IET ENERGY ENGINEERING SERIES 188

Coordination of Distributed Energy Resources in Microgrids

Other volumes in this series:

Volume 1	**Power Circuit Breaker Theory and Design** C.H. Flurscheim (Editor)
Volume 4	**Industrial Microwave Heating** A.C. Metaxas and R.J. Meredith
Volume 7	**Insulators for High Voltages** J.S.T. Looms
Volume 8	**Variable Frequency AC Motor Drive Systems** D. Finney
Volume 10	**SF_6 Switchgear** H.M. Ryan and G.R. Jones
Volume 11	**Conduction and Induction Heating** E.J. Davies
Volume 13	**Statistical Techniques for High Voltage Engineering** W. Hauschild and W. Mosch
Volume 14	**Uninterruptible Power Supplies** J. Platts and J.D. St Aubyn (Editors)
Volume 15	**Digital Protection for Power Systems** A.T. Johns and S.K. Salman
Volume 16	**Electricity Economics and Planning** T.W. Berrie
Volume 18	**Vacuum Switchgear** A. Greenwood
Volume 19	**Electrical Safety: A guide to causes and prevention of hazards** J. Maxwell Adams
Volume 21	**Electricity Distribution Network Design, 2nd Edition** E. Lakervi and E.J. Holmes
Volume 22	**Artificial Intelligence Techniques in Power Systems** K. Warwick, A.O. Ekwue and R. Aggarwal (Editors)
Volume 24	**Power System Commissioning and Maintenance Practice** K. Harker
Volume 25	**Engineers' Handbook of Industrial Microwave Heating** R.J. Meredith
Volume 26	**Small Electric Motors** H. Moczala *et al.*
Volume 27	**AC-DC Power System Analysis** J. Arrillaga and B.C. Smith
Volume 29	**High Voltage Direct Current Transmission, 2nd Edition** J. Arrillaga
Volume 30	**Flexible AC Transmission Systems (FACTS)** Y.-H. Song (Editor)
Volume 31	**Embedded Generation** N. Jenkins *et al.*
Volume 32	**High Voltage Engineering and Testing, 2nd Edition** H.M. Ryan (Editor)
Volume 33	**Overvoltage Protection of Low-Voltage Systems, Revised Edition** P. Hasse
Volume 36	**Voltage Quality in Electrical Power Systems** J. Schlabbach *et al.*
Volume 37	**Electrical Steels for Rotating Machines** P. Beckley
Volume 38	**The Electric Car: Development and future of battery, hybrid and fuel-cell cars** M. Westbrook
Volume 39	**Power Systems Electromagnetic Transients Simulation** J. Arrillaga and N. Watson
Volume 40	**Advances in High Voltage Engineering** M. Haddad and D. Warne
Volume 41	**Electrical Operation of Electrostatic Precipitators** K. Parker
Volume 43	**Thermal Power Plant Simulation and Control** D. Flynn
Volume 44	**Economic Evaluation of Projects in the Electricity Supply Industry** H. Khatib
Volume 45	**Propulsion Systems for Hybrid Vehicles** J. Miller
Volume 46	**Distribution Switchgear** S. Stewart
Volume 47	**Protection of Electricity Distribution Networks, 2nd Edition** J. Gers and E. Holmes
Volume 48	**Wood Pole Overhead Lines** B. Wareing
Volume 49	**Electric Fuses, 3rd Edition** A. Wright and G. Newbery
Volume 50	**Wind Power Integration: Connection and system operational aspects** B. Fox *et al.*
Volume 51	**Short Circuit Currents** J. Schlabbach
Volume 52	**Nuclear Power** J. Wood
Volume 53	**Condition Assessment of High Voltage Insulation in Power System Equipment** R.E. James and Q. Su
Volume 55	**Local Energy: Distributed generation of heat and power** J. Wood
Volume 56	**Condition Monitoring of Rotating Electrical Machines** P. Tavner, L. Ran, J. Penman and H. Sedding
Volume 57	**The Control Techniques Drives and Controls Handbook, 2nd Edition** B. Drury
Volume 58	**Lightning Protection** V. Cooray (Editor)
Volume 59	**Ultracapacitor Applications** J.M. Miller

Volume 62 **Lightning Electromagnetics** V. Cooray
Volume 63 **Energy Storage for Power Systems, 2nd Edition** A. Ter-Gazarian
Volume 65 **Protection of Electricity Distribution Networks, 3rd Edition** J. Gers
Volume 66 **High Voltage Engineering Testing, 3rd Edition** H. Ryan (Editor)
Volume 67 **Multicore Simulation of Power System Transients** F.M. Uriate
Volume 68 **Distribution System Analysis and Automation** J. Gers
Volume 69 **The Lightening Flash, 2nd Edition** V. Cooray (Editor)
Volume 70 **Economic Evaluation of Projects in the Electricity Supply Industry, 3rd Edition** H. Khatib
Volume 72 **Control Circuits in Power Electronics: Practical issues in design and implementation** M. Castilla (Editor)
Volume 73 **Wide Area Monitoring, Protection and Control Systems: The enabler for smarter grids** A. Vaccaro and A. Zobaa (Editors)
Volume 74 **Power Electronic Converters and Systems: Frontiers and applications** A.M. Trzynadlowski (Editor)
Volume 75 **Power Distribution Automation** B. Das (Editor)
Volume 76 **Power System Stability: Modelling, analysis and control** A.A. Sallam and Om P. Malik
Volume 78 **Numerical Analysis of Power System Transients and Dynamics** A. Ametani (Editor)
Volume 79 **Vehicle-to-Grid: Linking electric vehicles to the smart grid** J. Lu and J. Hossain (Editors)
Volume 81 **Cyber-Physical-Social Systems and Constructs in Electric Power Engineering** S. Suryanarayanan, R. Roche and T.M. Hansen (Editors)
Volume 82 **Periodic Control of Power Electronic Converters** F. Blaabjerg, K. Zhou, D. Wang and Y. Yang
Volume 86 **Advances in Power System Modelling, Control and Stability Analysis** F. Milano (Editor)
Volume 87 **Cogeneration: Technologies, optimisation and implementation** C.A. Frangopoulos (Editor)
Volume 88 **Smarter Energy: From smart metering to the smart grid** H. Sun, N. Hatziargyriou, H.V. Poor, L. Carpanini and M.A. Sánchez Fornié (Editors)
Volume 89 **Hydrogen Production, Separation and Purification for Energy** A. Basile, F. Dalena, J. Tong and T.N. Veziroğlu (Editors)
Volume 90 **Clean Energy Microgrids** S. Obara and J. Morel (Editors)
Volume 91 **Fuzzy Logic Control in Energy Systems with Design Applications in MATLAB®/Simulink®** İ.H. Altaş
Volume 92 **Power Quality in Future Electrical Power Systems** A.F. Zobaa and S.H.E.A. Aleem (Editors)
Volume 93 **Cogeneration and District Energy Systems: Modelling, analysis and optimization** M.A. Rosen and S. Koohi-Fayegh
Volume 94 **Introduction to the Smart Grid: Concepts, technologies and evolution** S.K. Salman
Volume 95 **Communication, Control and Security Challenges for the Smart Grid** S.M. Muyeen and S. Rahman (Editors)
Volume 96 **Industrial Power Systems with Distributed and Embedded Generation** R Belu
Volume 97 **Synchronized Phasor Measurements for Smart Grids** M.J.B. Reddy and D.K. Mohanta (Editors)
Volume 98 **Large Scale Grid Integration of Renewable Energy Sources** A. Moreno-Munoz (Editor)
Volume 100 **Modeling and Dynamic Behaviour of Hydropower Plants** N. Kishor and J. Fraile-Ardanuy (Editors)
Volume 101 **Methane and Hydrogen for Energy Storage** R. Carriveau and D.S-K. Ting
Volume 104 **Power Transformer Condition Monitoring and Diagnosis** A. Abu-Siada (Editor)
Volume 106 **Surface Passivation of Industrial Crystalline Silicon Solar Cells** J. John (Editor)

Volume 107 **Bifacial Photovoltaics: Technology, applications and economics** J. Libal and R. Kopecek (Editors)
Volume 108 **Fault Diagnosis of Induction Motors** J. Faiz, V. Ghorbanian and G. Joksimović
Volume 110 **High Voltage Power Network Construction** K. Harker
Volume 111 **Energy Storage at Different Voltage Levels: Technology, integration, and market aspects** A.F. Zobaa, P.F. Ribeiro, S.H.A. Aleem and S.N. Afifi (Editors)
Volume 112 **Wireless Power Transfer: Theory, technology and application** N. Shinohara
Volume 114 **Lightning-Induced Effects in Electrical and Telecommunication Systems** Y. Baba and V.A. Rakov
Volume 115 **DC Distribution Systems and Microgrids** T. Dragičević, F. Blaabjerg and P. Wheeler
Volume 116 **Modelling and Simulation of HVDC Transmission** M. Han (Editor)
Volume 117 **Structural Control and Fault Detection of Wind Turbine Systems** H.R. Karimi
Volume 119 **Thermal Power Plant Control and Instrumentation: The control of boilers and HRSGs, 2nd Edition** D. Lindsley, J. Grist and D. Parker
Volume 120 **Fault Diagnosis for Robust Inverter Power Drives** A. Ginart (Editor)
Volume 121 **Monitoring and Control using Synchrophasors in Power Systems with Renewables** I. Kamwa and C. Lu (Editors)
Volume 123 **Power Systems Electromagnetic Transients Simulation, 2nd Edition** N. Watson and J. Arrillaga
Volume 124 **Power Market Transformation** B. Murray
Volume 125 **Wind Energy Modeling and Simulation, Volume 1: Atmosphere and plant** P. Veers (Editor)
Volume 126 **Diagnosis and Fault Tolerance of Electrical Machines, Power Electronics and Drives** A.J.M. Cardoso
Volume 128 **Characterization of Wide Bandgap Power Semiconductor Devices** F. Wang, Z. Zhang and E.A. Jones
Volume 129 **Renewable Energy from the Oceans: From wave, tidal and gradient systems to offshore wind and solar** D. Coiro and T. Sant (Editors)
Volume 130 **Wind and Solar Based Energy Systems for Communities** R. Carriveau and D.S-K. Ting (Editors)
Volume 131 **Metaheuristic Optimization in Power Engineering** J. Radosavljević
Volume 132 **Power Line Communication Systems for Smart Grids** I.R.S. Casella and A. Anpalagan
Volume 134 **Hydrogen Passivation and Laser Doping for Silicon Solar Cells** B. Hallam and C. Chan (Editors)
Volume 139 **Variability, Scalability and Stability of Microgrids** S.M. Muyeen, S.M. Islam and F. Blaabjerg (Editors)
Volume 145 **Condition Monitoring of Rotating Electrical Machines** P. Tavner, L. Ran and C. Crabtree
Volume 146 **Energy Storage for Power Systems, 3rd Edition** A.G. Ter-Gazarian
Volume 147 **Distribution Systems Analysis and Automation, 2nd Edition** J. Gers
Volume 151 **SiC Power Module Design: Performance, robustness and reliability** A. Castellazzi and A. Irace (Editors)
Volume 152 **Power Electronic Devices: Applications, failure mechanisms and reliability** F Iannuzzo (Editor)
Volume 153 **Signal Processing for Fault Detection and Diagnosis in Electric Machines and Systems** M. Benbouzid (Editor)
Volume 155 **Energy Generation and Efficiency Technologies for Green Residential Buildings** D. Ting and R. Carriveau (Editors)
Volume 156 **Lithium-ion Batteries Enabled by Silicon Anodes** C. Ban and K. Xu (Editors)
Volume 157 **Electrical Steels, 2 Volumes** A. Moses, K. Jenkins, Philip Anderson and H. Stanbury
Volume 158 **Advanced Dielectric Materials for Electrostatic Capacitors** Q Li (Editor)
Volume 159 **Transforming the Grid Towards Fully Renewable Energy** O. Probst, S. Castellanos and R. Palacios (Editors)
Volume 160 **Microgrids for Rural Areas: Research and case studies** R.K. Chauhan, K. Chauhan and S.N. Singh (Editors)

Volume 161 **Artificial Intelligence for Smarter Power Systems: Fuzzy logic and neural networks** M.G. Simoes
Volume 166 **Advanced Characterization of Thin Film Solar Cells** N. Haegel and M. Al-Jassim (Editors)
Volume 167 **Power Grids with Renewable Energy: Storage, integration and digitalization** A.A. Sallam and Om P. Malik
Volume 169 **Small Wind and Hydrokinetic Turbines** P. Clausen, J. Whale and D. Wood (Editors)
Volume 170 **Reliability of Power Electronics Converters for Solar Photovoltaic Applications** F. Blaabjerg, A. Haque, H. Wang, Z.A. Jaffery and Y. Yang (Editors)
Volume 171 **Utility-scale Wind Turbines and Wind Farms** A. Vasel-Be-Hagh and D.S-K. Ting
Volume 172 **Lighting Interaction with Power Systems, 2 Volumes** A. Piantini (Editor)
Volume 174 **Silicon Solar Cell Metallization and Module Technology** T. Dullweber (Editor)
Volume 182 **Surge Protection for Low Voltage Systems** A. Rousseau (Editor)
Volume 184 **Compressed Air Energy Storage: Types, systems and applications** D. Ting and J. Stagner
Volume 193 **Overhead Electric Power Lines: Theory and practice** S. Chattopadhyay and A. Das
Volume 194 **Offshore Wind Power Reliability, Availability and Maintenance, 2nd Edition** P. Tavner
Volume 198 **Battery Management Systems and Inductive Balancing** A. Van den Bossche and A. Farzan Moghaddam
Volume 199 **Model Predictive Control for Microgrids: From power electronic converters to energy management** J. Hu, J.M. Guerrero and S. Islam
Volume 204 **Electromagnetic Transients in Large HV Cable Networks: Modeling and calculations** A. Ametani, H. Xue, T. Ohno and H. Khalilnezhad
Volume 212 **Battery State Estimation: Methods and models** S. Wang
Volume 905 **Power System Protection, 4 Volumes**

Coordination of Distributed Energy Resources in Microgrids

Optimisation, control, and hardware-in-the-loop validation

Yan Xu, Yu Wang, Cuo Zhang and Zhengmao Li

The Institution of Engineering and Technology

Published by The Institution of Engineering and Technology, London, United Kingdom

The Institution of Engineering and Technology is registered as a Charity in England & Wales (no. 211014) and Scotland (no. SC038698).

First published 2021

The Institution of Engineering and Technology
Michael Faraday House
Six Hills Way, Stevenage
Herts, SG1 2AY, United Kingdom

www.theiet.org

British Library Cataloguing in Publication Data
A catalogue record for this product is available from the British Library

ISBN 978-1-83953-268-9 (hardback)
ISBN 978-1-83953-269-6 (PDF)

Typeset in India by MPS Limited
Printed in the UK by CPI Group (UK) Ltd, Croydon

Contents

About the authors

Yan Xu received the B.E. and M.E degrees from South China University of Technology, Guangzhou, China in 2008 and 2011, respectively, and the Ph.D. degree from The University of Newcastle, Australia, in 2013. He conducted post-doctoral research with the University of Sydney Postdoctoral Fellowship, and then joined Nanyang Technological University (NTU) with the Nanyang Assistant Professorship. He is now an Associate Professor at School of Electrical and Electronic Engineering and a Cluster Director at Energy Research Institute @ NTU (ERI@N). His research focuses on power system stability, control, optimization, and data-analytics. His work has been funded by Singapore National Research Foundation (NRF), Ministry of Education (MOE), Energy Market Authority (EMA), Housing & Development Board (HDB), Rolls-Royce, Singtel, Infineon, Singapore Power (SP) Group, and technology start-up companies. Many of his research outcomes have been practically applied/licensed to the industry. As the first/corresponding author, Dr Xu has published 2 books, 90 IEEE Transactions papers and 30 IET journal papers. He has 14 "Web-of-Science highly cited papers" and received 10 IEEE/IET paper contest and conference best paper awards. Dr Xu is as an Editor for IEEE Transactions on Smart Grid, IEEE Transactions on Power Systems, and Associate Editor for IET Generation, Transmission & Distribution, etc. He is also serving as the Chairman for IEEE Power & Energy Society (PES) Singapore Chapter.

Yu Wang received his BEng degree in School of Electrical Engineering and Automation from Wuhan University, Wuhan, China, in 2011, and the MSc and PhD degree in Power Engineering from Nanyang Technological University, Singapore, in 2012 and 2017, respectively. He is now a research scientist with RollsRoyce@NTU Corporate Lab, Nanyang Technological University, Singapore. He was a recipient of the European Union Marie Skłodowska-Curie Action Individual Fellowship 2020. His research interests include distributed control and optimisation of energy storage systems, microgrids, and cyber-physical power systems. He has published 50 peer-reviewed journal and conference papers, including 3 'web-of-science highly cited papers'.

Cuo Zhang received his BE (Hons.) degree in electrical (power) engineering from the University of Sydney, Australia, in 2014 and the PhD degree in electrical engineering from the University of New South Wales, Australia, in 2018. He is now a research fellow with the University of New South Wales, Australia. His research interests include power system planning and operation, voltage stability and control, microgrids, multi-energy systems, and applications of optimisation theory and artificial intelligence in these areas. As the first author, he has published 10 IEEE Transactions journal papers including 2 'web-of-science highly cited papers' and received 3 IEEE/IET international conference best paper awards.

Zhengmao Li received the B.E. degree in information engineering, M.E. degree in electrical engineering from Shandong University, Jìnan, China, in 2013 and 2016, respectively. and the PhD. degree in electrical engineering with the School of Electrical and Electronic Engineering, Nanyang Technological University, Singapore, in 2020. He has been a research fellow with the Stevens Institute of Technology, Hoboken, NJ, USA during 2019–2021 and now he is a research fellow at Nanyang Technological University under the NTU-ETH project. His research interests include renewable energy integration, microgrid and multi-energy systems, and optimization techniques such as approximate dynamic programming, robust optimization, and stochastic optimization.

Foreword

A microgrid is composed of a medium/low-voltage distribution system with distributed energy resources (DERs) such as distributed generators (DGs), energy storage systems (ESSs), and controllable loads, which can operate either connected or disconnected from the main grid. Microgrids are an effective means to integrate renewable energy through smart grid technologies. In the past couple of decades, microgrids have been widely researched and developed and are starting to be widely deployed.

Due to the significantly diverse temporal and spatial characteristics of different DERs and the significant operational uncertainties from intermittent renewable energy-based DERs, the planning, operation, and control of DERs in microgrids present challenging problems. Thus, this book focuses on coordination aspects of DERs in the context of microgrids, presenting various advanced techniques for optimal planning, operation, and control of DERs for secure, economic, and robust microgrid operation. Major topics covered in this book include:

- Comprehensive mathematical modelling of DERs and microgrids.
- Optimal sizing and siting of DERs in microgrids.
- Robust and scholastics optimisation for active and reactive power dispatch of DERs in microgrids.
- Distributed coordinated control of DERs in microgrids in grid-connected and islanded modes.
- Power and controller hardware-in-the-loop tests for validation of control algorithms.

Written by a dedicated research team that has been working on DERs and microgrids for over ten years, this book is a systematic presentation of the authors' original research work and insights on these topics. With a balanced presentation of theory and practice, it can be a valuable reference for researchers, engineers, and graduate students in the areas of DERs and microgrids.

Claudio Cañizares, PhD, P.Eng., FIEEE, FRSC, FCAE
University Professor and Hydro One Endowed Chair
Executive Director, Waterloo Institute for Sustainable Energy (WISE)
University of Waterloo, Canada
Editor-in-Chief, *IEEE Transactions on Smart Grid*
IEEE Division VII Director-Elect

Preface

The distributed energy resource (DER) refers to the small-scale energy generation and storage units, as well as flexible loads. Microgrids are distribution systems with a group of interconnected loads and DERs within clearly defined electrical boundaries that act as a single controllable entity with respect to the main grid. A microgrid can connect and disconnect from the main grid to enable it to operate in both grid-connected and islanded modes.

This book focuses on three research problems of the DERs in the microgrid: *planning*, *operation*, and *control*. The planning aims to determine the optimal size and site as well as investment plan of the DER units in the microgrid over a long-time horizon, e.g., 5 years to 20 years; The operation of the microgrid focuses on the operational dispatch horizon over a short-term time interval, e.g., from 5 minutes to a few hours or one day. The objective is to dispatch the DERs for optimal energy management and bus voltage regulation; The control of the microgrid is concerned on the real-time timescale, i.e., milliseconds to seconds, and the objective is to maintain the voltages and frequency within desired operational ranges, achieve power sharing and balancing among DERs and loads.

Based on our research works in the past 10 years, this book is divided into the following four parts with 19 chapters:

Part I gives a general introduction to the DERs and the microgrid. We consider three types of DERs, which are distributed generators (DGs), energy storage systems (ESSs), and flexible loads. Their definitions, classification, and working principles are introduced. Then, microgrids are introduced and classified according to different architectures and technical characteristics. Finally, the research problems for planning, operation, and control of the DERs in the microgrid are discussed in depth, followed by a summary of our originally proposed methodologies that will be introduced in the rest of the book.

Part II focuses on the planning of the DER units in the microgrid. Four different methods are presented to optimise the deployment plan, including the sizing, siting, type selection, and installation timeline of the DER units in a microgrid. In particular, the DER investment phase and the operation phase are coordinately modelled so to minimise the expected total investment and operation costs over the entire project horizon.

Part III focuses on the operation of the DERs in the microgrid. To address the operational uncertainties in the microgrid, we adopt a temporally coordinated operation framework, where two operational stages are designed: the first stage aims to dispatch the slow-responding DERs over a relatively longer time horizon, e.g., hours to a day, and the second stage is for fast-responding and more flexible DERs over a relatively shorter time interval, e.g., 5 min to an hour. Optimisation methodologies for energy management and voltage regulation of the microgrid are respectively introduced.

Part IV focuses on the coordinated control of the DERs in the microgrid. We adopt a distributed control structure, where each DER unit only communicates with its neighbouring units so that the overall computational burden is shared amongst the individual units, and a more flexible and scalable microgrid control framework is allowed. Different distributed control systems are designed for DERs to support or control frequency and voltage in grid-connected and islanded mode, respectively. Besides, to validate the real-time performance of the control system, hardware-in-the-loop (HIL) tests are also introduced.

We hope this book can serve as a timely reference and guidebook for researchers, students, and engineers who seek to study distributed energy resource integration and microgrid operation and control as well as hardware-in-the-loop real-time simulation tests.

Yan Xu
Yu Wang
Cuo Zhang
Zhengmao Li

Part I

Distributed Energy Resources and Microgrids: Preliminaries

Chapter 1

Distributed energy resources: introduction and classification

1.1 Background

An electric power system is a network of the *generators* that supply the power, the *transmission and distribution grid* that transports and delivers the power, and the *loads* which consume the power, as illustrated in Figure 1.1. Conventionally, the generators are of large capacities (up to hundreds of megawatts), centralised, and mainly fuelled by fossil energy resources such as coal and gas.

In recent years, driven by the tremendous pressure of global climate changes and fossil-fuel crisis, modern power systems have been integrating high levels of renewable energy resources (RESs) at a fast-growing speed. Two types of RESs that stand out are solar photovoltaic (PV) and wind power. In addition to the centralised large-scale PV stations and wind power farms which are usually far from centres of the loads, smaller and modular-based RES units (typically less than 10 MW) such as rooftop PV and small wind turbines that are owned by consumers or utilities have also been widely deployed in the distribution networks close to the loads. Besides traditional electrical-focused components, electrical–thermal coupled devices such as on-site tri-generation of cooling, heat, and power (CHP) which can enable district cooling/heating for a higher energy efficiency are also gaining increasing popularity at the distribution network level. In view of the highly stochastic and intermittent nature of the RES units, energy storage systems (ESSs) are

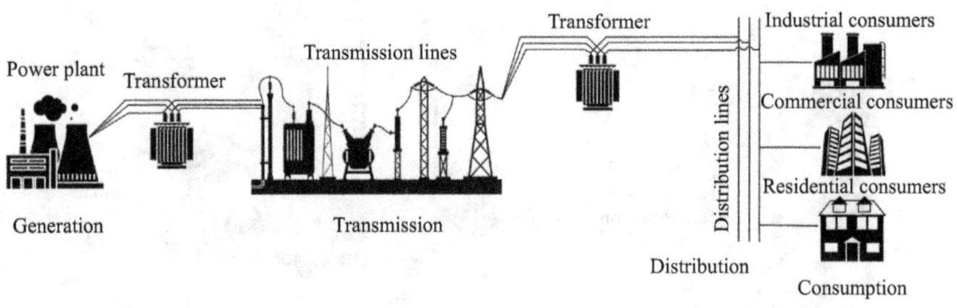

Figure 1.1 Illustration of a traditional electric power system

considered to be a necessary component to overcome the potential issues brought by the intermittency.

In the meantime, the demand side has been retrofitted by advanced metering infrastructure (AMI) which is a system of smart meters supported by bidirectional communication networks. The AMI can significantly facilitate the interactions between the consumers and the utilities by making the loads more flexible and controllable. For example, the loads can be directly or indirectly controlled via different demand response (DR) programmes where consumers can benefit from a time-varying tariff or gain economic compensations from their assistance in grid emergency situations.

A modernised power system with aforementioned new elements is also known as 'smart grid', as conceptually illustrated in Figure 1.2.

1.2 Definition and classification

The distributed energy resource (DER) refers to the above-mentioned *small-scale energy generation and storage units, as well as flexible loads that are connected to the distribution system*. The DERs can greatly benefit the society, consumers, and utilities:

1. DERs help reduce the dependence on fossil fuels and therefore reduce the overall carbon emissions.
2. DERs help save consumers' electricity bills.
3. DERs help boost reliability of the distribution networks given proper controls.
4. DERs provide trading options for consumers; for example, consumers can opt to trade amongst each other or with the utilities.
5. DERs help the utilities to achieve more flexible controls of the distribution grid.
6. DERs help reduce the cost of power system retrofitting and defer capital investments, which in turn reduce the overall cost incurred to consumers.

As shown in Figure 1.3, this book classifies the DERs into the following categories.

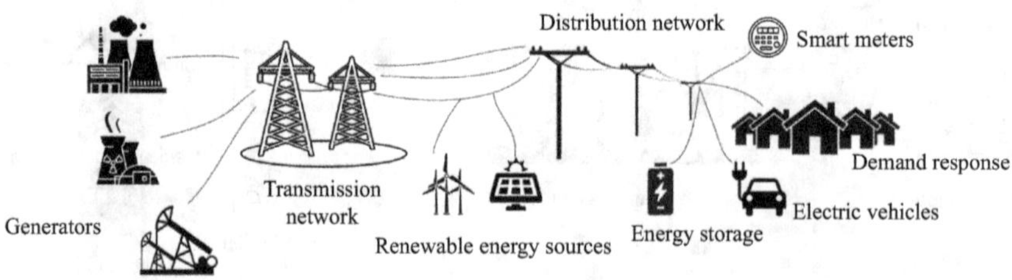

Figure 1.2 Illustration of 'smart grid'

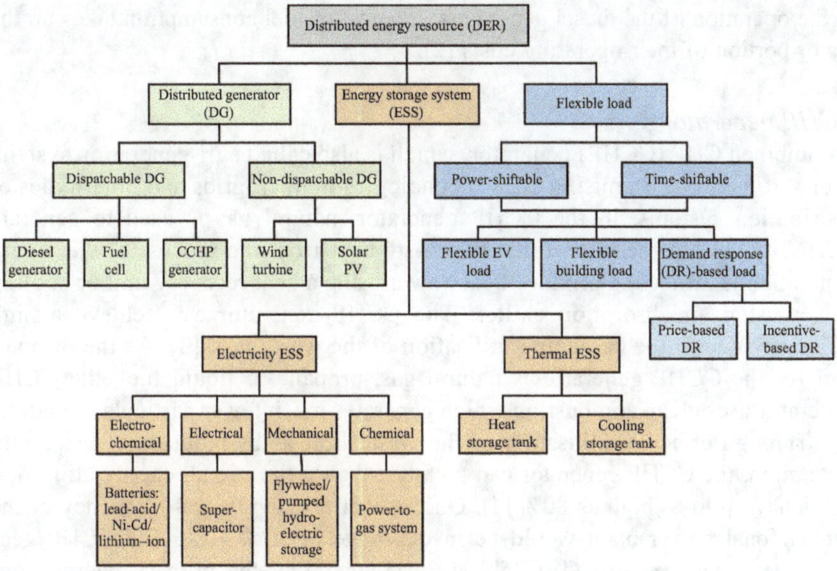

Figure 1.3 Classification of DERs

1.2.1 Distributed generator

The distributed generator (DG) is a small (typically less than 10 MW) and modular generation unit that is directly connected to the distribution network and produces electric power or other energy forms such as heat or cooling. Depending on the primary energy resources, the DG can be further classified into *dispatchable* DGs and *non-dispatchable* DGs.

1.2.1.1 Dispatchable DGs

This type of DG is fuelled by a stable primary energy supply, so its power output can be dispatched flexibly. The representative types of DGs are as follows.

Diesel generator

A diesel generator combines a diesel engine with an electric generator to generate electricity. A diesel engine is usually designed to run on diesel fuels with some types adapted for other liquid fuels or natural gas. A modern diesel generator can reach a fuel efficiency of around 30%. Given fuel-efficient engines, diesel generators are well suited for applications where continuous, reliable, and flexible electricity is required for extended periods of time. Besides, diesel generators are more economically viable in operations than other types of coal-fired generators due to the affordability of diesel as well as their nature of low maintenance [1]. However, the consumption of diesel would result in significant pollutions, which in some way prevents its wide applications in the distributed energy systems, especially for those with RESs which consider environmental impacts as a major factor.

In the operation of the diesel generators, the cost of fuel consumption takes up the major portion of their operating costs [2].

CCHP generator

A combined CHP (CCHP) generator, which is also called a tri-generation system, is one of the most promising DER technologies in microgrids and other types of distribution systems. In the CCHP generator, natural gas is used to generate electricity, and at the same time, the waste heat from the process of electricity generation is utilised to produce heat power using a heat recovery unit or cooling power using an absorption chiller. The CCHP generator can achieve a high efficiency due to the cascading utilisation of the waste heat [3]. As the primary fuel for the CCHP generator is natural gas, propane or liquid fuel, the CCHP generator uses clean combustion which generates much fewer particulate matters.

Through effective utilisation of the waste heat as by-products of electricity generation, the CCHP generator can greatly enhance the overall energy utilisation efficiency up to as high as 80% [4]. On the contrary, the typical efficiency of the conventional power plant would be in the range of 30%–40%. The CCHP generators can achieve around a 35%–40% reduction in the primary energy consumption as compared to conventional power plants and heat-only boiler as well as a 30%–35% reduction in gas emissions compared with coal-fired power plants. The CCHP generators are now widely used in district power, heating, or cooling systems for smart buildings, commercial stations, industrial plants, etc., with large thermal demands [5,6].

Under different conditions, the operating modes for CCHP generators vary. In some small-scale applications, it is inclined to use the CCHP generators for thermal demands, while electricity demands are met by power transactions with the main electricity grid or using other electrical power generation units such as diesel generators; whereas in large-scale applications where the electricity demands are of higher priorities, CCHP generators usually operate with the aim to follow the power demands.

Fuel cell

A fuel cell is an electrochemical cell that converts the chemical energy of a type of fuel (often hydrogen) into electricity with the aid of an oxidising agent (often oxygen) through an electrochemical process [7]. Fuel cells can produce electricity continuously as long as the fuel and the agent oxygen are supplied. As electricity and heat are both products of the fuel cell, they can also be used as tri-generation units. The electricity generation efficiency of the fuel cells can reach as high as 55%, which makes them sufficiently efficient for electricity generation purposes.

1.2.1.2 Non-dispatchable DGs

This type of DG is driven by intermittent energy resources such as wind or solar irradiance. Their power outputs are stochastic and non-dispatchable.

Wind turbine
Wind turbines generate electrical energy by converting the kinetic energy of wind flowing through rotor blades [8]. They are available in a wide range of sizes in the form of either horizontal or vertical axes. They are increasingly important renewable energy-based generators and are used worldwide to lower energy costs and reduce reliance on fossil fuels [9]. At the distribution network or the microgrid level, smaller wind turbines are generally adopted. To capture the maximum possible power, the maximum power point tracking (MPPT) is essential in the wind energy conversion systems mainly for the smaller wind turbines. The efficiency of wind turbines rises from around 30%–45% at normal times to around 50% at times of peak wind. It is also to be noted that due to the volatile nature of wind energy, the power outputs from the wind turbines are also intermittent and stochastic.

Solar PV
A solar PV system is an energy system designed to convert the solar radiance into the usable electricity by means of photovoltaics [10]. The efficiency of solar systems is around 15%–20%, and for some of high-quality ones, it exceeds 22%. The DC power from the solar panels can be inverted to AC power for real-world applications.

The solar PV systems range from small, rooftop-mounted or building-integrated systems with capacities from a few to several tens of kilowatts to large utility-scale power stations on the order of hundreds of megawatts. The solar PV system can be suitable for the different scales of applications depending on their sizes. Due to the advantages including silent operations with no moving parts and zero carbon emissions, solar PV systems have developed from being products used in niche markets into a mature technology used for mainstream renewable electricity generation [11]. Given the uncertain weather conditions, the solar irradiance can be intermittent which leads to fluctuating and stochastic power outputs from the solar PV systems.

1.2.2 Energy storage system

The ESS is a device that stores energy produced at one time for use in a later time. Through charging and discharging, it allows a bidirectional energy flow between the device and the power grid. As a result, the ESS can be used to assist reducing the power imbalance between energy demand and supply when needed.

Based on power or thermal energy grid integration, the ESS can be generally grouped as *electricity ESS* and *thermal ESS* which work for electric power and heat/cooling energies, respectively. For electricity ESS, it can be divided into four types: electrochemical, electromagnetic, mechanical, and chemical energy storage [12]. The thermal ESS comprises the heat or cooling storage tank. Typical ESSs utilised in microgrids are further introduced as follows.

1.2.2.1 Electrochemical energy storage

A battery is typical electrochemical energy storage consisting of one or more rechargeable cells. The non-rechargeable battery is called a primary battery, and the rechargeable battery is called a secondary battery [13]. In microgrid energy storage applications, there are several widely used types of batteries. The battery ESS size can range from several to thousands of kilowatts.

Lead–acid battery

It uses lead or lead oxide for the electrodes and sulphuric acid for the electrolyte. Among other candidates, lead–acid battery has relatively low cost and very mature technology. In addition, it has high reliability and high efficiency (80%–90%), which makes it suitable for microgrid applications. Currently, the lead–acid battery is still the most widely used rechargeable battery for small-to-medium-scale storage applications. The drawbacks of lead–acid batteries are short life cycles and low energy densities, which makes them unsuitable for portable applications.

Ni–Cd battery

It uses nickel oxide hydroxide and metallic cadmium for the electrodes and potassium hydroxide for the electrolyte. It has a long cycle life and good performances at low temperatures compared to other types of rechargeable batteries. One significant advantage of nickel–cadmium battery (Ni–Cd battery) is its high discharge rates and it can deliver current close to its rated capacity. However, the Ni–Cd battery has high self-discharge rates and is more costly than the lead–acid battery. Larger ventilated wet cell Ni–Cd batteries are used in emergency power, uninterruptible power supplies, and microgrid applications.

Lithium-ion battery

It uses intercalated lithium compound for the positive electrode and typically graphite for the negative electrode. During discharge, lithium ions move from the negative electrode to the positive electrode through the electrolyte, and reversely during charge. The lithium-ion battery has many advantages including a high energy density, very slow self-discharge rate, low maintenance, and no memory effects. One of the major drawbacks is its high manufacturing cost, which is a major barrier for large-scale applications. Another concern lies with the safety in applications. Direct damage or inappropriate use of lithium-ion batteries can lead to explosions and fire hazards. Nowadays, lithium-ion batteries are widely used for portable electronic devices, electric vehicles (EVs), renewable intensive microgrids, as well as military, maritime, and aerospace power systems.

1.2.2.2 Electrical energy storage

This type of ESS directly stores electrical energy. A supercapacitor or ultracapacitor is a type of capacitor with extra high capacities that store electricity in electrical fields [14]. Compared to rechargeable batteries, it has a much higher power density, which means it can be charged and discharged much faster. Supercapacitors are applied to applications where frequent rapid charge/discharge cycles are required. They can be widely found in electric or hybrid buses, trains, ships, cranes, as well as microgrids, where they are used for regenerative braking, propulsion and pulse load power supplies, and renewable energy intermittency mitigation. They are also widely used as hybrid ESSs in combination with rechargeable batteries, which possess advantages of both high power and energy densities.

1.2.2.3 Mechanical energy storage

This type of ESS stores the electrical energy in the form of mechanical energy, and generates electricity by releasing the mechanical energy to drive a generator.

Flywheel energy storage

It is a type of system that stores rotational energy which is a type of kinetic energy in proportion to the product of its moment of inertia and the square of its rotational speed by accelerating the rotor of flywheels to a very high speed [15]. When energy is extracted from the flywheel energy storage (FES), the rotational speed of the flywheel is reduced while rotational energy is released to generate electricity. On the contrary, the charging process of FES involves increasing the speed of the flywheel. The flywheel has a very long cycle life up to more than 20 years. It has charging/discharging efficiencies higher than 90%, which is higher than most of electrochemical ESSs. Application scenarios of flywheels are similar to those of ultracapacitors where high power density is required with the difference that flywheels can be built in much large capacity. Flywheels have been widely applied in UPS, transportation electrification, and renewable energy generation systems.

Pumped hydroelectric energy storage

It stores energy as the gravitational potential energy of water by pumping the water from a lower to a higher elevation reservoir [16]. The stored water can be released back to the lower reservoir through hydro-turbines during which electric power is generated. It is widely used as long-term energy storage for electric power systems. The small pumped hydroelectric energy storage (PHES) in buildings is another application scenario, although it is still not usually an economically viable solution. The round-trip energy efficiency of PHES is typically between 70% and 80%. The method is suitable for medium to long-term energy storage applications.

1.2.2.4 Chemical energy storage

The power-to-gas (P2G) system is a representative chemical energy storage technique. The P2G describes the process of converting electricity into gaseous energy carriers such as hydrogen or methane via water electrolysis. In a P2G system, the electrochemical process splits water into hydrogen and oxygen or converts it to methane. The hydrogen reacts with carbon dioxide in the presence of biocatalysts. The produced 'green' hydrogen or natural gas can be used in a range of pathways that promise to decouple renewable generation from electricity demand-helping to avoid surplus curtailment [17].

In the P2G system, there is a gas/hydrogen storage section where the gas/hydrogen is stored for a postponed use, and a fuel cell generation system, which reconverts the hydrogen to electricity during periods with power deficiencies. Much like the battery storage systems, the gas/hydrogen storage capacity is determined with the assumption that all the hydrogen produced by the hydrogen generators can be stored in the tanks without reaching a full charge [18].

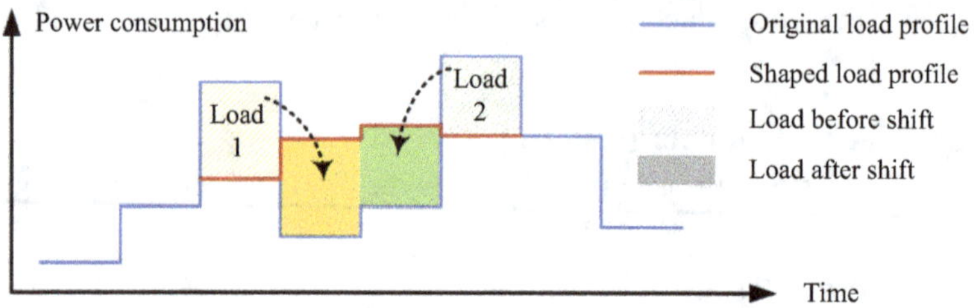

Figure 1.4 Illustration of time-shiftable loads

1.2.2.5 Thermal energy storage

Thermal energy storage which usually includes the heat storage tank and cooling storage tank uses various technologies. It stores available thermal energy in an insulated repository, which can be released for later use in different industrial and residential applications, such as space heating or cooling, hot water production, or electricity generation [19]. Depending on the specific technology, it allows excess thermal energy to be stored and used in hours, days, or even months later at scales covering the individual, building, multi-user building, district, town, or region. The storage media can be in different forms, such as liquid (water, molten salt, and thermal oil), solid (stone, concrete, metal, and ground), or liquid with a solid filler material (molten salt/stone) [20].

The thermal energy storage, both seasonal and short-term, is considered an important means for economically balancing high shares of variable renewable electricity production and integration of electricity and heating sectors in energy systems almost or completely fed by the renewable energy.

1.2.3 Flexible load

The flexible loads refer to those that have a high level of energy demand flexibility and can be controlled directly or indirectly by the utilities.

According to different energy demand flexibility, the loads can be generally classified to be ***time-shiftable*** and ***power-shiftable***. The former means the energy demand can be shifted from one time period to another without changing the total amount of the demand. The latter means the power demand can be increased or decreased at a particular time. In both scenarios, the inconvenience to consumers' energy usage should be kept at minimum and/or consumers should be economically compensated by the utilities who call for load shifting.

Figure 1.4 illustrates time-shiftable loads, where Load 1 is shifted to a later time period while Load 2 is shifted in advance. By doing so, the original load profile shown by the blue line is shaped into a new one as the red line, but the total amount of energy consumption remains unchanged.

On the other hand, power-shiftable loads are illustrated by Figure 1.5. During their corresponding time periods, Load 1 is increased, while Load 2 is decreased.

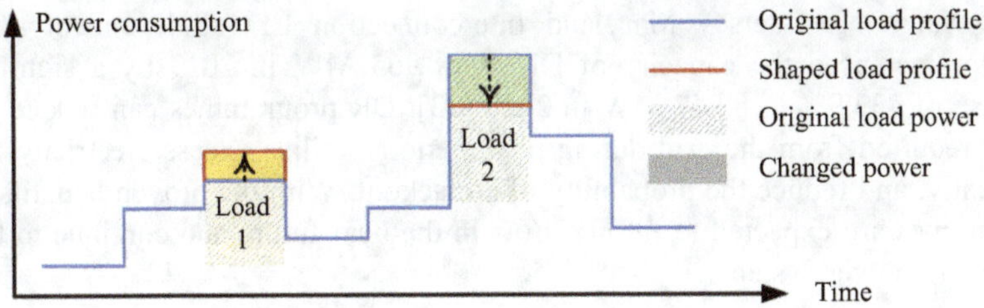

Figure 1.5 Demonstration of power-shiftable loads

With more power consumption of Load 1 and less of Load 2, the original load profile is reshaped to a new one.

From the practical implementation point of view, this book classifies the flexible loads into DR-based loads, flexible building loads, and flexible EV loads.

1.2.3.1 DR-based loads

DR programmes are applied by the utilities to achieve flexible loads and change load profiles. DR programmes can be classified into price based and incentive based.

Price-based DR programmes

Price-based DR programmes are set by utilities to change electricity prices for customers, such that the customers can take advantage of lower-priced periods and avoid consuming energy when the prices are higher [21]. In response to varying electricity prices, customers can schedule their loads in a flexible manner to participate in, thus providing DR programmes. Price-based DR programmes include time of use (ToU), critical peak pricing, extreme day pricing, and real-time pricing programmes.

Incentive-based DR programmes

On the other hand, incentive-based DR programmes influence customer behaviours of the electricity use through providing incentives. Incentives can be given as a form of payments from utilities to customers or through energy transactions in associated markets. For example, in direct load control, utilities remotely control certain customer appliances to increase or decrease power consumption with a financial compensation for the affected customers. Other incentive-based DR programmes include emergency DR, demand bidding/buy-back and capacity market programmes.

In this book, the DR programmes are targeted at small customers, which requires a larger-scale aggregation in order to make remarkable contributions in providing flexible load services for microgrids. DR programmes have been implemented in practice and demonstrated economic and technical benefits. In Australia, DR projects aimed to achieve 200 MW of capacity by 2020 based on various innovative customer behavioural change initiatives [22]. In the USA, the

Pennsylvania New Jersey Maryland Interconnection LLC independent system operator increased the capacity of DR from 965 MW in 2019 by a significant increase of 733.9% to 8,047 MW in 2020 [23]. DR programmes can reduce electricity required from the grid during peak periods, utilise excess electricity more efficiently, and reduce the probability of a blackout. With the proven benefits, DR programmes are expected to further grow in the near future and continue to bring benefits to power systems.

1.2.3.2 Flexible building loads

Buildings account for about 40% of global energy consumption [24]. In Singapore, buildings consume about half of the country's electricity [25]. In buildings, modern appliances can be operated in a way where varying power can be consumed without compromising expected performances, which gives room for high load flexibility.

In practice, smart appliances such as washing machines, clothes dryers, and pool pumps are interruptible or deferrable. These appliances can be efficiently and optimally scheduled as time-shiftable loads without sacrificing user convenience. On the other hand, thermostatically controlled appliances including heating, ventilation, and air conditioning (HVAC) systems and water heaters can be dynamically controlled as power-shiftable loads while ensuring comfort levels of customers.

Building automation plays an important role in aggregating and controlling the flexible time/power-shiftable loads in buildings. Through building automation, smart appliances can be aggregated to provide auxiliary services for the grid. As benefits, the flexible building loads can help save electricity bills for customers and improve energy utilisation efficiency. Given these economic and technical benefits, flexible building loads with building automation technologies are expected to be widely used in microgrid operation.

1.2.3.3 Flexible EV loads

EVs are interruptible and deferrable loads, which can be charged at individual households or at public charging stations. EVs can be treated as both time-shiftable and power-shiftable loads. Vehicle-to-gird (V2G) services which optimally schedule EV charging allow for high flexibility of power consumption and can benefit operation of power systems and microgrids.

V2G trials have been implemented widely especially in the US, the UK, and Germany [26]. The trials aim to validate potential of using EV charging/discharging operation for grid services including frequency response, power reserve, and time shifting of energy use.

To improve efficiency of V2G services, smart charging, where the EVs are monitored and controlled remotely through a reliable communication network, has been widely explored. A trial project conducted in the UK – 'Electric Nation' [27] – was launched to investigate impacts of residential EV charging on local electricity networks, as well as reliability and acceptability of smart charging amongst the EV owners. Based on the findings, smart charging can enable ToU-based charging for economic benefits while also mitigate potential risks of mass uptake of ToU incentives that may result in network congestions. With the aid of smart charging technologies and a robust communication network, charging service providers can effectively aggregate EVs as a type of flexible loads to provide V2G services in microgrids.

References

[1] M. Issa, H. Ibrahim, R. Lepage, *et al.*, 'A review and comparison on recent optimization methodologies for diesel engines and diesel power generators', *Journal of Power and Energy Engineering*, vol. 7, no. 6, p. 31, 2019.

[2] I. Baring-Gould and M. Dabo, 'Technology, performance, and market report of wind-diesel applications for remote and island communities', National Renewable Energy Lab. (NREL), Golden, CO (United States), 2009.

[3] D.W. Wu and R.Z. Wang, 'Combined cooling, heating and power: a review', *Progress in Energy and Combustion Science*, vol. 32, no. 5-6, pp. 459–495, 2006.

[4] H. Cho, A.D. Smith, and P. Mago, 'Combined cooling, heating and power: a review of performance improvement and optimization', *Applied Energy*, vol. 136, pp. 168–185, 2014.

[5] M. Liu, Y. Shi, and F. Fang, 'Combined cooling, heating and power systems: a survey', *Renewable and Sustainable Energy Reviews*, vol. 35, pp. 1–22, 2014.

[6] 'Global Combined Heat and Power (CHP) market – forecasts from 2020 to 2025', Report, Aug 2020, ID: 5174330.

[7] European Commission, 'Hydrogen energy and fuel cells: a vision of our future', The final report of the high-level group, 2003.

[8] D.A. Spera, *Wind Turbine Technology*, New York, NY: ASME Press, 1994.

[9] D. Corbus and M. Meadors, 'Small wind research technical report', NREL/TP-500-38550, Oct. 2005.

[10] V.V. Tyagi, N.A.A. Rahim, N.A. Rahim, *et al.*, 'Progress in solar PV technology: research and achievement', *Renewable and Sustainable Energy Reviews*, vol. 20, pp. 443–461, 2013.

[11] 'Solar PV', IEA, Paris, Available: www.iea.org/reports/solar-pv, 2020.

[12] 'Electrical energy storage', International Electrotechnical Commission White Paper, Dec. 2011.

[13] D. Linden, *Handbook of Batteries, Fuel and Energy Abstracts*, vol. 4, no. 36, 1995.

[14] M.S. Halper and J.C. Ellenbogen, 'Supercapacitors: a brief overview', The MITRE Corporation, McLean, Virginia, USA, pp. 1–34, 2006.

[15] California Energy Commission, 'Flywheel systems for utility scale energy storage: a transformative flywheel project for commercial readiness', Technical Report.

[16] B. Hadjerioua, K. Stewart, S. DeNeale, *et al.*, 'Pumped storage hydropower FAST commissioning technical analysis', No. PNNL-30199, Pacific Northwest National Lab.(PNNL), Richland, WA (United States), 2020.

[17] M. Götz, J. Lefebvre, F. Mörs, *et al.*, 'Renewable power-to-gas: a technological and economic review', *Renewable Energy*, vol. 85, pp. 1371–1390, 2016.

[18] C. Golling, R. Heuke, H. Seidl, and J. Uhlig, 'Component of an integrated energy transition, roadmap power to gas', German Energy Agency, Sept. 2019.

[19] U.S. Department of Energy, 'Energy storage grand challenge: energy storage market report', Technical Report, NREL/TP-5400-78461, DOE/GO-102020-5497, Dec. 2020.

[20] I. Sarbu and C. Sebarchievici, 'A comprehensive review of thermal energy storage', *Sustainability*, vol. 10, no. 1, p. 191, 2018.

[21] 'Benefits of demand response in electricity markets and recommendations for achieving them', U.S. Department of Energy, Feb. 2006.

[22] 'Demand response RERT Trial Year 1 report', Australian Renewable Energy Agency (ARENA), Mar. 2019.

[23] '2020–2021 winter reliability assessment report', North American Electric Reliability Corporation (NERC), Nov. 2020.

[24] 'Success stories in building energy efficiency', Center for Clean Air Policy.

[25] 'Building energy efficiency R&D roadmap', National Climate Change Secretariat.

[26] 'V2G global roadtrip: around the world in 50 projects', UK Power Networks, Innovate UK, Everoze and EVConsult, Oct. 2018.

[27] 'Electric Nation customer trial final report', Western Power Distribution, Oct. 2019.

Chapter 2

Microgrids: introduction and research problem descriptions

2.1 Definition

For most cases, the distributed energy resources (DERs) are connected to the low-voltage (LV) or medium-voltage (MV) distribution grid, which can form a *Microgrid* system. There are several definitions of a microgrid as follows:

The U.S. Department of Energy (DOE) defines a microgrid as '*a group of interconnected loads and DERs within clearly defined electrical boundaries that acts as a single controllable entity with respect to the main grid. A microgrid can connect and disconnect from the main grid to enable it to operate in both connected or island-mode*' [1].

The CIGRE C6.22 Working Group defines that '*Microgrids are electricity distribution systems containing loads and DERs, (such as distributed generators, storage devices, or controllable loads) that can be operated in a controlled, coordinated way either while connected to the main power network or while islanded*' [2].

Ref. [3] describes the microgrid as '*comprising low-voltage (LV) distribution systems with DERs. Such systems can operate either connected or disconnected from the main grid. The operation of DERs in the network can provide benefits to the overall system performance, if managed and coordinated efficiently*'.

To summarise, a microgrid has the following characteristics:

- Operated at LV to MV levels, typically in the range of 400 V to 69 kV.
- Comprises multiple DERs such as distributed generators (DGs), energy storage systems (ESSs), and flexible loads which are managed and dispatched by the microgrid control system.
- Exists in different sizes from hundreds of kilowatts to tens of megawatts, in the form of such as campus/institutional microgrid, building microgrid, commercial and industrial microgrid, military-based microgrid, or utility microgrid that involves a segment of the regulated grid.
- Deployed as an isolated system that has no physical connection to the main grid or coupled with the main grid through the point of common coupling (PCC) with a clearly defined electrical boundary.

- If coupled with the main grid, the microgrid should be able to operate in either grid-connected or islanded mode as a single controllable entity with respect to the main grid. For example, during a main grid failure, the microgrid should be able to disconnect from the main grid and operate autonomously, also known as the islanding capacity.

Compared with a traditional bulk power system, a microgrid is considerably different. Specifically, the microgrid system size is much smaller than that of a conventional large, interconnected power system. Besides, the microgrid networks are relatively short and have a higher resistance to reactance (R/X) ratio compared to the high-voltage power grids. As a result, the dynamic performance of microgrid and the physical relationships between voltage, angle, and active and reactive power flows are significantly different from the conventional grids. Another consequence of the small grid size is the higher-level operational uncertainty in the system, due to the large share of intermittent renewable energy sources. Moreover, since many DER units are interfaced with the grid via power-electronic converters, the system inertia is considerably lower in the microgrid compared to a traditional power grid. In the meantime, unlike the conventional bulk power system, where the generation plants and ESS stations are generally owned and centrally controlled by the utilities, the DER units in the microgrid can also be owned by end users. As a result, these DERs may not be controlled by a central controller, rather, they may be controlled in a decentralised or distributed manner. All of the above intrinsic differences between microgrids and bulk power systems require specialised methodologies.

This book focuses on three research areas of the microgrid: *planning*, *operation*, and *control*:

- The planning aims to determine the optimal size and site as well as investment plan of the DER units in the microgrid over a long-time horizon, e.g. 5–20 years.
- The operation of the microgrid focuses on the operational dispatch horizon over a short-term time interval, e.g. from 30 min to a few hours or one day ahead. The objective is to dispatch the DERs for optimal energy management and bus voltage regulation.
- The control of the microgrid is concerned with the real-time timescale, i.e. milliseconds to seconds, and the objective is to maintain the voltages and frequency within desired operational ranges, and achieve power sharing and balancing among DERs and loads.

2.2 Microgrid architecture and classification

Microgrids have multiple possible architectures depending on their size and functionalities. To reflect the recent trend of multiple energy integration, this book classifies the microgrid into two main architectures according to different energy carries: (1) *electric microgrid*, i.e. a power grid integrated with DERs that carries electric power only, and (2) *multi-energy microgrid* (*MEMG*), i.e. a coupled energy network of the power grid and the thermal grid integrated with DERs that carries both electric power and thermal power (heat and cooling).

2.2.1 Electric microgrid

Depending on the current format, the electric microgrid can be further classified into AC (alternative current) microgrid, DC (direct current) microgrid, and hybrid AC/DC microgrid.

2.2.1.1 AC microgrid

The AC power distribution network is most commonly used for microgrid development and implementation. In this architecture, the power is transmitted from the supply to the demand in AC form. DER units are directly connected to or interfaced via power-electronic converters with the distribution grid. In practice, the existing AC network infrastructure (distribution network, transformers, feeders, protection systems, etc.) can be retrofitted to be an AC microgrid with necessary communication and control system.

Figure 2.1 illustrates a typical AC microgrid. There are two networks, one is the power distribution network and the other is communication network.

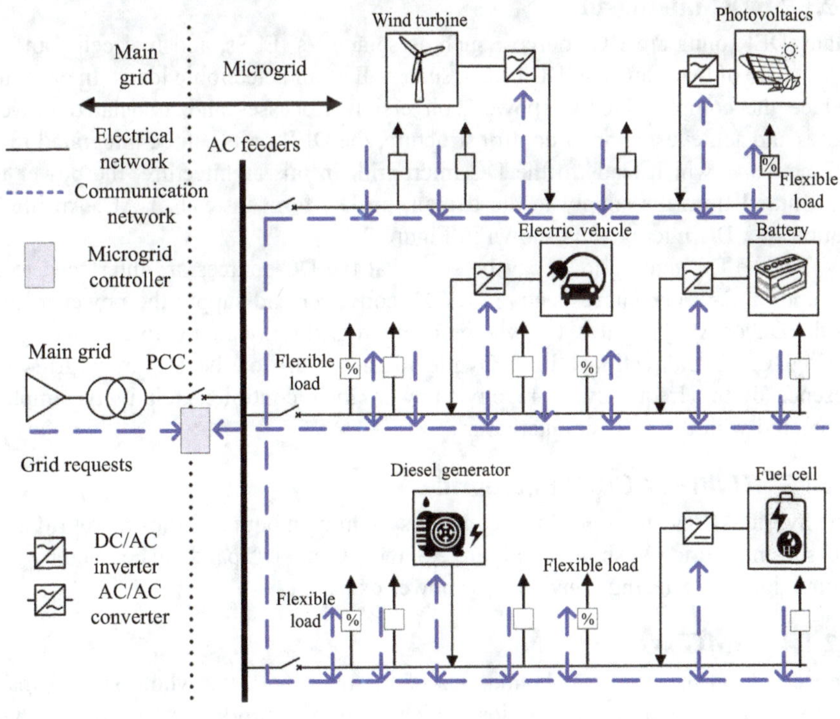

Figure 2.1 Architecture of a typical AC microgrid

In the power distribution network, various DER units, including photovoltaics (PVs), small wind turbines, electric vehicles (EVs), battery ESSs, diesel generators, fuel cells, and flexible loads, are directly connected to or interfaced via power-electronic converters with the network feeders. The whole microgrid is coupled with the main grid through a PCC, which can switch the microgrid between grid-connected or islanded operation mode. For cases where the main grid has a higher-voltage level, the PCC switch is connected to the LV side of the transformer, as shown in Figure 2.1.

The communication network allows information exchange between each component and the microgrid control system, based on which the microgrid operation and control are achieved. A microgrid communication system should be designed based on requirement and features of participating devices [4]. The communication networks can be categorised according to the coverage and functions, e.g. home area networks (HANs), neighbourhood area networks (NANs), and wide area networks (WANs). The communication signals can be transmitted by both wired communication (e.g. power line and optical fibre) and wireless communication (e.g. ZigBee, WiFi, and 5G). The Internet Protocol (IP) suite is a set of protocols to provide Internet services and is also used for microgrid applications. It can be applied to carry specific automation protocols for microgrids, such as Modbus, Distributed Network Protocol version 3.3 (DNP3), and IEC 61850.

2.2.1.2 DC microgrid

Many DER units are DC sources, such as solar PVs, ESSs, and fuel cells, and a large share of loads are also DC based, such as EVs and electronic loads. In order to reduce the complex DC/AC power conversion processes and associated power losses and achieve a simpler control structure, the DER units can be integrated to a DC network, which leads to the DC microgrid. In this architecture, the power is transmitted from the supply to the demand in DC form instead of AC. An illustration of a DC microgrid is shown in Figure 2.2.

For the DC microgrid, it can be seen that the DC sources are interfaced to a common DC feeder via respective DC/DC converter and supply the power to the loads (DC or AC). Besides, the whole DC microgrid is coupled with the main grid via a DC/AC bidirectional link. Owing to the nature of the DC microgrids in absence of the frequency or reactive power, their control is relatively simpler compared with their AC counterparts.

2.2.1.3 Hybrid AC/DC microgrid

The hybrid AC/DC microgrid is a coupled system with both AC sub-microgrid and DC sub-microgrid. As shown in Figure 2.3, the AC and DC parts are coupled by the bidirectional interfacing converter for power exchange.

2.2.1.4 NMG system

Depending on the electrical boundaries of the microgrid, the whole system may comprise multiple single microgrids, which forms a networked-microgrid (NMG) system. According to voltage levels, the NMG system can be formed by both LV

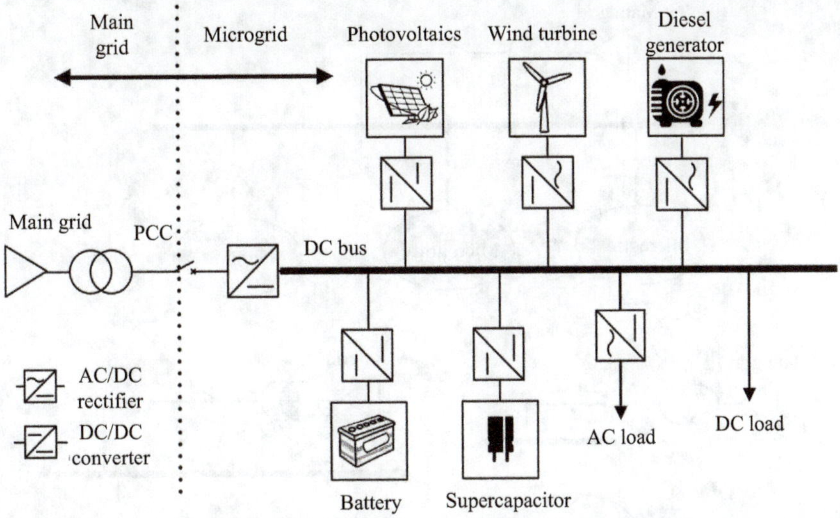

Figure 2.2 Architecture of a typical DC microgrid

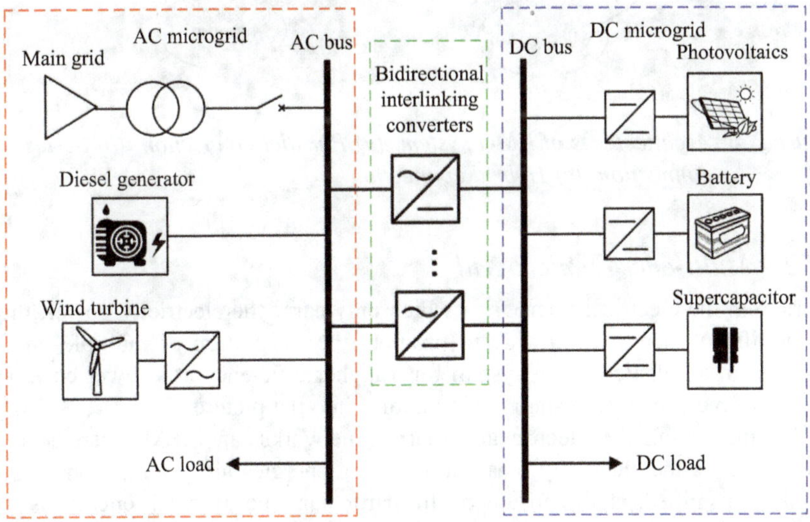

Figure 2.3 Architecture of a typical hybrid AC/DC microgrid

MGs and MV MGs. Each MG can be connected by tie lines or LV/MV transformers. From the structure perspective, the NMG system can be divided into three types: serial, parallel, and hybrid NMG system. Figure 2.4 shows the typical structures of NMG systems.

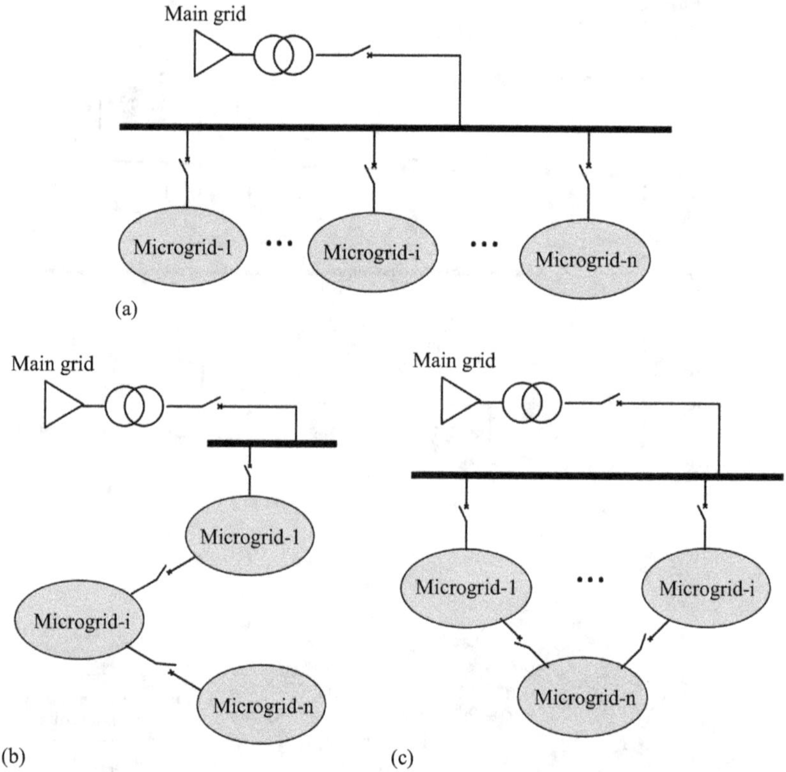

Figure 2.4 Architectures of NMG system. (a) Parallel connection. (b) Series connection. (c) Hybrid connection

2.2.2 Multi-energy microgrid

Apart from the electrical microgrids which only carry the electricity power, there are significant efforts worldwide to promote the integration of multiple energy resources in a holistic energy system for a higher efficiency at a lower cost. The concept of multi-energy system (MES) comes into the picture [5].

By integrating the electric and thermal networks, an MEMG can be constructed where the electricity, heat, and cooling energies are simultaneously generated, transmitted, and consumed. In particular, the thermal energy is not suitable for long-distance transmissions due to high energy transmission losses, and the MEMG offers a promising solution to introduce distributed heating and cooling.

The typical structure of an MEMG is shown in Figure 2.5.

Unlike the electrical microgrid, the MEMG also comprises the thermal energy flow and associated thermal inertia-based DERs. Usually in the MEMG, there are renewable energy-based generators such as wind turbines and PV cells, battery storage, combined cooling heat, and power (CCHP) units, fuel cells, power to thermal conversion units like electric boilers and electric chillers, and thermal

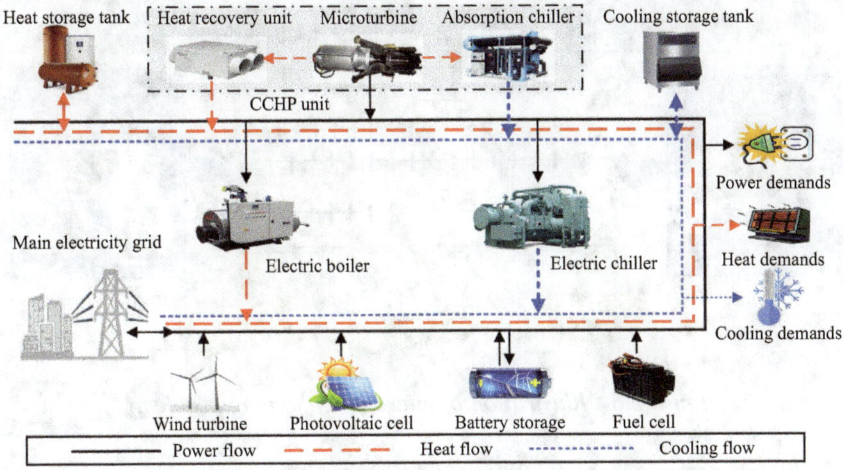

Figure 2.5 Architecture of the MEMG

storage, i.e. heat and cooling storage tanks. Besides, for the multi-energy loads, they can be flexible as usually the MEMG would sign long-term demand response contracts. Besides, there is a switch between the main grid and MEMG: when the switch is on, the microgrid operates in the grid-connected mode, and the main grid would provide power support for the MEMEG; while off, it is in the islanded mode and the MEMG has to sustain itself.

For the multi-energy interaction and coordination in the MEMG, first, the CCHP units could consume natural gas to produce electricity, and at the same time, the waste heat can be used by the heat recovery unit and absorption chiller to produce heat and cooling energy. In this regard, the holistic energy utilisation efficiency can be greatly improved. In addition, through the multi-energy conversion, the power-to-thermal conversion units significantly contribute to enhancing the interaction of different energy systems. Finally, the multi-energy storage and demand response scheme could further enhance the holistic energy dispatch flexibility to reduce the multi-energy supply cost.

2.3 Planning of DER units in microgrid

Microgrid can be implemented for different applications, such as campus/institutional microgrid, residential building and community microgrid, commercial and industrial microgrid, military-based microgrid, or utility-owned microgrid which is a segment of the regulated grid. Within the microgrid, the DER units can be owned by different parties, including the customers, utilities, as well as service providers or agents. Therefore, at the investment stage, one key research problem is to determine the deployment plan of the DER units in the microgrid towards a certain cost-effectiveness objective while satisfying the financial and operational constraints.

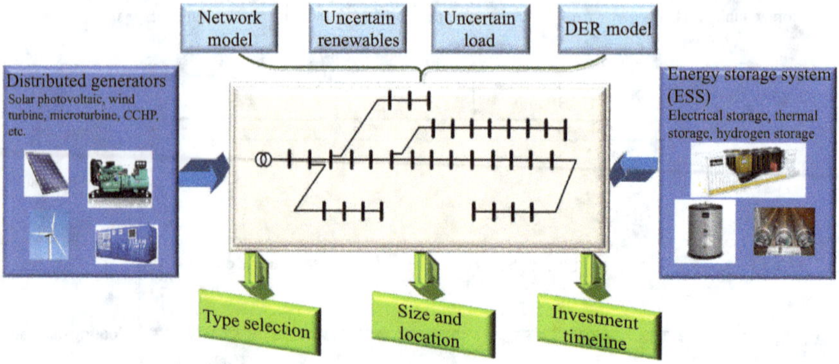

Figure 2.6 Illustration of microgrid planning problem

This research problem is illustrated in Figure 2.6. In general, there are two physical DER units which include DGs and ESS units to be installed in the microgrid. Specifically, the planner needs to select the types of the respective DER units, and decides their sizes and locations in the microgrid together with the installation timeline.

As a decision-making problem, the DER planning can be formulated as an optimisation model towards the economic and technical objectives considering constraints. Specifically, the key research problems are as follows:

- The microgrid network, including electrical microgrid and the MEMG, should be accurately modelled to reflect the power or thermal flow in the network. Given the highly non-linear and non-convex nature of the power and thermal flow, the network should be modelled in a mathematically tractable way without losing engineering accuracy.
- The DER units including DGs and ESSs should be accurately modelled to reflect their economic and technical characteristics. Similarly, the units should be modelled in a mathematically tractable way without losing engineering accuracy. Besides, for the non-dispatchable DER units, their stochastic power output nature should also be well modelled.
- The uncertainties of the renewable energy-based DER units and the load demands should be well modelled and addressed, including the short-term operational uncertainties such as the stochastic power output and load variation, and the long-term uncertainties such as the load demand growth and inflammation rate. The programming methods for optimisation under uncertainties should be employed in this regard, such as stochastic programming and robust optimisation, as well as their variants.
- The investment phase should holistically consider the operation phase; that is, once the DER units are installed, their operation and dispatch should be well accounted for. Therefore, the investment stage and the operation stage should be jointly modelled so as to minimise the total investment and operation costs over the entire planning horizon.

- The leading optimisation model can be of high dimensions and in large sizes. How to effectively solve this model within a reasonable amount of time is another research challenge. In this regard, efforts should be put into both problem formulations and solution algorithms.

In this book, the planning problem is covered in Part II.

Chapter 3 presents a composite sensitivity factor-based planning method for electrical microgrid. It aims to optimise the locations and sizes of the utility-owned renewable energy-based distributed generation units (such as wind turbines) to simultaneously decrease the network power losses and improve the voltage stability in microgrids.

Chapter 4 introduces a novel probability-weighted robust optimisation method to allocate dispatchable DG such as microturbines and non-dispatchable DG such as wind turbines in the electrical microgrid. It aims to maximise the total operating profit under uncertainties of renewable power generation and loads throughout a long-term planning horizon.

Chapter 5 presents a two-stage coordinated stochastic programming method for optimally placing heterogeneous DGs in an MEMG considering the uncertainties from renewable energy sources.

Chapter 6 introduces a risk-averse multi-stage adaptive stochastic optimisation approach for heterogeneous energy storage deployment in a residential MEMG, considering the diverse uncertainties and multi-energy demand-side management.

2.4 Microgrid operation

The operation of the microgrid focuses on the operational dispatch horizon over a short-term time interval, e.g. from 30 min to a few hours or one day. The objective is to dispatch the DERs for optimal energy management and bus voltage regulation.

Depending on whether there is a physical connection with a main grid and practical needs, the microgrid can operate in either grid-connected or islanded mode. In the grid-connected operation mode, the main grid can be viewed as a slack bus (i.e. an infinite generation source) to balance the power demand and supply of the microgrid through the PCC. Besides, the microgrid as a whole entity can be managed to trade electric/thermal energy with the main grid. For the islanded operation mode, since there is not a slack bus, the microgrid should be able to balance the energy supply and the demand within the grid. Under emergency cases, e.g. when the supply is insufficient, the load curtailment is needed. Such load shedding can be voluntary under the demand response programme.

For most of the cases, the microgrid operation problem is also formulated as an optimisation model with predefined objective function(s) and operational constraints. Specifically, the key research problems are as follows:

- Similar to the planning problem, the operation of the DERs is also based on accurate modelling of the microgrid network and the DER components. For different DERs, their operational cost and benefit characteristics, as well as the uncertainty in power generation and load demand should be well modelled.

- For the energy management problem, different operation objectives can be considered, such as minimising the total operation costs or maximising the total revenue of the microgrid.
- Due to the high R/X ratio of the distribution network, the real power has a considerable impact on the bus voltage magnitude. Because of the highly intermittent and uncertain power output of the non-dispatchable DGs, the microgrid will experience fast and stochastic voltage fluctuations and deviations. Traditional volt/var regulation devices such as on-load-tap-changer and capacitor bank are mechanical equipment and are not fast enough to respond to the voltage variations. In this case, the power-electronic inverters that interface the DER units (such as solar PV) should be used to provide fast and flexible volt/var support. Therefore, proper coordination of the inverters and mechanical volt/var regulation devices is needed.
- Different DERs, i.e. generators, storage units, and loads, have substantially different intrinsic features. Even for the same DER category, different units could have significantly operational characteristics. First, the operational response time of the DERs can be very different. For instance, the price-based DR needs a longer lead time to be dispatched, e.g. one day ahead, while the direct load control (DLC) can be much faster since the loads can be directly shed by the operators. Besides, compared with electric power, thermal energy is of slow response speed due to the thermal inertia. Therefore, for MEMG, the thermal-related units need a longer lead time than the electric units. Second, the operational flexibility of the DERs is also significantly different. For example, the ESSs can be flexibly dispatched to charge and discharge power/thermal energy within a short time horizon. By contrast, the microturbine, once turned on, is not so flexible to be turned off considering the shut-up/shutdown cost.
- Unlike the planning problem which aims at a long-term decision process and not time critical, the optimisation of the DER operation should be solved within a short time window, so that the dispatch signal can be timely delivered to the DERs ahead of each dispatch interval. Therefore, the solution speed of the operational optimisation model should be fast enough.
- One critical research challenge for the DER operation problem is to address multiple operational uncertainties in the microgrid, such as stochastic variations of power output of non-dispatchable DGs, power and thermal energy loads, electricity prices, as well as uncertain demand response behaviours to the electricity price, etc. Therefore, the programming methods for optimisation under uncertainties should be employed in this regard, such as stochastic programming and robust optimisation, as well as their variants.

In this book, the DER operation problem is covered in Part III. In particular, to address the diverse uncertainties and different operational characteristics of the DERs, we adopt a temporally coordinated operation framework (see Figure 2.7), where two operational stages are designed: the first stage aims to dispatch the slow-responding DERs over a relative longer time horizon, e.g. hours to a day, and the second stage is for fast-responding and more flexible DERs over a relatively shorter time interval, e.g. 15 min to an hour. Formulated as a two-stage coordinated optimisation model, the second stage can be viewed as a recourse action to compensate for decisions in the first stage to address the uncertainties.

Part III-1 focuses on the energy management of the microgrid.

Chapter 7 introduces a robust coordinated operation approach for ESS and DLC. They are, respectively, scheduled in hourly and 15-min intervals for the two stages. They are optimally coordinated through a two-stage robust optimisation model considering the worst case of the uncertain variations of renewable energy resources.

Chapter 8 presents a similar robust optimisation approach for price-based DR and dispatchable DG. The daily price for customers is scheduled one day ahead and microturbine outputs are dispatched hourly. The two operation stages are robustly coordinated considering uncertainties from renewables and loads.

Chapter 9 introduces a system-wide optimal coordinated energy dispatch method for an MEMG in both the grid-connected and islanded modes. The proposed energy dispatch method aims to minimise the microgrid net operating cost and enhance the dispatch flexibility in supplying power, heat, and cooling in the day-ahead energy market.

Chapter 10 presents a two-stage coordinated stochastic programming method for an MEMG under diverse uncertainties from renewable energy sources, power loads, and electricity transaction prices between the microgrid and the utility grid.

In Chapter 11, the price-based DR and indoor temperature control are utilised to flexibilise the electric and thermal loads, respectively. A two-stage operation approach is proposed to optimally coordinate the CCHP units, flexible electric and thermal loads, and thermal storage units. The coordination is optimised by the robust optimisation method to guarantee solution robustness under multiple uncertainties.

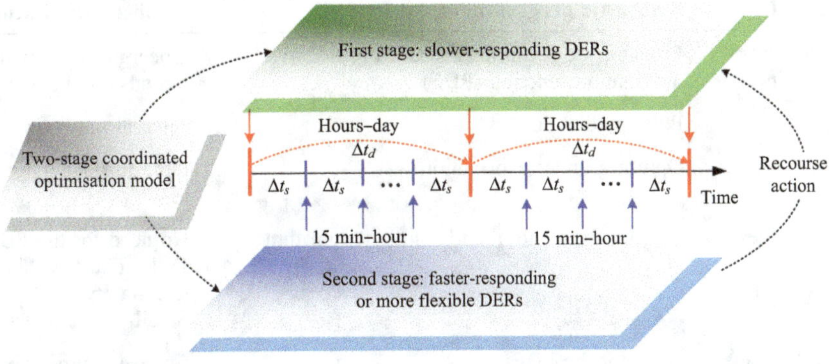

Figure 2.7 Temporally coordinated operation framework for the DERs in microgrid

Part III-2 focuses on the voltage regulation of the microgrid.

Chapter 12 presents a two-stage coordinated stochastic programming method for volt/var control of the microgrid. In the hourly timescale (first stage), capacitor banks and transformer tap changers are scheduled before stochastic renewable outputs and load variations are realised. In the 15-min timescale (second stage), inverters that interface with the renewable energy resources provide var support to supplement the first-stage decision after uncertainties are observed.

Chapter 13 presents a three-stage robust inverter-based voltage/var control approach for high PV penetrated microgrids. In the first stage, capacitor banks and an on-load tap changer are scheduled hourly in a rolling horizon. In the second stage, PV inverters are dispatched in a short time window. In the third stage, the inverters respond to real-time voltage violations through local droop controllers. To address uncertain PV output and loads, the robust optimization method is used to coordinate the three operation stages and guarantee robust solutions.

It is important to note that to solve the optimisation problems including the planning and the operation of the microgrid, this book mainly adopts two optimisation methods, stochastic programming and robust optimisation. A general comparison between them is given in Table 2.1, and detailed formulation and algorithms can be found in the aforementioned chapters.

2.5 Microgrid control

The control of the microgrid is concerned with the real-time timescale, i.e. milliseconds to seconds, and the objective is to maintain the voltages and frequency within desired operational ranges, and achieve power sharing and balancing among DERs and loads. Note that this book only targets at the electrical microgrid, which means the control only concerns the frequency and voltage controls. Inheriting

Table 2.1 Comparison of stochastic programming and robust optimisation

Method	Stochastic programming	Robust optimisation
Uncertainty modelling	Probabilistic scenarios based on probability distribution function (PDF)	Uncertainty set with bounds and budgets
Input	Point prediction	Interval prediction
Model	Optimise under expectation	Optimise under worst case
Advantages	Simpler formulation and solution algorithm	No need for the PDFs Fully robust within uncertainty sets Small problem size
Disadvantages	Need for the PDFs Probabilistic robustness Large problem size	Complex formulation and solution process Tend to be conservative

from bulk power systems, the microgrid can also be controlled in a hierarchical framework [6], as illustrated in Figure 2.8.

- **Primary control:** implemented at the individual DER units, it aims at localised frequency and voltage stabilisation and regulation. The primary control includes the inner control loops of power converters and generators. Besides, the primary control should control power converters and generators to provide instantaneous active/reactive power response to the frequency/voltage variations. Typically, the droop-based control law is used in this regard, and its response speed is very fast (on the order of milliseconds), which helps to stabilise the system. However, due to its proportional control nature, the frequency and voltage deviations cannot be eliminated.
- **Secondary control:** the main purpose of secondary control is to eliminate the frequency/voltage deviations and achieve accurate power sharing. Besides, the synchronisation control and power quality improvement can also be included in this control level.
- **Tertiary control:** the main purpose of the tertiary control is to increase the economy and resilience of microgrids. The tertiary control contains the economic dispatch, optimal power flow, unit commitment functions of microgrids, as well as the energy exchange with external systems. The business model and market level design of microgrids also belong to this control level. Note that the tertiary control essentially corresponds to the operation phase of the microgrid, which is covered in Part III of this book.

It is important to point out that depending on practical microgrid deployment, the above secondary control objectives can be integrated with the energy management and economic dispatch function within a microgrid. In this case, the secondary control is referred to as the microgrid energy management system, and the tertiary control is defined for coordinating the operation of multiple microgrids and communicating needs or requirements from the host grid [7]. Many recent research works have been reported that follow this control hierarchy definition, e.g., [8–10].

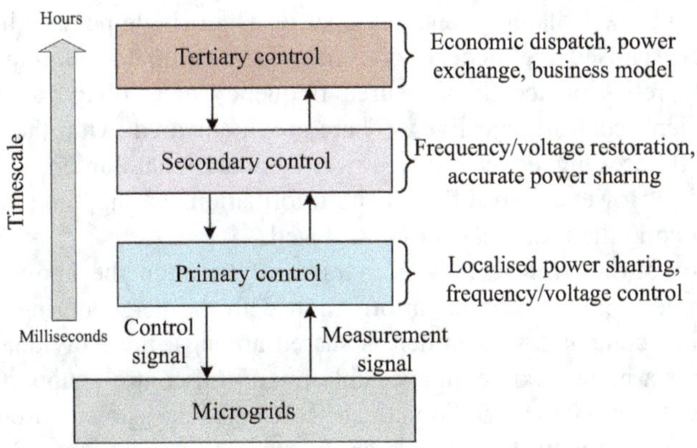

Figure 2.8 Illustration of the hierarchical control framework of the microgrid

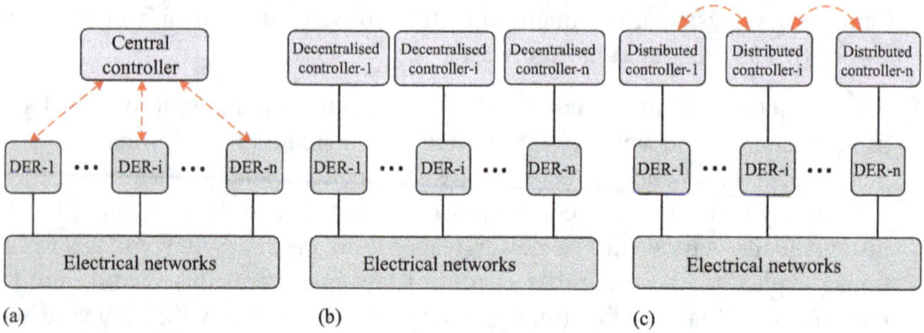

Figure 2.9 Illustration of three microgrid control system architectures. (a) Centralized architecture. (b) Decentralized architecture. (c) Distributed architecture

In this book, we categorise that all the energy management and optimal dispatch functions of microgrids into the tertiary level, aiming to differentiate it from the real-time coordinated (distributed) control in secondary level, i.e., eliminating frequency/voltage deviations and achieving accurate power sharing without a central controller, which is the focus of Part IV. However, it should be noted that the control hierarchy definitions in [7] could better reflect most realistic microgrids, which rely on a centralised control system other than distributed control.

For the microgrid control system, the architecture can be generally classified into three topologies: centralised, decentralised, and distributed, as illustrated in Figure 2.9.

- The **centralised architecture** relies on a single controller which communicates with all the other components in the network. The centralised control is most mature and enables a global coordination in the microgrid system. However, it requires a one-to-all communication infrastructure which may not be available in the microgrid and/or conflict with the autonomous management of customer-owned DERs. Besides, it has a significant computational complexity and low scalability, and could suffer from single point failure risks.
- The **decentralised architecture** does not need a communication infrastructure and only relies on locally measured frequency and voltage to achieve the independent control objectives. Therefore, compared with the centralised control, it has a higher reliability, lower computational burden, and increased scalability. However, since there is no information sharing, the global coordination among the DERs cannot be achieved.
- The **distributed architecture** is a trade-off between the above two. Each control agent only exchanges information with the neighbouring agents, and the overall computational burden is shared amongst the individual agents. It allows for a more flexible and scalable microgrid control, supports the plug-and-play function (e.g. additional DERs can be integrated through a low-bandwidth communication link to an existing DER, with the communication topology being a tunable design variable), protects customer data privacy, and

is less prone to single point failure risk. This book mainly focuses on this type of control architecture.

Depending on different operation modes, the microgrid control objective is different.

- In the **grid-connected mode** of operation, the frequency of the microgrid and the voltage at the PCC are dominantly determined by the main grid. In this case, the DERs can be controlled to provide frequency support to the entire power system and provide reactive power support for microgrid voltage regulation purpose.
- In the **islanded mode**, the microgrid operates as an independent entity. In this case, the voltages and frequency of the microgrid are no longer supported by a main grid and should be controlled by the DER units. The control system should ensure active and reactive power to be accurately balanced within the microgrid.

The microgrid operational optimisation is an open-loop decision process, where the output is the dispatch signal to each DERs, e.g. the power output of a DG for the next one hour. In contrast, the microgrid control system sends real-time control signals to the DERs and collects the feedbacks of the DERs and the whole system, e.g. frequency and voltage, which is a closed-loop. Therefore, the real-time control performance of the microgrid control system should be sufficiently tested and validated. In this case, in addition to traditional simulation tests, the hardware-in-the-loop (HIL) validation becomes popular. Typically, an HIL platform consists of a real-time simulator (RTS) and the interconnected hardware facilities. It can be classified into two types:

- **Controller HIL (CHIL):** The power system is simulated in real time within the RTS while the complete or a proportion of its controls is embedded within an external micro-controller interconnected with the RTS.
- **Power HIL (PHIL):** A portion of the power system is simulated in real time within the RTS while the remainder of the power system is represented by physical power equipment interconnected to the RTS.

Compared with the digital simulation, the HIL platform emulates a more realistic environment for the evaluation of the controllers.

2.6 Microgrid stability

Given the significant differences between the microgrid and the conventional bulk power system (as discussed in Section 2.1), the dynamic performance of microgrids and the intrinsic mathematical relationships among bus voltages, rotor angles, and power flows are also significantly different. For example, in a conventional bulk power system, transient and voltage stability problems typically occur more often than frequency stability ones, whereas in islanded microgrids, maintaining frequency stability is more challenging due to the low system inertia and high-level penetration of intermittent RES. Besides, some stability problems observed in large interconnected systems, such as inter-area oscillations and voltage collapse, are not observed in microgrid systems. As a result, the microgrid stability should be specifically defined [11].

In [11], the microgrid stability is generally categorised into Control System Stability (i.e., phenomena pertaining to the equipment control systems) and Power Supply and Balance Stability (i.e., phenomena pertaining to active and reactive power sharing and balance).

- **Power supply and balance stability:** it refers to the ability of the system to maintain power balance and effectively share the demand power among DERs, so that the system satisfies operational requirements. These types of stability problems are associated with the loss of a generation unit, violation of DERs limits, poor power sharing among multiple DERs, wrong selection of slack(s) resources, and/or involuntary no-fault load tripping. In addition, certain type of loads, such as constant power loads or induction motors, may trigger certain types of instability in the system, such as voltage and harmonic problems. This class of stability can be subcategorised into Frequency and Voltage Stability [11].
- **Control system stability:** this stability issues may arise due to inadequate control schemes (e.g., harmonic resonance of parallel DERs) and/or poor tuning of one or more pieces of equipment controllers. In the latter case, the poorly tuned controller(s) is the primary source of instability, and the system cannot be stabilised until the controller is re-tuned or the associated piece of equipment is disconnected. This type of stability is concerned on electric machines and inverters control loops, LCL filters, and phase locked loops (PLLs). This category of stability is subcategorised into Electric Machine and Converter Stability [11].

The detailed classification of the microgrid stability by [11] is shown in Figure 2.10.

Table 2.2 summarises the characteristics of various types of stability issues in microgrids [11].

The primary and secondary control of the microgrid are covered in Part IV of this book, where the secondary control is based on a distributed architecture. Part IV-1 focuses on the frequency control support from the DERs to a grid-connected microgrid. In Chapter 14, the distributed ESS units are aggregated together to provide frequency control support to the power system, and a finite-time leader-follower consensus controller is designed to control the small-scale ESS units via sparse communication networks. In Chapter 15, the grid-interactive smart buildings are used to support power system frequency control. A distributed sliding mode controller with a leader-follower communication structure is designed, where the leader provides power and comfort/energy-level references for the smart building aggregator based on the area frequency deviation, and references are tracked by each smart building (follower).

Part IV-2 focuses on the voltage control support from the DER inverters. Chapter 16 presents a hybrid control strategy based on DER inverters for voltage

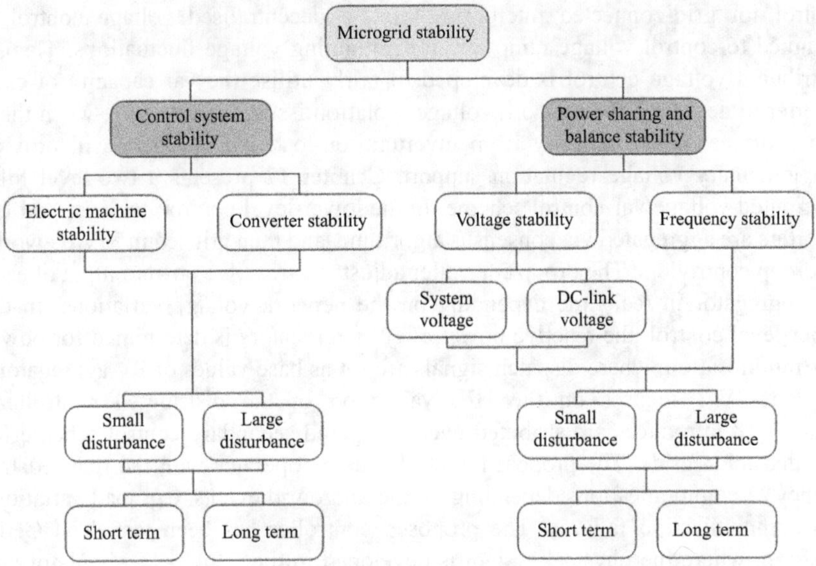

Figure 2.10 Classification of microgrid stability [11]

Table 2.2 Characteristics of microgrid stability issues [11]

Category	**Control system stability**		**Power supply and balance stability**	
Subcategory	**Electric machine stability**	**Converter stability**	**Voltage stability**	**Frequency stability**
Root cause	Poor controller tuning	Poor controller tuning, PLL bandwidth, PLL synchronisation failure, harmonic instability	DERs power limits, inadequate reactive power supply, poor reactive power sharing, load voltage sensitivities, DC-link capacitor	DERs active power limits, inadequate active power supply, poor active power sharing
Phenomenon	Undamped oscillations, aperiodic voltage and/or frequency increase/drop	Undamped oscillations, low steady-state voltages, high-frequency oscillations	Low steady-state voltages, large power swings, high dc-link voltage ripples	High rate of change of frequency, low steady-state frequency, large power and frequency swings

control for grid-connected microgrid. First, a decentralised voltage control is designed to control voltage ramp rate for mitigating voltage fluctuations. Then, a distributed voltage control is developed to fairly utilise the var capacity of each inverter to deal with the network voltage violation issue. Furthermore, when there is a shortage of var capacity from inverters, on-load tap changers will provide supplementary voltage regulation support. Chapter 17 presents a two-level fully distributed voltage/var control scheme. In the lower-level control, the user-end PV inverters are aggregated via consensus algorithms and then utilised in MV networks by droop controllers. The droop controller adjusts the reactive power output of each PV aggregator in real time depending on the network voltage variations. In the upper-level control, the reactive power of PV aggregators is determined for power loss minimisation, where dispatch signals are set as base values of PV aggregators.

Part IV-3 focuses on the HIL validation of the distributed controllers. Chapter 18 introduces a distributed event-triggered secondary control scheme for islanded microgrids. The proposed control scheme operates with the reduced frequency of communications depending on the microgrid events, e.g. load variations and communication failures. The proposed controller has been tested on CHIL platform, where the physical system is developed in the OPAL-RT simulator and the cyber system is realised in embedded systems 'Raspberry Pis'. Chapter 19 presents a new three-level coordinated control method for PV inverters for voltage control support. The proposed control method can smooth the voltage profiles, restrain the voltage rise/drop problems, and coordinate all PV inverters in real time when there is no feasible local solution. The PHIL experiment has been conducted for validating the proposed control method under various scenarios.

References

[1] 'DOE Microgrid Workshop Report', pp. 1–32, San Diego, California, August 30–31, 2011.

[2] C. Marnay, S. Chatzivasileiadis, C. Abbey, *et al.*, 'Microgrid evolution roadmap', *2015 International Symposium on Smart Electric Distribution Systems and Technologies (EDST)*, pp. 139–144, 2015, doi:10.1109/SEDST.2015.7315197.

[3] N. Hatziargyriou, *Microgrids: Architectures and Control*, UK: Wiley-IEEE Press, 2014, ISBN: 978-1-118-72068-4.

[4] S. Kumar, S. Islam, and A. Jolfaei, 'Microgrid communications – protocols and standards', in *Variability, Scalability and Stability of Microgrids*, UK: IET, pp. 139–144, 2019.

[5] P. Mancarella, 'MES (multi-energy systems): an overview of concepts and evaluation models', *Energy*, vol. 65, pp. 1–17, 2014.

[6] A. Bidram and A. Davoudi, 'Hierarchical structure of microgrids control system', *IEEE Transactions on Smart Grid,* vol. 3, no. 4, pp. 1963–1976, 2012.

[7] D. E. Olivares, A. Mehrizi-Sani, A. H. Etemadi, *et al.*, 'Trends in microgrid control', *IEEE Transactions on Smart Grid*, vol. 5, no. 4, pp. 1905–1919, 2014.

[8] C. Zhang and Y. Xu, 'Hierarchically-coordinated voltage/VAR control of distribution networks using PV inverters', *IEEE Transactions on Smart Grid*, vol. 11, no. 4, pp. 2942–2953, 2020.
[9] B. Alghamdi and C. A. Cañizares, 'Frequency regulation in isolated microgrids through optimal droop gain and voltage control', *IEEE Transactions on Smart Grid*, vol. 12, no. 2, pp. 988–998, 2021.
[10] S. Cordova, C. A. Cañizares, A. Lorca, and D. E. Olivares, 'An energy management system with short-term fluctuation reserves and battery degradation for isolated microgrids', *IEEE Transactions on Smart Grid*, vol. 12, no. 6, pp. 4668–4680, 2021.
[11] IEEE-PES Task Force on Microgrid Stability Analysis and Modeling, 'Microgrid stability definitions, analysis, and examples', *IEEE Transactions on Power Systems*, vol. 35, no. 1, pp. 13–29, 2020.

Part II

Coordinated Planning of DERs in Micogrids: Optimal Sizing and Siting

Chapter 3

Composite sensitivity factor-based method for DG planning

Distributed generation (DG) can provide multiple benefits to distribution systems such as network power loss reduction and voltage stability enhancement. In this chapter, a novel composite sensitivity factor (CSF)-based method is introduced for optimal placement of network-owned DG units to simultaneously decrease losses and improve voltage stability in a distribution network. This method prioritises the buses which are more sensitive to power loss and voltage stability, and then it applies sensitivity factors to settle DG units iteratively. Besides, uncertainties of renewable DG outputs are fully considered with a discrete Monte Carlo simulation. Via tests on two distribution systems with different scenarios, this chapter presents high computing efficiency and unique satisfying solutions of the CSF-based method for DG planning.

Nomenclature

A Acronyms

CSF	Composite sensitivity factor
DG	Distributed generation
GA	Genetic algorithm
LLRI	Line loss reduction index
MINLP	Mixed-integer non-linear programming
MOGA	Multi-objective genetic algorithm
PDF	Probability density function
VCPI	Voltage collapse proximity indicator
VCPIRI	Voltage collapse proximity indicator reduction index
WT	Wind turbine

B Indices, numbers, and sets

i/j	Index of network buses
k	Index of wind speed level

l	Index of network lines
N	Number of network buses
N_b	Number of network lines
N_{PQ}	Set of PQ buses

C Parameters and variables

CF_i	Capacity factor of the DG on bus i (%)
CSF	Composite sensitivity factor
G_{ij}	Conductance of network line from i to j (S)
I_{ij}	Current from bus i to bus j (A)
$I_{ij,\max}$	Allowed maximum line current (A)
$J_{(.)}$	Sub-matrix of the Jacobian matrix
LSF	Power loss sensitivity factor
MP	Allowed maximum DG penetration percentage (%)
$P_{\mathrm{DG},i}$	Rated active power output of the DG on bus i (W)
P_{Li}	Active power load on bus i (W)
P_l	Active power transmitted by line l (W)
$P_{l,\max}$	Maximum active power which can be transmitted by line l (W)
P_{loss}	Active power loss (W)
P_{w}	WT output power (W)
P_{wr}	WT rated output power (W)
V_i	Voltage magnitude of bus i (p.u.)
$V_{i,\min/\max}$	Allowed minimum/maximum bus voltage (p.u.)
VSF	Voltage sensitivity factor
v	Wind speed (m/s)
$v_{\mathrm{ci}}/v_{\mathrm{co}}$	Cut-in/cut-off wind speed of WT (m/s)
v_k	Mean wind speed of the kth wind speed level (m/s)
v_{kl}/v_{ku}	Lower/upper bounds of the kth wind speed level (m/s)
v_{m}	Mean wind speed (m/s)
v_{r}	Rated wind speed of WT (m/s)
ε, σ	Pre-designed criteria for the objectives
ω	Sensitivity weighting factor
θ_i	Voltage angle of bus i (°)
$(.)_0$	Value in pre-change status
$(.)_{\mathrm{ave}}$	Average value
$(.)_{\mathrm{w}}$	Value with a DG unit added
$(.)_{\mathrm{wo}}$	Value without a DG unit added
$\Delta(.)$	Value of deviation

3.1 Introduction

With increased installation of DG since the beginning of this century, traditional centralised generation systems are evolving towards decentralised systems. The DG growth and the drivers behind it are summarised in [1]. One of the key reasons is the nature of the flexible allocation of DG. DG is a kind of generation unit which can be installed close to loads and especially mid-scale DG units can be invested in and planned by network companies. These network companies own the mid-scale DG units and get benefits through optimising their locations and controlling their operation. Thus, the network-owned DG contributes significantly to converting a distribution network from a passive system to an active one, i.e. an active distribution network, resulting in immense benefits in terms of techniques and economics. In [1–4], the authors summarised the benefits of DG as shown in Table 3.1.

However, as DG proliferates, challenges occur as well, in particular for power system planners [1]. For example, it is concluded that as the penetration rises, the system power loss falls first to the minimum then increases as a negative impact of DG [5]. Thus, DG is expected to be with an optimal penetration level to the system, which is required to be optimised. References [6,7] suggest different heuristic methodologies of DG planning to reduce network power losses. Tabu search with neighbour solution movements to find the optimal placements of DG is developed in [6]. Reference [7] proposes an iterative heuristic approach to add DG units in larger loss areas to reduce power loss with the help of B-loss coefficients. On the other hand, some mathematical approaches are also applied; Reference [8] utilises mixed-integer non-linear programming (MINLP) to minimise power loss with network-owned mid-scale DG, while a Hereford Ranch algorithm and a genetic algorithm (GA) are applied in [9]. Meanwhile, many researchers take into account the voltage stability issue when they optimise the DG allocation. Reference [10] applies binary particle swarm optimisation and [11] suggests MINLP to improve the voltage stability by optimising the network-owned DG allocation.

Table 3.1 Technical and economic advantages of distributed generation

Major technical advantages	
1	Reduction of energy loss
2	Improvement of voltage stability
3	Improvement of power quality
4	Relief of distribution feeders
5	Reduction of transmission and distribution lines congestion
6	More sensitive track and response to local loads
Major economic advantages	
1	Short lead time and low investment risk
2	Reduction in maintenance costs of transmission and distribution lines
3	Reduction of economic investment for environmental cares
4	Low operation costs due to peak hour control

Considering the importance of both the power loss reduction and the voltage stability improvement, a multi-objective planning problem has received significant research interests. In [2,12], two composite multi-objective indices with weighting factors which can be modified with the predefined planning priority are derived for assessing the impacts of DG on the multiple objectives. Herein, the authors of [12] suggest utilising GA to achieve an optimal index meaning that the DG allocation is optimised. Furthermore, a multi-objective genetic algorithm (MOGA) is applied in [13] to reduce the network power loss and improve the voltage profile. It is worth mentioning that the MOGA is a widely accepted solution algorithm for the multi-objective planning problems. However, the MOGA has some shortcomings like providing uncertain solutions, slow procedures, and a predefined and fixed priority of the multiple objectives which is not practical for a long-term planning problem.

In order to solve the DG planning problem more effectively, a novel CSF-based method is proposed in this chapter for optimising placements of DG units to simultaneously reduce the power loss and improve the voltage stability. This approach can perform a determinate solution efficiently at a fast computing speed, and the objective priority can be dynamically adjusted during the iterative solution process.

The proposed approach is based on sensitivity analysis. A voltage sensitivity factor (VSF) is derived in [11] for selecting candidate buses to enhance the voltage profile, which is utilised in this chapter to form a composite sensitivity index. For the power loss reduction purpose, a new power loss sensitivity factor is derived. Since the ranks of the power loss and the voltage sensitivities may conflict, a CSF is developed using a weighting factor which can be dynamically modified to adjust the planning priority between the power loss reduction and the voltage stability improvement objectives. The CSF converts the multi-ranking conflicting issue into a simple single ranking to obtain the optimal location of a DG unit. Thus, the buses with large CSFs can be prioritised for DG installation to achieve an overall benefit which is the combination of the power loss reduction and the voltage stability enhancement.

On the other hand, it is important that two assessment indices are considered to quantify the power loss reduction and voltage stability improvement, respectively. In [2,10], a power loss reduction index is proposed, and it is modified in this chapter to assess impacts of an installed DG unit on the power loss reduction. For the voltage stability assessment, voltage profile [14] and voltage deviation index [15] are usually accepted assessment indices. However, these indices utilise only bus voltage magnitudes to quantify the voltage stability, which cannot reflect a stability nature of the system. In fact, causes of the voltage stability include not only the voltage magnitude, but also active and reactive power balances as well as branch line capacity. Considering this issue, first, a voltage collapse proximity indicator (VCPI) [16] is applied in this chapter to assess the voltage stability. The VCPI indicates impacts of the active and reactive power on the voltage stability, which the voltage profile and the voltage deviation index cannot. Then, a VCPI reduction index (VCPIRI) is proposed in this chapter.

Finally, based on the proposed assessment indices and the CSF of each candidate bus, an iterative algorithm is proposed with verification of network operating constraints. The proposed method also takes into account uncertainties of DG power generation and loads. To achieve this purpose, Monte Carlo simulation, where the wind speed is simulated with a Rayleigh probability density function (PDF) and the loads follow hourly load probability distribution, is proposed to simulate these uncertainties in power flow calculation.

3.2 Sensitivity factors

3.2.1 Power loss sensitivity factor

When DG supplies active power into the network, the power loss of all the branch lines has corresponding variations. The ratio of all the line power loss change summation to the DG injected power increment of a specific bus is defined as a power loss sensitivity factor for this bus and it can be derived as follows.

For a stable power system, the Jacobian matrix can be used for sensitivity analysis by using the following linearised model:

$$\begin{bmatrix} \Delta P \\ \Delta Q \end{bmatrix} = \begin{bmatrix} J_{P\theta} & J_{PV} \\ J_{Q\theta} & J_{QV} \end{bmatrix} \begin{bmatrix} \Delta\theta \\ \Delta V \end{bmatrix} \tag{3.1}$$

where $J_{P\theta}$, J_{PV}, $J_{Q\theta}$, and J_{QV} are four decomposed sub-matrices of the Jacobian matrix.

Assuming that DG generates active power only, i.e. its power factor is unity, the reactive power incremental change is zero. Then, the incremental change of bus active power injection can be calculated by either of the followings:

$$\Delta P = \left(J_{P\theta} - J_{PV}{J_{QV}}^{-1}J_{Q\theta}\right)\Delta\theta = J_{MP\theta}\Delta\theta \tag{3.2}$$

$$\Delta P = \left(J_{PV} - J_{P\theta}{J_{Q\theta}}^{-1}J_{QV}\right)\Delta V = J_{MPV}\Delta V \tag{3.3}$$

where $J_{MP\theta}$ and J_{MPV} are two modified Jocobian matrices.

For all the branch lines, active power loss is given by

$$P_{\text{loss}} = f([\theta, V]) = 0.5\sum_{i=1}^{N}\sum_{j=1}^{N}\left\{G_{ij} \times \left[V_i^2 + V_j^2 - 2V_iV_j\cos\left(\theta_j - \theta_i\right)\right]\right\} \tag{3.4}$$

where N is the total number of the network buses and G_{ij} is the real part of the corresponding element in the network line admittance matrix.

In the pre-change status, the line power loss and the network bus voltage are defined as $P_{\text{loss},0}$ and $[\theta_0, V_0]$, respectively. After a change of network bus active power injection, the corresponding line active power loss can be calculated by the corresponding variations of bus voltage as

$$P_{\text{loss},0} + \Delta P_{\text{loss}} = f([\theta_0 + \Delta\theta, V_0 + \Delta V]) \tag{3.5}$$

Equation (3.5) can be further expressed by Taylor's series expansion of $[\theta, V]$ as

$$P_{\mathrm{loss},0} + \Delta P_{\mathrm{loss}} = f([\theta_0, V_0]) + \left.\frac{\mathrm{d}f([\theta, V])}{\mathrm{d}\theta}\right|_{\theta_0} \Delta\theta + \left.\frac{\mathrm{d}f([\theta, V])}{\mathrm{d}V}\right|_{V_0} \Delta V + h.o.t. \tag{3.6}$$

Assuming the variations of bus voltage are small enough to ignore the higher-order terms (h.o.t.), the increment of the line power loss is obtained by

$$\Delta P_{\mathrm{loss}} = \left.\frac{\mathrm{d}f([\theta, V])}{\mathrm{d}\theta}\right|_{\theta_0} \Delta\theta + \left.\frac{\mathrm{d}f([\theta, V])}{\mathrm{d}V}\right|_{V_0} \Delta V \tag{3.7}$$

Substituting (3.2) and (3.3) into (3.7), the proposed power loss sensitivity factor is derived as a one-row matrix of sensitivities for the total network power loss to each bus DG active power output increment as

$$LSF = \frac{\Delta P_{\mathrm{loss}}}{\Delta P} = \left.\frac{\mathrm{d}f([\theta, V])}{\mathrm{d}\theta}\right|_{\theta_0} {J_{MP\theta}}^{-1} + \left.\frac{\mathrm{d}f([\theta, V])}{\mathrm{d}V}\right|_{V_0} {J_{MPV}}^{-1} \tag{3.8}$$

These power loss sensitivity factors can be positive for power loss increasing or negative for loss decreasing when DG generates more active power. Thus, the small negative power loss sensitivity factors (large absolute values with negative signs) can be applied in selecting the candidate buses for DG installation to reduce more branch line power losses.

3.2.2 *Voltage sensitivity factor*

On the other hand, the impacts of the DG active power outputs on the network voltage stability should also be considered. It is assumed that all the DG units operate PQ control and all the buses should be modelled as PQ buses except the root bus. Thus, the bus voltages vary due to the DG active power injection with the unity power factor. Similar to the power loss sensitivity factor, a VSF of a certain bus is defined as the ratio of each bus corresponding voltage magnitude variation to the DG injected active power increment at this bus. Given (3.3), a matrix of sensitivities for each PQ bus voltage magnitude to each bus DG active power output increment is derived as

$$VSF = \frac{\Delta V}{\Delta P} = {J_{MPV}}^{-1} \tag{3.9}$$

The VSF has positive values for all the buses, since DG active power generation always increases the bus voltage magnitudes. Thus, the buses with the large VSFs are considered as the candidate buses for the DG installation.

3.2.3 Composite sensitivity factor

When a distribution network is planned with DG, locations of DG units are expected to be optimised to reduce the network power losses and improve the voltage stability simultaneously, which is considered as a multi-objective optimisation problem. As described in Sections 3.2.1 and 3.2.2, the buses with the small negative power loss sensitivity factors can be selected to install DG units to reduce the network power losses, meanwhile those with the large VSFs can be prioritised for enhancing the voltage stability through DG active power supply. However, the priority to install DG units of the candidate buses with small negative power loss sensitivity factors can conflict with the rank of the large VSFs. This issue occurs in most multi-objective optimisation problems. To avoid the confliction between the power loss reduction and the voltage stability improvement objectives, a CSF is developed in this chapter to quantify the overall sensitivity as follows.

First, a parameter for scalarising the power loss sensitivity factors and the VSFs is defined as

$$\alpha = \frac{LSF_{\text{ave}}}{VSF_{\text{ave}}} \tag{3.10}$$

where LSF_{ave} is the average value of all the PQ bus power loss sensitivity factors, and VSF_{ave} is the average value of all the PQ bus VSF summations.

Then, the proposed CSF is derived as a one-row matrix of composite sensitivities of each PQ bus as

$$CSF = \omega\frac{-LSF}{\alpha} + (1-\omega)VSF^* \tag{3.11}$$

where ω is a sensitivity weighting factor, $0 \le \omega \le 1$, and VSF^* is a one-row matrix where each element is the summation of the corresponding PQ bus VSFs.

This CSF has two advantages in DG planning:

1. A multi-sensitivity ranking with a confliction issue can be converted into a single sensitivity ranking.
2. The sensitivity weighting factor can be modified to adjust the planning priority between the power loss reduction and the voltage stability improvement objectives.

Therefore, the buses with large CSFs can be prioritised for DG installation to achieve an overall benefit which is a combination of the power loss reduction and the voltage stability enhancement.

3.3 Power loss and voltage stability assessment indices

3.3.1 Line loss reduction index

When a certain amount of DG active power is injected into the distribution network, it is obvious that the network power loss is decreased because of reduction of some branch power flows. However, it is concluded in [5] that as DG penetration increases, the network power loss decreases first to the minimum level, then the power loss begins to increase. This means that an optimal penetration level of DG exists for the minimum network power loss. For a purpose of obtaining this optimal penetration, a line loss reduction index (LLRI) is defined as the ratio of the total line power loss in the network with a DG unit added ($Loss_{\mathrm{w}}$) to the total line power loss without this added DG unit ($Loss_{\mathrm{wo}}$), and it can be represented as

$$LLRI = \frac{Loss_{\mathrm{w}}}{Loss_{\mathrm{wo}}} \tag{3.12}$$

This index can assess the impacts of each single DG unit on the line power loss as following:

$LLRI < 1$: the installed DG unit can reduce the total line power loss;
$LLRI = 1$: the installed DG unit cannot impact on the total line power loss;
$LLRI > 1$: the installed DG unit can rise the total line power loss.

Besides, when the LLRI is less than 1, the smaller values mean the more positive effects can be achieved by the DG unit on the power loss reduction.

3.3.2 Voltage collapse proximity indicator reduction index

A VCPI is developed in [16] to assess the network voltage stability. In [16], a single line VCPI is defined as the ratio of the active power transmitted through a specific line, P_l, to the maximum active power which can be transmitted, $P_{l,\max}$, and it can be expressed for a specific line l as

$$VCPI_l = \frac{P_l}{P_{l,\max}} \tag{3.13}$$

In this chapter, a network overall VCPI is proposed by averaging all the line VCPIs of the network as

$$VCPI_n = \frac{1}{N_b}\sum_{l=1}^{N_b} VCPI_l \tag{3.14}$$

where N_b is the total number of the network lines. If the proposed VCPI is close to 1, it is indicated that the network voltage instability may occur. On the contrary, if it is close to 0, the network voltage can be considered as strongly stable. Thus, this VCPI is expected to be decreased by installing DG units. Similar to the LLRI, a VCPIRI is proposed in this chapter to quantify the reduction of the network VCPI, in other words, to assess the improvement of the network voltage stability. The proposed VCPIRI is defined as the ratio of the network VCPI with a DG unit added ($VCPI_{\mathrm{w}}$) to the network VCPI without this added DG unit ($VCPI_{\mathrm{wo}}$):

$$VCPIRI = \frac{VCPI_{\mathrm{w}}}{VCPI_{\mathrm{wo}}} \tag{3.15}$$

Similarly, the following characteristic explains the usage of the VCPIRI in the DG planning:

VCPIRI < 1: the installed DG unit can improve the network voltage stability;
VCPIRI = 1: the installed DG unit cannot impact on the network voltage stability;
VCPIRI > 1: the installed DG unit can worsen the network voltage stability.

3.4 Composite sensitivity factor-based method

3.4.1 Distributed generation and load modelling

DG includes many types which can be classified into controllable and uncontrollable distributed resources, or rotating and non-rotating generation structures. Herein, the renewable DG attracts more attention, due to its sustainable and environmentally friendly benefits. However, the renewable resources, like wind and solar energy, are uncontrollable distributed resources with power generation uncertainties. Considering the renewable DG attractive prospects and its uncertain output power characteristics, the proposed DG type applied in chapter is distributed wind turbines (WTs) whose output power depends on wind speed uncertainties. However, it is worth noting that other renewable DG units like photovoltaic (PV) panels with solar irradiance uncertainties can be considered when necessary. The details of PV modelling can be found in [17].

3.4.1.1 WT output power modelling with wind speed uncertainty

In [8], the authors develop a novel WT power generation modelling where the output power is divided into different levels with their corresponding occurrence probabilities. Furthermore, these power levels are applied into a power flow process to simulate different wind speed scenarios with their corresponding occurrence probabilities. This probabilistic simulation can be regarded as a discrete Monte Carlo simulation and applied in this CSF-based method for DG allocation. The proposed WT power generation modelling is described as follows.

First, it is noticed that the uncertain and varying wind speed is the root reason that WT output power cannot be fully controlled. Thus, the modelling of the wind speed should be completed. References [18,19] support the detailed analysis of the wind speed characteristics and modelling. In addition, it is recommended that the Rayleigh PDF can be applied in simulation of wind speed most accurately. This function can be expressed as

$$f(v) = \frac{2v}{c^2}\exp\left[-\left(\frac{v}{c}\right)^2\right] \tag{3.16}$$

$$c = \frac{2}{\sqrt{\pi}}v_{\mathrm{m}} \tag{3.17}$$

where v_{m} is the mean wind speed of a certain period.

Then, based on Rayleigh PDF, the simulated wind speed is divided into specified levels with their own probabilities which can be calculated as

$$\text{Proba}(v_k) = \int_{v_{kl}}^{v_{ku}} f(v)\text{d}v \tag{3.18}$$

where v_{kl} and v_{ku} are the lower and upper bounds of the kth wind speed level, v_k is the mean of v_{kl} and v_{ku}, and v_k stands for the kth wind speed level.

Finally, each wind speed level v_k has a corresponding output power level, $P_{\text{w}}(v_k)$, which can be obtained by

$$P_{\text{w}}(v_k) = \begin{cases} 0, & 0 \leq v_k \leq v_{\text{ci}} \text{ or } v_k \geq v_{\text{co}} \\ P_{\text{r}} \times \dfrac{v_k - v_{\text{ci}}}{v_{\text{r}} - v_{\text{ci}}}, & v_{\text{ci}} < v_k < v_{\text{wr}} \\ P_{\text{r}}, & v_{\text{wr}} \leq v_k < v_{\text{co}} \end{cases} \tag{3.19}$$

where P_{r} is the WT rated power, and v_{ci}, v_{co}, and v_{r} are the WT cut-in, cut-off, and rated wind speeds, respectively.

Table 3.2 indicates the modelling result with the 3.8 m/s mean wind speed, which is applied in the case study in Section 3.5. Since the proposed WT modelling is dependent on the division of the wind speed, other modelling results can be obtained as well when necessary.

It is worth mentioning that the WT output power is provided in the specified values which are discrete, and the probabilities of these values make up probability distribution which can simulate the uncertain WT power generation efficiently. If the discrete levels are divided more enough, even to be continuous values, the model is a de facto Monte Carlo simulation model. Thus, the proposed simulation using the discrete levels with the occurrence probabilities can be regarded as a discrete Monte Carlo simulation.

3.4.1.2 Load modelling

Similarly, the load of each bus can be modelled in the discrete Monte Carlo simulation concept as well. In [8], by a clustering method, the load is divided into ten levels with different occurrence probabilities. Each level has a specified load percentage of the peak load to fit the hourly load distribution given by [20]. This load modelling is applied in the power flow calculation to involve all uncertain loads with the probabilities. Table 3.3 shows the different load levels with their corresponding occurrence probabilities.

3.4.2 Composite sensitivity factor-based method

A novel planning method based on the CSF for optimising the DG allocation to reduce the network power losses and to improve the system voltage stability simultaneously is developed in this chapter. The CSF is applied in this method to select the most satisfying location where the new DG installation can both reduce the power losses and improve the voltage stability most efficiently. The sensitivity weighting factor is initialised as 0.5, since neither of the two objectives is

Table 3.2 Wind turbine output power levels (v_{ci} = 4 m/s, v_{co} = 25 m/s, and v_r = 14 m/s)

Level	Wind speed (m/s)	Output power (% of rated power)	Probability (%)
1	0–4	0	18.61
2	4–5	5	8.90
3	5–6	15	9.57
4	6–7	25	9.69
5	7–8	35	9.34
6	8–9	45	8.62
7	9–10	55	7.65
8	10–11	65	6.54
9	11–12	75	5.40
10	12–13	85	4.31
11	13–14	95	3.33
12	14–25	100	7.99
13	>25	0	0.00

Table 3.3 Load levels

Level	Percentage of peak load	Probability (%)
1	100	1.00
2	85.3	5.60
3	77.4	10.57
4	71.3	16.54
5	65	16.54
6	58.5	16.30
7	51	16.30
8	45.1	9.12
9	40.6	4.73
10	35.1	3.30

prioritised at the beginning. After one DG unit is assumed to be settled on the best location, the power flow is calculated again with probabilistic WT and load uncertainty models to obtain new network power loss and voltage stability results. These results are verified with LLRI and VCPIRI to assess whether the effects of the new installed DG unit can reach the planning requirement as follows:

$$LLRI \le \varepsilon \quad \text{and} \quad VCPIRI \le \sigma \tag{3.20}$$

where ε and σ are pre-designed criteria for the objectives.

If the new DG unit shows the acceptable positive impacts on the network and all the network constraints are satisfied, the method continues to find the next most satisfying location and install the next DG unit. If the impacts are negative, the assumed installed DG unit is withdrawn and the sensitivity weighting factor in CSF is modified to improve the overall impacts. Here, the weighting factor must be changed between lower and upper limits, ω_l and ω_u, which are designed on the requirement of the planners' objective priority. Then, the planning method continues with the modified sensitivity weighting factor. The proposed method continues iteratively until one of the following termination conditions occurs:

1. Neither LLRI nor VCPIRI can be acceptable.
2. The sensitivity weighting factor cannot be modified within the limits.
3. Some network constraints violate.

Herein, the network constraints include the following.

Bus voltage magnitude limit:

$$0.95 \leq V_i \leq 1.05, \forall i \in N_{PQ} \tag{3.21}$$

where V_i is the bus voltage magnitude of the PQ bus i, and N_{PQ} is the set of all the PQ buses.

Branch line current capacity limit:

$$0 \leq I_{ij} \leq I_{ij\max}, \forall i,j \in N \tag{3.22}$$

where I_{ij} is the current of the branch line ij from the bus i to the bus j, $I_{ij\max}$ is the current capacity of the branch line ij, and N is the set of all the buses in the network.

Maximum DG penetration limit:

$$\sum_{i=1}^{N_{PQ}} CF_i \times P_{DGi} \leq MP \sum_{i=1}^{N_{PQ}} P_{Li} \tag{3.23}$$

where CF_i is the capacity factor of the DG at the bus i, P_{DGi} is the rated active power output of the DG at the bus i, MP is the maximum allowed DG penetration percentage, and P_{Li} is the load at the bus i.

The flow chart of the proposed DG allocation method is shown in Figure 3.1.

3.5 Case study

3.5.1 Test system description

A 1.1 MW rated power WT is considered as a DG unit to be installed, and the proposed WT output power model described in Section 3.4.1.1 is applied. First, the characteristics of the proposed power loss sensitivity factors are verified on a 33-bus radial distribution system whose data can be found in [21]. Second, a 5-year single-stage WT allocation problem with the maximum installation number of the WTs as 7 is solved by both the proposed CSF-based method and the conventional MOGA, respectively. This problem is tested on Ontario's standard offer programme system which is a practical distribution system given by [8]. Finally, a multi-stage WT allocation test is brought in on the Ontario's system and the results of the proposed CSF-based method and the conventional MOGA are compared.

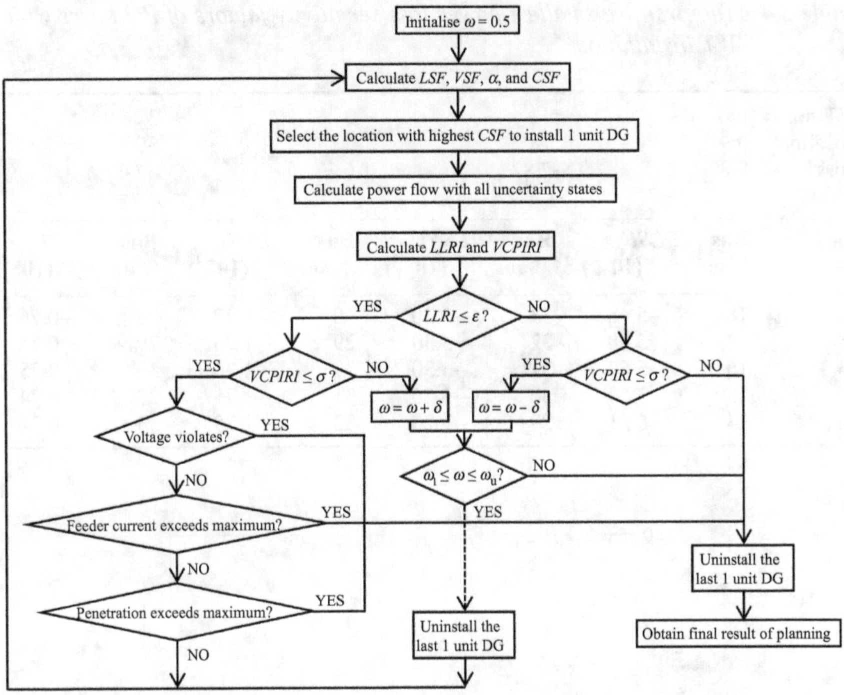

Figure 3.1 A CSF-based method for distributed generation allocation optimisation

This multi-stage allocation problem has a 20-year total planning horizon which is divided into four stages, and each stage has 5 years.

The simulations in this chapter are conducted on a 64-bit PC with 3.30-GHz CPU and 8 GB RAM, and they are programmed on MATLAB® with the help of MATPOWER 5.1 [22].

3.5.2 Power loss sensitivity factor analysis

To verify the effectiveness of the power loss sensitivity factor, for the 33-bus system, the power loss sensitivity factor of each bus is calculated, and a WT is installed at the location with the smallest power loss sensitivity factor. With this WT installed, the total network power loss is calculated. Then, all the power loss sensitivity factors are calculated again to install the next WT. Table 3.4 shows the results of the first five smallest power loss sensitivity factors for installation of three WTs, and the total network power loss. Figure 3.2 displays changes of all the power loss sensitivity factors for each WT installation process. It is seen from Table 3.4, when there is no WT, bus 18 has the smallest power loss sensitivity factor as -5.68×10^{-4} so that the first WT should be located at this bus. After the first WT allocated, the smallest power loss sensitivity factor, -3.30×10^{-4}, comes to bus 33. It indicates that the smallest power loss sensitivity factor location is

Table 3.4 The first five smallest power loss sensitivity factors of PQ buses with WT installation

WT no.	0		1		2		3	
Location	0		18		33		30	
Loss (kW)	9.89		7.82		6.27		5.66	
Rank	**Bus no.**	***LSF* (10^{-4})**	**Bus no.**	***LSF* (10^{-4})**	**Bus no.**	***LSF* (10^{-4})**	**Bus no.**	***LSF* (10^{-4})**
1	18	−5.68	33	−3.30	30	−1.77	28	−0.76
2	17	−5.68	32	−3.30	29	−1.76	29	−0.75
3	16	−5.67	31	−3.30	33	−1.76	30	−0.75
4	15	−5.64	30	−3.26	31	−1.75	27	−0.74
5	14	−5.61	29	−3.20	32	−1.75	33	−0.74

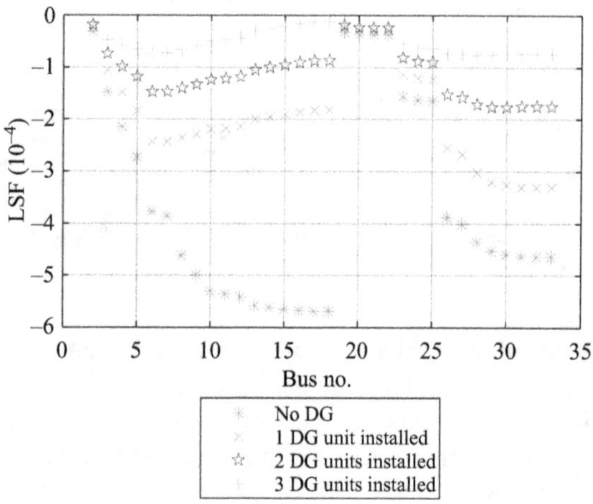

Figure 3.2 Power loss sensitivity factors of all PQ buses with WT installation

changed and the value is increased after a WT installed, which also leads to the installation of the second and third WTs. It states that the bus active power injection is much less sensitive to reduce network power losses after a WT installed. In addition, as the power loss sensitivity factors trend to be zero, less power loss reduction is achieved by the installed WTs. On the other hand, from Figure 3.2, it is argued that the power loss sensitivity factors of the buses can be grouped based on the network topology and the power loss sensitivity factors in the same group have similar values. It is considered that when a WT is located at a bus, it can help reducing power losses in the area around it, which leads to increased power loss sensitivity factors in the same group by similar rates. In a short conclusion, the power loss sensitivity factors can stand for how much power loss can be reduced with a WT installed, and they rise continuously when more WTs are allocated in the network.

3.5.3 Single-stage planning results

Table 3.5 presents the single-stage 5-year planning results of applying the proposed CSF-based method and the MOGA, respectively. Since the MOGA cannot provide determinate solutions, the solutions shown in this table are the best ones among 10 times simulations. The MOGA shows less network energy loss than that of the CSF-based method, and the VCPI results are similar. But it is noticed that the allowed maximum seven WTs are all allocated by the MOGA and only four by the CSF-based method. It is obvious that this MOGA solution is not accepted with consideration of economic investment. Then, the allowed WT installation number for the MOGA planning is decreased to four, same with the CSF-based method solution. The solution of the MOGA with four WTs is obtained and shown in the third column of Table 3.5. Significantly similar planning results are obtained by the CSF-based method and the MOGA for the network power/energy loss and the VCPI, which means the CSF-based method can achieve a satisfying solution as the MOGA does when the total installed WTs are same in the single-stage WT allocation planning problem. However, the tremendous average run-time difference draws attention, 2,115 s for the MOGA while only 77 s for the CSF-based method. In brief, the CSF-based method provides higher computation efficiency.

3.5.4 Multi-stage planning with load growth results

Tables 3.6 and 3.7 display the optimal allocation results of the WTs obtained by the CSF-based method and the MOGA, respectively. Similarly, for a comparison

Table 3.5 Single-stage planning solutions of CSF-based method and MOGA

Metdods		CSFBM	MOGA	
Power loss (MW)		0.0213	0.015	0.0211
Energy loss (MWh)		186.7	131.1	185.3
VCPI		0.0782	0.078	0.0776
Candidate buses installed with WT units	40	3	2	2
	39	0	0	0
	38	0	0	0
	37	0	0	0
	28	1	1	1
	26	0	0	0
	24	0	0	0
	35	0	0	0
	33	0	1	1
	32	0	0	0
	23	0	0	0
	19	0	3	0
Total installed WTs		4	7	4
Average run time (s)		77	1,946	2,115

Table 3.6 Multi-stage planning solutions of CSF-based method

Stage		1	2	3	4
Power loss (MW)		0.0213	0.0336	0.0497	0.0821
Energy loss (MWh)		933	1,472	2,176	3,598
Total energy loss (MWh)		8,179			
VCPI		0.0782	0.0998	0.1258	0.1591
Average of VCPIs		0.1157			
Candidate buses installed with WT units	40	3	4	5	6
	39	0	0	0	0
	38	0	0	0	0
	37	0	0	0	0
	28	1	1	2	2
	26	0	0	0	0
	24	0	0	0	0
	35	0	0	0	0
	33	0	0	0	0
	32	0	0	0	0
	23	0	0	0	0
	19	0	0	0	0
Total installed WT units		4	5	7	8
Average run time (s)		269			

Table 3.7 Multi-stage planning solutions of MOGA

Stage		1	2	3	4
Power loss (MW)		0.0220	0.0335	0.0474	0.0789
Energy loss (MWh)		962	1,469	2,077	3,454
Total energy loss (MWh)		7,963			
VCPI		0.0818	0.0999	0.1253	0.1573
Average of VCPIs		0.1161			
Candidate buses installed with WT units	40	4	4	4	4
	39	0	0	1	1
	38	0	0	0	0
	37	0	0	0	0
	28	0	0	0	0
	26	0	1	2	2
	24	0	0	0	0
	35	0	0	0	0
	33	0	0	0	1
	32	0	0	0	0
	23	0	0	0	0
	19	0	0	0	0
Total installed WT units		4	5	7	8
Average run time (s)		6,016			

purpose, the maximum numbers of the WT installation in the MOGA for all the stages are set same as the CSF-based method solution. The result shown in Table 3.7 is also the best one from ten solutions, considering the MOGA uncertain result characteristic. From these two tables, it is observed that the total energy loss of the CSF-based method solution is 8,179 MWh, which is slightly higher than that of the MOGA as 7,963 MWh. Besides, for the VCPI reduction purpose, the CSF-based method accomplishes similar results for all the stages as the MOGA does. Thus, it can be noted that the CSF-based method solution is not only acceptable and quite close to the conventional planning optimal solution through the MOGA, but also determinate, which is the shortcoming of the MOGA. Furthermore, for the run time 269 s used by the CSF-based method is significantly less than that by the MOGA as 6,016 s, which indicates high computing efficiency of the CSF-based method.

3.6 Conclusion

In this chapter, a novel CSF-based method is proposed for optimising the allocation of the DG units. This approach aims to utilise the DG units to simultaneously reduce the network power losses and improve the network voltage stability, which is considered as a multi-objective planning problem. The proposed CSF-based method is tested on the 33-bus radial distribution system and a real planning system of the Ontario's standard offer programme, and it is compared with the planning results of the conventional MOGA. The test results show the following advantages of the proposed CSF-based method: (1) it can provide a determinate solution in contrast to inconsistent solutions from the MOGA; (2) the iterative process performs efficiently at a significantly fast speed; and (3) the objective priority can be adjusted depending on planners' priority, which is more practical for the real planning project. Therefore, the proposed CSF-based method can achieve an efficient and determinate optimisation planning solution to decrease the network power losses and enhance the voltage stability, simultaneously.

References

[1] J.A.P. Lopes, N. Hatziargyriou, J. Mutale, P. Djapic, and N. Jenkins, 'Integrating distributed generation into electric power systems: a review of drivers, challenges and opportunities', *Electric Power Syst. Research*, vol. 77, no. 9, pp. 1189–1203, 2008.

[2] P. Chiradeja and R. Ramakumar, 'An approach to quantify the technical benefits of distributed generation', *IEEE Trans. Energy Conv.*, vol. 19, no. 4, pp. 764–773, 2004.

[3] S. Porkar, P. Poure, A. Abbaspour-Tehrani-fard, and S. Saadate, 'A novel optimal distribution system planning framework implementing distributed generation in a deregulated electricity market', *Electric Power Syst. Research*, vol. 80, no. 7, pp. 828–837, 2010.

[4] W. El-Khattam, K. Bhattacharya, Y. Hegazy, and M.M.A. Salama, 'Optimal investment planning for distributed generation in a competitive electricity market', *IEEE Trans. Power Syst.*, vol. 19, no. 3, pp. 1674–1684, 2004.

[5] V.H.M. Quezada, J.R. Abbad, and T.G.S. Roman, 'Assessment of energy distribution losses for increasing penetration of distributed generation', *IEEE Trans. Power Syst.*, vol. 21, no. 2, pp. 533–540, 2006.

[6] K. Nara, Y. Hayashi, K. Ikeda, and T. Ashizawa, 'Application of tabu search to optimal placement of distributed generators', *Proc. IEEE PES Winter Meeting*, vol. 2, pp. 918–923, Columbus, OH, USA, 2001.

[7] T. Griffin, K. Tomsovic, D. Secrest, and A. Law, 'Placement of dispersed generation systems for reduced losses', *Proc. 33rd Annual Hawaii Int. Conf. Syst. Sciences*, pp. 1–9, Maui, HI, USA, Jan. 2000.

[8] Y.M. Atwa and E.F. El-Saadany, 'Probabilistic approach for optimal allocation of wind-based distributed generation in distribution systems', *IET Renew. Power Gen.*, vol. 5, no. 1, pp. 79–88, 2011.

[9] J. Kim, S. Nam, S. Park, and C. Singh, 'Dispersed generation planning using improved Hereford Ranch algorithm', *Electric Power Syst. Research*, vol. 47, no. 1, pp. 47–55, 1998.

[10] S.Y. Su, C.N. Lu, R.F. Chang, and G. Gutierrez-Alcaraz, 'Distributed generation interconnection planning: a wind power case study', *IEEE Trans. Smart Grid*, vol. 2, no. 1, pp. 181–189, 2011.

[11] R.S. Al Abri, E.F. El-Saadany, and Y.M. Atwa, 'Optimal placement and sizing method to improve the voltage stability margin in a distribution system using distributed generation', *IEEE Trans. Power Syst.*, vol. 28, no. 1, pp. 326–334, 2013.

[12] D. Singh, D. Singh, and K.S. Verma, 'Multiobjective optimization for DG planning with load models', *IEEE Trans. Power Syst.*, vol. 24, no. 1, pp. 427–436, 2009.

[13] M. Sedighizadeh and A. Rezazadeh, 'Using genetic algorithm for DG allocation to reduce losses and improve voltage profile', *World Academy of Science, Engineering and Technology*, vol. 37, pp. 251–256, 2008.

[14] R. Caire, N. Retiere, S. Martino, C. Andrieu, and N. Hadjsaid, 'Impact assessment of LV distributed generation on MV distribution network', *IEEE PES Summer Meeting*, vol. 3, pp. 1423–1428, Chicago, IL, USA, 2002.

[15] B. Venkatesh, R. Ranjan, and H.B. Gooi, 'Optimal reconfiguration of radial distribution systems to maximise loadability', *IEEE Trans. Power Syst.*, vol. 19, no. 1, pp. 260–266, 2004.

[16] M. Moghavvemi and O. Faruque, 'Real-time contingency evaluation and ranking technique', *IEE Proc. Gen., Transm. and Distrib.*, vol. 145, no. 5, pp. 517–524, 1998.

[17] Z.M. Salameh, B.S. Borowy, and A.R.A. Amin, 'Photovoltaic module-site matching based on the capacity factors', *IEEE Trans. Energy Conv.*, vol. 10, no. 2, pp. 326–332, 1995.

[18] J. Seguro and T. Lambert, 'Modern estimation of the parameters of the Weibull wind speed distribution for wind energy analysis', *J. Wind Eng. Ind. Aerod.*, vol. 85, no. 1, pp. 75–84, 2000.

[19] J. Hetzer, D.C. Yu, and K. Bhattarai, 'An economic dispatch model incorporating wind power', *IEEE Trans. Energy Conv.*, vol. 23, no. 2, pp. 603–611, 2008.
[20] J.M.S. Pinheiro, C.R.R. Dornellas, M.T. Schilling, A.C.G. Melo, and J.C.O. Mello, 'Probing the new IEEE Reliability Test System (RTS-96): HL-II assessment', *IEEE Trans. Power Syst.*, vol. 13, no. 1, pp. 171–176, 1998.
[21] M.E. Baran and F.F. Wu, 'Network reconfiguration in distribution systems for loss reduction and load balancing', *IEEE Trans. Power Del.*, vol. 4, no. 2, pp. 1401–1407, 1989.
[22] R.D. Zimmerman, C.E. Murillo-Sanchez, and R.J. Thomas, 'MATPOWER: steady-state operations, planning, and analysis tools for power systems research and education', *IEEE Trans. Power Syst.*, vol. 26, no. 1, pp. 12–19, 2011.

Chapter 4

Probability-weighted robust optimisation method for DG planning

Distributed generation (DG) has been rapidly integrated into microgrids. However, uncertain power generation of intermittent DG such as wind turbines (WTs) brings challenges to DG planning problems. This chapter introduces a novel probability-weighted robust optimisation (PRO) method to allocate DG units including microturbines and WTs in microgrids, aiming to maximise the total profit over a long-term planning horizon. First, probability-weighted uncertainty sets are proposed to model uncertainties including wind power outputs and loads during a long-term planning horizon, which can fully cover the uncertainty spectrum with uncertainty probability distribution. Then, the PRO method optimises DG locating and sizing under the worst uncertainty cases considering their occurrence probabilities. Furthermore, a modified column-and-constraint generation algorithm is developed to solve the PRO problem. Simulation results show that the DG planning obtained by the proposed PRO method can achieve full operating robustness against any possible uncertainty realisation.

Nomenclature

A Acronyms

C&CG	Column and constraint generation
DG	Distributed generation
MILP	Mixed-integer linear programming
NPV	Net present value
O&M	Operation and maintenance
PDF	Probability density function
PRO	Probability-weighted robust optimisation
PV	Photovoltaic
RES	Renewable energy source
RO	Robust optimisation
SO	Stochastic optimisation
WT	Wind turbine

B Sets and indices

I, i	Set and index of system buses
J, j	Set and index of DG installation candidate buses
M, m	Set and index of DG units
T, t	Set and index of operation periods
U	Combined uncertainty set formed by U_v^{LD} and U_w^{WT}
$U_l^{\mathrm{LD}}, U_w^{\mathrm{WT}}$	Probability-weighted uncertainty set l, w for loads and WT output power, respectively
Y, y	Set and index of years in planning horizon
Φ	Set of probability-weighted uncertainty sets

C Parameters

c	Annual cost increasing rate (%)
d	Annual discount rate (%)
g	Annual load growth rate (%)
G_y^{MT}	Generation cost of microturbine in year y (\$/kWh)
h	Time interval of one operation period (h)
H_y	Total operation hours in year y (h)
$Inv^{\mathrm{MT/WT}}$	Investment cost of one microturbine/WT (\$/kVA)
m	Number of DG units
n, n_l, n_w	Numbers of uncertainty sets U, U_l^{LD}, U_w^{WT}
$OM_y^{\mathrm{MT/WT}}$	Yearly operation and maintenance cost of one microturbine/WT in year y (\$/kVA/year)
P_i^{cap}	Capacity of branch following bus i (kW)
P^{r}	Rated power of one WT (kW)
$\widehat{P}_{i,t,y}^{\mathrm{LD}}$	Predicted peak active loads at bus i during period t in year y (kW)
$\underline{P}, \bar{P}$	Lower and upper bounds of uncertain renewable power generation or active loads (kW)
$Pr_y^{\mathrm{buy}}, Pr_y^{\mathrm{sell}}$	Electricity price for buying energy from and selling energy to main grid in year y (\$/kWh)
Pr_y^{cus}	Electricity price for selling energy to customers in year y (\$/kWh)
R_i, X_i	Resistance/reactance of branch following bus i
$S^{\mathrm{MT/WT}}$	Capacity of one microturbine/WT (kVA)
$s^{\mathrm{ci}}, s^{\mathrm{co}}, s^{\mathrm{r}}$	Cut-in, cut-off, and rated wind speeds of WT (m/s)
s^{m}	Mean wind speed (m/s)
V_0	Voltage magnitude of microgrid substation (p.u.)
$V^{\min}, V^{\max}$	Allowed minimum and maximum bus voltage magnitude (p.u.)
$\underline{\mu}, \bar{\mu}$	Lower and upper bounds of uncertain renewable power generation or active loads for all operation periods in one year (kWh)

ε	Excess reserve margin percentage for microgrid islanded operation
ρ	Occurrence probability
τ	Transformation factor for yearly cost
φ_i	Ratio between active and reactive loads

D Variables

$P_{i,t,y}$, $Q_{i,t,y}$	Active and reactive power through main branch following bus i during period t in year y (kW/kVar)
$P^{\mathrm{LD}}_{i,t,y(,l)}$, $Q^{\mathrm{LD}}_{i,t,y(,l)}$	Uncertain active and reactive loads at bus i during period t in year y (in uncertainty set l) (kW/kVar)
$P^{\mathrm{lat}}_{i,t,y}$, $Q^{\mathrm{lat}}_{i,t,y}$	Active and reactive power through lateral branch following bus i during period t in year y (kW/kVar)
$P^{\mathrm{def}}_{t,y}$, $P^{\mathrm{sur}}_{t,y}$	Deficient and surplus power of microgrid during period t in year y (kW)
$P^{\mathrm{MT}}_{j,t,y}$	Output power of microturbines at bus j during period t in year y (kW)
$P^{\mathrm{WT}}_{j,t,y(,w)}$	Uncertain output power of WTs at bus j during period t in year y (in uncertainty set w) (kW)
$V_{i,t,y}$	Bus voltage magnitude of bus i during period t in year y (p.u.)
$\alpha^{\mathrm{MT/WT}}_{j,m}$	Binary decision of installing m microturbines/WTs at bus j

4.1 Introduction

DG is a key element in a microgrid. In general, DG can be divided into dispatchable units, e.g. microturbines, and non-dispatchable units which are based on renewable energy sources (RESs), e.g. WTs and solar photovoltaics (PVs). DG can bring considerable technical and economic benefits such as reduction of energy losses, improvement of power quality as well as a low investment risk, low operation costs, and so forth. More detailed advantages of the DG are described in Section 3.1.

DG planning aims to determine the optimal locating and sizing of different DG units in active distribution networks or microgrids.

For dispatchable DG, Ref. [1] proposes a multi-stage planning structure for dispatchable DG and distribution network expansion. In [2], a multi-objective DG planning optimisation model is proposed to minimise real and reactive power losses and voltage deviations simultaneously by applying a genetic algorithm and four indices are derived to assess the objectives. Reference [3] applies a net present value (NPV) criterion to optimally allocate DG units considering investment, operation and maintenance (O&M), and energy purchase costs.

For non-dispatchable DG, since the RESs are naturally intermittent and stochastic, the uncertainty of the non-dispatchable DG output power can significantly complicate its planning problem. To address the uncertainty impacts over a long-term planning horizon, stochastic optimisation (SO) methods which model the uncertainties by scenario sampling have been used in DG planning problems [4,5]. To accurately reflect uncertainty probability distribution, a large number of uncertainty scenarios need to be generated in these SO methods. However, the optimisation with such many scenarios incurs considerably heavy computational burdens. To improve the uncertainty modelling efficiency, Ref. [6] proposes a probabilistic optimisation method for a DG allocation problem, where the probability density of renewable power generation and load uncertainties is modelled as different states. In each state, the uncertainty like wind speed is within specific limits and the average value between the limits is used to represent the uncertainty state. Each state is assumed to occur with a probability over a long-term planning horizon. Thus, the uncertainties are modelled effectively and their impacts during the planning horizon are considered in the optimisation. Compared to the SO, only a small number of uncertainty states are generated in the probabilistic optimisation to present the uncertainties with relatively accurate probability distribution, which enhances the computational efficiency. This probabilistic optimisation method has been applied in recent DG planning works [7–9]. Herein, Ref. [7] considers the system voltage stability margin as the DG planning optimisation objective. Plug-in electric vehicle penetration level and allocation of different types of DG are optimised simultaneously to achieve multiple objectives in [8]. Reference [9] applies the probabilistic method to model the operating variability due to intermittent renewable power generation and loads in a multi-stage DG planning problem. This probabilistic optimisation method has also been applied in other planning problems, such as transmission expansion problems [10]. However, these states presented by the average values cannot fully cover the spectrum of the uncertainties. Thus, short-term operating constraint violations may still occur, that is, planning solutions obtained by the probabilistic optimisation method are not robust over the whole planning horizon.

In recent years, to improve operating robustness against uncertainties, robust optimisation (RO) approaches have been applied for power system operation and planning problems, such as unit commitment problems [11], voltage/VAR control problems [12], and transmission expansion problems [13]. For microgrid DG planning, Ref. [14] applies the RO method to maximise economic benefits by optimising DG locations and sizes. In this RO-based planning problem, uncertain RES power generation and loads are modelled within uncertainty sets based on a single-day profile. Within the uncertainty sets, the worst case of RES power generation and load realisation in the single-day profile is searched and the DG units are allocated under this worst case. Therefore, the planning solution can guarantee operating constraint satisfaction for any realisation of the uncertain RES power generation and loads in this single-day profile. In recent works, the RO approach has been applied to DG allocation problems considering uncertainty of market prices and microgrid islanding conditions [15], voltage control and reactive power compensation [16], operating resilience against natural disasters [17], and interaction among provisional microgrids, coupled microgrids, and main grid [18]. However, a single-day profile is inadequate to represent uncertainties in a long-term planning horizon, e.g. 10 years. Thus, the planning solution is only robust for the single-day profile, yet the operating constraint violations may still occur for other RES power generation and load profiles. Besides, the obtained planning solution would vary depending on selection of profiles, and it is difficult to obtain a typical profile for a long-term planning horizon.

To ensure full robustness for short-term operation under uncertainties over a long-term planning horizon, this chapter proposes a PRO method which takes advantages of both the probabilistic optimisation and the RO methods. First, uncertainties are modelled in probability-weighted uncertainty sets which can fully cover the uncertainty spectrum and accurately reflect the probability distribution over the long-term horizon. Within the probability-weighted uncertainty sets, DG allocation considering present capital investment costs and future operating costs is robustly optimised under the worst cases taking their occurrence probabilities into account. Furthermore, a modified column and constraint generation (C&CG) algorithm is developed to solve the proposed PRO-based DG planning problem.

The proposed PRO-based DG planning method is tested and compared with the RO-based method. The simulation results show that the DG planning optimised by the PRO method is fully robust against the uncertainties and no operating constraint violation would occur.

4.2 Mathematical formulation

This chapter considers one prevailing type of non-dispatchable renewable DG, i.e. WTs, and one prevailing type of dispatchable DG, i.e. microturbines. Both of them are widely invested and owned by the microgrid owners in practice. Note that other types of DG including PVs and fuel cells can be also applied in microgrids when necessary.

The DG planning optimisation model is formulated as follows:

$$\min C_{\text{inv}} + C_{\text{OM}} + C_{\text{MT}} + C_{\text{grid}} - C_{\text{rev}} \tag{4.1}$$

$$\text{s.t.} \quad C_{\text{inv}} = \sum_{j \in J} \sum_{m \in M} \left(\alpha_{j,m}^{\text{MT}} mS^{\text{MT}} Inv^{\text{MT}} + \alpha_{j,m}^{\text{WT}} mS^{\text{WT}} Inv^{\text{WT}} \right) \tag{4.2}$$

$$C_{\text{OM}} = \sum_{y \in Y} \sum_{j \in J} \sum_{m \in M} \delta_y \left(\alpha_{j,m}^{\text{MT}} mS^{\text{MT}} H_y OM_y^{\text{MT}} + \alpha_{j,m}^{\text{WT}} mS^{\text{WT}} H_y OM_y^{\text{WT}} \right) \tag{4.3}$$

$$C_{\text{MT}} = \sum_{y \in Y} \sum_{t \in T} \sum_{j \in J} \delta_y P_{j,t,y}^{\text{MT}} h\tau G_y^{\text{MT}} \tag{4.4}$$

$$C_{\text{grid}} = \sum_{y \in Y} \sum_{t \in T} \delta_y \left(P_{t,y}^{\text{def}} h\tau Pr_{t,y}^{\text{buy}} - P_{t,y}^{\text{sur}} h\tau Pr_{t,y}^{\text{sell}} \right) \tag{4.5}$$

$$C_{\text{rev}} = \sum_{y \in Y} \sum_{t \in T} \sum_{i \in I} \delta_y P_{i,t,y}^{\text{LD}} h\tau Pr_{t,y}^{\text{cus}} \tag{4.6}$$

$$\delta_y = \frac{1}{(1+d)^y}, \forall y \tag{4.7}$$

$$\alpha_{j,m}^{\mathrm{MT}} \in \{0,1\}, \alpha_{j,m}^{\mathrm{WT}} \in \{0,1\}, \forall j, m \tag{4.8}$$

$$\sum_{m \in M} \alpha_{j,m}^{\mathrm{MT}} = 1, \sum_{m \in M} \alpha_{j,m}^{\mathrm{WT}} = 1, \forall j \tag{4.9}$$

$$\varepsilon \sum_{i \in I} \widehat{P}_{i,t,y}^{\mathrm{LD}} \leq \sum_{j \in J} \sum_{m \in M} \left(\alpha_{j,m}^{\mathrm{MT}} m S^{\mathrm{MT}} + \alpha_{j,m}^{\mathrm{WT}} m S^{\mathrm{WT}} \right), \forall t, y \tag{4.10}$$

$$P_{i+1,t,y} = P_{i,t,y} + P_{i+1,t,y}^{G} - P_{i+1,t,y}^{\mathrm{LD}} - P_{i+1,t,y}^{\mathrm{lat}}, \forall i, t, y \tag{4.11}$$

$$Q_{i+1,t,y} = Q_{i,t,y} - Q_{i+1,t,y}^{\mathrm{LD}} - Q_{i+1,t,y}^{\mathrm{lat}}, \forall i, t, y \tag{4.12}$$

$$V_{i+1,t,y} = V_{i,t,y} - \frac{R_i P_{i,t,y} + X_i Q_{i,t,y}}{V_0}, \forall i, t, y \tag{4.13}$$

$$V^{\min} \leq V_{i,t,y} \leq V^{\max}, \forall i, t, y \tag{4.14}$$

$$-P_i^{\mathrm{cap}} \leq P_{i,t} \leq P_i^{\mathrm{cap}}, \forall i, t \tag{4.15}$$

$$P_{1,t,y} = P_{t,y}^{\mathrm{def}} - P_{t,y}^{\mathrm{sur}}, P_{t,y}^{\mathrm{def}} \geq 0, P_{t,y}^{\mathrm{sur}} \geq 0, \forall t, y \tag{4.16}$$

$$P_{j,t,y}^{G} = P_{j,t,y}^{\mathrm{MT}} + P_{j,t,y}^{\mathrm{WT}}, \forall j, t, y \tag{4.17}$$

$$\sum_{m \in M} \alpha_{j,m}^{\mathrm{WT}} m \underline{P}_{j,t,y}^{\mathrm{WT}} \leq P_{j,t,y}^{\mathrm{WT}} \leq \sum_{m \in M} \alpha_{j,m}^{\mathrm{WT}} m \bar{P}_{j,t,y}^{\mathrm{WT}}, \forall j, t, y \tag{4.18}$$

$$0 \leq P_{j,t,y}^{\mathrm{MT}} \leq \sum_{m \in M} \alpha_{j,m}^{\mathrm{MT}} m S^{\mathrm{MT}}, \forall j, t, y \tag{4.19}$$

$$\underline{P}_{i,t,y}^{\mathrm{LD}} \leq P_{i,t,y}^{\mathrm{LD}} \leq \bar{P}_{i,t,y}^{\mathrm{LD}}, \forall i, t, y \tag{4.20}$$

$$Q_{i,t,y}^{\mathrm{LD}} = \varphi_i P_{i,t,y}^{\mathrm{LD}}, \forall i, t, y \tag{4.21}$$

This planning optimisation problem aims to minimise the overall cost consisting of the investment cost as well as the operational costs and revenue. In (4.1), C_{inv} is the capital cost of DG units at the investment stage; C_{OM}, C_{MT}, C_{grid}, and C_{rev} are the operational costs and revenue which are realised by optimal operation after DG installation, and their values are adversely impacted by uncertainties such as the WT power generation and loads. Note that the operational costs and revenue are minimised together with the investment cost in (4.1). By doing so, considering the capital cost in the investment stage as well as the operational costs and revenue in the operation stage, long-term economic benefits for microgrid operators can be maximised.

The investment cost which is determined by DG installation decisions $\alpha_{j,m}^{\mathrm{MT}}$ and $\alpha_{j,m}^{\mathrm{WT}}$ is calculated by (4.2). The O&M costs of the microturbines and the WTs are computed by (4.3), and the microturbines also have the fuel cost depending on the amount of energy generation as (4.4). Equation (4.5) calculates the energy transaction cost for the microgrid exchange energy with the main grid. Besides, the revenue made for the microgrid operators by selling electric energy to customers is calculated by (4.6). In (4.4)–(4.6), τ is used to transform the costs and revenue in each operation period to yearly values.

To practically evaluate the economic benefits of the DG planning, the NPV criterion is employed for (4.3)–(4.6) to compare the expected operational costs and revenue with the investment cost over the whole planning horizon. The NPV discounts each year's cash flow back to the present, giving a net value of the planning in today's currency. Thus, the costs and revenue in the operation stage converted by an NPV factor and the capital cost in the investment stage have the identical present currency value, such that they can be optimised simultaneously in (4.1). The calculation of the NPV factor for the yth year is given as (4.7).

In (4.8), binary variables are used to denote DG installation decisions for different capacities. Constraint (4.9) guarantees that only one DG installation decision can be made for each type of DG and for each DG installation candidate bus. In addition, the total installed DG capacity should be not less than a certain percentage (excess reserve margin) of the total loads for a reliable microgrid islanded operation requirement [14] which is constrained by (4.10).

To achieve robust operation, operating constraints of each short-term operation period must be taken into account in the long-term planning problem. A linearised distribution load flow (Dist-Flow) model has been widely applied in the microgrid operation and planning problems [19]. The linearised Dist-Flow model is used as the network operating constraints as (4.11)–(4.13).

Constraint (4.14) shows that the voltage magnitude of each bus must be kept within the allowed range. To avoid overflow, the power flow of each branch is constrained within the capacity as (4.15).

Moreover, the relationship between the power flow from the substation and the electricity transaction is provided by (4.16). Herein, $P^{\mathrm{def}}_{t,y} > 0$ means that the microgrid buys electricity from the main grid and $P^{\mathrm{sur}}_{t,y} > 0$ means that the microgrid sells electricity to the main grid. With this constraint, the power exchange for the electricity transaction calculation can be easily obtained in the optimisation problem.

For the DG unit operation, (4.17) computes the total active power generated at each DG candidate bus $P^{G}_{j,t,y}$. Herein, the WT output power of each candidate bus is constrained within the predicted range as (4.18). In terms of the microturbines, their output power can be optimally dispatched within the installed capacity as (4.19).

In terms of the loads, the active power load at each bus is limited within the predicted range for each planning year as (4.20), and the reactive power load is computed by (4.21) based on the assumption that the power factor is fixed for each bus.

The objective function (4.1) subject to the constraints (4.2)–(4.21) makes up a mixed-integer linear programming (MILP) optimisation problem with uncertainty variables $P^{\mathrm{LD}}_{i,t,y}$ and $P^{\mathrm{WT}}_{j,t,y}$. The control variables include the microturbine and WT installation decisions $\alpha^{\mathrm{MT}}_{j,m}$ and $\alpha^{\mathrm{WT}}_{j,m}$ in the investment stage and the microturbine power generation dispatch decision $P^{\mathrm{MT}}_{j,t,y}$ in the operation stage.

4.3 Probability-weighted robust optimisation

To achieve robust operation in the microgrid during the whole planning horizon, the uncertainties including the WT output power and the loads must be effectively modelled, and the DG planning problem must be optimised with full consideration of the uncertainties. This chapter proposes a PRO method to optimise the DG planning in the microgrid. The uncertainties are modelled in probability-weighted uncertainty sets which can reflect probability distribution of the uncertainties accurately. With the probability-weighted uncertainty sets, the PRO method can obtain a robust and optimal planning solution.

4.3.1 Probabilistic modelling of uncertainties

In [6], a probability density function (PDF) is used to model uncertainty variables. For example, for WTs, the uncertain output power is modelled according to the wind speed into different states accompanied by the corresponding occurrence probabilities. Thus, the modelling of the WT output power is derived with the uncertain wind speed characteristics and its state-based modelling. It is normally recommended that *Rayleigh* PDF which is subtype of *Weibull* PDF can express long-term behaviours of wind speed accurately [20]. In this chapter, *Rayleigh* PDF is applied to generate several wind speed states accompanied by their corresponding occurrence probabilities. This PDF of the wind speed s is shown as

$$f_{WS}(s) = \frac{2s}{c^2}\exp\left[-\left(\frac{s}{c}\right)^2\right] \tag{4.22}$$

$$c = \frac{2}{\sqrt{\pi}}s^{\mathrm{m}} \tag{4.23}$$

where s^{m} is the mean wind speed.

Then, the wind speed can be divided and clustered into several states (s_{w}) which have respective lower and upper bounds ($\underline{s}_{\mathrm{w}}$ and $\bar{s}_{\mathrm{w}}$). The occurrence probability of each state $\rho(s_{\mathrm{w}})$ is calculated as the integral of the PDF from the lower bound to the upper bound as

$$\rho(s_{\mathrm{w}}) = \int_{\underline{s}_{\mathrm{w}}}^{\bar{s}_{\mathrm{w}}} f(s)\mathrm{d}s \tag{4.24}$$

Finally, the WT output power of each state $P_{\mathrm{w}}^{\mathrm{WT}}$ is calculated according to the wind speed states and the WT power generation function as

$$P^{\mathrm{WT}}(s_{\mathrm{w}}) = \begin{cases} 0, & 0, s_{\mathrm{w}} \le s^{\mathrm{ci}} \quad \text{or} \quad s_{\mathrm{w}} \ge s^{\mathrm{co}} \\ P^{\mathrm{r}} \times \dfrac{s_{\mathrm{w}} - s^{\mathrm{ci}}}{s^{\mathrm{r}} - s^{\mathrm{ci}}}, & s^{\mathrm{ci}} < s_{\mathrm{w}} < s^{\mathrm{r}} \\ P^{\mathrm{r}}, & s^{\mathrm{r}} \le s_{\mathrm{w}} < s^{\mathrm{co}} \end{cases} \tag{4.25}$$

Table 4.1 Probabilities for different load states

State	Peak load (%)	Probability (%)
1	90–100	44.73
2	80–90	18.27
3	70–80	9.23
4	60–70	16.38
5	50–60	11.39

Herein, P^{r} is the rated output power of one WT and s^{ci}, s^{co}, and s^{r} are the WT cut-in, cut-off, and rated wind speeds, respectively.

Similarly, the uncertain loads can be divided and clustered into several different states with the average value P_l^{LD} and the occurrence probability ρ_l. The probability distribution can be characterised based on the hourly load data in the IEEE RTS-1996 [21]. Each load state has a percentage range of the predicted peak load. Table 4.1 shows the five designed load states which are utilised in this chapter. Note that other clustered load states can also be used when available, and they do not affect effectiveness of the probabilistic modelling method.

4.3.2 Probability-weighted uncertainty sets

In the RO method, uncertainty sets are used to model uncertainty variables, which consist of prediction interval bounds of uncertainties and uncertainty budgets measuring overall uncertainty levels. The uncertainty sets can provide the worst case of uncertainty realisation for obtaining a robust solution during the optimisation process. In traditional works, an uncertainty set is designed based on a single prediction profile [11–14].

To reflect the uncertainties realistically and effectively during the long-term planning horizon and ensure full robustness of the planning solution, this chapter proposes probability-weighted uncertainty sets to model the uncertainties. First, different load and WT output power states are generated with their probability distribution characteristics by the probabilistic modelling method as described in Section 4.3.1. Then, the uncertainty set for each load or WT output power state is constructed as

$$U_l^{\mathrm{LD}} = \left\{ \underline{\mu}_{y,l}^{\mathrm{LD}} \le \sum_{t\in T}\sum_{i\in I} h \times P_{i,t,y,l}^{\mathrm{LD}} \le \bar{\mu}_{y,l}^{\mathrm{LD}}, \forall y \right.$$

$$\left. \underline{P}_{i,t,y,l}^{\mathrm{LD}} \le P_{i,t,y,l}^{\mathrm{LD}} \le \bar{P}_{i,t,y,l}^{\mathrm{LD}}, \forall i,t,y \right\}, \forall l = 1,2,\ldots,n_l \tag{4.26}$$

$$U_w^{\mathrm{WT}} = \left\{ \underline{\mu}_{y,w}^{\mathrm{WT}} \le \sum_{t\in T}\sum_{i\in I} h \times P_{i,t,y,w}^{\mathrm{WT}} \le \bar{\mu}_{y,w}^{\mathrm{WT}}, \forall y \right.$$

$$\left. \underline{P}_{i,t,y,w}^{\mathrm{WT}} \le TP_{i,t,y,w}^{\mathrm{WT}} \le \bar{P}_{i,t,y,w}^{\mathrm{WT}}, \forall i,t,y \right\}, \forall w = 1,2,\ldots,n_w \tag{4.27}$$

where i, t, and y indicate the bus number, the operation period, and the planning year, respectively. For each set, there are T operation periods which have equal time length h. $\underline{P}$ and $\bar{P}$ are power limits for each period. $\underline{\mu}$ and $\bar{\mu}$ are total energy limits of all the periods for each state. U_l^{LD} and U_w^{WT} stand for the probability-weighted uncertainty sets of the loads and the WT power generation, respectively, and their corresponding occurrence probabilities are ρ_l and ρ_w. Two sets are constructed by containing all U_l^{LD} and U_w^{WT} as elements, respectively, and these two sets of the probability-weighted uncertainty sets are expressed as

$$\Phi^{\mathrm{LD}} = \{U_l^{\mathrm{LD}}\}, \quad \Phi^{\mathrm{WT}} = \{U_w^{\mathrm{WT}}\} \tag{4.28}$$

Last, combined probability-weighted uncertainty sets $U_n \in \Phi$ are formed by Cartesian product of Φ^{LD} and Φ^{WT} and the occurrence probability ρ_n of each combined set U_n is calculated by multiplying ρ_l and ρ_w as follows:

$$\begin{aligned} &\Phi = \Phi^{\mathrm{LD}} \times \Phi^{\mathrm{WT}} = \{(U_l^{\mathrm{LD}}, U_w^{\mathrm{WT}})\} \\ &\forall U_n \in \Phi: \rho_n = \rho_l \times \rho_w \end{aligned} \tag{4.29}$$

The combined probability-weighted uncertainty sets U_n with their occurrence probabilities can cover all uncertainty realisation which follows the probability distribution for the long-term planning horizon, rather than a single profile. These sets are applied in the PRO method to provide the worst cases of uncertainty realisation where a robustly optimal solution can be obtained.

4.3.3 Probability-weighted robust optimisation model

With the probability-weighted uncertainty sets, a PRO model for the proposed DG planning problem is formulated in a matrix compact form as follows:

$$\min_x a^T x + \sum_{n \in N} \rho_n \left(\max_{u_n \in U_n} \min_{y_n} b^T y_n + c^T u_n \right) \tag{4.30}$$

$$\text{s.t.} \quad Dx \geq e \tag{4.31}$$

$$Fx + Gy_n + Hu_n \leq i, \forall n \tag{4.32}$$

$$Jx + Ky_n + Lu_n = m, \forall n \tag{4.33}$$

where x denotes the DG installation decision variables, y denotes the microturbine power generation dispatch decision variables, energy transaction-dependent variables, and power flow dependent variables, u denotes the uncertain loads $P_{i,t,y,l}^{\mathrm{LD}}$, and the uncertain WT output power $P_{i,t,y,w}^{\mathrm{WT}}$. The first objective item $a^T x$ is the investment cost (4.2), and the second objective item $b^T y_n + c^T u_n$ is the microgrid operational costs and revenue over the whole planning horizon (4.4)–(4.7).

Constraints (4.31)–(4.33) represent the DG investment constraints and the microgrid operating constraints (4.8)–(4.21).

In the proposed PRO model, '*max–min*' searches the worst case of uncertainty realisation in each combined probability-weighted uncertainty set U_n and meanwhile minimises the operational costs under the worst case. Then, all the minimised operational costs are summed up with the weights of their corresponding uncertainty sets, i.e. the occurrence probabilities. Thus, the operation under the uncertainties is fully taken into account in the PRO model. Since the DG planning is optimised under the worst cases, the DG installation decision would be fully robust against all uncertainty realisation during the long-term planning horizon, and no operating constraint violation would occur.

4.3.4 Solution algorithm

The C&CG algorithm has been widely applied to solve conventional two-stage RO problems with a high convergence speed [22]. Unlike the conventional two-stage RO problem, the PRO problem proposed in this chapter has multiple probability-weighted uncertainty sets, which cannot be solved by the original C&CG algorithm. To address this issue, this chapter develops a modified C&CG algorithm which can solve the PRO problem under the worst case of each probability-weighted uncertainty set.

The proposed modified C&CG algorithm is demonstrated in Figure 4.1. It decomposes the PRO problem into a master problem which is an MILP problem and a sub-problem in the form of '*max–min*'. The C&CG process solves the master problem and the sub-problem iteratively until the gap between the lower bound given by the master problem and the upper bound given by the sub-problem is reduced to a predefined termination threshold. This is regarded as the outer loop process of the proposed modified C&CG algorithm.

Unlike the original C&CG algorithm, the modified one solves the sub-problem with the probability-weighted uncertainty sets iteratively, which is regarded as the inner loop process.

In the inner loop, for each probability-weighted uncertainty set, the '*max–min*' sub-problem is solved efficiently by the strong duality theory [19] and an outer approximation algorithm [23] which can transform the sub-problem into a linear programming equivalent model which can be solved efficiently. If a finite solution is obtained, the corresponding uncertainty variables construct the worst case for this probability-weighted uncertainty set. This worst case is added into the master problem as a non-optimal cut. Otherwise, some certain values of the uncertainty variables can cause an infeasible '*min*' problem. This means that despite any operation effort, an operating constraint violation would occur under this specific case that the uncertainty variables realise as these values. These values of the uncertainty variables are identified to construct the worst case. In this condition, the worst case is added into the master problem as an infeasible cut. If the infeasible cuts do not arise, the upper bound is updated; otherwise, the upper bound is kept as same with that in the last iteration of the outer loop.

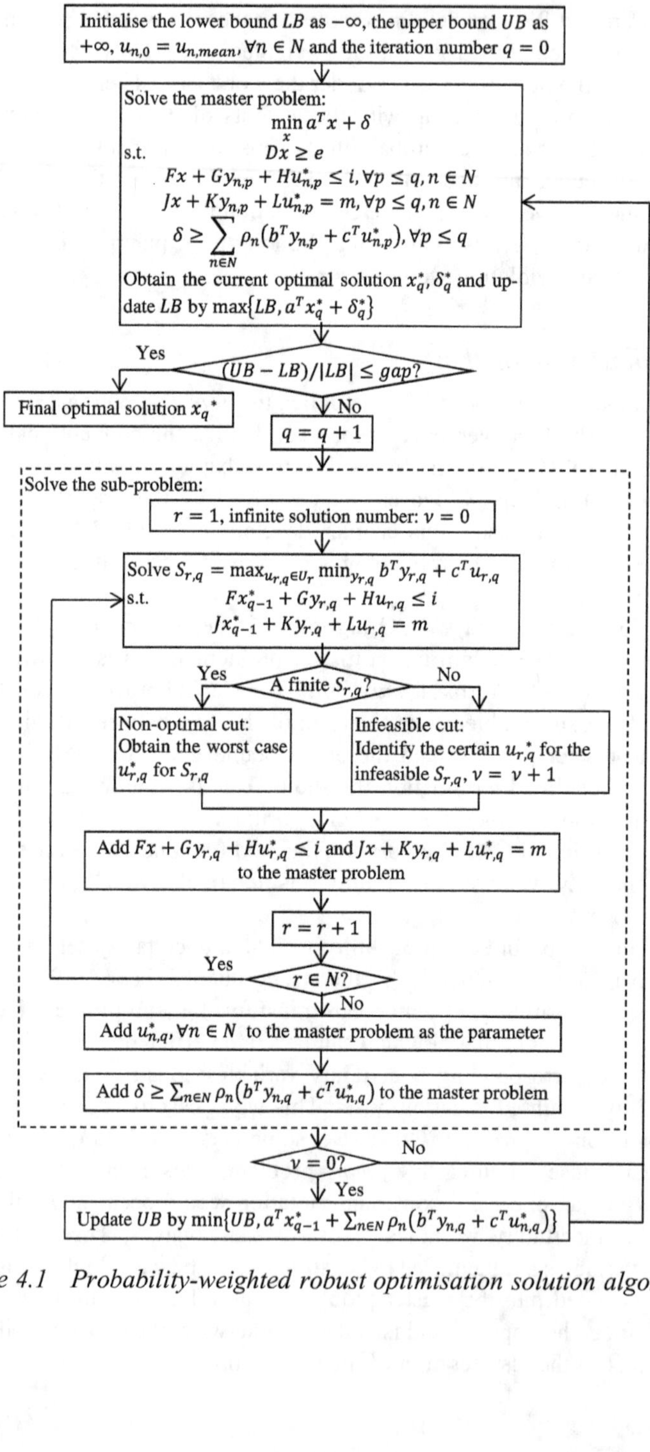

Figure 4.1 Probability-weighted robust optimisation solution algorithm

4.4 Case study

4.4.1 Test system description

The proposed DG planning method is tested on the 33-bus radial distribution system which has been modified to a microgrid [24,25], and its data can be obtained from [26]. The system topology is shown in Figure 4.2. Buses 3, 6, 9, 12, 18, 22, 25, 27, 30, and 33 are selected as DG installation candidate buses based on the sensitivity analysis given by Chapter 3 and they are marked as red in Figure 4.2.

The planning horizon is 10 years. Each probability-weighted uncertainty set has four 15-min operation periods (i.e. $h = 0.25$ h) and τ is 8,760. The other parameters used in this case study are shown in Table 4.2. Only the first-year values of the generation cost, the O&M costs, and the electricity prices are shown. For the remaining years, the values can be calculated based on the first-year values and the annual cost increasing rate, c. For example, the yth-year generation cost can be calculated as follows:

$$G_y^{\mathrm{MT}} = (1+c)^{y-1} G_1^{\mathrm{MT}}, \forall y \tag{4.34}$$

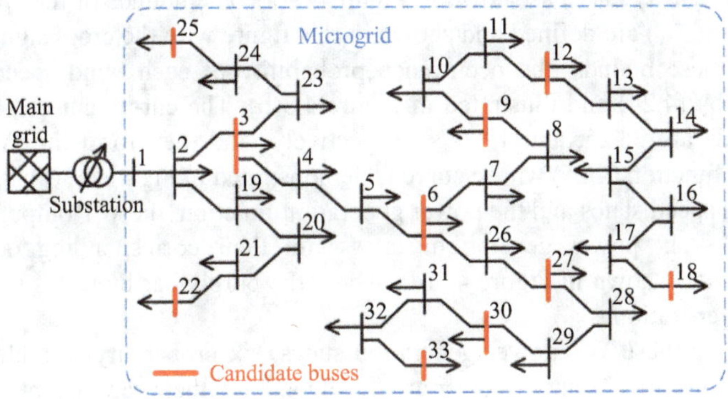

Figure 4.2 Topology of test microgrid

Table 4.2 Parameters in distributed generation planning

Parameter	Value	Parameter	Value
c	6%	OM_1^{MT}	\$0.008/kWh
d	5%	OM_1^{WT}	\$0.007/kWh
g	3%	Pr_1^{buy}	\$0.038/kWh
G_1^{MT}	\$0.036/kWh	Pr_1^{cus}	\$0.142/kWh
Inv^{MT}	\$2,293/kVA	$S^{\mathrm{MT/WT}}$	30 kVA
Inv^{WT}	\$1,882/kVA	ε	115%

It is worth noting that any practical data can be used and does not affect the effectiveness of the proposed DG planning method.

The simulations are conducted on a 64-bit PC with 3.30-GHz CPU and 8 GB RAM using MATLAB® platform and YALMIP interface [27]. The proposed optimisation problems are solved by GUROBI solver [28].

4.4.2 Probability-weighted uncertainty sets

In the first year, the predicted peak load is 3.715 MW for this microgrid. Based on the load states given by Table 4.1, five probability-weighted uncertainty sets for the loads are developed and their parameters are shown in Table 4.3. For the lower and upper bounds and the uncertainty budgets, only the first-year parameters are given in this table.

For the remaining years, the mean loads, the lower and upper bounds, as well as the uncertainty budgets can be calculated based on the first-year values and the annual load growth rate, g. For example, the mean loads can be calculated as follows:

$$P_{i,t,y,l}^{\mathrm{LD}} = (1+g)^{y-1} P_{i,t,1,l}^{\mathrm{LD}}, \forall i, t, y, l \tag{4.35}$$

The mean wind speed is 5 m/s for (4.22) and (4.23) to obtain the wind speed probability density curve as shown in Figure 4.3(a). The bounds of the wind speed states ($\underline{s}_{\mathrm{w}}$ and $\bar{s}_{\mathrm{w}}$) are defined and shown in this figure with the cross signs as well. Based on these bounds, the occurrence probability of each wind speed state is calculated by (4.24) and illustrated in Figure 4.3(b). The cut-in, cut-off, and rated wind speeds are 3, 25, and 12 m/s, respectively, and they form the WT power generation function (4.25) whose curve is demonstrated in Figure 4.3(c). According to the wind speed states and the power generation function, the WT output power is modelled as six power generation states with their corresponding occurrence probabilities as shown in Figure 4.3(d), where the purple bar indicates the state of no power generation.

Based on these WT power generation states, six probability-weighted uncertainty sets for the WT output power are developed and their parameters are shown in Table 4.4. It is assumed that the mean wind speed does not change during the whole planning horizon. However, if long-term forecasting information is

Table 4.3 Probability-weighted uncertainty sets for loads (first year)

Uncertainty set l	**1**	**2**	**3**	**4**	**5**
$\underline{\mu}_{1,l}^{\mathrm{LD}}$ (MWh)	3.344	2.972	2.601	2.229	1.858
$\bar{\mu}_{1,l}^{\mathrm{LD}}$ (MWh)	3.715	3.344	2.972	2.601	2.229
$\underline{P}_{i,t,1,l}^{\mathrm{LD}}/\widehat{P}_{i,t,1}^{\mathrm{LD}}$ (%)	81	72	63	54	45
$\bar{P}_{i,t,1,l}^{\mathrm{LD}}/\widehat{P}_{i,t,1}^{\mathrm{LD}}$ (%)	110	99	88	77	66
Probability ρ_l (%)	44.73	18.27	9.23	16.38	11.39

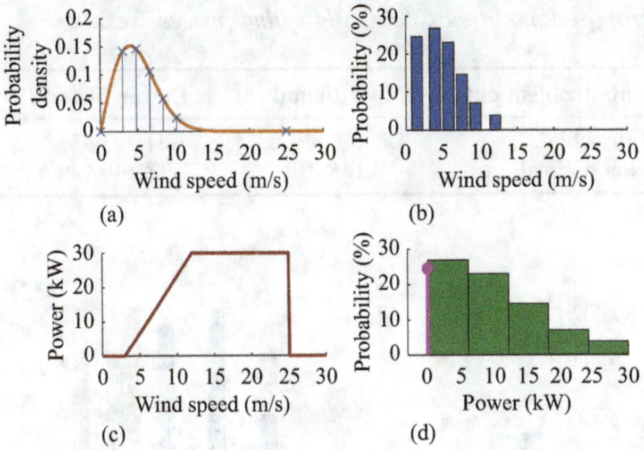

Figure 4.3 (a) Wind speed probability density; (b) wind speed probability; (c) WT power generation function; (d) WT power generation probability

Table 4.4 Probability-weighted uncertainty sets for WT output power

Uncertainty set *w*	**1**		**2**	**3**	**4**	**5**	**6**
$\underline{s}_w$ (m/s)	0	25	3	4.8	6.6	8.4	10.2
$\bar{s}_w$ (m/s)	3	∞	4.8	6.6	8.4	10.2	25
$\underline{\mu}_{y,w}^{\mathrm{WT}}$ (kWh)	0		0	6	12	18	24
$\bar{\mu}_{y,w}^{\mathrm{WT}}$ (kWh)	0		6	12	18	24	30
$\underline{P}_{j,t,y,w}^{\mathrm{WT}}/S^{\mathrm{WT}}$ (%)	0		0	16	32	48	64
$\bar{P}_{j,t,y,w}^{\mathrm{WT}}/S^{\mathrm{WT}}$ (%)	0		24	48	72	96	100
Probability ρ_w (%)	24.63		26.88	23.04	14.55	7.09	3.81

available, the parameters of the probability-weighted uncertainty sets can be easily modified.

According to these load and WT output power uncertainty sets, 30 combined probability-weighted uncertainty sets applied in the PRO method are formed and their corresponding occurrence probabilities are calculated by using (4.29).

4.4.3 Distributed generation planning results

Using the proposed modified C&CG algorithm, a DG planning solution is obtained after only two iterations with a significantly small gap. The proposed modified C&CG algorithm process results are shown in Table 4.5. In the first iteration, some infeasible cuts are produced, while only non-optimal cuts are generated in the second iteration. This indicates that the proposed modified C&CG algorithm can find the worst cases where the operating constraint violations occur in the first iteration. Then, the master problem is solved according to the worst case of each

Table 4.5 Proposed modified C&CG algorithm process results

Iteration	Sub-problem cut	Upper bound ($)	Lower bound ($)	Gap (%)
1	Infeasible	$+\infty$	−21,793,937	Inf
2	Non-optimal	−18,146,129	−18,146,149	0.0001

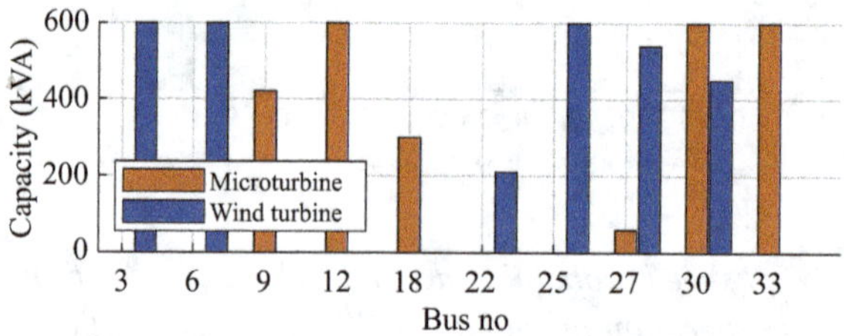

Figure 4.4 DG planning solution

combined probability-weighted uncertainty set to obtain a robust planning solution which leads to no violation. In the second iteration, there are only the non-optimal cuts, which means that robust operation without any operating constraint violation can be achieved with the robust planning solution. With the non-optimal cuts, the master problem optimises the DG planning again based on the worst cases. Thus, after two iterations, the DG planning solution is robustly and optimally obtained.

The DG planning results are shown in Figure 4.4. The microturbines with a total capacity of 2.58 MVA are installed on buses 9, 12, 18, 27, 30, and 33. The WTs with a total capacity of 3 MVA are installed on buses 3, 6, 22, 25, 27, and 30. The total investment cost is 11.56 million dollars ($11.56M). The operating profit in the NPV under the worst cases is about $18.15M.

4.4.4 Operation results with Monte Carlo simulation

To comprehensively examine the operating robustness of the PRO-based DG planning solution, 10,000 random scenarios of WT output power and loads are generated by Monte Carlo sampling (1,000 scenarios for each year). The sampled wind speeds follow the *Rayleigh* PDF and the WT output power has a range from 0 to 30 kWh for each WT. The loads follow the *Gaussian* PDF of each load state with a range from 50% to 110% of $\widehat{P}^{LD}$ in each year. For a comparison purpose, DG planning solutions obtained by the RO-based planning approach [14] with four different cases are also examined with these random scenarios. These four cases have different uncertainty profiles as follows:

RO-1: wind power profile used in [14];

RO-2: low wind power profile (average output power below 50% of the rated WT capacity);

Table 4.6 Comparison between different distributed generation planning methods

DG planning method	PRO	RO-1	RO-2	RO-3	RO-4
Voltage violation rate (%)	0	8.12	0	3.12	11.44
Operating profit in NPV ($M)	23.51	24.71	23.43	24.50	24.74

RO-3: high wind power profile (average output power above 50% of the rated WT capacity);

RO-4: large wind power variation range (from 0% to 100% of the rated WT capacity).

The comparison results are shown in Table 4.6. It can be seen in this table that the proposed PRO method can ensure a 100% satisfaction of the operating constraints, while the RO method reports 8.12%, 3.12%, and 11.44% voltage violations of all the scenarios for cases RO-1, RO-3, and RO-4, respectively. Although the operating profit in the NPV is sacrificed slightly, the PRO-based DG planning solution can guarantee the robustness for the short-term operation under the uncertainty realisations in the long-term planning horizon. For the case RO-2 which uses a low wind power profile, although the RO solution can also satisfy the operating constraints for all the random scenarios, its operating profit in the NPV is less than that of the proposed PRO solution. Therefore, the proposed PRO-based DG planning method is overall superior.

Besides, it is worth pointing out that for the RO method, given the different single-day profiles, the obtained planning solutions are different from each other. This adds more difficulty and complexity to DG planning decision makers. In contrast, the PRO method can always provide a determinate solution, which allows more robust decision-making.

4.5 Conclusion

This chapter proposes a PRO method to optimise the DG allocation in the microgrid. The uncertainties are modelled in the probability-weighted uncertainty sets which fully take the occurrence probability distribution of the uncertainties into account. Within each probability-weighted uncertainty set, the proposed PRO method searches the worst case of uncertainty realisation and optimises the DG allocation based on the worst case. Moreover, this chapter develops a modified C&CG algorithm to effectively solve the PRO-based DG planning problem. The simulation results show that the proposed PRO method can achieve a robust DG planning solution efficiently without any operating constraint violation during the whole planning horizon, while the RO method cannot guarantee a robust solution. Thus, taking the advantages of both the probabilistic optimisation method and the RO one, the PRO method can guarantee a robust and optimal solution for the DG planning in the microgrid.

References

[1] S. Haffner, L.F.A. Pereira, L.A. Pereira, and L.S. Barreto, 'Multistage model for distribution expansion planning with distributed generation – Part I: Problem formulation', *IEEE Trans. Power Del.*, vol. 23, no. 2, pp. 915–923, 2008.

[2] D. Singh, D. Singh, and K.S. Verma, 'Multiobjective optimization for DG planning with load models', *IEEE Trans. Power Syst.*, vol. 24, no. 1, pp. 427–436, 2009.

[3] E. Naderi, H. Seifi, and M.S. Sepasian, 'A dynamic approach for distribution system planning considering distributed generation', *IEEE Trans. Power Del.*, vol. 27, no. 3, pp. 1313–1322, 2012.

[4] V.F. Martins and C.L.T. Borges, 'Active distribution network integrated planning incorporating distributed generation and load response uncertainties', *IEEE Trans. Power Syst.*, vol. 26, no. 4, pp. 2164–2172, 2011.

[5] K. Zou, A.P. Agalgaonkar, K.M. Muttaqi, and S. Perera, 'Distribution system planning with incorporating DG reactive capability and system uncertainties', *IEEE Trans. Sustain. Energy*, vol. 3, no. 1, pp. 112–123, 2012.

[6] Y.M. Atwa, E.F. El-Saadany, M.M.A. Salama, and R. Seethapathy, 'Optimal renewable resources mix for distribution system energy loss minimization', *IEEE Trans. Power Syst.*, vol. 25, no. 1, pp. 360–370, 2010.

[7] R.S. Al Abri, E.F. El-Saadany, and Y.M. Atwa, 'Optimal placement and sizing method to improve the voltage stability margin in a distribution system using distributed generation', *IEEE Trans. Power Syst.*, vol. 28, no. 1, pp. 326–334, 2013.

[8] M.F. Shaaban, and E.F. El-Saadany, 'Accommodating high penetrations of PEVs and renewable DG considering uncertainties in distribution systems', *IEEE Trans. Power Syst.*, vol. 29, no. 1, pp. 259–270, 2014.

[9] S.F. Santos, D.Z. Fitiwi, A.W. Bizuayehu, *et al.*, 'Novel multi-stage stochastic DG investment planning with recourse', *IEEE Trans. Sustain. Energy*, vol. 8, no. 1, pp. 164–178, 2017.

[10] F. Ugranli and E. Karatepe, 'Transmission expansion planning for wind turbine integrated power systems considering contingency', *IEEE Trans. Power Syst.*, vol. 31, no. 2, pp. 1476–1485, 2016.

[11] R. Jiang, J. Wang, M. Zhang, and Y. Guan, 'Two-stage minimax regret robust unit commitment', *IEEE Trans. Power Syst.*, vol. 28, no. 3, pp. 2271–2282, 2013.

[12] C. Zhang, Y. Xu, Z.Y. Dong, and J. Ravishankar, 'Three-stage robust inverter-based voltage/var control for distribution networks with high-level PV', *IEEE Trans. Smart Grid*, vol. 10, no. 1, pp. 782–793, 2019.

[13] B. Chen, J. Wang, L. Wang, Y. He, and Z. Wang, 'Robust optimization for transmission expansion planning: minimax cost vs. minimax regret', *IEEE Trans. Power Syst.*, vol. 29, no. 6, pp. 3069–3077, 2014.

[14] Z. Wang, B. Chen, J. Wang, J. Kim, and M.M. Begovic, 'Robust optimization based optimal DG placement in microgrids', *IEEE Trans. Smart Grid*, vol. 5, no. 5, pp. 2173–2182, 2014.

[15] A. Khodaei, S. Bahramirad, and M. Shahidehpour, 'Microgrid planning under uncertainty', *IEEE Trans. Power Syst.*, vol. 30, no. 5, pp. 2417–2425, 2015.

[16] S. Wang, S. Chen, L. Ge, and L. Wu, 'Distributed generation hosting capacity evaluation for distribution systems considering the robust optimal operation of OLTC and SVC', *IEEE Trans. Sustain. Energy*, vol. 7, no. 3, pp. 1111–1123, 2016.

[17] W. Yuan, J. Wang, F. Qiu, C. Chen, C. Kang, and B. Zeng, 'Robust optimization-based resilient distribution network planning against natural disasters', *IEEE Trans. Smart Grid*, vol. 7, no. 6, pp. 2817–2826, 2016.

[18] A. Khodaei, 'Provisional microgrid planning', *IEEE Trans. Smart Grid*, vol. 8, no. 3, pp. 1096–1104, 2017.

[19] C. Zhang, Y. Xu, Z.Y. Dong, and J. Ma, 'Robust operation of microgrids via two-stage coordinated energy storage and direct load control', *IEEE Trans. Power Syst.*, vol. 32, no. 4, pp. 2858–2868, 2017.

[20] G. Boyle, *Renewable Energy*, UK: Oxford University Press, 2004.

[21] C. Grigg, P. Wong, P. Albrecht, *et al.*, 'The IEEE reliability test system-1996. A report prepared by the reliability test system task force of the application of probability methods subcommittee', *IEEE Trans. Power Syst.*, vol. 14, no. 3, pp. 1010–1020, 1999.

[22] B. Zeng and L. Zhao, 'Solving two-stage robust optimization problems using a column-and-constraint generation method', *Operations Research Letters*, vol. 41, no. 5, pp. 457–461, 2013.

[23] D. Bertsimas, E. Litvinov, X.A. Sun, J. Zhao, and T. Zheng, 'Adaptive robust optimization for the security constrained unit commitment problem', *IEEE Trans. Power Syst.*, vol. 28, no. 1, pp. 52–63, 2013.

[24] M.V. Kirthiga, S.A. Daniel, and S. Gurunathan, 'A methodology for transforming an existing distribution network into a sustainable autonomous micro-grid', *IEEE Trans. Sustain. Energy*, vol. 4, no. 1, pp. 31–41, 2013.

[25] C. Zhang, Y. Xu, Z.Y. Dong, and K.P. Wong, 'Robust coordination of distributed generation and price-based demand response in microgrids', *IEEE Trans. Smart Grid*, vol. 9, no. 5, pp. 4236–4247, Sept. 2018.

[26] M.E. Baran and F.F. Wu, 'Network reconfiguration in distribution systems for loss reduction and load balancing', *IEEE Trans. Power Del.*, vol. 4, no. 2, pp. 1401–1407, 1989.

[27] J. Lofberg, 'YALMIP: a toolbox for modelling and optimization in MATLAB', *Proc. IEEE Inter. Conf. on Robotics and Automation*, pp. 284–289, Taipei, Taiwan, 2004.

[28] GUROBI, Available: www.gurobi.com, accessed March 2018.

Chapter 5

Multi-stage stochastic programming method for multi-energy DG planning

The multi-energy microgrid (MEMG) comprises heterogeneous distributed generators (DGs) such as wind turbines (WTs), diesel generators (DEs), and combined cooling, heat, and power (CCHP) plants. Proper placement of these DGs is critical for the system energy efficiency and network reliability performance. This chapter proposes a two-stage coordinated method for optimally placing heterogeneous DGs in an MEMG project considering the uncertainties from renewable energy sources (RESs). Apart from optimising the traditional DG size and location, this method considers the optimal DG type and investment year simultaneously by maximising the project net present value (NPV), which consists of investment costs and operation costs. The whole problem is modelled as a two-stage coordinated stochastic optimisation model, where the long-term DG investment is determined at the first stage and operation decisions are determined at the second stage. The proposed method is verified on a test MEMG system. The simulation results show that its NPV is positive, which means the method is effective and should be implemented. Compared with the conventional DG placement approaches, the proposed method is more robust against the RES uncertainties and can better coordinate the heterogeneous energies with higher dispatch flexibility and economic profits.

Nomenclature

A Abbreviations

AC	Absorption chiller
CCHP	Combined cooling, heat, and power
DE	Diesel generator
DG	Distributed generator
EB	Electric boiler
EC	Electric chiller
GB	Gas boiler
HRU	Heat recovery unit
MG/MEMG	Microgrid/multi-energy microgrid

MILP	Mixed-integer linear programming
MT	Microturbine
NPV	Net present value
PV	Photovoltaic cell
RES	Renewable energy source
SEMG	Single-energy microgrid
WT	Wind turbine

B Superscript

Λ	Set for $\{p, q, d, t, i\}$
b	Index of branches, $b \in \mathrm{Br}(i, i+1)$
$d/t/i$	Index of days/dispatch periods/buses
$p/q/c/s$	Index of years/DG types/seasons/scenarios

C Constants and parameters

a^q, b^q	Cost coefficients for DEs
$C^{\mathrm{U}}_{\mathrm{DG}}/C^{\mathrm{D}}_{\mathrm{DG}}$	Unit DG start-up/shutdown cost
dr	Discount rate
$N_{\mathrm{D}}/N_{\mathrm{T}}$	Number of days in a season/dispatch periods in a day
N_{P}	Project lifetime
N_{Q}	Total number of DG types
N_{B}	Total number of buses in the MEMG
$N_{\mathrm{CB}}/N_{\mathrm{NB}}$	Total number of candidate/non-candidate buses
$N_{\mathrm{CQ}}/N_{\mathrm{VQ}}$	Total number of DGs with discrete unit/continuous capacity
$P_{\mathrm{DG,REAL}}$	Available active power of RES
$P_{\mathrm{L}}/H_{\mathrm{H}}/H_{\mathrm{C}}$	Power/heat/cooling loads of the MEMG
$R_{\mathrm{DG,DOWN}}/R_{\mathrm{DG,UP}}$	Maximal ramping down/up limits
$R^{\mathrm{b}}/X^{\mathrm{b}}$	Branch resistance/reactance
$S_{\mathrm{DG,\ ALL}}$	Maximal DG investment capacity for the MEMGs
$S_{\mathrm{DG,MIN}}/S_{\mathrm{DG,MAX}}$	Minimal/maximal DG investment capacity for each bus
$S_{\mathrm{DG,UN}}$	The unit capacity of DGs in discrete size
V_0	Substation voltage
$\gamma_{\mathrm{DG,MIN}}/\gamma_{\mathrm{DG,MAX}}$	Minimal/maximal DG output ratio to its capacity
Δt	Unit dispatch interval
$\Delta V^{\max}_{\mathrm{BUS}}$	Maximal bus voltage deviation
$\varepsilon_{\mathrm{DG,SAL}}$	The salvage factor, which is the ratio of DG salvage value to its initial investment cost
$\eta_{\mathrm{MT}}/\eta_{\mathrm{DE}}$	Electrical efficiency of MTs/DEs

η_{EB}/η_{EC}	Coefficient of performance of EBs/ECs
η_{MH}/η_{MC}	Power to heat/cooling ratios of MTs
$\kappa_{MT}/\kappa_{B}/\kappa_{DE}$	Emission conversion factor of MTs/grid/DEs
χ_{B}/χ_{S}	Electricity purchasing/selling prices between the MEMGs and utility grid
χ_{CO2}	Unit CO_2 penalty price
$\chi_{DG,INV}$	Unit DG investment cost
$\chi_{DG,F}/\chi_{DG,V}$	The fixed/variable unit maintenance cost of DGs
χ_{NG}/χ_{DE}	Price for the natural gas/diesel
$\chi_{PS}/\chi_{HS}/\chi_{CS}$	The power/heat/cooling retail price to the local customers
χ_{RES}	The unit RES curtailment cost
χ_{SUB}	Unit subsidy price for RES

D Variables

C_{CT}	RES curtailment cost
C_{FC}/C_{TC}	Cost from fuel consumption/power transaction
C_{INV}	The investment cost for DGs
C_{MC}/C_{EM}	Cost from DG maintenance/gas emission
C_{REV}/C_{SUB}	Revenue from the local customers/RES subsidy
C_{ST}/C_{SD}	Start-up/shutdown cost
C_{SAL}^{Np}	DG salvage value
H_{EB}/H_{EC}	Heat/cooling outputs for EBs/ECs
N_{DG}	Number of DGs in discrete size
P_{WT}/P_{PV}	Power outputs of WTs/PVs
P_{BX}/P_{SX}	Purchasing/selling power from/to the utility grid
P_{DG}/P_{DE}	Power outputs of DGs/DEs
P_{EB}/P_{EC}	Power inputs for EBs/ECs
P_{F}	Diesel inputs for DEs
P_{G}	Gas inputs for MTs
$P_{MT}/H_{CMT}/H_{HMT}$	Power/cooling/heat outputs of MTs
P_{PF}/Q_{PF}	Active/reactive power flow
$P^{p,d,t,0,b}{}_{PF}/Q^{p,d,t,0,b}{}_{PF}$	Parallel power flow of branch *b*
S_{DG}	DG capacity
U_{DG}/U_{DE}	Binary variable to indicate whether a DG/DE is on: 0 is off, 1 is on
U_{DGI}	Binary variable to indicate whether a DG is installed: 1 is installed, 0 is not
V_{BUS}	Bus voltage

5.1 Introduction

Compared with the conventional centralised coal-fired generators, heterogeneous DGs such as WTs, DEs, and photovoltaic cells (PVs) are decentralised, module-based, flexible, and located much closer to the loads. For the practical application, they are usually deployed and operated within a microgrid (MG) [1], providing advantages for emission reduction, reliability enhancement, etc.

For the DG application in MGs, its placement in the network is critical for the system techno-economic performance [2,3]. Specifically, DG placement refers to deciding their type, location, size, investment year, generation mix, etc. In the literature, the DG placement has been optimised for the purpose of power loss minimisation [4], reliability enhancement [5], system voltage stability improvement [6], and cost–benefit balancing [7], etc. Reference [8] proposes an optimal placement method for DG type and size selection in an MG for minimisation of annual cost and pollution from the combustion processes. Reference [9] first decides the locations for the DGs by identifying the loss sensitivity index of each bus, and then determines their size based on the cost–benefit analysis. Ref. [10] presents an optimisation model that determines the optimal DG size and generation mix for an isolated MG. All of the above studies in [4–10] assume that there is perfect generation information for the power outputs of WTs and PVs. However, the RES outputs are stochastic in nature and their uncertainties should be tackled for a robust system planning solution, so as to avoid violation of the operational limits if the significant uncertainties occur.

To handle the uncertainties in the DG placement, Ref. [11] employs the robust optimisation method to maximise the system long-term cost under the RES uncertainties. Reference [12] presents a model for DG placement in an MG with uncertain physical information and the model is then solved by the robust optimisation method. Reference [13] describes a DG placement method for minimising the cost for supplying a remote telecommunication station exclusively by the RES, and the robust optimisation method is also used. Reference [14] compares the stochastic and robust optimisation methods in the optimal DG placement scheme under the objective of minimising system annualised cost. The simulation results show that the robust optimisation method tends to be conservative over a long planning period than the stochastic method and highly depends on the predefined uncertainty budgets.

It is worth noting that the DG placement in [4–14] only focuses on the single-energy microgrids (SEMGs), i.e. the MGs with only the electricity. However, with the fast development of the MEMGs [15–18], multiple energy carriers such as the power and thermal energy are interacted and coordinated by the heterogeneous DGs. Thus, the above traditional separate planning methods would not be sufficient for the long-term planning of interdependent energy systems. Further, 'synergies' among the heterogeneous energies in the MEMGs would introduce an additional energy supply degree of freedom; in this regard, the system cost/profit would be potentially reduced/increased.

In related literature, Ref. [19] presents a reliability-constrained optimisation approach to determine the size of CCHP plants. The RES uncertainties are handled by the stochastic optimisation method in the operation. References [20,21] obtain the best DG alternatives and control strategies by considering the uncertainties in a long-term time frame based on the decision theory. Nonetheless, for the DG placement research in [19–21], the distribution network power flow model is not considered, so the planning decisions may not be viable for practical application. In Ref. [22], a DG planning method for a residential MEMG with the RES, battery storage, and demand response scheme is proposed. The stochastic model takes the RES intrinsic fluctuation behaviours into account. Reference [23] proposes a multi-objective optimisation model and an evidential reasoning approach to solve the DG sizing problem. In this model, the interval variables are introduced to tackle the RES uncertainties. However, in the above research work [22,23], the DG placement only considers the power-related DGs, while the placement of other energy-type of DGs such as CCHP units is not studied sufficiently (their placement is predefined). Consequently, 'synergies' among heterogeneous energies have not been optimally achieved in their studies.

Moreover, most of the above works focus only on determining the size and location of DGs [4,14,19–23], but a more comprehensive study on simultaneously deciding the DG type as well as investment year under the multi-energy coordination scheme has not been studied sufficiently yet. The DG type and investment year optimisation are also critical for MEMGs for the following reasons:

1. Prior to the project, choosing the right types of DGs is important because different types of DGs have different economic and technical properties. The DG type optimisation would bring a new degree of freedom for the DG placement and potentially reduce/increase the system cost/profit.
2. For the investment year optimisation, Refs. [4–14,19–23] have not taken the annual load growth [21] into account, and they only consider the one-time DG investment at the beginning of the project. However, with the practical ever-growing loads, their DG placement decisions would probably be far from sufficient. Besides, the investment year optimisation will bring more flexibility for the DG placement and hence further reduce the system cost. In the literature, few studies have considered the investment year optimisation: Ref. [24] proposes a tri-level expansion planning framework for an isolated MG under uncertainty, and Ref. [25] presents a stochastic multi-objective approach for optimal dynamic planning of interconnected MGs under uncertainty. Though the investment year is taken into account, the studies in [24,25] are still limited in the SEMG research scope. To sum up, for the DG placement, it is necessary to consider different DG investment phases and multiple DG types in the MEMGs.

As for the solution algorithms, the DG placement in some studies is divided into problems of different levels [8,12]. In such cases, the globally optimal solution is hard to be obtained. Further, many previous works apply evolutionary algorithms such as the Genetic Algorithm or Particle Swarm Optimisation [6,8,9,23]. Though effective, these algorithms usually suffer from problems such as excessive computational burden, inconsistent solutions, and non-rigorous convergence criterion.

To fill the inadequacies in the literature, this chapter proposes an optimal two-stage placement method for the heterogeneous DGs in a grid-connected MEMG considering RES uncertainties. Compared with existing works, the heterogeneous

DGs in this chapter include the WTs, DEs, CCHP plants, PVs, electric boilers (EBs), and electric chillers (ECs). The main novel contributions are summarised as follows:

1. A two-stage stochastic placement method for heterogeneous DGs in the MEMG under RES uncertainties is proposed. Compared with the conventional operation research, the DG planning problem is investigated in the multi-energy coordination scheme.
2. Apart from the DG size and location planning, this study considers the type and investment year optimisation for a more comprehensive and flexible DG placement to optimally minimise/maximise the project cost/profit.
3. To handle the RES uncertainties and avoid any conservative solutions, the proposed DG placement model is formulated as a multi-stage stochastic optimisation problem.
4. The model is converted into a mixed-integer linear programming (MILP) problem via linearisation processing. Then comprehensive case studies are done to verify its effectiveness and advantages.

The remainder of this chapter is organised as follows: Section 5.2 introduces the models for the MEME system and multi-energy conversion block. Section 5.3 presents the DG placement model and provides the linearisation method. The solution methodology is outlined in Section 5.4. Section 5.5 describes the test system set-up and case studies for evaluating the effectiveness of the proposed method. Section 5.6 lists the simulation results and the corresponding analysis. Section 5.7 presents the conclusions. Besides, the supplementary contents are given in Appendix A.

5.2 MEMG modelling

An MEMG carries the heterogeneous energy flows including power, heat, and cooling. The heterogeneous DGs are the generation components for converting fuels to different types of energy. In this section, all the DGs are modelled to match the system loads via the multi-energy conversion block.

5.2.1 System configuration

Figure 5.1 shows a typical structure of a grid-connected MEMG. The input energy is transferred into the output side by multi-energy conversion block. For the multi-energy supply, the system thermal (heat or cooling) loads are provided by the EBs, CCHP plants, and ECs; the system power loads are supplied by the DEs, CCHP plants, WTs, PVs, as well as the transaction power from the utility grid.

5.2.2 DG classification

The heterogeneous DGs in MEMGs can be classified as follows:

1. Non-dispatchable DGs, i.e. their power outputs are intermittent and non-dispatchable such as the WTs and PVs. However, their power outputs can be curtailed if needed.

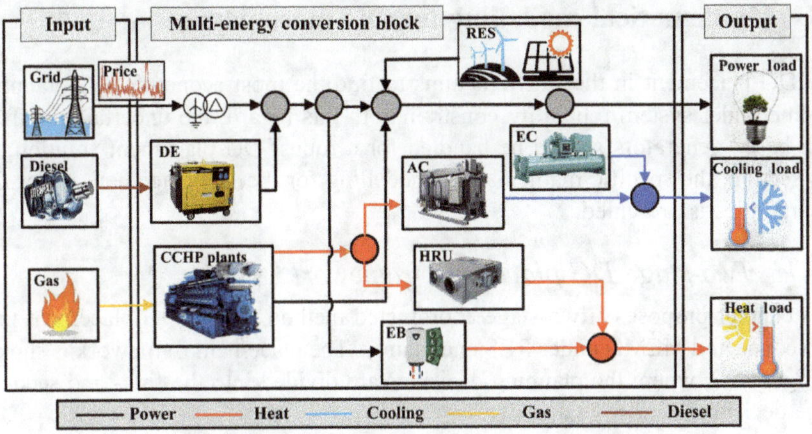

Figure 5.1 Typical structure of an MEMG

2. Dispatchable DGs, i.e. their power outputs can be dispatched flexibly, such as the DEs, CCHP plants, EBs, and ECs.

In the MEMG, the system uncertainty comes from the non-dispatchable DGs and should be handled to maintain the system operation reliability and economy. In addition, the DGs can also be classified according to their energy carriers; in this way, the CCHP plant is the tri-generation unit while others can only generate the single type of energy.

5.2.3 Multi-energy conversion

In the multi-energy conversion block of Figure 5.1, the microturbine (MT) is used as the main generator for the CCHP plants. It consumes natural gas to produce power, and the waste heat is utilised by the heat recovery unit (HRU) or absorption chiller (AC) to generate heat or cooling energy, respectively [18,26]. Each MT is specified by its electrical efficiency and thermal-to-power ratio, which are used to calculate its power and thermal outputs. To enhance the flexibility of supplying multiple energies, EBs and ECs can be utilised to convert the power to thermal energy. For DEs, they consume the diesel to generate power. Based on the above, the mathematical model of multi-energy conversion block is given in (5.1):

$$\underbrace{\begin{bmatrix} P_{\mathrm{DE}} \\ P_{\mathrm{MT}} \\ H_{\mathrm{HMT}} + H_{\mathrm{EB}} \\ H_{\mathrm{CMT}} + H_{\mathrm{EC}} \end{bmatrix}}_{\text{Output}} = \underbrace{\begin{bmatrix} \eta_{\mathrm{DE}} & 0 & 0 & 0 \\ 0 & \eta_{\mathrm{MT}} & 0 & 0 \\ 0 & \eta_{\mathrm{MH}} & \eta_{\mathrm{EB}} & 0 \\ 0 & \eta_{\mathrm{MC}} & 0 & \eta_{\mathrm{EC}} \end{bmatrix}}_{\text{Multi-energy conversion matrix}} \times \underbrace{\begin{bmatrix} P_{\mathrm{F}} \\ P_{\mathrm{G}} \\ P_{\mathrm{EB}} \\ P_{\mathrm{EC}} \end{bmatrix}}_{\text{Input}} \tag{5.1}$$

5.3 Mathematical modelling for DG placement

The DG placement in the MEMGs aims to find the most economic DG planning scheme under system reliability constraints. In this regard, the uncertainties from RES-based generators should be handled for a robust DG placement solution. In this section, the specific mathematical modelling for the DG placement under the uncertainties is presented.

5.3.1 Two-stage DG placement framework

This chapter proposes a two-stage coordinated method for the DG placement in a grid-connected MEMG under RES uncertainty. The placement framework is shown as Figure 5.2, where the planning decisions are divided into the first- and second-stage decisions:

1. First- or investment-stage decisions, which refer to the investment plan, namely the type, location, size, and installation phase of heterogeneous DGs. In this chapter, 4 years are taken as a unit investment phase. The DG capacity to be installed at the beginning year of a phase (i.e. the first, fifth, ninth, etc., years) is set as the separate decision variable sets. During each phase, the capacity is kept the same for the whole phase.
2. Second- or operation-stage decisions, which refer to the optimal operating dispatch of all the DGs after they are installed.

The mathematical models of the above two-stage planning framework will be detailed in the next subsection.

5.3.2 Two-stage mathematical modelling

In this study, the objective function is the project NPV which should be maximised. The two-stage DG placement problem in Figure 5.2 can be modelled as follows:

1. Objective function:

$$NPV_s = \operatorname*{Max}_{z \in CF_z, x \in CG_x} \{ \underbrace{-F(z)}_{\text{Investment stage}} + \underbrace{G(x)}_{\text{Operation stage}} \} \tag{5.2}$$

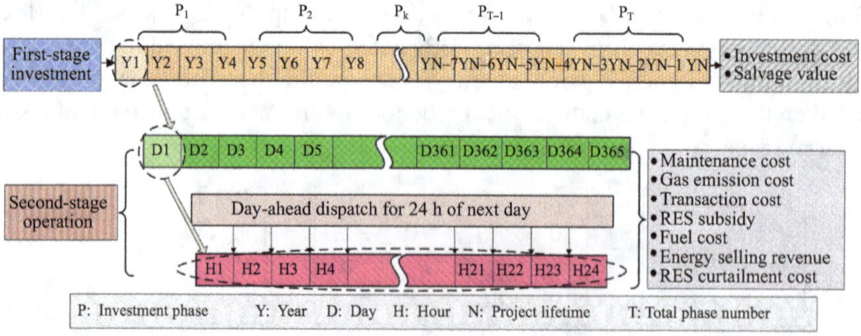

Figure 5.2 Proposed two-stage DG placement structure

where in (5.2), NPV_s denotes the NPV of the project. z is the decision variable of the planning stage, i.e. DG type, size, location, and investment phase. CF_z denotes the constraints for the DG placement, which are shown in (5.4)–(5.12). $F(z)$ is the objective of the investment stage, and its mathematical model is shown in (5.3). x denotes the decision variable at the operation stage. CG_x and $G(x)$ are the constraints and objective of the operation stage.

2. Constraints for the investment stage:

$$F(z) = \sum_{p \in N_P} \frac{C_{\text{INV}}^{p} - C_{\text{SAL}}^{N_P}}{(1+dr)^{p-1}} \tag{5.3}$$

$$C_{\text{INV}}^{p} = \sum_{q \in N_Q} \sum_{i \in N_{\text{CB}}} \left[\chi_{\text{DG,INV}} \cdot \left(S_{\text{DG}}^{p,q,i} - S_{\text{DG}}^{p-1,q,i} \right) \right] \tag{5.4}$$

$$C_{\text{SAL}}^{N_P} = \varepsilon_{\text{DG,SAL}}^{p,q} C_{\text{INV}}^{p} \quad \forall p, q \tag{5.5}$$

$$S_{\text{DG}}^{p,q,i} = N_{\text{DG}}^{p,q,i} \cdot S_{\text{DG,UN}}^{q} + S_{\text{DG}}^{p-1,q,i} \quad \forall p; q \in N_{\text{CQ}}; i \in N_{\text{CB}} \tag{5.6}$$

$$S_{\text{DG}}^{p,q,i} \geq S_{\text{DG}}^{p-1,q,i} \quad \forall p; q \in N_{\text{VQ}}; i \in N_{\text{CB}} \tag{5.7}$$

$$S_{\text{DG}}^{p,q,i} \geq 0 \quad \forall p, q; i \in N_{\text{CB}} \tag{5.8}$$

$$S_{\text{DG}}^{p,q,i} = S_{\text{DG}}^{p-1,q,i} \quad \forall q; i \in N_{\text{CB}}; \ p \neq \text{1st year of each phase} \tag{5.9}$$

$$U_{\text{DGI}}^{p,q,i} \cdot S_{\text{DG,MIN}}^{q} \leq S_{\text{DG}}^{p,q,i} \leq U_{\text{DGI}}^{p,q,i} \cdot S_{\text{DG,MAX}}^{q} \ \forall p, q; i \in N_{\text{CB}} \tag{5.10}$$

$$\sum_{i \in N_{\text{CB}}} S_{\text{DG}}^{p,q,i} \leq S_{\text{DG,ALL}}^{q} \ \forall p, q \tag{5.11}$$

$$\sum_{i \in N_{\text{NB}}} U_{\text{DGI}}^{p,q,i} = 0 \quad \forall p, q \tag{5.12}$$

where (5.3) is the mathematical model for the NPV from the investment stage. Equation (5.4) shows the DG investment cost. Equation (5.5) is the salvage value, which means that the owner of the project can get some cash back by selling the used DGs at the end of the project [24]. Equation (5.6) denotes the capacity limits of DGs that have a discrete unit size, such as the WTs, DEs, and CCHP plants. Equation (5.7) denotes the capacity limits for the DGs such as PVs, EBs, and ECs, which are available in all variable sizes. Equation (5.8) means that all the capacity should be larger than or equal to 0, and (5.9) denotes that only the DG capacity to be installed at the beginning year of a phase is set as separate decision variables. Equation (5.10) denotes the DG capacity limits for each node and (5.11) denotes the total DG capacity limits of the MEMG. Equation (5.12) denotes that the DGs cannot be installed at the non-candidate buses.

3. Constraints for the operation stage:

$$G(x) = \sum_{p \in N_P} \left[\begin{matrix} C_{\mathrm{REV}}^p + C_{\mathrm{SUB}}^p - C_{\mathrm{FC}}^p - C_{\mathrm{TC}}^p \\ -C_{\mathrm{CT}}^p - C_{\mathrm{MC}}^p - C_{\mathrm{EM}}^p - C_{\mathrm{SD}}^p - C_{\mathrm{ST}}^p \end{matrix} \right] \Big/ (1+dr)^p \tag{5.13}$$

$$C_{\mathrm{REV}}^p = \sum_{q \in N_{\mathrm{Q}}} \sum_{d \in N_{\mathrm{D}}} \sum_{t \in N_{\mathrm{T}}} \sum_{i \in N_{\mathrm{B}}} (\chi_{\mathrm{PS}} P_{\mathrm{L}}^{\Lambda} + \chi_{\mathrm{HS}} H_{\mathrm{H}}^{\Lambda} + \chi_{\mathrm{CS}} H_{\mathrm{C}}^{\Lambda}) \ \forall p \tag{5.14}$$

$$C_{\mathrm{SUB}}^p = \sum_{q \in N_{\mathrm{Q}}} \sum_{d \in N_{\mathrm{D}}} \sum_{t \in N_{\mathrm{T}}} \sum_{i \in N_{\mathrm{CB}}} \chi_{\mathrm{SUB}} \cdot (P_{\mathrm{WT}}^{\Lambda} + P_{\mathrm{PV}}^{\Lambda}) \ \ \forall p \tag{5.15}$$

$$C_{\mathrm{FC}}^p = \sum_{q \in N_{\mathrm{Q}}} \sum_{d \in N_{\mathrm{D}}} \sum_{t \in N_{\mathrm{T}}} \sum_{i \in N_{\mathrm{CB}}} \left[\begin{matrix} \chi_{\mathrm{NG}} P_{\mathrm{MT}}^{\Lambda} / \eta_{\mathrm{MT}} + \\ \chi_{\mathrm{DE}} (a^q U_{\mathrm{DE}}^{\Lambda} + b^q P_{\mathrm{DE}}^{\Lambda}) \end{matrix} \right] \ \ \forall p \tag{5.16}$$

$$C_{\mathrm{TC}}^p = \sum_{q \in N_{\mathrm{Q}}} \sum_{d \in N_{\mathrm{D}}} \sum_{t \in N_{\mathrm{T}}} (\chi_{\mathrm{B}} P_{\mathrm{BX}}^{p,q,d,t,1} - \chi_{\mathrm{S}} P_{\mathrm{SX}}^{p,q,d,t,1}) \ \ \forall p \tag{5.17}$$

$$C_{\mathrm{MC}}^p = \sum_{q \in N_{\mathrm{Q}}} \sum_{i \in N_{\mathrm{CB}}} (\chi_{\mathrm{DG,F}} S_{\mathrm{DG}}^{p,q,i} + \sum_{d \in N_{\mathrm{D}}} \sum_{t \in N_{\mathrm{T}}} \chi_{\mathrm{DG,V}} P_{\mathrm{DG}}^{\Lambda}) \ \ \forall p \tag{5.18}$$

$$C_{\mathrm{EM}}^p = \sum_{q \in N_{\mathrm{Q}}} \sum_{d \in N_{\mathrm{D}}} \sum_{t \in N_{\mathrm{T}}} \sum_{i \in N_{\mathrm{CB}}} \chi_{\mathrm{CO2}} \left[\begin{matrix} (\kappa_{\mathrm{MT}} P_{\mathrm{MT}}^{\Lambda}) + (\kappa_{\mathrm{B}} P_{\mathrm{BX}}^{\Lambda}) + \\ \kappa_{\mathrm{DE}} (a^q U_{\mathrm{DE}}^{\Lambda} + b^q P_{\mathrm{DE}}^{\Lambda}) \end{matrix} \right] \ \ \forall p \tag{5.19}$$

$$C_{\mathrm{ST}}^p = \sum_{q \in N_{\mathrm{Q}}} \sum_{d \in N_{\mathrm{D}}} \sum_{t \in N_{\mathrm{T}}} \sum_{i \in N_{\mathrm{CB}}} \max\{0, U_{\mathrm{DG}}^{\Lambda} - U_{\mathrm{DG}}^{p,q,d,t-1,i}\} C_{\mathrm{DG}}^{\mathrm{U}} \ \forall p \tag{5.20}$$

$$C_{\mathrm{SD}}^p = \sum_{q \in N_{\mathrm{Q}}} \sum_{d \in N_{\mathrm{D}}} \sum_{t \in N_{\mathrm{T}}} \sum_{i \in N_{\mathrm{CB}}} \max\{0, U_{\mathrm{DG}}^{p,q,d,t-1,i} - U_{\mathrm{DG}}^{\Lambda}\} C_{\mathrm{DG}}^{\mathrm{D}} \ \ \forall p \tag{5.21}$$

$$C_{\mathrm{CT}}^p = \sum_{q \in N_{\mathrm{Q}}} \sum_{d \in N_{\mathrm{D}}} \sum_{t \in N_{\mathrm{T}}} \sum_{i \in N_{\mathrm{CB}}} \chi_{\mathrm{RES}} (P_{\mathrm{DG,REAL}}^{\Lambda} - P_{\mathrm{DG}}^{\Lambda}) \ \ \forall p \tag{5.22}$$

$$\gamma_{\mathrm{DG,MIN}} S_{\mathrm{DG}}^{p,q,i} \cdot U_{\mathrm{DG}}^{\Lambda} \leq P_{\mathrm{DG}}^{\Lambda} \leq \gamma_{\mathrm{DG,MAX}} S_{\mathrm{DG}}^{p,q,i} \cdot U_{\mathrm{DG}}^{\Lambda} \ \ \forall p,q,d,t; i \in N_{\mathrm{CB}} \tag{5.23}$$

$$P_{\mathrm{DG}}^{\Lambda} \leq P_{\mathrm{DG,REAL}}^{\Lambda} \ \ \forall p,q,d,t; i \in N_{\mathrm{CB}} \tag{5.24}$$

$$R_{\mathrm{DG,DOWN}}^q \Delta t \leq P_{\mathrm{DG}}^{\Lambda} - P_{\mathrm{DG}}^{p,q,d,t-1,i} \leq R_{\mathrm{DG,UP}}^q \Delta t \quad \forall p,q,d,t; i \in N_{\mathrm{CB}} \tag{5.25}$$

$$H_{\mathrm{H}}^{p,d,t,i} = H_{\mathrm{HMT}}^{\Lambda} + H_{\mathrm{EB}}^{\Lambda} \ \forall p,q,d,t,i \tag{5.26}$$

$$H_{\mathrm{C}}^{p,d,t,i} = H_{\mathrm{CMT}}^{\Lambda} + H_{\mathrm{EC}}^{\Lambda} \ \forall p,q,d,t,i \tag{5.27}$$

$$P_{\mathrm{PF}}^{p,d,t,b+1} = P_{\mathrm{PF}}^{p,d,t,b} - P_{\mathrm{PF}}^{p,d,t,0,b+1} - P_{\mathrm{L}}^{p,d,t,b} + P_{\mathrm{DG}}^{p,d,t,b} \ \ \forall p,q,d,t,i \tag{5.28}$$

$$Q_{\mathrm{PF}}^{p,d,t,b+1} = Q_{\mathrm{PF}}^{p,d,t,b} - Q_{\mathrm{PF}}^{p,d,t,0,b+1} - Q_{\mathrm{L}}^{p,d,t,b} \ \forall p,q,d,t,i \tag{5.29}$$

$$V_{\mathrm{BUS}}^{p,d,t,i+1} = V_{\mathrm{BUS}}^{p,d,t,i} - (R^b P_{\mathrm{PF}}^{p,d,t,b} + X^b Q_{\mathrm{PF}}^{p,d,t,b})/V_0 \ \forall p,q,d,t,i \tag{5.30}$$

$$1 - \Delta V_{\mathrm{BUS}}^{\max} \le V_{\mathrm{BUS}}^{p,d,t,i} \le 1 + \Delta V_{\mathrm{BUS}}^{\max} \ \forall p,q,d,t,i \tag{5.31}$$

where (5.13) denotes the NPV from the operation stage. Equation (5.14) is the revenue from selling power, heat, and cooling energy to the local customers. Equation (5.15) is the subsidy for the utilisation of RES. Equation (5.16) means that the system fuel cost comes from the DEs and CCHP plants. Equation (5.17) is power transaction revenue with the utility grid (the first bus is directly connected to the utility grid). Equation (5.18) denotes the DG maintenance cost. Equation (5.19) is the emission cost from CCHP plants, utility grid, and DEs. Equations (5.20) and (5.21) denote the start-up and shutdown cost of the dispatchable DGs. Equation (5.22) is the cost of RES power curtailment if needed. Equation (5.23) means that the DG power output is limited by the installed capacity times the generation efficiency. Equation (5.24) denotes that for non-dispatchable DGs, their power outputs are limited by the available power generation capacity. Note that in (5.22) and (5.24), the DG in the subscript refers merely to RES-based generators. As the RES-based generators are subset of the DGs, to avoid any confusion from the different abbreviations or variables, no new abbreviations are set solely for them. Equation (5.25) shows the ramp-up/down limits for the CCHP plants, DEs, ECs, and EBs. Equations (5.26) and (5.27) mean that the heat and cooling generation should be equal to its consumption. Equations (5.28)–(5.30) model the power flow of the distribution network, which is based on the linearised *Dist-Flow* equations [27]. Equation (5.31) constrains the bus voltage magnitude. To illustrate (5.28)–(5.31) in a more intuitive way, a simple power network model is shown in Figure 5.3.

In Figure 5.3, p and q correspond to P_{PF} and Q_{PF} in (5.28)–(5.30). P and Q denote the net active and reactive loads, which are equal to the subtraction of the loads with the generation. The details can be referred to our former work in [27,28].

It should be noted that there are a number of non-linear items in (5.20), (5.21), and (5.23). To reduce the computation burdens and improve solution quality, those non-linear items are linearised in Appendix A. Then, the proposed DG placement model becomes an MILP one.

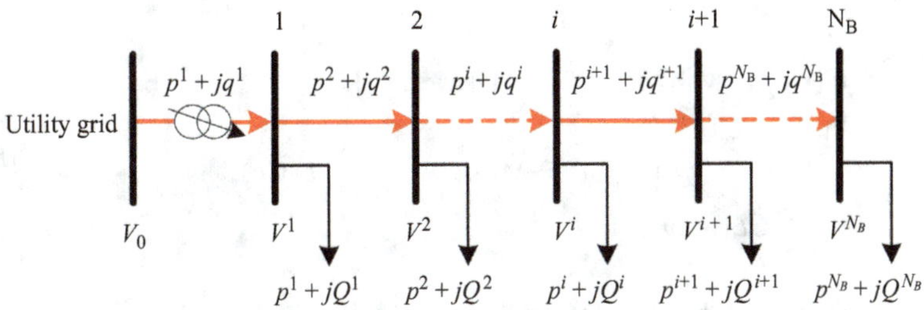

Figure 5.3 Schematic diagram of a radial distribution network

5.4 Solution method

The dimension of DG placement problem in Section 5.3 can be very high, due to the large number of DG types, system buses, planning years, dispatch periods, and operating days in each year. Besides, uncertainties from RES-based generators should be handled in the MEMG operation. This section reduces the problem dimension first, and then proposes the stochastic optimisation method for tackling the uncertainties.

5.4.1 Typical seasonal day selection

In the two-stage DG placement framework, there are 365 dispatch days in a year, which makes the problem size excessively high. To reduce the computation burden, a set of typical seasonal days is selected to represent the whole year [26]. Then, the two-stage planning model can be converted as

$$NPV_s = \underset{z \in CF_z, x \in CG_x}{\text{Max}} \{ \underbrace{-F(z)}_{\text{Investment stage}} + \underbrace{\sum_{c \in N_C} N_D^c \cdot G(x|c)}_{\text{Operation stage}} \} \tag{5.32}$$

$$\text{s.t.} \quad (5.3)\text{–}(5.31) \tag{5.33}$$

where N_D^c is the number of days in season c; N_C is the number of seasons; and $G(x|c)$ is the objective of the operation stage in season c.

Note that all the 365 dispatch days can still be fully modelled if the computational burden is not a concern.

5.4.2 Stochastic optimisation model

To handle the RES uncertainties, the operation stage is solved by a two-stage stochastic programming method. At its first stage, dispatch decisions for all the heterogeneous DGs before the realisation of RES uncertainties are determined. They are called the '*here-and-now*' *decisions*. At the second stage when the real-time RES outputs are revealed, the power exchange between the MEMG and utility grid compensates the first-stage decisions to maintain the energy balance. The transaction power is the '*wait-and-see*' decision.

A Stochastic model: In this study, the stochastic optimisation model can be denoted as

$$\begin{aligned} &\text{Min } G(x|c) = \underset{w}{\text{Min}}\{S(w|c) + E[Q(w|c,\omega)]\} \\ &\text{s.t.} \quad w \in CS_w|z \\ &\qquad\qquad Q(w|c,\omega) = \underset{y}{\text{Min}} \qquad L(y|c) \\ &\qquad y \in CL(w,\omega) \end{aligned} \tag{5.34}$$

where $S(w|c)$ is the objective of DG operation in (5.13)–(5.22); w denotes the decision variables for the DG dispatch; CS_w is constraints for w in (5.23)–(5.27);

ω is stochastic variation of the RES outputs; $Q(w|c, \omega)$ is the optimal value of transaction cost under the uncertainty; y is the decision variable which denotes the transaction power; $L(y|c)$ is the mathematical formulation for transaction cost in (5.17); $CL(w, \omega)$ means the constraints for y in (5.28)–(5.31); and $E[Q(w|c, \omega)]$ denotes the expectation of transaction cost.

B Deterministic equivalent: Assuming that ω has a finite number of possible realisations which are called the scenarios and denoted as $\omega_1, \ldots, \omega_n$ with respective possibilities of $\chi_1, \ldots, \chi_n$, then the model in (5.34) can be converted into a deterministic one as follows:

$$\begin{aligned} \min_x G(x|c) = &\min_{w, y_1, y_2, \ldots, y_n} [S(w|c) + \sum_{n \in N_S} \chi_n L(y_n|c)] \\ \text{s.t.} \quad & w \in CS_w|z \\ & y_n \in CL(w, \omega_n), \ \forall n \end{aligned} \tag{3.35}$$

where n is the index of the forecasted scenarios, and N_s is the total number of scenarios. All the scenarios can be obtained by the scenario generation and reduction techniques, which are commonly utilised in literature as [28].

In (5.35), the objective and constraints are all linear; thus, our proposed two-stage DG placement model can be efficiently solved by some existing solvers such as the CPLEX.

5.5 Test system set-up and case study

For the proposed DG placement method, its effectiveness is justified on a test MEMG system. Besides, since there are already some traditional DG placement methods in the literature, the superiority of our proposed method should be further verified in comparison with them. In this section, the corresponding case studies are done in the following subsections.

5.5.1 Test system set-up

The proposed DG placement method is tested on a grid-connected MEMG as shown in Figure 5.4, which is structured on the IEEE 33-bus radical distribution system [21,27,28].

In this study, the data from an MG system in Tianjin, China, is utilised to model this MEMG. Practically, in Tianjin, thermal energies are mainly supplied locally for a region of residential buildings or communities (denoted as the thermal groups 1–3 in Figure 5.4), as the long-distance thermal transmission will result in high time delay and great thermal loss. Given this fact, the heat network is not modelled. The project lifetime is 12 years with 4 years as one investment phase; that is, the new DG installation only happens at the beginning of the first, fifth, and ninth years. The discount rate is 6% [8] and load growth rates for power and thermal energy are set as 6.7% and 5%, respectively [21,28]. The salvage factors

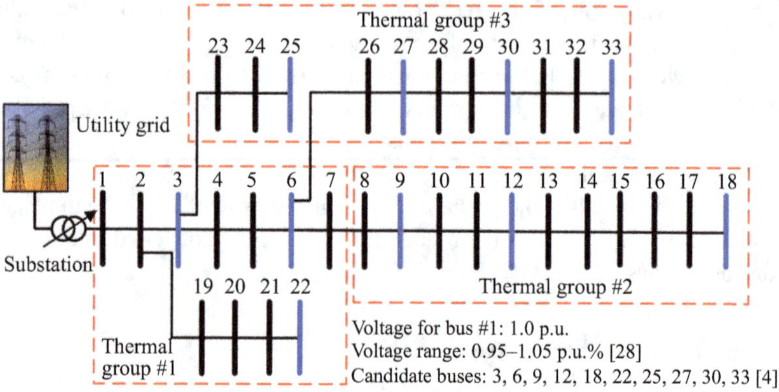

Figure 5.4 The 33-bus radial distribution system-based MEMG

for three phases are set as 0.05, 0.15, and 0.25, respectively. Other technical details for heterogeneous DGs are listed in Table 5.1, which summarises the synthesised figures from available literature, reports, and commercial catalogues [8,18,27–31]. Other technical details regarding the MEMG can be seen in Figure 5.4.

The dispatch horizon for the operation stage is set as 1 day and the dispatch interval is 1 h [15].

Tianjin has four distinct seasons as the spring (91 days), summer (91 days), autumn (91 days), and winter (92 days) [32]. Based on the data of the year 2017 (Jan.–Dec.), Figure 5.5 shows the total forecasted RES generation and multi-energy demands of four typical seasonal days in the year 2018. The energy tariffs are shown in Table 5.2 [8,18], and the prices for the power transaction, natural gas, and diesel are extracted from Tianjin Development and Reform Commission [33]. For the thermal and power energy supplied to the customers, the fixed pricing scheme is applied as in our former study [18].

The load ratio of each group to the total thermal loads is set as 35%, 30%, and 35%, respectively. The capacity of the substation linking the MEMG and utility grid is set as 1.5 times of its peak power loads [4]. The forecasting errors for the RES generation are set as 25% deviation to their prediction values [28].

5.5.2 Methods comparison

To fully demonstrate the advantages of the proposed method, three DG placement methods that are commonly employed in the literature are compared as follows:

Method #1: Separately DG placement in SEMGs [8]: In this method, under the RES uncertainties, the gas boilers (GBs) are utilized instead of the CCHP plants, EBs, and ECs for the thermal loads. The WTs, DEs, and transaction power are coordinated to satisfy the system power loads. The parameters for GBs are shown in Table 5.3.

Method #2: Single-type single-investment-phase DG placement: From the simulation results mentioned earlier in this section, only the Cap-65, DE-B, WT-B,

Table 5.1 Parameters for different DGs [8,15,21,26–28]

Type	Unit size (kW)	Investment cost ($/kW)	Maintenance cost ($/kWh)	System size limits (kW)	Bus size limits (kW)	Power-to-heat ratio	Power-to-cooling ratio	Electric efficiency (%)
Microturbine (or CCHP plants)								
ICE-250	250	1,452	0.0159	3,000	1,000	1.48	1.78	28
Cap-200	200	1,795	0.0121	3,000	1,000	1.57	1.89	30
Cap-65	65	2,134	0.0078	3,000	1,000	1.8	2.16	32
Diesel generators						a	b	
DE-A	80	1,521	0.16	3,500	800	0.0841	0.246	–
DE-B	120	1,121	0.24	3,500	800	0.0161	0.252	–
Wind turbines								
WT-A	80	2,134	0.24	1,800	320	–	–	–
WT-B	120	2,012	0.32	1,800	320	–	–	–
Photovoltaic cells								
PV-A	–	2,039	0.32	1,500	200	–	–	–
PV-B	–	1,792	0.45	1,500	200	–	–	–
Electric boilers								
EB-A	–	231	0.2	2,000	700	3	–	–
EB-B	–	195	0.23	2,000	700	2.75	–	–
EB-C	–	158	0.25	2,000	700	2.5	–	–
Electric chillers								
EC-A	–	489	0.2	1,800	600	–	4	–
EC-B	–	395	0.25	1,800	600	–	3.5	–

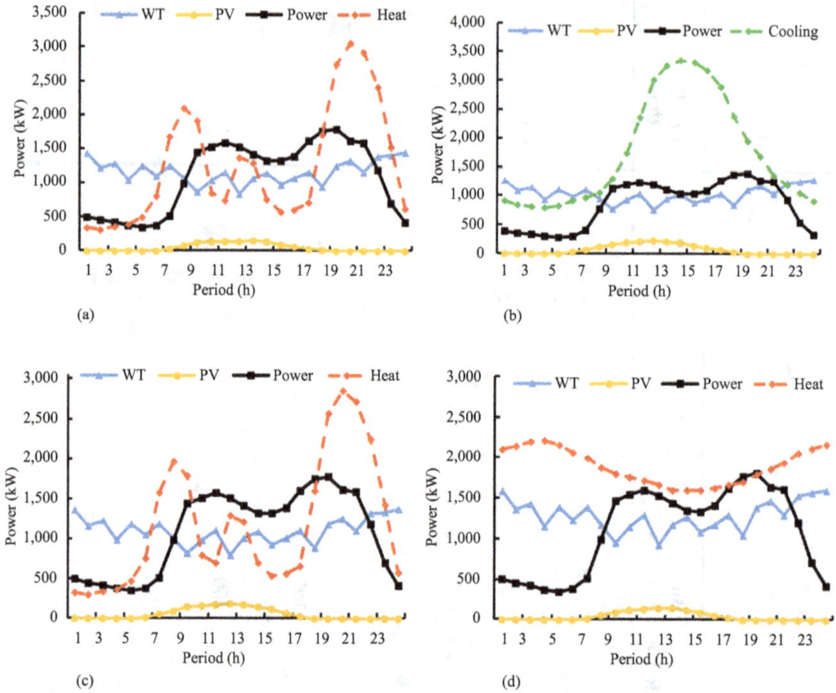

Figure 5.5 Typical daily RES generation and multi-energy loads. (a) Spring. (b) Summer. (c) Autumn. (d) Winter

Table 5.2 Energy tariffs and CO_2 conversion factors [8,15,29–33]

Type	**Price ($/kWh)**	**Time period**
Local power	0.17	All day
Local thermal	0.022	All day
Power transaction	0.0768	0.00–6.00, 22.00–24.00
	0.1429	6.00–8.00, 11.00–17.00
	0.1785	8.00–11.00, 17.00–22.00
Natural gas	0.0569	All day
Diesel	1.0 ($/L)	All day
Unit RES curtailment	0.005	All day
CO_2 emission	0.0312 ($/kg)	All day
Emission conversion factors	**Value**	**Unit**
CCHP plants	0.184	kg/kWh
Utility grid	0.889	kg/kWh
DEs	2.6	kg/L

Table 5.3 Parameters for GBs [8]

Type parameters	GB-A		GB-B		GB-C	
	Heat	Cooling	Heat	Cooling	Heat	Cooling
Efficiency	0.75	0.89	0.8	0.95	0.82	0.98
Maintenance ($/kW)	0.29		0.33		0.36	
Investment ($/kW)	35		42		52	
Emission factors (kg/kWh)			0.226			
System/bus size limits (kW)			8,000/1,000			

EB-A, EC-A, and PV-B are considered and installed at the beginning of the first project year [4,11,16–20]. Other placement aspects remain the same as our proposed method.

Method #3: Deterministic DG placement: The uncertainties from RES generation are not considered and other placement aspects remain the same.

5.5.3 Assumptions for case study

In the case study, to verify the effectiveness of the proposed DG placement method, some assumptions are made in the following. Though those assumptions make different methods simple, they are reasonable and effective in reducing the problem dimensions.

1. All the DGs are assumed to have the same lifetime as the project. In fact, different DGs should have different lifetime; however, considering DG lifetime in detail would make the model much more complicated without adding novel scientific contributions. As this chapter aims to focus more on the DG planning and multi-energy coordination in MEMGs, the above assumption is made. Besides, the system performance considering different DG lifetime is listed as the future work.
2. The predefined candidate buses, the number of the thermal groups, and some simulation data such as the salvage factors can also be regarded as parameter assumptions. However, as the proposed model aims to provide a general DG placement method, any other real-world reasonable data or configuration for the DGs or system structure can also be applied, which won't influence its effectiveness.
3. It is assumed that initial sizes for all DGs are 0. In practice, it can be other real-world values which won't affect the effectiveness of the proposed method.

The simulation is conducted on an Intel(R) Xeon(R) E5-1630 3.70 GHz 64-bit PC with 16G RAM. The CPLEX solver is used to solve the model via the General Algebraic Modelling System (GAMS).

5.6 Simulation results and discussions

For the DG placement method, the simulation results of the project NPV are used to verify its effectiveness. If the NPV is positive, it means that the project should be invested as it gains value, or it should be definitely rejected [27]. Compared with other methods, the NPV difference denotes the indicator for the superiority of the proposed method. In this section, NPV of the proposed method and traditional methods are compared and analysed.

5.6.1 Investment-stage simulation results and discussions

In the simulation, 2,000 random scenarios are initially generated and then reduced to 10 representative ones with corresponding possibilities using the simultaneous backward method [27]. The scenarios and possibilities are shown in Figure A.1 of Appendix A.

The optimisation results for the investment stage are listed in Table 5.4. Note that in Table 5.4, first, the grey blocks denote that there are investment decisions at this phase and its subtraction with the DG size at the last phase is the newly installed DG capacity at the new phase. For example, at Phase 1 (year 1–4), there is a 90.7 kW installation of EB-A at bus 9, at Phase 2 (year 5–8), the newly installed EB-A is (128.1 − 90.7) = 37.4 kW, and at Phase 3 (year 9–12), an additional capacity of (129.6 − 128.1) = 1.5 kW of the EB-A is installed. Second, if no DG

Table 5.4 The DG placement results (kW)

	Bus									
Year	**3**	**6**	**9**	**12**	**18**	**22**	**25**	**27**	**30**	**33**
Cap-65										
1–4	65	195	65	130	0	0	130	65	0	65
5–8	65	195	65	195	0	65	130	65	65	65
9–12	65	195	65	195	65	65	130	65	65	130
EB-A										
1–4	123.2	96	90.7	0	146.8	41.5	167.8	92.9	0	0
5–8	123.2	96	128.1	0	146.8	88.5	214.8	92.9	0	0
9–12	138.5	96	129.6	0	146.8	120.4	216.2	92.9	0	0
EB-B										
9–12	0	0	0.0	57.1	0	66.7	74.1	0	0	0
EC-A										
1–4	200.2	0	68.4	0	105.8	0	141.3	0	0	48.25
EC-B										
1–4	0	0	25.0	0	0	50.7	74.3	0	0	0
5–8	0	0	50.3	0	0	103.8	115.2	0	0	0
9–12	0	0	90.9	0	0	143.0	115.2	0	0	0
PV-B										
1–12	137.6	137.6	136.2	122.2	118.9	137.7	181.6	181.6	177.8	169
WT-A										
1–12	80	0	0	0	0	80	80	0	80	0
WT-B										
1–12	0	240	120	0	240	0	240	0	120	0

installation at a certain phase, the phase index is omitted; for example, for EB-B at bus 12, there is the only installation at Phase 3 (year 9–12). Finally, if no more DG installation at later phases, the phases are combined, e.g. the WT-A at bus 6, where there is the only new installation at the first phase.

From the investment-stage simulation results, it can be seen the following:

1. The investment-related parameters, mainly the unit investment cost and the operation-related parameters such as the energy efficiencies and energy conversation ratios would exert different effects on the DG placement. For the CCHP plants, though the type Cap-65 has the highest unit investment cost, its high energy efficiency and energy conversion ratio result in much lower operation cost (i.e. fuel cost) in the long run; thus, it is installed rather than other two types. For EB, the unit investment cost of EB-A is higher than EB-B, but the operation efficiency of EB-B and EB-C is lower, which leads to the investment of EB-A at earlier stages and larger capacity (there is no investment of EB-C at all). Similar results are obtained for the EC investment. For the WTs and PVs, as their operation performance (mainly maintenance cost) are similar, the investment cost plays a leading role in their placement.
2. The unit size will affect the placement of DGs in discrete size. Except for higher efficiency, Cap-65 has a smaller unit size, which means that it is more flexible for deployment; thus, the Cap-65 type of MT is chosen. As for WTs, though WT-B is the majority, the smaller unit size of WT-A is more economical than a unit of WT-B, when not so large capacity is needed.
3. Though the unit investment cost for DEs is lower than the CCHP plants, there is still none DEs installed; this is because the CCHP plants can provide multiple energies with higher efficiency and lower cost than the DEs. The results verify the effectiveness and practicability of the placement of co- or tri-generation units in the MEMGs.
4. The investment phase optimisation plays a role in the placement of some DGs such as the Cap-65, EB-A, and EC-B. The reasons lie in two aspects: on the one hand, with the increasing multi-energy loads, their capacities which are invested at the earlier phases are not enough; on the other hand, the DGs which are invested at later phases tend to get more salvage value. The specific case study will be done in the later section.
5. For the RES-based generators, i.e. WTs and PVs, there is an only one-time investment at the Phase 1 and their installed capacity is larger than the maximal RES available active power, which results in no RES curtailment at all. This is because the RES outputs are the energy generation with relatively low investment and operation cost and they tend to be utilised as much as possible.

5.6.2 Operation-stage simulation results and discussions

After getting the investment-stage decisions, the operation-stage decisions for all the DGs as well as power transaction can also be obtained.

The dispatch results for the units in the year 2028 are shown in Figures 5.6 and 5.7. In these figures, ‘EX’ means the net power transaction between the MEMG and

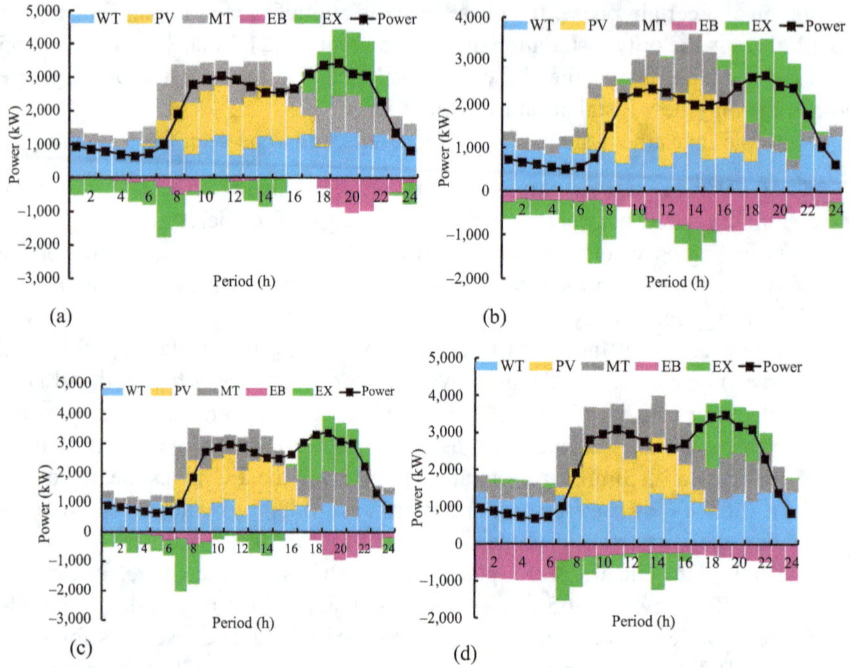

Figure 5.6 Power balance condition in the year 2028. (a) Spring. (b) Summer. (c) Autumn. (d) Winter

utility grid. It is equal to the amount of purchasing power minus the selling power. 'Power', 'Heat', and 'Cooling' mean the system power, heat, and cooling loads, respectively.

From Figures 5.6 and 5.7, all the heterogeneous DGs coordinate with one another to supply the multi-energy demands. The RES outputs follow the maximal available active power as there is no RES curtailment. When the transaction prices are at valley periods, i.e. 0.00–6.00, 22.00–24.00, more power would be transferred to heat via EBs or cooling via ECs to save money for supplying thermal demands. When the transaction prices are at peak periods, i.e. 8.00–11.00, 17.00–22.00, more natural gas is consumed via CCHP plants to generate relatively cheaper power and thermal energy for the system multi-energy loads as well as sell the excess electricity out to earn more from the utility grid. When the transaction prices are at medium periods, i.e. 6.00–8.00, 11.00–17.00, the programme will decide the most economical way for the multi-energy coordination. The final NPV for the project plan is $15.02M, which means that this project should be implemented since its NPV is positive.

In terms of the solution speed, there are 1,832,653 continuous and 93,096 discrete variables in the mathematical model, while the solution time is only 21.5 h, which is fully computationally acceptable for a 12-year planning project.

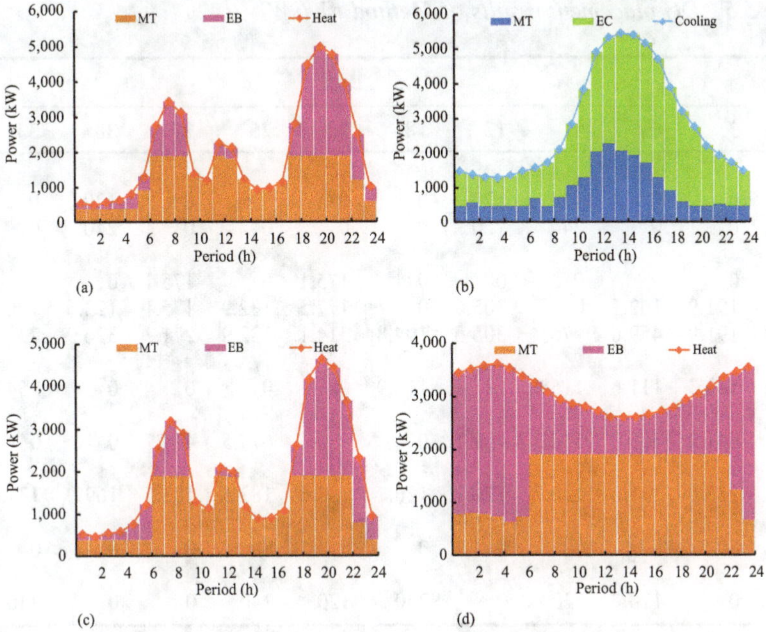

Figure 5.7 Thermal balance condition in the year 2028: (a), (c), and (d) are for the heat and (b) is for the cooling. (a) Spring. (b) Summer. (c) Autumn. (d) Winter

5.6.3 Methods comparison results

After obtaining the simulation results for the proposed DG placement method, to further verify its effectiveness, the comparisons with the traditional DG placement methods as stated in Section 5.5.2. are done.

The simulation results of **Method #1** are given in Table 5.5.

It can be seen from Table 5.5 that, due to the similar operation performance with DE-A but lower unit investment cost, DE-B is installed to maintain the voltage reliability and supply of the power loads. For thermal supply, all the three types of GBs are installed but given the superior economic operation performance of GB-C, it has the largest invested capacity. However, when supplying the multi-energy loads, the DEs and GBs are of lower efficiency but higher operating cost. Besides, there is no interaction between different energies at all; thus, the system cannot take the advantage of the price difference to effectively transfer power to thermal energy to save money. Overall, the NPV for **Method #1** is much lower, which is only \$9.95M. The simulation results demonstrate the advantages of our proposed DG placement in '*multi-energy*' scheme.

The simulation results for **Method #2** are shown in Table 5.6.

The NPV of **Method #2** is \$14.15M, which is also lower than our method. The reason is with the one-time investment, all the DGs have to be installed prior to the

Table 5.5 DG placement results of ***Method #1*** *(kW)*

	Bus									
Year	**3**	**6**	**9**	**12**	**18**	**22**	**25**	**27**	**30**	**33**
DE-B										
5–8	0	0	120	120	0	0	0	0	120	0
9–12	0	0	240	120	0	0	0	0	240	0
GB-A										
1–4	0	0	0	0	0	175.1	0	175.1	0	0
5–8	191.9	162.2	0	305.6	0	177.5	132.9	175.1	123.6	100.0
9–12	191.9	459.0	176.7	305.6	194.8	314.1	132.9	271.7	326.9	233.3
GB-B										
1–12	143.3	111.6	115.4	128.9	114.9	0	0	0	0	254.9
GB-C										
1–12	641.2	463.0	537.3	409.1	0	0	679.5	424.7	0	0
PV-B										
1–12	137.6	137.6	126.7	126.7	126.7	137.6	183.5	183.5	169.0	171.0
WT-A										
1–12	80	80	80	0	0	0	80	0	0	0
WT-B										
1–12	0	120	120	0	240	120	240	0	0	240

Table 5.6 DG placement results of ***Method #2*** *(kW)*

	Bus									
Year	**3**	**6**	**9**	**12**	**18**	**22**	**25**	**27**	**30**	**33**
Cap-65										
1–12	0	260	0	65	130	0	65	0	0	195
EB-A										
1–12	90.8	239.9	250.0	25.0	131.7	124.3	250.0	89.5	25.0	90.5
EC-A										
1–12	183.6	37.6	200.0	0	124.1	139.3	250.0	0	110.5	0
PV-B										
1–12	136.2	136.2	128.8	126.7	122.2	137.6	181.6	181.6	169.0	180.3
WT-B										
1–12	0	240	240	0	120	120	360	0	0	240
DE-B										
1–12	0	0	0	0	240	0	0	120	0	0

project with large sizes, and it would be a waste of capacity at early years and also result in lower salvage value. Besides, with much less choice for DG types, the solution space of the programming model is limited. Thus, it can be seen that enabling the type and investment phase optimisation benefits the project economic performance.

The simulation results for **Method #3** are shown in Table 5.7.

*Table 5.7 DG placement results of **Method #3** (kW)*

Year	Bus 3	6	9	12	18	22	25	27	30	33
Cap-65										
1–4	195	0	65	130	0	65	0	130	0	130
5–9	195	0	65	195	0	130	0	130	0	195
9–12	195	0	65	195	65	130	0	130	0	195
EB-A										
1–4	81.7	68.4	138.9	0	98.6	112.1	122.2	0	0	140.0
5–9	81.7	87.7	176.3	0	98.6	138.3	167.7	0	0	140.0
9–12	81.7	128.1	177.0	0	98.6	138.3	208.0	0	0	140.0
EB-B										
9–12	0	0	58.0	0	0	74.1	74.1	0	0	0
EC-A										
1–12	110.1	0	127.9	0	32.9	71.8	113.6	0	0	66.0
EC-B										
1–4	54.1	0.0	25.0	0	32.0	32.0	97.7	0	0	0
5–9	72.8	28.2	65.0	0	32.0	32.0	157.7	0	0	0
9–12	103.7	28.2	74.3	0	32.0	32.0	166.5	0	0	0
PV-B										
1–12	133.7	133.7	133.7	128.7	128.7	133.7	178.2	178.2	178.2	173.4
WT-A										
1–12	0	80	0	0	80	80	0	0	0	80
WT-B										
1–12	120	0	0	240	120	120	240	0	240	0

The NPV of **Method #3** is \$15.52M, which is slightly higher than our method. This is because the solution space is reduced and hence the NPV of our stochastic approach is lower considering the RES uncertainties in terms of scenarios. The cost between the deterministic approach (**Method #3**) and the stochastic approach is called the value of perfect information and it is used to help the system immune against the uncertainties.

However, when a set of 1,000 random scenarios for the RES outputs of four typical days every year is used to check the robust performance of the planning decisions, the average NPV of all the scenarios of **Method #3** becomes \$14.55M but the average NPV of our method is \$14.93M. It means that with the real-time RES outputs (i.e. the practical condition), our (stochastic) method has a higher potential in gaining more profits than the deterministic method. This result justifies the effectiveness of our stochastic DG placement method in immunising against the practical RES fluctuations to reduce the system cost.

5.7 Conclusion

This chapter develops an optimal two-stage placement method for the heterogeneous DGs in an MEMG under the uncertainties from the RES. The DG placement scheme is decomposed into three investment phases with the project NPV to be maximized. By employing the stochastic optimisation method, the RES uncertainties are incorporated and then handled in the system two-stage decision-making structure. Simulation results show that the project NPV of our method is positive, which means that this project can gain values and thus should be implemented. Besides, to further demonstrate the effectiveness of our proposed method, three conventional DG placement methods are then compared, i.e. separate DG placement method, single-type-single-investment-phase DG placement method, and deterministic DG placement method. The results show that our method, which considers the multi-energy coordination, multiple investment phase, and type optimisation in the DG placement scheme, would add extra flexibility for the system holistic energy supply and obtain a more economic and robust DG placement plan.

Appendix A

To make the overall chapter structure clear, additional contents which support the study are presented in this section. The specific contents can be seen in the following subsections.

A.1 Model linearisation

In this chapter, the non-linear items in the DG placement model are linearised as follows:

1. *Linearisation of the max function*, i.e. (5.20) and (5.21): Separating the max function into two individual parts will do, and the general form $A \geq \max(B,C)$ is taken as an example as

$$A \geq B; A \geq C \tag{A.1}$$

2. *Linearisation of the bilinear item*, i.e. (5.23): Take $z = Ax$ as an example in which $\bar{A}$ is a continuous variable which is bounded below by zero and above by A (or any Big M) and x is a binary variable, then z can be linearised as

$$z \leq \bar{A} \cdot x; \quad z \leq A; \quad z \geq A - (1 - x) \cdot \bar{A}; \quad z \geq 0 \tag{A.2}$$

A.2 Scenarios and corresponding possibilities

In this study, the ten representative scenarios with their corresponding possibilities are shown in Figure A.1.

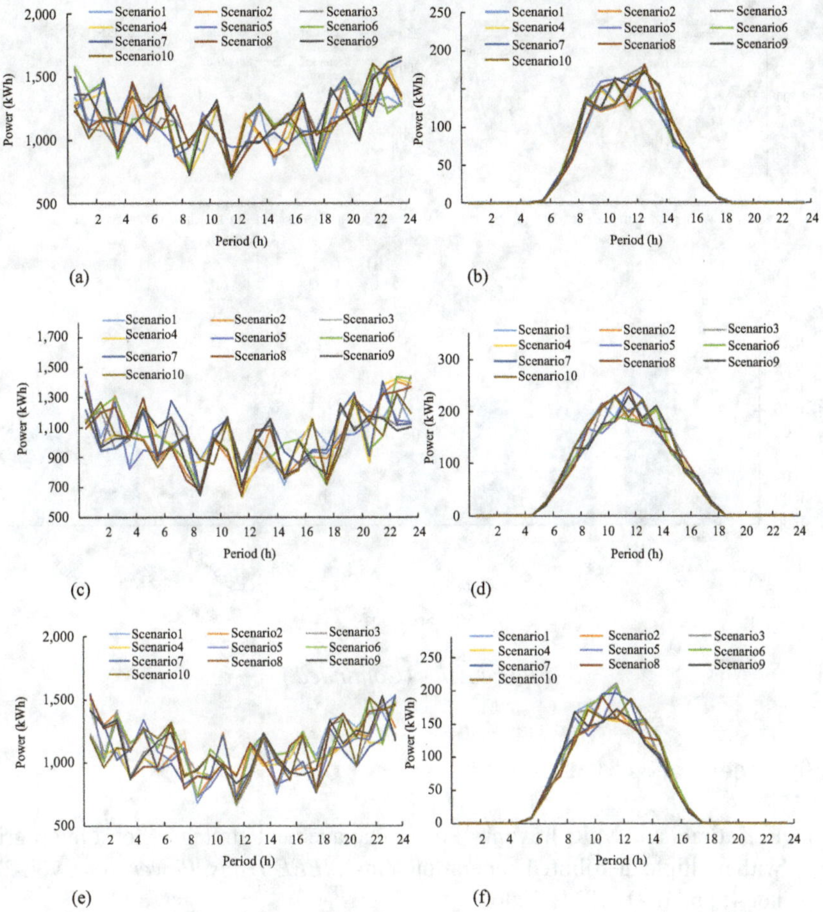

Figure A.1 The scenarios and corresponding possibilities: (a) wind sources in spring; (b) solar source in spring; (c) wind sources in summer; (d) solar source in summer; (e) wind sources in autumn; (f) solar source in autumn; (g) wind sources in winter; (h) solar source in winter; (i) the corresponding possibilities for ten scenarios

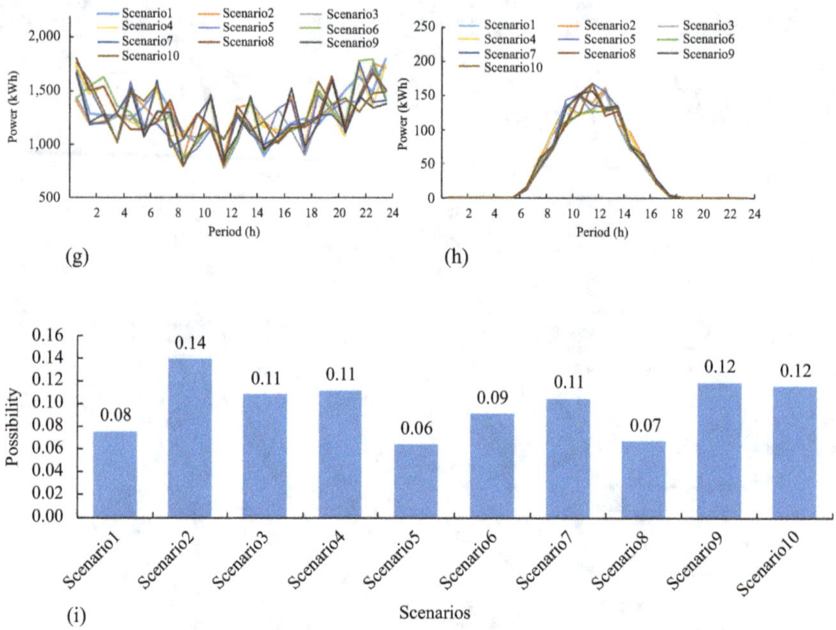

Figure A.1 (Continued)

References

[1] F. Katiraei and M.R. Iravani, 'Power management strategies for a microgrid with multiple distributed generation units', *IEEE Trans. Power Syst.*, vol. 21, no. 4, pp. 1821–1831, 2006.

[2] H. Cho, A.D. Smith, and P. Mago, 'Combined cooling, heating and power: a review of performance improvement and optimization', *Appl. Energy*, vol. 36, no. 1, pp. 168–185, 2014.

[3] W. Gu, Z. Wu, R. Bo, *et al.*, 'Modeling, planning and optimal energy management of combined cooling, heating and power microgrid: a review', *Int. J. Elec. Power*, vol. 54, no. 1, pp. 26–37, 2014.

[4] C. Zhang, Y. Xu, Z.Y. Dong, *et al.*, 'A composite sensitivity factor based method for networked distributed generation planning', *Proc. Power Syst. Comput. Conf. (PSCC)*, pp. 1–7, Genoa, Italy, June 2016.

[5] M.F. Shaaban, Y.M. Atwa, and E.F. El-Saadany, 'DG allocation for benefit maximization in distribution networks', *IEEE Trans. Power Syst.*, vol. 28, no. 2, pp. 639–649, 2013.

[6] R.S. Al Abri, E.F. El-Saadany, and Y.M. Atwa, 'Optimal placement and sizing method to improve the voltage stability margin in a distribution system using distributed generation', *IEEE Trans. Power Syst.*, vol. 28, no. 1, pp. 326–334, 2013.

[7] J. Chen, W. Zhang, J. Li, *et al.*, 'Optimal sizing for grid-tied microgrids with consideration of joint optimization of planning and operation', *IEEE Trans. Sustain. Energy*, vol. 9, no. 1, pp. 237–248, 2018.

[8] L. Guo, W. Liu, J. Cai, *et al.*, 'A two-stage optimal planning and design method for combined cooling, heat and power microgrid system', *Energy Convers. Manag.*, vol. 74, no. 1, pp. 433–445, 2013.

[9] A.K. Basu, A. Bhattacharya, S. Chowdhury, *et al.*, 'Planned scheduling for economic power sharing in a CHP-based micro-grid', *IEEE Trans. Power Syst.*, vol. 27, no. 1, pp. 30–38, 2012.

[10] S. Mashayekh, M. Stadler, G. Cardoso, *et al.*, 'Security-constrained design of isolated multi-energy microgrids', *IEEE Trans. Power Syst.*, vol. 33, no. 3, pp. 2452–2462, 2018.

[11] Z. Wang, B. Chen, J. Wang, *et al.*, 'Robust optimization based optimal DG placement in microgrids', *IEEE Trans. Smart Grid*, vol. 5, no. 5, pp. 2173–2182, 2014.

[12] A. Khodaei, S. Bahramirad, and M. Shahidehpour, 'Microgrid planning under uncertainty', *IEEE Trans. Power Syst.*, vol. 30, no. 5, pp. 2417–2425, 2015.

[13] T. Dragičević, H. Pandžić, D. Škrlec, *et al.*, 'Capacity optimization of renewable energy sources and battery storage in an autonomous telecommunication facility', *IEEE Trans. Sustain. Energy*, vol. 5, no. 4, pp. 1367–1378, 2014.

[14] G.G. Moshi, C. Bovo, A. Berizzi, *et al.*, 'Optimization of integrated design and operation of microgrids under uncertainty', *Proc. Power Syst. Comput. Conf. (PSCC)*, pp. 1–7, Genoa, Italy, June 2016.

[15] P. Mancarella, 'MES (multi-energy systems): an overview of concepts and evaluation models', *Energy*, vol. 65, no. 1, pp. 1–17, 2014.

[16] 'NTU to build an offshore integrated system of renewable energy sources', http://media.ntu.edu.sg/NewsReleases/Pages/newsdetail.aspx?news=830aa91e-aa88-4c7b-8f9500c171c17560&from=timeline&isappinstalled=0, accessed October 2016.

[17] Z. Li and Y. Xu, 'Temporally-coordinated optimal operation of a multi-energy microgrid under diverse uncertainties', *Appl. Energy*, vol. 240, no. 1, pp. 719–729, 2019.

[18] Z. Li and Y. Xu, 'Optimal coordinated energy dispatch of a multi-energy microgrid in grid-connected and islanded modes', *Appl. Energy*, vol. 210, no. 1, pp. 974–986, 2018.

[19] M.R. Benam, S.S. Madani, S.M. Alavi, *et al.*, 'Optimal configuration of the CHP system using stochastic programming', *IEEE Trans. Power Del.*, vol. 30, no. 3, pp. 1048–1056, 2015.

[20] E. Carpaneto, G. Chicco, P. Mancarella, *et al.*, 'Cogeneration planning under uncertainty: Part I: Multiple time frame approach', *Appl. Energy*, vol. 88, no. 4, pp. 1059–1067, 2011.

[21] E. Carpaneto, G. Chicco, P. Mancarella, *et al.*, 'Cogeneration planning under uncertainty. Part II: Decision theory-based assessment of planning alternatives', *Appl. Energy*, vol. 88, no. 4, pp. 1075–1083, 2011.

[22] R. Atia and N. Yamada, 'Sizing and analysis of renewable energy and battery systems in residential microgrids', *IEEE Trans. Smart Grid*, vol. 7, no. 3, pp. 1204–1213, 2016.
[23] F. Wei, Q.H. Wu, Z.X. Jing, *et al.*, 'Optimal unit sizing for small-scale integrated energy systems using multi-objective interval optimization and evidential reasoning approach', *Energy*, vol. 111, no. 1, pp. 933–946, 2016.
[24] Z. Wang, Y. Chen, S. Mei, *et al.*, 'Optimal expansion planning of isolated microgrid with renewable energy resources and controllable loads', *IET Renew. Power. Gen.*, vol. 11, no. 7, pp. 931–940, 2017.
[25] F.S. Gazijahani and J. Salehi, 'Stochastic multi-objective framework for optimal dynamic planning of interconnected microgrids', *IET Renew. Power. Gen.*, vol. 11, no. 14, pp. 1749–1759, 2017.
[26] M. Alipour, B. Mohammadi-Ivatloo, and K. Zare, 'Stochastic scheduling of renewable and CHP-based microgrids', *IEEE Trans. Ind. Informat.*, vol. 11, no. 5, pp. 1049–1058, 2015.
[27] Y. Xu, Z.Y. Dong, K.P. Wong, *et al.*, 'Optimal capacitor placement to distribution transformers for power loss reduction in radial distribution systems', *IEEE Trans. Power Syst.*, vol. 28, no. 4, pp. 4072–4079, 2013.
[28] Y. Xu, Z.Y. Dong, R. Zhang, *et al.*, 'Multi-timescale coordinated voltage/var control of high renewable-penetrated distribution systems', *IEEE Trans. Power Syst.*, vol. 32, no. 6, pp. 4398–4408, 2017.
[29] L. Ma, N. Liu, and J. Zhang, *et al.*, 'Energy management for joint operation of CHP and PV prosumers inside a grid-connected microgrid: a game theoretic approach', *IEEE Trans. Ind. Informat.*, vol. 12, no. 5, pp. 1930–1942, 2016.
[30] W. Wei, F. Liu, and S. Mei, 'Distributionally robust co-optimization of energy and reserve dispatch', *IEEE Trans. Sustain. Energy*, vol. 7, no. 1, pp. 289–300, 2015.
[31] M. Tasdighi, H. Ghasemi, and A. Rahimi-Kian, 'Residential microgrid scheduling based on smart meters data and temperature dependent thermal load modeling', *IEEE Trans. Smart Grid*, vol. 5, no. 1, pp. 349–357, 2014.
[32] 'Tianjin', https://en.wikipedia.org/wiki/Tianjin, accessed May 2019.
[33] 'Tianjin Development and Reform Commission', www.tjdpc.gov.cn/gzcx/syjgcx/gd/201307/t2013070530044.shtml, accessed May 2019.

Chapter 6

Stochastic planning of heterogeneous energy storage (HES) in residential MEMG

The deployment of heterogeneous energy storage (HES), which mainly consists of the power and thermal storage (TS), is critical for improving the overall energy utilisation efficiency of the multi-energy systems. Thus, a risk-averse method for HES deployment in a residential multi-energy microgrid (MEMG) is proposed taking diverse uncertainties and multi-energy demand-side management (DSM) into account. Besides the size and location planning of the HES, its optimal investment phase is optimised by maximising the system equivalent daily profit (EDP) and minimising the risk at the same time. To deal with the diverse uncertainties from renewable energy sources (RESs), power loads, outdoor temperature, and hot water demands, a multi-stage adaptive stochastic optimisation (SO) approach is used. Through the linearisation approaches and scenario sampling, the original non-linear deployment model is transformed into a mixed-integer linear programming (MILP) one and tested on an IEEE 33-bus distribution network-based residential MEMG. The effectiveness of the proposed method is verified in numerous case studies. All simulation results indicate that our proposed method can effectively improve the system EDP and is more immune to heterogeneous uncertainty sources. Besides, our method can be practically applied for the emerging residential MEMGs, such as smart buildings and green homes, with long-term DSM contracts.

Nomenclature

A Abbreviations

CCHP	Combined cooling, heat, and power
CVaR	Conditional value-at-risk
DSM	Demand-side management
EB	Electric boiler
EC	Electric chiller
EDP	Equivalent daily profit
EES	Electrical energy storage
GAMS	General Algebraic Modelling System
GB	Gas boiler

HES	Heterogeneous energy storage
MILP	Mixed-integer linear programming
PBDR	Price-based demand response
PV	Photovoltaic cell
RES	Renewable energy source
MEMG	Residential MEMG
SO	Stochastic optimisation
TS	Thermal storage
TST	Thermal storage tank
VaR	Value-at-risk
WT	Wind turbine

B Superscript

b	Index for branches, $b \in \mathrm{Br}(i, i+1)$
j	Index for the price levels
m	Index for the first year of each investment phase, $m \in p$
$p/t/d/i$	Index for years/dispatch periods/days/buses
Λ	Set for $\{p, d, t, i\}$

C Constants and parameters

α_{BH}	Correction factor for envelopes like walls, doors, etc.
α_{CL}	Confidential level
A_{CL}	Air volume that infiltrates into the room
C_{AV}	Equivalent specific heat of the room
$C_{\mathrm{WR}}/C_{\mathrm{Air}}$	Specific heat capacity of water/air
dr	Discount rate
I_{LE}	Gap length of the door or window crack
K_{HT}	Heat transfer coefficient of the room
$L_{\mathrm{PL}}/L_{\mathrm{QL}}$	Active/reactive PBDR levels
N_{F}	The normalised factor for Beta Distribution
N_{M}	The total number of investment phases
$N_{\mathrm{P}}/N_{\mathrm{B}}/N_{\mathrm{T}}/N_{\mathrm{D}}$	Number of years/buses/dispatch periods/days a year
$P_{\mathrm{DG,REAL}}$	Available energy resource of RES
$P_{\mathrm{DL}}/H_{\mathrm{HL}}/H_{\mathrm{CL}}$	Power/heat/cooling demands
$P_{\mathrm{WT}}/P_{\mathrm{PV}}$	The power output of wind turbines/photovoltaic cells
$P_{\mathrm{DL,Pr}}/Q_{\mathrm{DL,Pr}}$	Active/reactive power demands before the DSM
R/X	Branch impedance
$S_{\mathrm{ES}}^{\min}/S_{\mathrm{ES}}^{\max}$	Minimal/maximal HES capacity for each bus/group

S_{BH}	Area of the building envelope
$S_{\mathrm{ES}}^{\mathrm{ALL}}$	Maximal energy capacity for the HES system
T_{IW}	The temperature of inlet water for water tanks
$T_{\mathrm{IN}}^{\mathrm{set}}/T_{\mathrm{SW}}^{\mathrm{set}}$	Temperature set-point for room/hot water
$V_0/\Delta V_{\mathrm{BUS}}^{\max}$	Substation voltage/maximal voltage deviation
$V_{\mathrm{RB}}/V_{\mathrm{HW}}$	The volume of the room/hot water needed
x_s	The generalised symbol for power outputs of WTs and PVs
y_s	The generalised symbol for the outdoor temperature and multi-energy demands
β_{cx}/β_{jx}	Correction factor due to solar irradiation/wind blow
$\gamma_{\mathrm{DG}}^{\min}/\gamma_{\mathrm{DG}}^{\max}$	Minimal/maximal DG output ratio
$\gamma_{\mathrm{ES}}^{\max}$	The maximal charging/discharging power ratio of the energy storage to its capacity
γ_{it}	The ratio of intrusion loss to basic thermal transfer loss
γ_{LL}	The leakage loss ratio of the energy storage
$\gamma_{\mathrm{SOC}}^{\min}/\gamma_{\mathrm{SOC}}^{\max}$	Minimal/maximal state of charge for energy storage
Δt	Unit dispatch interval
$\Delta T_{\mathrm{V}}/\Delta T_{\mathrm{W}}$	Maximal temperature deviation for room/hot water
$\varepsilon_{\mathrm{SAL}}$	The ratio of HES salvage value to its investment cost
$\eta_{\mathrm{BC}}/\eta_{\mathrm{BD}}$	Charging/discharging efficiency of battery storage
η_{MT}	Electrical efficiency of CCHP plants
$\eta_{\mathrm{TC}}/\eta_{\mathrm{TD}}$	Charging/discharging efficiency of thermal storage
ρ_{RK}	The weighted risk factor
θ/ψ	Scale parameters for beta distribution
$\kappa_{\mathrm{MT}}/\kappa_{\mathrm{B}}$	Emission penalty price of CCHP plants/utility grid
μ/σ	Mean/standard deviation for normal distribution
$\rho_{\mathrm{WR}}/\rho_{\mathrm{Air}}$	Water/air density
$\tau_{\mathrm{T}}/\tau_{\mathrm{B}}$	The decay rate of thermal/battery storage
$\chi_{\mathrm{B}}/\chi_{\mathrm{S}}$	Power purchasing/selling price between MEMG and utility grid
$\chi_{\mathrm{DG}}^{\mathrm{M}}/\chi_{\mathrm{ES}}^{\mathrm{M}}$	The unit maintenance cost of generators/storage
$\chi_{\mathrm{ES}}^{\mathrm{INV}}$	Unit investment cost of energy storage
χ_{NG}	Nature gas price
$\chi_{\mathrm{PS}}/\chi_{\mathrm{HS}}/\chi_{\mathrm{CS}}$	Power/heat/cooling retail prices to the local customers
χ_{SUB}	The unit subsidy price for RES

D Variables

η_{VaR}	The project VaR
a	Binary indicator for the state of price or load levels
C_{EDP}	The project EDP

C_{FC}/C_{TC}	Fuel/power transaction cost
C_{INV}/C_{SAL}	Investment cost/salvage value
C_{MC}/C_{EM}	Maintenance/gas emission cost
C_{REV}/C_{SUB}	Revenue from local customers/RES subsidy
C_{ST}/C_{SD}	Start-up/shut-down cost
E_T/E_B	Energy stored in the thermal/battery storage
H_{BH}/H_{AH}	Basic/additional thermal transfer loss
H_{CN}/H_{CR}	Infiltration/intrusion thermal loss
H_{EB}/H_{EC}	Heat/cooling output of electric boilers/chillers
H_{MH}/H_{MC}	Heat/cooling output of CCHP plants
H_{TC}/H_{TD}	Charging/discharging power of thermal storage
H_{TL}/H_{NET}	Total space thermal demands/net thermal
P_{MT}	The power output of CCHP plants
P_{BC}/P_{BD}	Charging/discharging power of battery storage
P_{BX}/P_{SX}	Purchasing/selling power between MEMG and utility
P_{DG}/P_{ES}	Generalised power output for generators/storage
P_{EB}/P_{EC}	Power input for electric boilers/chillers
P_{PF}/Q_{PF}	Active/reactive power flow
P_{RL}/Q_{RL}	Active/reactive power loads after demand response
$P^{p,d,t,0,b}{}_{PF}/Q^{p,d,t,0,b}{}_{PF}$	Parallel active/reactive power of branch b
$R_{DG,DN}/R_{DG,UP}$	Maximal ramping down/up limits
S_{ES}	The capacity of the energy storage
$S_{ES,Eq}$	The effective capacity of energy storage considering the leakage loss
T_{IN}/T_{OT}	Indoor/outdoor temperature
T_{SW}	The temperature of the water inside the water tank
U_{DG}	Binary state to indicate the on-off status of generators
U_{ES}	Binary state to denote investment decisions for HES
U_{ESC}/U_{ESD}	Binary charging/discharging states of HES
V_{BUS}	Bus voltage

6.1 Introduction

Energy storage is regarded as a critical unit in the energy system which could store energy for later usage [1]. The deployment of energy storage system is vital as it contributes to handling the stochastic power, increasing the penetration levels of RES, shaving peak energy demands, etc. [2,3].

In the literature, the optimal deployment of the electrical energy storage (EES) is more investigated. Reference [4] formulates an optimal deployment model for the EES in an electricity system to shift the power demands and reduce the energy supplying cost. In [5], an economic dispatch approach for EES sizing in the power system is proposed. This approach decides the optimal EES capacity regarding its operation cost under various RES penetration levels. Reference [6] introduces a planning method for integrating EES in a microgrid (MG) to study its merits over diverse conceivable regulatory services. Reference [7] decides the best EES capacity in the RES-combined power system to get the optimal operating cost. Nonetheless, in the above research works, the EES only facilitates the RES integration without coordinating all the units at the system level. Therefore, their placement solutions may not be optimal from the system view. In this light, Ref. [3] studies the EES deployment in an MG based on cost–benefit analysis. The time series and neural network techniques are employed to predict the wind and solar power. In [8], to decide the best EES capacity and location, the probabilistic load model is adopted to combine more load scenarios. The above research work has used the deterministic operation approaches by assuming that there can be perfect information about the generation from renewables and end-user energy consumption. However, the RES and energy consumption are intrinsically stochastic, which makes all the solutions obtained from the deterministic approaches unreliable.

To deal with heterogeneous uncertainties in the planning of the energy storage system, various means are used. In [9], the best EES deployment scheme is obtained by the deterministic operation tool with the uncertainties dealt with the model predictive control method. Reference [10] studies the optimal EES capacity optimisation problem to minimise the annual cost by a bi-level programming method. The uncertainties are managed by the system reserve. Though the diverse uncertainties are taken into account, both works still use deterministic approaches, which fail to provide a reliable or robust deployment plan. In this sense, the SO and robust optimisation (RO) approaches have been employed. The study in [11] proposes an SO method for EES capacity in a wind–diesel-based isolated grid. In [12], the optimal EES sizing problem in MGs with highly penetrated wind power is solved by the unit-commitment-based RO method. The research in [13] investigates an efficient cost-reliability optimisation method to obtain the optimal EES location and size in an MG considering the DSM for electricity. To handle the uncertainties from wind turbines (WTs), the SO method is utilised. Reference [14] introduces the SO approach for EES deployment to reduce the energy supply operating cost. Authors in [15] propose a combined robust capacity optimisation method for EES and distributed generators to save the operating cost under the intermittent RES outputs. For the long-term planning research, the worst-case-oriented RO method can be overly conservative compared with the SO method, while in fact, the occurrence probability of the worst case is definitely low. As for the SO method, currently, the majority of the research is risk-neutral [13,14], nevertheless, this shall pose a great threat to the reliable operation of the energy systems. Consequently, the decision makers would be willing to choose a solution with a bit higher operational costs but much lower risk [16].

In addition, most of the studies focus merely on the deployment of EES in the electricity network. However, recently, there is a dramatic development of multi-energy systems which integrate a large amount of power, heat, and cooling energy carriers [17]. At the distribution network level, multi-energy systems are usually realised in forms of residential MEMGs, such as smart buildings and intelligent homes [18]. These multi-energy systems usually require a large quantity of thermal demand. Thus, to enhance the overall energy efficiency and operation flexibility, the EES deployment can be transformed into the HES deployment in the emerging residential MEMGs. Accordingly, the uncertainties from only the power system should be changed to from the multi-energy systems, such as the outdoor temperature or thermal demands [19]. Regarding this aspect of research, very limited research work has been done in the literature. Reference [20] plans the HES size aiming to minimise the planning cost and gas emissions by a two-stage approach. The first stage uses the genetic algorithm to solve the planning problem and the second stage applies the MILP solvers to obtain the operating solution. In [21], both the system structure and HES capacity configuration are investigated in a deployment framework for a smart MG. Then, an MILP-based planning method is presented to reduce the net operating cost. The research in [22] introduces an optimal design model for an MG that involves the HES. To reduce the computational burden, the typical days are used to represent the yearly horizon. Though the above research work achieves the efficient HES deployment, the network constraints and the effects of multi-energy uncertainty sources are neglected. As a result, a reliable operation scheme for the energy system cannot be guaranteed, and their solutions are not practical for real-world applications.

Moreover, as the DSM is a promising and effective solution to support the system operation as well as save energy bills for residents, there is a widespread installation of smart meters in residential buildings which has signed long-term DSM contracts with the main grid [18]. Compared with the electricity, thermal energy (heat or cooling) gets a higher inertia because residents can be still comfortable within a tolerable temperature range [23]. Though the DSM for electricity is widely applied in different fields like power grid transmission planning [24], robust operation [25], low-carbon sustainability [26], etc., few research has been done to address its coordination with the thermal energy especially for the HES deployment of residential MEMGs.

Finally, the majority of the existing deployment works, either for the EES or HES, only takes the one-off installation at the beginning of the project into account, which ignores the increasing load growth. In this regard, their deployment decisions would be far from enough, given the annual load growth in reality [27]. Hence, the multi-phase deployment, where the HES can be invested at different years of the project, can be considered the more comprehensive and flexible HES planning research.

To fill the existing research gap, in this chapter, an optimal risk-averse stochastic HES deployment method is presented in a grid-tied residential MEMG under heterogeneous uncertainties. Compared with the existing research works, the main contributions are summarised as follows:

1. A novel risk-averse HES deployment model in the grid-tied residential MEMGs is proposed. In spite of merely focusing on the EES, its integration with the TS and different energy properties are also studied.

2. For achieving the more comprehensive and flexible HES deployment, besides the traditional capacity and location planning, the investment phase selection is also taken into consideration, given in reality that there is an annual load growth.
3. Multiple uncertainty sources not only from the RES and power loads of the power system but from the outdoor temperature and hot water needs of the thermal system are also taken into full account. Then, a scenario-based adaptive SO approach is applied to tackle all uncertainties.
4. The DSM for multiple energies is integrated in the HES deployment work, given its merits in flexibilising the system operation and reducing energy bills. Especially for the thermal system, the DSM in terms of the specific thermal modelling for local residential buildings is built with thermal dynamics and thermal inertia.

For the real-world implications, with the increasing number of green buildings, smart homes, etc., that are partially powered by the RES and have signed long-term multi-energy DSM contracts, our proposed method can be directly used for deploying their HES system.

The rest of this chapter is organised as follows. Section 6.2 describes the basic structure of the residential MEMG with the specific modelling of all its units. Section 6.3 introduces the proposed risk-averse HES deployment model and its scenario-based multi-stage adaptive SO form. Section 6.4 describes the solution method for the HES deployment and Section 6.5 presents a detailed case study and simulation results of the HES deployment. At last, Section 6.6 concludes the whole chapter.

6.2 Modelling of the residential MEMG

6.2.1 Basic structure of the residential MEMG

The typical structure of an residential MEMG is given in Figure 6.1. The main components are the distributed generators, HES system, and multi-energy demands. The residential MEMG is grid-connected, which indicates that the power can be

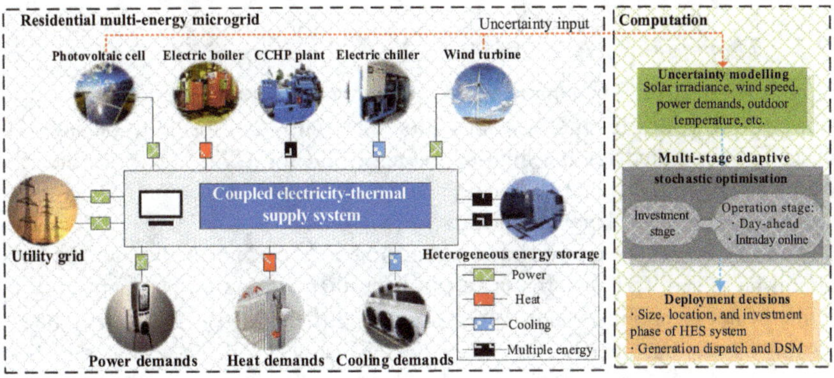

Figure 6.1 The basic structure for an residential MEMG

exchanged freely with the utility grid [18]. Further, the residents are assumed to have signed a long-term contract with the residential MEMG to save their energy bills.

The distributed generators in the residential MEMG can be categorised as follows [27]:

1. Dispatchable unit: This refers to the generators which are in small sizes, flexible control, and high ramp rates: electric boilers (EBs), combined cooling, heat, and power (CCHP) plants with the main generator as micro-turbines, and electric chillers (ECs).
2. Non-dispatchable unit: This refers to the generator whose energy outputs are highly volatile, such as the WTs and photovoltaic cells (PVs).

For the HES system, it is mainly composed of the power energy type, i.e. battery storage (BS), and thermal energy type, i.e. thermal storage tank (TST).

All the components in the residential MEMG are coordinated to satisfy the multi-energy demands to maintain the reliability and economy [19]. The data for the RES generation, power demands, outdoor temperature, residential hot water needs, and demand response pattern can be collected by the smart meters in the building. The collected data are inserted in the computation procedure as optimisation parameters.

The computation procedures for the HES deployment are described in Sections 6.4 and 6.5. The deployment decisions can be resulted.

6.2.2 Battery storage

The BS is widely used in the power system for shifting the loads and shaving the peak demands. Compared with other types of EESs such as the pumped energy storage and compressed air energy storage, its power density ranges from a few kilowatts to several megawatts. This makes BS much more suitable for its application in the MG or community level [2,6].

The relationship between energy and power in BS is given as follows:

$$E_{\mathrm{B}}^{t} = E_{\mathrm{B}}^{t-1}(1 - \tau_{\mathrm{B}}) + P_{\mathrm{BC}}^{t}\eta_{\mathrm{BC}}\Delta t - P_{\mathrm{BD}}^{t}\eta_{\mathrm{BD}}\Delta t \tag{6.1}$$

6.2.3 Thermal storage tank

The typical TST model is demonstrated in Figure 6.2. Its walls are composed of the stainless steel, concrete, and heat insulation materials from inside to outside. The tank is well insulated for the high operating efficiency [17]. The temperature in the tank can be roughly stratified into the hot water layer (temperature, T_{H}; volume, V_{H}) and cooling water layer (temperature, T_{C}; volume, V_{C}).

In both layers, the temperature is the same as the connected water networks and constant. The TST absorbs or releases the thermal energy by controlling the volume of hot or cooling water. When it produces heat energy, the temperature of the supply water becomes higher – thus, the TST now is called a heat storage tank; while for the cooling supply, the TST is called a cooling storage tank whose return water temperature is much higher than the supply one.

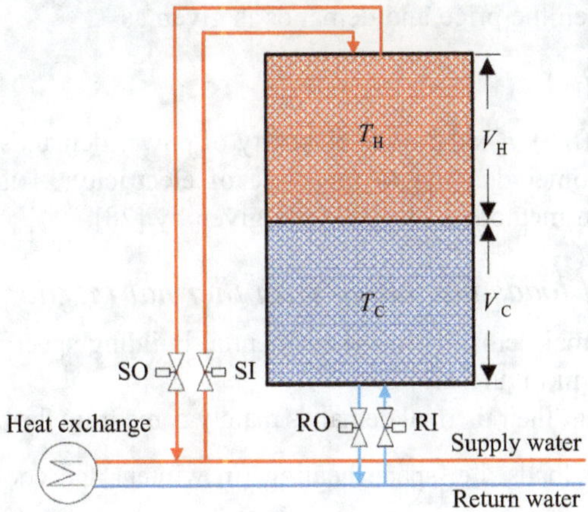

Figure 6.2 Basic structure of a TST. SI, SO: inlet and outlet switch for the supply water; RI, RO: inlet and outlet switch for return water

The absorbing and releasing power of TST is shown as follows:

$$H_{TC}^t = C_{WR} \cdot \rho_{WR} \cdot V_H^t \cdot (T_H - T_C) \quad (6.2)$$

$$H_{TD}^t = -C_{WR} \cdot \rho_{WR} \cdot V_C^t \cdot (T_C - T_H) \quad (6.3)$$

The relationship between the energy and power of TST is indicated as

$$E_T^t = E_T^{t-1}(1 - \tau_T) + H_{TC}^t \eta_{TC} \Delta t - H_{TD}^t \eta_{TD} \Delta t \quad (6.4)$$

6.2.4 Price-based demand response (PBDR) for the electricity

The price-based demand response (PBDR) is adopted for electricity DSM in the residential MEMG. The reasons are the following:

1. It allows customers to manage demand freely without partial loads being controlled directly by the system operator. In the PBDR, electricity prices of the following day are designed and delivered to customers a day ahead, and then the customers adjust their demands based on the prices. When the prices are high, the customers would reduce their demands to save bills. When they are low, customers would be incented to increase their demands.
2. It contributes to the system reliability enhancement when there is a large penetration level of fluctuating RES [25], that is, it could help to consume the large percentage of renewables, and also when there is lack of enough energy, it could reduce the demands which are not critical.

Currently, the PBDR has been widely applied in the power system planning, power market regulation, and economic dispatch, etc. In the PBDR scheme, the relationship between the price and demands is given as

$$P_{DL}^{t} = A \cdot (\chi_{PS})^{\varepsilon} \tag{6.5}$$

In Constraint (6.5), ε is the price elasticity of power demands, representing the sensitivity of customer demand to the price of electricity; A is a constant. The specific calculation methods of ε and A are given by [28].

6.2.5 Thermal loads modelling with thermal inertia

To model the thermal demands in the residential building specifically, the typical structure is shown in Figure 6.3 [29].

For the residents, their thermal demands mainly come from the below two aspects:

1. Space thermal loads, i.e. space heating (in winter) and cooling (in summer) [19];
2. Residential hot water demands (in spring, autumn, and winter) [30].

All the above thermal loads can be modelled with thermal dynamics which will be detailed in the following.

6.2.5.1 Thermal demand classification

For space thermal loads, the thermal generation should compensate for any thermal loss to maintain the suitable room temperature. The total thermal loads are calculated as follows [31]:

$$H_{TL}^{t} = H_{BH}^{t} + H_{AH}^{t} + H_{CN}^{t} + H_{CR}^{t} \tag{6.6}$$

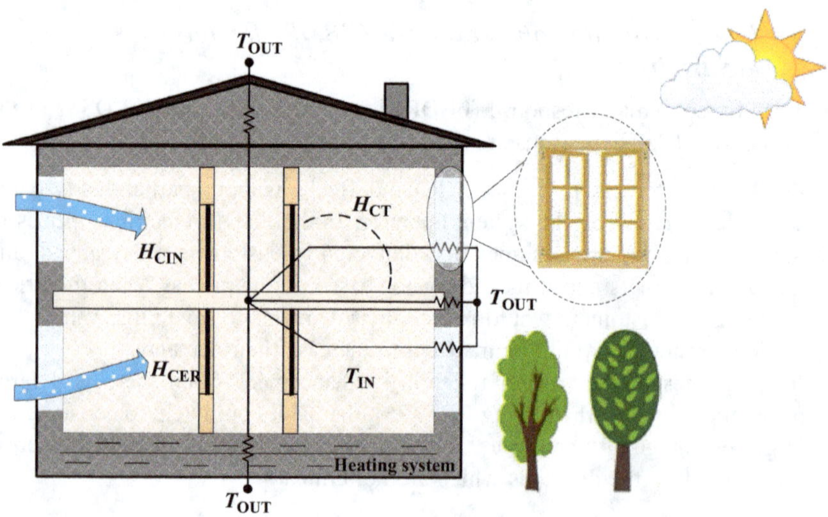

Figure 6.3 The structure of a typical room in a residential building

The basic thermal transfer loss includes the thermal transfer from the building envelope like doors, windows, walls, and floors. It can be formed as (6.7):

$$H_{\mathrm{BH}}^{t} = \alpha_{\mathrm{BH}} K_{\mathrm{HT}} S_{\mathrm{BH}} (T_{\mathrm{IN}}^{t} - T_{\mathrm{OT}}^{t}) \tag{6.7}$$

The additional thermal loss considers the loss brought by the wind blow and height of the building. It is formulated as

$$H_{\mathrm{AH}}^{t} = H_{\mathrm{BH}}^{t} \cdot (\beta_{cx} + \beta_{jx}) \tag{6.8}$$

The infiltration thermal loss means the thermal demands to heat the cold air (in winter) or cool the hot air (in summer) from outside to inside via the door or window crack. It is formed as

$$H_{\mathrm{CN}}^{t} = 0.364 \cdot I_{\mathrm{LE}} A_{\mathrm{CL}} (T_{\mathrm{IN}}^{t} - T_{\mathrm{OT}}^{t}) \tag{6.9}$$

The intrusion thermal loss describes the thermal demands that cool the hot air (in summer) or heat the cold air (in winter) from the casual opening of doors or windows. It can be demonstrated as

$$H_{\mathrm{CR}}^{t} = H_{\mathrm{BH}}^{t} \gamma_{it} \tag{6.10}$$

All the above thermal loss is basically related to the difference between the indoor and outdoor temperature, and then the total thermal demands can be reformulated in a general form as

$$H_{\mathrm{TL}}^{t} = C_{\mathrm{AV}} (T_{\mathrm{IN}}^{t} - T_{\mathrm{OT}}^{t}) \tag{6.11}$$

6.2.5.2 Dynamic thermal demand modelling

According to the first law of thermodynamics, the thermal loads in (6.6)–(6.11) can be reformulated as a dynamic model with the temperature change as

$$C_{\mathrm{Air}} \rho_{\mathrm{Air}} V_{\mathrm{RB}} dT_{\mathrm{IN}}^{t} / dt = H_{\mathrm{NET}}^{t} \tag{6.12}$$

Due to the thermal inertia, the indoor temperature varies slowly and the corresponding variations during each period, e.g. 1 h, can be neglected [19,29]. As a result, thermal dynamics model in (6.12) can be converted into an hourly state model as

$$T_{\mathrm{IN}}^{t} = H_{\mathrm{NET}}^{t} \Delta t / (C_{\mathrm{Air}} \rho_{\mathrm{Air}} V_{\mathrm{RB}}) + T_{\mathrm{IN}}^{t-1} \tag{6.13}$$

Note that some other factors like the human activities and power appliances may also influence the thermal demands, but as the influence can be minor, they are ignored here [2,31].

6.2.5.3 Hot water demands

The hot water is utilised for the showering or washing needs of all residents. The thermal demands from the hot water can be calculated as follows [29]:

$$H_{\mathrm{HW}}^{t} = C_{\mathrm{WR}} \rho_{\mathrm{WR}} V_{\mathrm{HW}}^{t} (T_{\mathrm{SW}}^{t} - T_{\mathrm{IW}}^{t}) \tag{6.14}$$

6.2.6 The demand side management (DSM) for thermal energy

Apart from the slow variation rate of thermal energy, the fact that residents can be comfortable within an adjustable indoor or hot-water temperature range constitutes another aspect of the thermal inertia [30]. This can be seen as the DSM for the thermal energy. The model is given as follows:

$$T_{\mathrm{IN}}^{\mathrm{set}} - \Delta T_{\mathrm{V}} \leq T_{\mathrm{IN}}^{\Lambda} \leq T_{\mathrm{IN}}^{\mathrm{set}} + \Delta T_{\mathrm{V}} \tag{6.15}$$

$$T_{\mathrm{SW}}^{\mathrm{set}} + \Delta T_{\mathrm{W}} \leq T_{\mathrm{SW}}^{\Lambda} \leq T_{\mathrm{SW}}^{\mathrm{set}} + \Delta T_{\mathrm{W}} \tag{6.16}$$

6.3 Mathematic modelling for HES deployment

6.3.1 Two-stage HES deployment framework

In this chapter, the HES deployment can be decomposed into two stages as in Figure 6.4:

1. **Investment stage:** Decisions at this stage are for the investment, i.e. the HES location, capacity, and investment phase. As shown in Figure 6.4, a period of 3 years is taken as a unit investment phase, and every 3 years, there can be new HES investment decisions.
2. **Operation stage:** Decisions at this stage involve the dispatch of the HES system, distributed generators, and DSM scheme. The heterogeneous uncertainty sources are tackled via a further two-substage-coordinated operation framework. The first substage is the day-ahead dispatch stage; the second substage is the intraday online dispatch stage. The details for the two substages are given in Section 6.4.

6.3.2 Mathematical formulation

In this study, it is assumed there is a steady annual growth rate for multi-energy demands. To evaluate the effectiveness of the method, the maximisation of

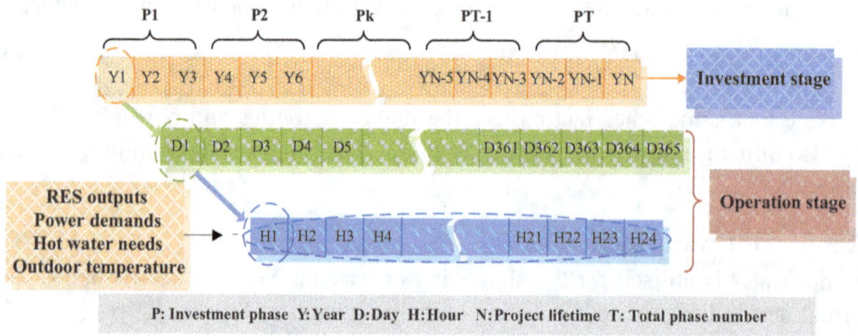

Figure 6.4 Two-stage HES deployment framework

'mean-risk', which is equal to the subtraction of EDP and risk cost is set as the objective. In finance, the EDP is defined as the profit per day of owning and operating an asset over its entire lifespan. It is calculated by dividing the net present value of a project by the present value of the daily factor. The risk in this chapter is measured by the conditional value-at-risk (CVaR) [16,32]. Given the above, the risk-averse HES deployment model can be formed as

1. *Objective:*

$$\max_{z \in Z, x \in X, \eta_{\text{VaR}} \in \Re} \quad \mathbf{E}(C_{\text{EDP}}) - \rho_{\text{RK}} CVaR_{\alpha}(C_{\text{EDP}}) \tag{6.17}$$

$$\text{s.t.,} \quad C_{\text{EDP}} = \frac{1}{365 \times N_{\text{P}}} \cdot \frac{dr(1+dr)^{N_{\text{P}}}}{(1+dr)^{N_{\text{P}}} - 1} \cdot [\underbrace{-F(z)}_{\text{Investment stage}} + \underbrace{G(x)}_{\text{Operation stage}}] \tag{6.18}$$

$$CVaR_{\alpha}(C_{\text{EDP}}) = \eta_{\text{VaR}} + \frac{1}{1-\alpha_{\text{CL}}} \mathbf{E}[\max(C_{\text{EDP}} - \eta_{\text{VaR}}, 0)] \tag{6.19}$$

$$F(z) = \sum_{p \in N_{\text{P}}} \frac{C_{\text{INV}}^{p}}{(1+dr)^{p-1}} - \frac{C_{\text{SAL}}^{N_{\text{P}}}}{(1+dr)^{N_{\text{P}}-1}} \tag{6.20}$$

$$G(x) = \sum_{p \in N_{\text{P}}} \frac{(C_{\text{REV}}^{p} + C_{\text{SUB}}^{p} - C_{\text{FC}}^{p} - C_{\text{TC}}^{p} - C_{\text{MC}}^{p} - C_{\text{EM}}^{p} - C_{\text{ST}}^{p} - C_{\text{SD}}^{p})}{(1+dr)^{p}} \tag{6.21}$$

$$C_{\text{SAL}}^{N_{\text{P}}} = \sum_{p \in N_{\text{P}}} \varepsilon_{\text{SAL}}^{p} C_{\text{INV}}^{p} \tag{6.22}$$

$$C_{\text{INV}}^{p} = \sum_{i \in N_{\text{B}}} \chi_{\text{ES,INV}}(S_{\text{ES}}^{p,i} - S_{\text{ES}}^{p-1,i}) \tag{6.23}$$

$$C_{\text{REV}}^{p} = \sum_{d \in N_{\text{D}}} \sum_{t \in N_{\text{T}}} \sum_{i \in N_{\text{B}}} (\chi_{\text{PA}}^{\Lambda} P_{\text{DL}}^{\Lambda} + \chi_{\text{HS}} H_{\text{HL}}^{\Lambda} + \chi_{\text{CS}} H_{\text{CL}}^{\Lambda}) \tag{6.24}$$

$$C_{\text{SUB}}^{p} = \sum_{d \in N_{\text{D}}} \sum_{t \in N_{\text{T}}} \sum_{i \in N_{\text{B}}} \chi_{\text{SUB}}(P_{\text{WT}}^{\Lambda} + P_{\text{PV}}^{\Lambda}) \tag{6.25}$$

$$C_{\text{FC}}^{p} = \sum_{d \in N_{\text{D}}} \sum_{t \in N_{\text{T}}} \sum_{i \in N_{\text{B}}} (\chi_{\text{NG}} P_{\text{MT}}^{\Lambda} / \eta_{\text{MT}}) \tag{6.26}$$

$$C_{\text{TC}}^{p} = \sum_{d \in N_{\text{D}}} \sum_{t \in N_{\text{T}}} \sum_{i \in N_{\text{B}}} (\chi_{\text{B}} P_{\text{BX}}^{\Lambda} - \chi_{\text{S}} P_{\text{SX}}^{\Lambda}) \tag{6.27}$$

$$C_{\text{MC}}^{p} = \sum_{i \in N_{\text{B}}} \sum_{d \in N_{\text{D}}} \sum_{t \in N_{\text{T}}} \chi_{\text{DG}}^{\text{M}} P_{\text{DG}}^{\Lambda} + \chi_{\text{ES,M}} P_{\text{ES}}^{\Lambda} \tag{6.28}$$

$$C_{\text{EM}}^{p} = \sum_{d \in N_{\text{D}}} \sum_{t \in N_{\text{T}}} \sum_{i \in N_{\text{B}}} \kappa_{\text{MT}} P_{\text{MT}}^{\Lambda} + \kappa_{\text{B}} P_{\text{BX}}^{\Lambda} \tag{6.29}$$

$$C_{\mathrm{ST}}^{p} = \sum_{d \in N_{\mathrm{D}}} \sum_{t \in N_{\mathrm{T}}} \sum_{i \in N_{\mathrm{B}}} \max\{0, U_{\mathrm{DG}}^{\Lambda} - U_{\mathrm{DG}}^{p,d,t-1,i}\} C_{\mathrm{DG}}^{\mathrm{U}} \tag{6.30}$$

$$C_{\mathrm{SD}}^{p} = \sum_{d \in N_{\mathrm{D}}} \sum_{t \in N_{\mathrm{T}}} \sum_{i \in N_{\mathrm{B}}} \max\{0, U_{\mathrm{DG}}^{p,d,t-1,i} - U_{\mathrm{DG}}^{\Lambda}\} C_{\mathrm{DG}}^{\mathrm{D}} \tag{6.31}$$

2. *Constraints:*

$$\text{Constraints (6.1)–(6.16)} \tag{6.32}$$

$$S_{\mathrm{ES}}^{p,i} \geq S_{\mathrm{ES}}^{p-1,i} \tag{6.33}$$

$$U_{\mathrm{ES}}^{p,i} \cdot S_{\mathrm{ES}}^{\min} \leq S_{\mathrm{ES}}^{p,i} \leq U_{\mathrm{ES}}^{p,i} \cdot S_{\mathrm{ES}}^{\max} \tag{6.34}$$

$$\sum_{i \in N_{\mathrm{B}}} S_{\mathrm{ES}}^{p,i} \leq S_{\mathrm{ES}}^{\mathrm{ALL}} \tag{6.35}$$

$$S_{\mathrm{ES,Eq}}^{p,i} = \sum_{m \in N_{\mathrm{M}}} (S_{\mathrm{ES}}^{m,i} - S_{\mathrm{ES}}^{m-1,i})(1 - \gamma_{\mathrm{LL}})^{p-m}, \quad p \geq m \tag{6.36}$$

$$0 \leq P_{\mathrm{ESC}}^{\Lambda} \leq \gamma_{\mathrm{ES}}^{\max} U_{\mathrm{ESC}}^{\Lambda} S_{\mathrm{ES,Eq}}^{p,i} \tag{6.37}$$

$$0 \leq P_{\mathrm{ESD}}^{\Lambda} \leq \gamma_{\mathrm{ES}}^{\max} U_{\mathrm{ESD}}^{\Lambda} S_{\mathrm{ES,Eq}}^{p,i} \tag{6.38}$$

$$U_{\mathrm{ESC}}^{\Lambda} + U_{\mathrm{ESD}}^{\Lambda} \leq 1 \tag{6.39}$$

$$\sum_{i \notin N_{\mathrm{B}}} U_{\mathrm{ES}}^{p,i} = 0 \tag{6.40}$$

$$\gamma_{\mathrm{SOC}}^{\min} S_{\mathrm{ES,Eq}}^{p,i} \leq E_{\mathrm{ES}}^{\Lambda} \leq \gamma_{\mathrm{SOC}}^{\max} S_{\mathrm{ES,Eq}}^{p,i} \tag{6.41}$$

$$E_{\mathrm{ES}}^{p,d,0,i} = E_{\mathrm{ES}}^{p,d,N_{\mathrm{T}},i} \tag{6.42}$$

$$\gamma_{\mathrm{DG}}^{\min} S_{\mathrm{DG}}^{p,i} \leq P_{\mathrm{DG}}^{\Lambda} \leq \gamma_{\mathrm{DG}}^{\max} S_{\mathrm{DG}}^{p,i} \tag{6.43}$$

$$P_{\mathrm{DG}}^{\Lambda} \leq P_{\mathrm{DG,REAL}}^{\Lambda} \tag{6.44}$$

$$R_{\mathrm{DG,DN}}^{\Lambda} \Delta t \leq P_{\mathrm{DG}}^{\Lambda} - P_{\mathrm{DG}}^{p,d,t-1,i} \leq R_{\mathrm{DG,UP}}^{\Lambda} \Delta t \tag{6.45}$$

$$\begin{aligned} P_{\mathrm{PF}}^{p,d,t,b+1} &= P_{\mathrm{PF}}^{p,d,t,b} - P_{\mathrm{PF}}^{p,d,t,0,b+1} - P_{\mathrm{DL}}^{\Lambda} + P_{\mathrm{MT}}^{\Lambda} + P_{\mathrm{BD}}^{\Lambda} \\ &\quad - P_{\mathrm{BC}}^{\Lambda} + P_{\mathrm{WT}}^{\Lambda} + P_{\mathrm{PV}}^{\Lambda} - P_{\mathrm{EB}}^{\Lambda} - P_{\mathrm{EC}}^{\Lambda} \end{aligned} \tag{6.46}$$

$$Q_{\mathrm{PF}}^{p,d,t,b+1} = Q_{\mathrm{PF}}^{p,d,t,b} - Q_{\mathrm{PF}}^{p,d,t,0,b+1} - Q_{\mathrm{DL}}^{p,d,t,i} \tag{6.47}$$

$$V_{\mathrm{BUS}}^{p,d,t,i+1} = V_{\mathrm{BUS}}^{p,d,t,b} - (R^{b} P_{\mathrm{PF}}^{p,d,t,b} + X^{b} Q_{\mathrm{PF}}^{p,d,t,b}) / V_{0} \tag{6.48}$$

$$1 - \Delta V_{\mathrm{BUS}}^{\max} \leq V_{\mathrm{BUS}}^{p,d,t,i} \leq 1 + \Delta V_{\mathrm{BUS}}^{\max} \tag{6.49}$$

$$H_{\mathrm{TL}}^{\Lambda} + H_{\mathrm{WR}}^{\Lambda} = H_{\mathrm{MH}}^{\Lambda} + H_{\mathrm{EB}}^{\Lambda} + H_{\mathrm{MC}}^{\Lambda} + H_{\mathrm{EC}}^{\Lambda} + H_{\mathrm{TD}}^{\Lambda} - H_{\mathrm{TC}}^{\Lambda} \tag{6.50}$$

Constraint (6.17) is the objective of the risk-averse HES deployment model [32]. Constraint (6.18) denotes the mathematical formulation of the EDP in which z and x are the set of investment- and operation-stage decisions; Constraint (6.19) means the system risk cost in terms of CVaR. $F(z)$ in Constraint (6.20) and $G(z)$ in Constraint (6.21) denote the cost from the investment- and operation-stage, respectively; Constraint (6.22) is the salvage value of the HES system; Constraint (6.23) indicates the investment cost of the HES system; Constraint (6.24) is the revenue from selling power and thermal energy to residents; Constraint (6.25) is the subsidy from the utilisation of RES; Constraints (6.26)–(6.29) are the CCHP fuel cost, power transaction cost, maintenance cost, as well as emission cost from CCHP plants and the main grid; Constraints (6.30)–(6.31) are the start-up and shutdown cost for dispatchable generators; Constraint (6.33) means that the size of the HES system in the present year should be no less than that of the last year. It is assumed that HES can be available in all variable sizes [19]; Constraint (6.34) denotes HES size limits for each bus or thermal group. Constraint (6.35) describes the HES capacity limits for residential MEMG; Constraint (6.36) is the effective capacity of the energy storage considering the leakage loss. The capacity of the energy storage is decreasing gradually after the investment [33]; Constraints (6.37)–(6.42) denote all the HES operating constraints (subscript ES denotes the generalised abbreviation for BS and TST); Constraints (6.37) and (6.38) mean that the charging (absorbing) or discharging (releasing) power of the HES system are within the capacity limits; Constraint (6.39) shows that HES cannot charge (absorb) and discharge (release) at the same time; Constraint (6.40) denotes that the HES system can only be installed at candidate buses or thermal groups; Constraint (6.42) denotes that starting energy is equal to its ending energy of the HES system for the same scheduling flexibility [19]; Constraint (6.43) denotes that the power output of each generator should be within the limits (subscript DG is the generalised abbreviation for generators), while Constraint (6.44) shows that the RES power output should be less than the available sources; Constraint (6.45) means that the ramp up and down rate limits for dispatchable generators; Constraints (6.46)–(6.48) are the linearised *Dist-Flow* [2,19]; Constraint (6.49) shows that the bus voltage should be maintained in a secure range; Constraint (6.50) denotes that the thermal generation should be equal to its consumption [17,27].

Note that the battery degradation cost [18,34] is ignored in this long-term deployment work as we aim to find the rough planning decisions for the HES system [12,20,22].

Further, there are a number of non-linear items such as Constraints (6.5), (6.19), (6.30), (6.31), (6.37), and (6.38). To reduce the computation burden and improve solution quality, those non-linear items are linearised in Appendix A. After the linearisation, the proposed HES deployment model becomes an MILP problem.

6.4 Solution method

6.4.1 Typical seasonal day selection approach

Given 365 days in a year, the HES deployment model is in excessively high dimension and computationally demanding. To prune the model's size, a set of typical seasonal days is selected to represent the yearly horizon. Regarding this, the EDP can be transformed as

$$C_{\mathrm{EDP}} = \frac{1}{365 \times N_{\mathrm{P}}} \cdot \frac{dr(1+dr)^{N_{\mathrm{P}}}}{(1+dr)^{N_{\mathrm{P}}}-1} \cdot [\underbrace{-F(z)}_{\text{Investment stage}} + \underbrace{\sum_{c \in N_{\mathrm{C}}}}_{\text{Operation stage}} N_{\mathrm{D}}^{c} \cdot G(x|c)] \tag{6.51}$$

where N_{D}^{c} is the number of days in season c; N_{C} is the number seasons; and $G(x|c)$ is the objective of operation-stage in season c.

6.4.2 Two-substage stochastic operation model

To handle the heterogeneous uncertainties, a two-substage adaptive SO method is proposed as Figure 6.5.

In Figure 6.5, the operation stage can be decomposed into the day-ahead and intraday dispatch substages. Given that the HES system and PBDR are dynamic and sensitive to the holistic price information, their 24-h dispatch decisions are made in the day-ahead dispatch substage. At this substage, diverse uncertainties are not revealed in real time. In the intraday online dispatch substage, where the diverse uncertainties are realised, all the dispatchable generators that have faster response speeds are coordinated to keep the real-time balance of multi-energy generation and consumption.

6.4.2.1 Uncertainty modelling

For energy outputs of WTs and PVs, their stochastic variations typically follow the Beta distribution as

$$f_{x_s} = (x_s)^{\theta-1} \cdot (1-x_s)^{\psi-1} \cdot N_{\mathrm{F}} \tag{6.52}$$

As for power demands, outdoor temperature, and residential hot water needs, the normal distribution can be used to represent their forecasting errors [19]:

$$f_{y_s} = 1/(\sqrt{2\pi}\sigma) \cdot \exp[-(y_s-\mu)^2/(2\sigma^2)] \tag{6.53}$$

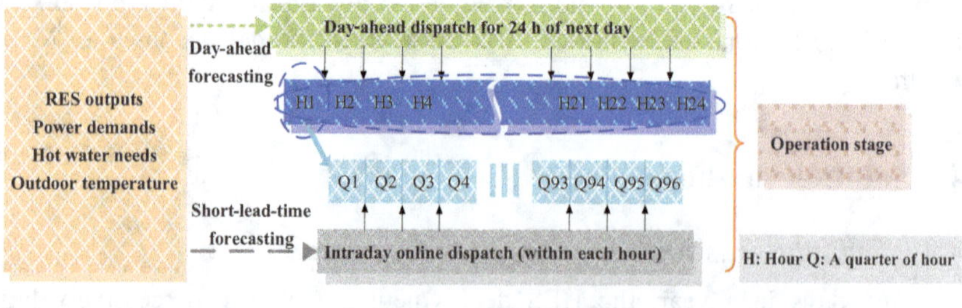

Figure 6.5 Two-substage coordinated SO framework

6.4.2.2 Scenario generation and reduction

Based on probability distributions of all the uncertainty sources in Constraints (6.52) and (6.53), the Latin hypercube sampling method is utilised to produce plenty of near-random samples of parameters for the SO model [27]. Nonetheless, the large number of scenarios would lead to a high computational burden. Thus, the simultaneous backward reduction method is applied to choose a much smaller but still representative set for practical applications. The scenario generation and reduction method can be referred to [13,19,27].

After the scenario generation and reduction, the SO model can be formulated as an MILP problem as Constraints (6.54) and (6.55):

$$\underset{x}{\text{Min}}\, G(x) = \underset{w, y_1, y_2, \ldots, y_q}{\text{Min}} \left[S(w) + \sum_{q \in N_Q} c_q L(y_q)\right] \tag{6.54}$$

$$\begin{aligned} \text{s.t.} \quad & w \in CS_w | z \\ & y_q \in CL(w, \omega_q), \ \forall q \end{aligned} \tag{6.55}$$

where q is the index of forecasting scenarios; N_Q is the total number of scenarios; c_q is the scenario probability; $S(w)$ is objective of investment stage and day-ahead dispatch substage; w is decision variables for size, location, investment phase, and corresponding dispatch of HES system and PBDR scheme; CS_w is constraints associated with w; $\sum_{q \in N_Q} c_q L(y_q)$ is the expectation of cost from the dispatchable generators and power transaction between the residential MEMG and utility grid; y_q denotes the corresponding decision variables for power flow, dispatchable generators, and temperature for the rooms and hot water; ω_q is stochastic variation of system uncertain sources; $CL(w, \omega_q)$ is the constraints associated with y_q.

6.5 Simulation results

6.5.1 *Set-up of the system*

The test grid-connected residential MEMG is built with reference to the real conditions in Tianjin (38°N, 117°E), China, as shown in Figure 6.6. Its lifespan is 9 years with 3 years as a unit investment phase. The multiple energy is served mainly for neighbouring aggregated residential buildings, which are all in three regions. The heat networks among different thermal groups are not considered as the long-distance thermal transmission will result in the substantial thermal loss [35]. The predefined configuration of all the generators is shown in Table 6.1, while the substation capacity is set as 1.5 times of the peak power in residential MEMG [2] to guarantee that there is enough power supply when faults occur. The technical details for the HES system are listed in Table 6.2. The studied residential MEMG in Figure 6.6 is built based on the IEEE 33-bus standard test system in [36].

Three typical days from transition seasons (spring and autumn), summer, and winter are employed to represent the whole year. Based on the data of the year 2018 (Jan.–Dec.), Figure 6.7 demonstrates the forecasting value of all uncertainties in the year 2019.

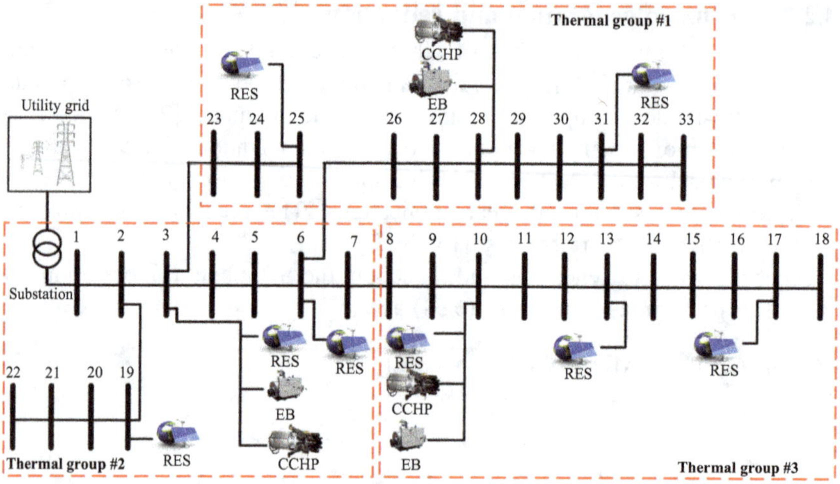

Figure 6.6 The IEEE 33-bus distribution network-based residential MEMG

Table 6.1 Configuration of distributed generators (kW)

Node	WT	PV	CCHP	EB	EC
3	150	40	1,200	650	500
6	187.5	80	0	0	0
10	187.5	64	1,000	500	450
13	112.5	64	0	0	0
17	187.5	96	0	0	0
19	150	128	0	0	0
25	187.5	144	0	0	0
28	0	0	800	500	350
31	187.5	80	0	0	0

The system energy tariffs are shown in Table 6.3, which are extracted from [17], [19,20,37]. For the thermal and power energy supplied to residents, the fixed pricing scheme is used [2]. According to the electricity price and usage data in [38], the price elasticity of electricity demands is estimated as −0.75.

In the residential MEMG, there are 16 buildings oriented south with 12 floors in each building; On each floor, there are two identical apartments, each with a floor area of 150 (15 × 10) m^2. The area of north/south and east/west facing walls are 42 (15 × 2.8) m^2 and 28 (10 × 2.8) m^2, respectively. The south and north walls are equipped with double-glazed windows to the outside environment with areas of 10.5 (3.75 × 0.7) m^2 and 8.4 (3 × 0.56) m^2, respectively. All the windows are of casement types without any blinds or shading devices. The area of the door on the east side is 2 (2 × 1) m^2. All the walls and flat roofs of the room are comprised of the same structural insulated panels [31].

Table 6.2 Technical details for the HES system [19,20]

Energy storage	$\chi_{ES,INV}$	$\chi_{ES,M}$	$S_{ES,MIN}/S_{ES,MAX}$
Battery storage	100 ($/kWh)	0.03 ($/kWh)	0/1,500 (kWh)
Heat storage tank	36.7 ($/kWh)	0.025 ($/kWh)	0/1,800 (kWh)
Cooling storage tank	36.7 ($/kWh)	0.025 ($/kWh)	0/1,800 (kWh)

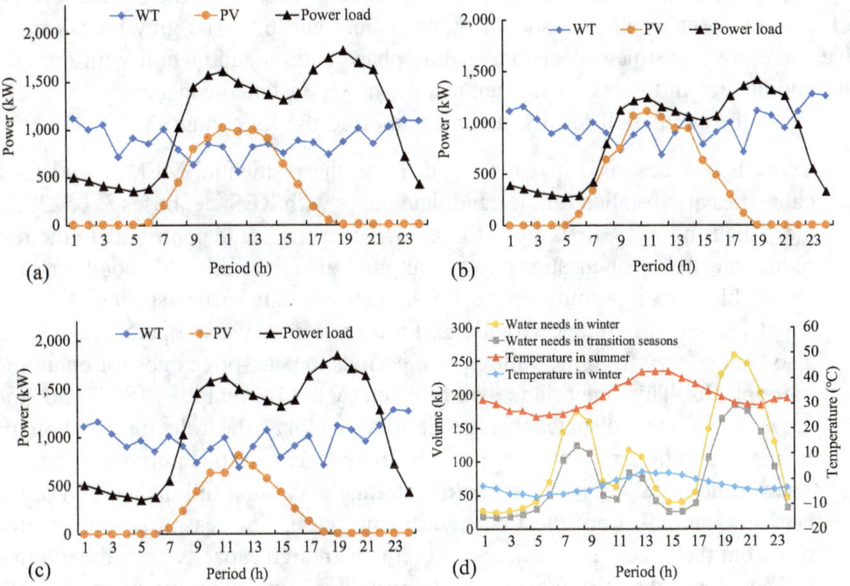

Figure 6.7 Forecast of the RES output, outdoor temperature, and multi-energy demands. (a) Transition seasons. (b) Summer. (c) Winter. (d) Temperature and water needs

Table 6.3 Energy tariffs and emission prices [17,19,27]

Type	Price ($/kW)	Type	Price ($/kWh)	Period (h)
Local power	0.18	Power transaction	0.0468	0.00–6.00, 23.00–24.00
Local thermal	0.022		0.1029	6.00–8.00, 11.00–17.00
Natural gas	0.039		0.1785	8.00–11.00, 17.00–22.00
Emission price	CCHP	0.0057 ($/kW)	Utility grid	0.028 ($/kW)

The discount rate is 6% and load growth rates for power and thermal demands are set as 6.7% and 5%, respectively [5,27].

Note that in reality, any other real-world data can also be used in the proposed method without violating its effectiveness. The mathematical model is implemented in General Algebraic Modelling System (GAMS) and solved by the CPLEX solver.

6.5.2 Investment-stage simulation results and discussion

In the simulation, 2,000 scenarios are generated first, and then they are reduced to 5 representative ones with the corresponding probabilities. After the simulation, the HES deployment results are shown in Tables 6.4 and 6.5. The grey blocks mean that there are investment decisions at this phase and its subtraction with the corresponding size in the last phase denotes the newly installed capacity.

From the simulation results, it can be inferred the following:

1. There is a widespread investment of BS in the residential MEMG, and most capacities are installed on the candidate buses with RES, i.e. buses 3, 6, 22, 25, etc. This is because, first, the sufficient BS capacity could contribute to the risk reduction from stochastic power outputs of renewables. Second, as the renewables are intermittent, the BS investment can ensure sufficient power supply and avoid voltage violations. Further, in the system operation, BS can also shift energy from the valley-price periods to peak-price ones for obtaining more profits. This function is mainly taken by the BS on buses 18, 27, 30, and 33; however, the dispatchable generators can play the role for the voltage regulation, and there is not so much BS investment for this purpose.
2. There is no installation of the cooling storage tank but a full installation of the heat storage tank from the first investment phase. The reason is to avoid the risk from the uncertainty sources and remain enough capacity after the internal leakage loss; the system chooses to install enough capacity of the energy

Table 6.4 Deployment results for BS (kWh)

Year/Bus	3	6	18	22	25	27	30	33
1–3	1,500	0	0	1,500	1,500	0	0	0
4–6	1,500	466.0	101.7	1,500	1,500	0	0	473.2
7–9	1,500	466.0	101.7	1,500	1,500	378.0	81.19	473.2

Table 6.5 Deployment results for TS (kWh)

Year 1–9	Group 1	Group 2	Group 3
Cooling storage tank	0	0	0
Heat storage tank	1,800	1,800	1,800

storage, i.e. BS and heat storage. However, the cooling storage tank works only in the summer while the heat storage tank contributes to supplying heat both in transition seasons as well as winter days. As a result, given the larger amount of heat demands, more utilisation rate but the same unit installation cost with the cooling storage tank, the invested capacity of the heat storage tank is much larger and more widespread.

3. The investment phase selection plays a vital role in the HES deployment scheme. For instance, on buses 6, 18, 27, and 33, the BS is installed at different phases. This contributes to avoiding any capacity redundancy at the early stage of the project, and thus enables a more economical and flexible HES deployment.

6.5.3 Operation-stage simulation results and discussions

To show the effectiveness of the proposed method in the operation stage, the dispatch results in the year 2026 are demonstrated in Figure 6.8:

From Figures 6.6–6.8, all the units are coordinated to supply multi-energy demands to maximise the EDP and reduce the risk at the same time: when the transaction prices are low, the thermal energy is converted by the electricity via EBs or ECs instead of the CCHP plants, to save the fuel cost; when the prices are high, the system thermal energy is supplied by CCHP plants, and the power surplus is sold to the main grid for gaining profits. As for the HES system, the BS charges at the valley-price periods and discharges at peak-price periods to gain more profits from main grid. For the TS, it makes the dispatch of the CCHP plants more flexible by absorbing its surplus thermal energy when there is more power generation than required and releasing thermal energy when not so much power is needed from CCHP plants. In addition, the DSM scheme shifts the multi-energy loads to other periods for increasing the system profits. Its function for the cost saving is verified in the next section.

The final result for the mean-risk model is \$417.9, and the EDP is positive as \$712.1 per day. It means that this project should be implemented as it gains profit.

For the solution performance, there are 1,128,586 continuous and 138,915 discrete variables, and the solution time is 11.63 h, which is efficient enough for a 9-year project.

6.5.4 Case comparison

To fully show the advantage of the proposed method, three conventional deployment methods for the energy storage are compared:

Method #1: Separate deployment method, i.e. the energy storage system is installed in the respective single-energy MGs. The gas boilers (GBs) are utilised instead of CCHP plants, ECs, and EBs for supplying the thermal demands, while the WTs, PVs, and exchanging power from the main grid serve the total power demands [17].

The corresponding simulation results show that there is no TS installation and the deployment results for the BS are shown in Table 6.6.

From the simulation results, under the mean-risk model, the BS is largely installed in the early phases to gain more profit and avoid risk from heterogeneous uncertainties. Further, the units for electricity generation in the MG are only RES-based ones, whose outputs are intermittent and fluctuate; thus, the BS is mainly installed on the buses with RES to maintain the bus voltage in a secure range. Meanwhile, the BS could shift energy to gain profits. For the TS investment, as the GBs can burn fuels to produce the thermal energy flexibly, it is unnecessary to install TS given that it will bring extra installation and operating cost.

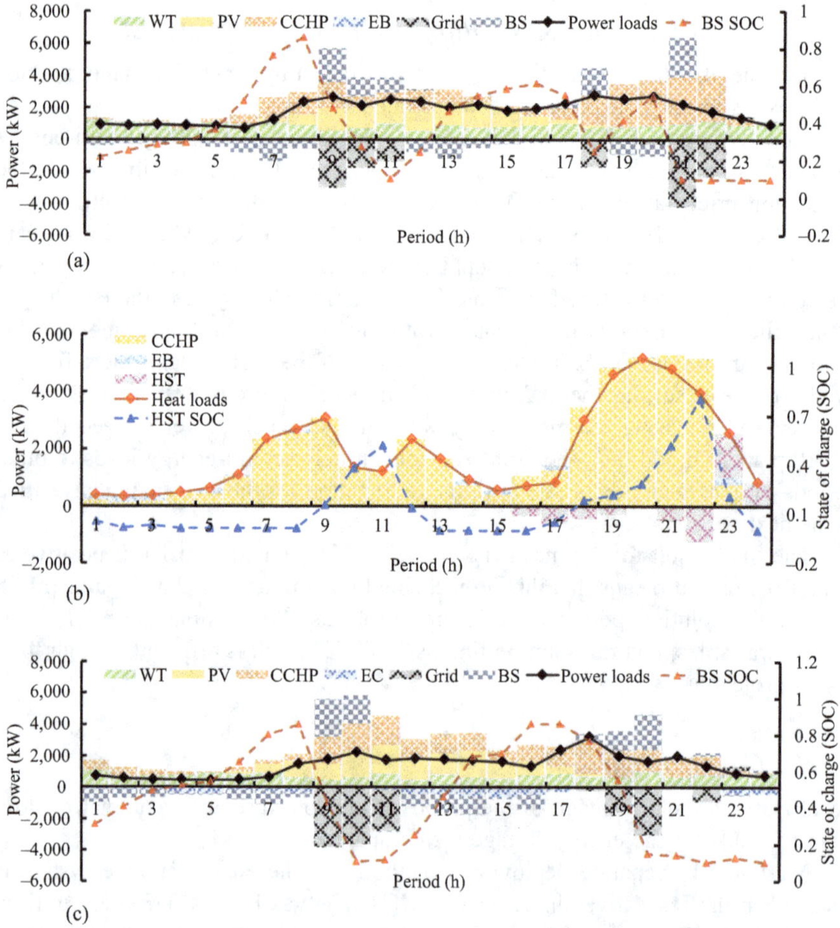

Figure 6.8 Energy balance condition in the year 2026. (a) Power balance in transition seasons. (b) Heat balance in transition seasons. (c) Power balance condition in summer. (d) Cooling balance condition in summer. (e) Power balance in winter. (f) Heat balance in winter

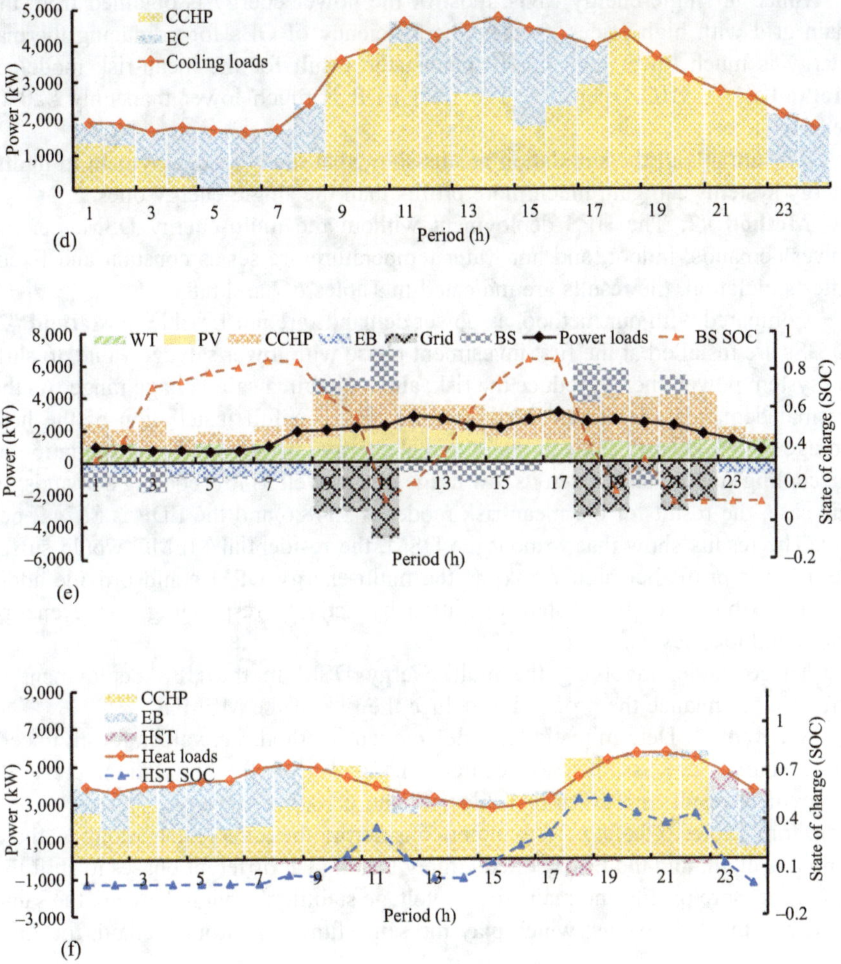

Figure 6.8 (Continued)

*Table 6.6 Deployment results for battery of **Method #1** (kW)*

Year/Bus	3	6	18	22	25	33
1–3	1,500	535.4	0	1,500	1,500	0
4–9	1,500	535.4	124.3	1,500	1,500	964.6

Since in single-energy MGs, most of the power energy is obtained from the main grid with high prices, further, the efficiency of GBs for producing thermal energy is much lower than CCHP plants; the result for the mean-risk model of **Method #1** is $102.4 per day. The final EDP is much lower than only $203.7 per day.

To sum up, all the simulation results show that the HES deployment in multi-energy systems can gain much more profits than the single-energy ones.

Method #2: The HES deployment without the multi-energy DSM, i.e. the power demands, indoor, and hot water temperature are set as constant and fixed. After simulation, the results are indicated in Tables 6.7 and 6.8.

Compared with our method, as power demands are not flexible in **Method #2**, all BSs are installed at the first investment phase with lower salvage value, to shift the system power energy, reduce the risk, and maintain a safe voltage range. As the thermal demands are not as flexible, there is still a full installation of the heat storage tank to flexbilise the overall thermal energy operation and no installation of the cooling storage tank given its low usage rate as well as low cooling demands. In this case, the result for the mean-risk model is $348.6 and the EDP is $674.2 per day. The results show that without the DSM, the residential MEMG would suffer from lower profit but higher risk, as the multi-energy DSM could provide additional flexibility for the system operation by actively responding to the energy prices and load levels.

To conclude, involving the multi-energy DSM in the HES deployment is effective to enhance the profits and reduce the risk of the MEMG.

Method #3: Deterministic HES deployment method, i.e. values for all uncertainty sources are assumed as accurate enough. After the simulation, the HES deployment results are given in Tables 6.9 and 6.10.

From Tables 6.9 and 6.10, without considering the diverse uncertainties, there is no risk at all, and the BS is stilled largely installed at different phases to shift the power for more profits and maintain the voltage stability; though there are the same TS deployment decisions, which play the same function as our method, the final

*Table 6.7 Deployment results for BS of **Method #2** (kWh)*

Year/Bus	3	6	22	25	27	33
7–9	1,500	562.7	1,500	1,500	428.4	508.9

*Table 6.8 Results for TS of **Method #2** (kWh)*

Year 1–9	Group 1	Group 2	Group 3
Cooling storage tank	0	0	0
Heat storage tank	1,800	1,800	1,800

*Table 6.9 Deployment results for BS of **Method #3** (kWh)*

Year/Bus	3	6	18	22	25	27	33
1–3	1,500	0	0	1,500	678.4	0	0
4–6	1,500	428.6	0	1,500	1,137.2	0	487.5
7–9	1,500	578.3	179.3	1,500	1,500	254.8	487.5

*Table 6.10 Results for TS of **Method #3** (kWh)*

Year 1–9	Group 1	Group 2	Group 3
Cooling storage tank	0	0	0
Heat storage tank	1,800	1,800	1,800

EDP for **Method #3** is $729.7 per day, which is higher than our method. The reason is that considering the heterogeneous uncertainties will result in the risk. Hence, the corresponding EDP is lower. The cost difference between the proposed method and the deterministic approach is called *Value of Perfect Information*, which is used to immune against the system uncertainties as well as the potential risk.

Nonetheless, to check the performance of **Method #3** when uncertainties are realised, a set of 1,000 random scenarios of typical days every year are used. The average EDP of **Method #3** is $682.6 per day, while our method is $701.3. It shows that when the residential MEMG is exposed to uncertainty sources, the proposed adaptive SO method has a higher potential in gaining more profits than the deterministic one as it already immunises against the diverse uncertainties through the risk-averse SO process.

6.6 Conclusion and future work

This chapter introduces a risk-averse multi-stage SO HES deployment approach in an residential MEMG considering multi-energy DSM and diverse uncertainties. The deployment framework is decomposed into three investment phases with the 'mean-risk' as the objective. By using the scenario-based multi-stage adaptive SO method, the uncertainties are effectively tackled. The case study shows that the risk-averse HES deployment can help gain profits while avoiding the risk. Thus, it can be implemented. Moreover, comparisons of the proposed method with three conventional deployment methods are done. The results mean that the system expected profit is around 3 times the separate energy storage deployment method. This indicates that deploying the HES in the multi-energy scheme is necessary as the CCHP plants are highly efficient. Compared with the HES deployment approach without the multi-energy DSM, there is a rate of increment 5.62% in the system profit. This demonstrates the effectiveness of the DSM in the HES deployment; further, compared with the deterministic optimisation method, though there is a 2.47% rate of decrement in the expected profit, nonetheless, when the system is exposed to the real-life uncertainties, there is a rate of increment of 2.74% in the profit. This shows that our method is effective in dealing with uncertainties and avoiding the risk.

For the practical application, with the increasing penetration level of renewables and widespread installation of smart sensors, this work could provide references for deploying the HES in the smart homes, green buildings, etc.

For the future work, for one thing, to conduct more comprehensive operation research for energy storage, the degradation model [18,34] can be incorporated. For another, as the solution dimension for the multi-energy planning and operation scheme is usually high, more advanced algorithms can be explored to improve the solution efficiency.

Appendix A

A.1 Model linearisation

In this chapter, all the non-linear items can be linearised for higher efficiency:

1. *Linearisation of the power demand under the PBDR:* The stepwise price-elastic demand curve uses different price levels to represent the price elasticity of power demands; with these levels, the power demands can be calculated/linearised as (assuming power factors are fixed) follows [2]:

$$P^t_{\mathrm{DL}} = P^t_{\mathrm{DL,Pr}} \cdot \sum_{j=1}^{N_J} \alpha^{t,j} \cdot L^{t,j}_{\mathrm{PL}} \tag{A.1}$$

$$Q^t_{\mathrm{DL}} = Q^t_{\mathrm{DL,Pr}} \cdot \sum_{j=1}^{N_J} \alpha^{t,j} \cdot L^{t,j}_{\mathrm{QL}} \tag{A.2}$$

2. *Linearisation of the max function*, i.e. Constraints (6.19, 6.30, 6.31): separating the max function into two individual parts will work, and we take $A \geq \max(B, C)$ as an example as

$$A \geq B; A \geq \mathrm{C} \tag{A.3}$$

3. *Linearisation of the bilinear item*, i.e. Constraints (6.35, 6.36): Take $z = Ax$ as an example in which $\bar{A}$ is a continuous variable and x is a binary variable, and A is bounded below by zero and above by $\bar{A}$ (or any big M), then it can be linearised as

$$z \leq \bar{A} \cdot x; \quad z \leq A; \quad z \geq A - (1 - x) \cdot \bar{A}; \quad z \geq 0 \tag{A.4}$$

References

[1] P. Poonpun and W.T. Jewell, 'Analysis of the cost per kilowatt-hour to store electricity', *IEEE Trans. Energy. Convers.*, vol. 23, no. 2, pp. 529–534, 2008.

[2] C. Zhang, Y. Xu, Z. Li, and Z.Y. Dong, 'Robustly coordinated operation of a multi-energy microgrid with flexible electric and thermal loads', *IEEE Trans. Smart. Grid.*, vol. 10, no. 3, pp. 2765–2775, 2018.

[3] S. Chen, H.B. Gooi, and M. Wang, 'Sizing of energy storage for microgrids', *IEEE Trans. Smart. Grid.*, vol. 3, no. 1, pp. 142–151, 2012.

[4] C. Thrampoulidis, S. Bose, and B. Hassibi, 'Optimal placement of distributed energy storage in power networks', *IEEE Trans. Automat. Contr.*, vol. 61, no. 2, pp. 416–429, 2016.

[5] Y. Yang, H. Li, A. Aichhorn, J. Zheng, and M. Greenleaf, 'Sizing strategy of a distributed battery storage system with high penetration of photovoltaic for voltage regulation and peak load shaving', *IEEE Trans. Smart. Grid.*, vol. 5, no. 2, pp. 982–991, 2014.

[6] G. Carpinelli, G. Celli, S. Mocci, F. Mottola, F. Pilo, and D. Proto, 'Optimal integration of distributed energy storage devices in smart grids', *IEEE Trans. Smart. Grid.*, vol. 4, no. 2, pp. 985–995, 2013.

[7] P. Xiong and C. Singh, 'Optimal planning of storage in power systems integrated with wind power generation', *IEEE Trans. Sustain. Energ.*, vol. 7, no. 1, pp. 232–240, 2016.

[8] A. S. A. Awad, T. H. M. EL-Fouly, and M. M. A. Salama, 'Optimal ESS allocation for a load management application', *IEEE Trans. Power. Syst.*, vol. 30, no. 1, pp. 327–336, 2015.

[9] I. Miranda, N. Silva, and H. Leite, 'A holistic approach to the integration of battery energy storage systems in island electric grids with high wind penetration', *IEEE Trans. Sustain. Energ.*, vol. 7, no. 2, pp. 775–785, 2016.

[10] J. Zhang, K. Li, M. Wang, *et al.*, 'A bi-level program for the planning of an islanded microgrid including CAES', *IEEE Trans. Ind. Appl.*, vol. 52, no. 4, pp. 2768–2777, 2016.

[11] C. Abbey and G. Joós, 'A stochastic optimization approach to the rating of energy storage systems in wind-diesel isolated grids', *IEEE Trans Power Syst*, vol. 24, no. 1, pp. 418–426, 2009.

[12] H. Khorramdel, J. Aghaei, B. Khorramdel, and P. Siano, 'Optimal battery sizing in microgrids using probabilistic unit commitment', *IEEE Trans Ind. Informat.*, vol. 12, no. 2, pp. 834–843, 2016.

[13] S. Nojavan, M. Majidi, and N.N. Esfetanaj, 'An efficient cost-reliability optimization model for optimal siting and sizing of energy storage system in a microgrid in the presence of responsible load management', *Energy*, vol. 139, no. 15, pp. 89–97, 2017.

[14] E. Hajipour, M. Bozorg, and M. Fotuhi-Firuzabad, 'Stochastic capacity expansion planning of remote microgrids with wind farms and energy storage', *IEEE Trans. Sustain. Energ.*, vol. 6, no. 2, pp. 491–498, 2015.

[15] P. Yang and A. Nehorai, 'Joint optimization of hybrid energy storage and generation capacity with renewable energy', *IEEE Trans. Smart. Grid.*, vol. 5, no. 4, pp. 1566–1574, 2014.
[16] M. Roustai, M. Rayati, A. Sheikhi, and A. Ranjbar, 'A scenario-based optimization of smart energy hub operation in a stochastic environment using conditional-value-at-risk', *Sustain. Cities. Soc.*, vol. 39, pp. 309–316, 2018.
[17] Z. Li and Y. Xu, 'Optimal coordinated energy dispatch of a multi-energy microgrid in grid-connected and islanded modes', *Appl. Energy*, vol. 210, pp. 974–986, 2017.
[18] S. Sharma, Y. Xu, A. Verma, and B.K. Panigrahi, 'Time-coordinated multi-energy management of smart buildings under uncertainties', *IEEE Trans. Ind. Informat.*, vol.15, no.8, 2019.
[19] Z. Li and Y. Xu, *et al.*, 'Temporally-coordinated optimal operation of a multi-energy microgrid under diverse uncertainties', *Appl. Energy*, vol. 240, pp. 719–729, 2019.
[20] L. Guo, W. Liu, J. Cai, B. Hong, and C. Wang, 'A two-stage optimal planning and design method for combined cooling, heat and power microgrid system', *Energy Convers. Manag*, vol. 74, pp. 433–445, 2013.
[21] T. Ma, J. Wu, L. Hao, *et al.*, 'The optimal structure planning and energy management strategies of smart multi-energy systems', *Energy*, vol. 160, pp. 122–141, 2018.
[22] P. Gabrielli, M. Gazzani, E. Martelli, and M. Mazzotti, 'Optimal design of multi-energy systems with seasonal storage', *Appl. Energy*, vol. 219, pp. 408–424, 2018.
[23] W. Gu, J. Wang, S. Lu, Z. Luo, and C. Wu, 'Optimal operation for integrated energy system considering thermal inertia of district heating network and buildings,' *Appl. Energy*, vol. 199, pp. 234–246, 2017.
[24] C. Li, Z.Y. Dong, G. Chen, F. Luo, and J. Liu, 'Flexible transmission expansion planning associated with large-scale wind farms integration considering demand response', *IET Gener. Transm. Dis.*, vol. 9, no. 15, pp. 2276–2283, 2015.
[25] C. Zhang, Y. Xu, Z.Y. Dong, and K.P. Wong, 'Robust coordination of distributed generation and price-based demand response in microgrids,' *IEEE Trans. Smart. Grid.*, vol. 9, no. 5, pp. 4236–4247, 2018.
[26] B. Zeng, J. Zhang, X. Yang, J. Wang, J. Dong, and Y. Zhang, 'Integrated planning for a transition to the low-carbon distribution system with renewable energy generation and demand response', *IEEE Trans Power Syst*, vol. 29, no. 3, pp. 1153–1165, 2014.
[27] Z. Li, Y. Xu, S. Fang, and S. Mazzoni, 'Optimal placement of heterogeneous distributed generators in a grid-connected multi-energy microgrid under uncertainties', *IET Renew. Power. Gener.*, vol.13, no.14, 2019.
[28] P.R. Thimmapuram, J. Kim, A. Botterud, and Y. Nam, 'Modeling and simulation of price elasticity of demand using an agent-based model', *Proc. IEEE ISGT*, pp. 1–8, Jan. 2010.

[29] M. Tasdighi, H. Ghasemi, and A. Rahimi-Kian, 'Residential microgrid scheduling based on smart meters data and temperature-dependent thermal load modeling', *IEEE Trans. Smart. Grid.*, vol. 5, no. 1, pp. 349–357, 2014.
[30] W. Gu, S. Lu, Z. Wu, *et al.*, 'Residential CCHP microgrid with load aggregator: operation mode, pricing strategy, and optimal dispatch', *Appl. Energy*, vol. 205, pp. 173–186, 2017.
[31] S. Yao, W. Gu, S. Zhou, S. Lu, C. Wu, and G. Pan, 'Hybrid timescale dispatch hierarchy for combined heat and power system considering the thermal inertia of heat sector', *IEEE Access*, vol. 6, pp. 63033–63044, 2018.
[32] R. Schultz and S. Tiedemann, 'Conditional value-at-risk in stochastic programs with mixed-integer recourse', *Math. Program.*, vol. 105, no.1, pp. 365–386, 2006.
[33] 'Battery capacity', www.sciencedirect.com/topics/engineering/, 2017.
[34] A. Ahmadian, M. Sedghi, B. Mohammadi-ivatloo, A. Elkamel, M. Aliakbar Golkar, and M. Fowler, 'Cost-benefit analysis of V2G implementation in distribution networks considering PEVs battery degradation', *IEEE Trans. Sustain. Energ.*, vol. 9, no. 2, pp. 961–970, 2017.
[35] J. Wang, H. Zhong, and C. Tan, 'Economic benefits of integrating solar-powered heat pumps into a CHP system', *IEEE Trans. on Sustain. Energ.*, vol. 9, no. 4, pp. 1702–1712, 2018.
[36] M.E. Baran and F.F. Wu, 'Network reconfiguration in distribution systems for loss reduction and load balancing', *IEEE Trans. Power. Deliver.*, vol. 4, no. 2, pp. 1401–1407, 1989.
[37] 'Power prices for the Tianjin Grid', http://fzgg.tj.gov.cn/gzcx/syjgcx/gd/201307/t20130705_30044.shtml#, 2019.
[38] A. Faruqui and S. Sergici, 'Household response to dynamic pricing of electricity: a survey of 15 experiments', *J. Regul. Econ.*, vol. 38, no. 2, pp. 193–225, 2010.

Part III

Coordinated Operation of DERs in Microgrids: Energy Management and Voltage Regulation

Chapter 7

Hourly coordination of energy storage and direct load control

This chapter introduces a robust optimisation (RO) approach for optimal operation of microgrids. The uncertain power generation of renewable energy sources (RESs) is addressed by collaborative schedule of energy storage (ES) and direct load control (DLC) through a two-stage complementary framework: an hour-ahead charging/discharging of ES and a quarter-hour-ahead activation of DLC. The objective is to maximise the total operating profit of the microgrid considering operation and maintenance (O&M) costs as well as energy transaction costs. Assuming the renewable power generation randomly varies within a bounded uncertainty set, the operation problem is modelled to a two-stage robust optimisation (TSRO) model and solved by a column-and-constraint generation algorithm. Compared with conventional operation methods, the ES and DLC are coordinated in different timescales, and RES uncertainties are fully addressed during operation decision-making, ensuring robust solutions.

Nomenclature

A Acronyms

C&CG	Column and constraint generation
DLC	Direct load control
DG	Distributed generation
ESS	Energy storage system
O&M	Operation and maintenance
PV	Photovoltaic
RES	Renewable energy source
RO	Robust optimisation
SO	Stochastic optimisation
TSRO	Two-stage robust optimisation
WT	Wind turbine

B Sets and indices

$Br(i)$	Set of all branches that connect to bus i
$Br(i,j)$	Branch between buses i and j
b	Index of branches
HQ	Set of all quarters in the current operation hour
J_{ch}, J_{dis}	Set of all ES charging/discharging levels
$N_{\mathrm{L/WT/PV/ES}}$	Set of buses that have loads/WTs/PVs/ESSs, respectively
T	Set of the operation periods
$U_{\mathrm{WT/PV}}$	Uncertainty set for WT/PV output power

C Parameters

$C_{\mathrm{WT,OM}}, C_{\mathrm{PV,OM}}$	O&M cost of WTs/PVs (\$/MWh)
$C_{\mathrm{sell}}, C_{\mathrm{buy}}$	Price for selling/buying electricity to/from the main grid (\$/MWh)
$C_{\mathrm{L,con}}, C_{\mathrm{L,unc}}$	Price for selling electricity to controllable/uncontrollable load (\$/MWh)
$C_{\mathrm{ES,ch}}, C_{\mathrm{ES,dis}}$	O&M cost of ESSs during charging/discharging operation (\$/MWh)
$E_{0,m}$	Initial energy stored in the ESSs at bus m (MWh)
$E_{r,m}$	Energy capacity of the ESSs at bus m (MWh)
$L_{\mathrm{ch},m,j}, L_{\mathrm{dis},m,j}$	Charging/discharging power rate (% of rated power) of the ESSs at bus m on level j
$K_{\mathrm{L,con}}$	Ratio of controllable load to total load, same as the maximum load cutting rate during DLC
$P^{\mathrm{fc}}_{\mathrm{WT},n,t}, P^{\mathrm{fc}}_{\mathrm{PV},n,t}$	Forecasted output power of the WTs/PVs at bus n during period t (MW)
$P^{\max}_{\mathrm{ch},m}, P^{\max}_{\mathrm{dis},m}$	Rated charging/discharging power of the ESSs at bus m (MW)
$P_{\mathrm{L,con}}, P_{\mathrm{L,unc}}$	Total controllable and uncontrollable loads in microgrid (MW)
$P^{\min/\max}_{\mathrm{WT},n,t}, P^{\min/\max}_{\mathrm{PV},n,t}$	Minimum/maximum forecasted output power of the WTs/PVs at bus n during period t (MW)
$P^{\mathrm{rated}}_{\mathrm{WT},n}, P^{\mathrm{rated}}_{\mathrm{PV},n}$	Rated output power of the WTs/PVs at bus n (MW)
R_b, X_b	Resistance/reactance of branch b
$\eta_{\mathrm{ch}}, \eta_{\mathrm{dis}}$	ESS energy conversion efficiency during charging/discharging (ESS overall energy conversion efficiency, $\eta = \eta_{\mathrm{ch}}/\eta_{\mathrm{dis}}$)
$\mu_{\mathrm{WT},l}, \mu_{\mathrm{WT},u}, \mu_{\mathrm{PV},l}, \mu_{\mathrm{PV},u}$	Lower/upper uncertainty budget of WT/PV output power
π	Time length of a quarter in the current operation hour, i.e. 0.25 h

D Variables

$K_{\mathrm{DLC},q}$	Load cutting rate (% of total load) for DLC during quarter hour q
$P_{\mathrm{WT},n,t}$, $P_{\mathrm{PV},n,t}$	Uncertain output power of the WTs/PVs at bus n during period t (MW)
$P_{b,t}$, $Q_{b,t}$	Active/reactive power through branch b during period t (MW/MVar)
$P_{0,b,t}$, $Q_{0,b,t}$	Active/reactive power through the lateral branch of branch b during period t (MW/MVar)
$P_{\mathrm{def},t}$, $P_{\mathrm{sur},t}$	Power deficiency/surplus of microgrid during period t (MW)
$P_{\mathrm{DLC},i,q}$, $Q_{\mathrm{DLC},i,q}$	Active/reactive load at bus i during quarter hour q under DLC (MW/MVar)
$V_{i,t}$	Voltage at bus i during period t (p.u.)
$\alpha_{\mathrm{ch},m,j}$, $\alpha_{\mathrm{dis},m,j}$	Binary charging/discharging decision of the ESSs at bus m on charging/discharging level j

7.1 Introduction

With large-scale installations of distributed generation (DG), today's distribution networks are now evolving from conventional passive systems to active decentralised systems such as microgrids. In order to reduce greenhouse gas emission and alleviate dependence on fossil fuels, RESs such as wind and solar energy sources are dominantly adopted in the microgrids. However, unlike conventional fossil-fuel-based generation, wind turbines (WTs) and solar photovoltaics (PVs) can only generate uncertain, intermittent, and non-dispatchable power, which causes significant difficulties for microgrid operation [1], e.g. when the power generation of the WTs and PVs is excessively high, it may be curtailed; when the renewable power generation is low, the power from the microgrid itself may not fulfil local loads.

To alleviate these problems, Refs. [1,2] suggest integrating energy storage systems (ESSs) into microgrids to achieve 'time-shifting' of energy, which allows the redundant energy produced by the WTs and the PVs to be stored during low-load periods and released during peak-load periods. With this energy shifting, microgrid operators can make more operating profits. Besides, as discussed in [2,3], ESSs can also provide other benefits such as mitigating renewable power generation intermittency, improving power system reliability, and so forth. As prices of ESSs continue to drop, it is now becoming common to deploy ESSs in a microgrid for a more efficient energy management purpose. However, ESSs also have clear drawbacks which hinder their effects, such as limited power/energy capacities and reduced lifetimes due to frequent charging/discharging operation. In the literature, ESS operation can be optimised based on historical or predicted renewable power generation [4–10]. As one major difficulty caused by RESs is uncertain and stochastic power generation, uncertainty analysis and optimisation methods are applied for the ESS operation. For example, [11] suggests a probabilistic approach where different wind power states are generated with corresponding occurrence probabilities to simulate uncertain wind power conditions. In [12], Monte Carlo simulations are utilised to simulate wind power uncertainty data to improve solution robustness. Reference [13] proposes a finite prediction error concept to involve RES uncertainties, so that the optimisation process can be achieved with predictable renewable power generation profiles.

On the other hand, with increased controllability of loads in smart grids, demand response is another solution to address the uncertainties arisen by RESs [1]. Demand response programmes can be classified, based on how loads are changed, into price-based demand response and incentive-based demand response methods [14]. Among demand response programmes, DLC is an incentive-based demand response method, which directly shuts down remote controllable and non-essential customer appliances such as air conditioners and water heaters, to maintain power balance in a microgrid. Thus, DLC has stronger controllability and can respond faster to mitigate the uncertainties than the other demand response methods.

Considering the complementary characteristics of ESSs and DLC, this chapter proposes a two-stage coordinated operation method for microgrids in presence of uncertain renewable power generation. In the first stage, ESSs are scheduled to charge/discharge on an hourly basis. The optimisation is based on 1-h ahead RES output power predictions. In the second stage, DLC is then implemented in each hour to complement the ESS operation, when the RES output power deviates significantly from the predictions. In this method, the ESSs are operated to deal with slow variations of the renewable power generation and improve economic benefits which the DLC cannot achieve alone. On the other hand, the second-stage DLC aims to balance power generation and loads in the microgrid when the ESS charging/discharging power is deficient, which can overcome the capacity limitation of the ESSs. Besides, in particular, the DLC can be activated in a relatively short time interval, such that it can compensate the relatively long response time of the ESSs to address fast renewable power generation variations.

The proposed coordinated operation method of the ESSs and the DLC is modelled as an optimisation problem with the uncertainties of renewable power generation, and then it is solved by a TSRO method. Compared to conventional stochastic optimisation (SO) methods which can only provide probabilistic guarantees for constraint satisfaction [11,12], RO methods can obtain an optimal solution by considering the worst case of uncertainty realisation within uncertainty sets [15,16]. In the literature, RO methods have been successfully applied for unit commitment [15–17], microgrid planning [18], and ESS planning [19] problems to handle RES uncertainties. However, its application in microgrid operation is relatively limited. In general, RO methods have three major benefits. First, they only require modest data of uncertainty, such as mean predictions and ranges of possible uncertainty realisation. It is a significant advantage, considering that SO methods are unable to provide reliable solutions when probability density functions are partially available or even not available. Second, RO methods are immune against any realisation of uncertainties presented by uncertainty sets. Since RO solutions are obtained under the worst case of uncertainty realisation, operating constraints for any uncertainty realisation are satisfied and the solutions are regarded robust. Third, the uncertainty sets support an efficient way in uncertainty modelling and RO methods are computationally efficient, while in SO methods, scenario sampling is significantly time-consuming. Furthermore, the TSRO method can improve solution robustness through the second-stage compensation operation against uncertainty realisation.

The major technical contents in this chapter to be introduced are summarised as follows.

A new coordinated microgrid operation framework is proposed, which schedules ESSs and activates DLC in different timescales to complement each other under uncertainties of renewable power generation.

A TSRO model is developed, which maximises the total operating profit of the microgrid while satisfying microgrid operating constraints with any uncertainty realisation.

A Lagrange-dual-based column and constraint generation (C&CG) algorithm is developed to solve the proposed TSRO problem, considering a variety of testing scenarios.

7.2 Two-stage coordination of ES operation and DLC

7.2.1 Principle of multi-stage coordination

With different purposes, distributed energy resources (DERs) can response in different timescales to support flexible power generation or consumption on different levels.

For example, the price-based responsive loads require more time to respond for a daily peak shaving purpose; that is, customers need to know optimised electricity prices and schedule their loads with a long leading time. On the other hand, microturbines and directly controlled loads can be dispatched to achieve adjustable power generation and consumption at a short time interval. For ESSs, they are expected to be scheduled to provide an energy time-shifting function; otherwise, they can also charge and discharge frequently to compensate renewable power generation fluctuations at a short time interval.

Generally, slow-responsive DERs such as the price-based responsive loads and the energy-shifting ESSs have large capacities of power or energy. On the other hand, other DERs such as the loads enabled with DLC and the microturbines require a fast responding speed but not a large capacity.

To improve overall operating efficiency and guarantee network operating constraints under uncertainties of renewable power generation and loads, a multi-stage coordination framework is proposed and suggested.

In this framework, the slow-responsive and large-capacity DERs are scheduled for a relatively long period, to achieve overall high operating efficiency, e.g. minimise the total operating cost. Accordingly, long-timescale scheduling of DERs is required as first-stage operation. The scheduling timescale of this first stage can be commonly from 1 h to 1 day, even to a week.

Then, during a scheduled time window, second-stage operation is applied to compensate the first-stage operation to track uncertainty realisation. Herein, the fast-responsive DERs are dispatched or controlled at a short time interval to provide frequently adjustable power generation or consumption which can eliminate negative impacts of the uncertainty realisation. The second-stage dispatch intervals can be up to 1 h. If real-time control is available, a third stage to support real-time compensation to further track fluctuations of the uncertainties can also be involved in a multi-stage coordination framework.

The multi-stage coordination framework should be designed with full consideration of the characteristics of multiple DERs, such as responding speeds and capacities.

In this chapter, ESSs are used to be scheduled hourly in the first operation stage and DLC is adjusted every 15 min to track the uncertainty realisation as the second operation stage, forming an hourly two-stage coordination framework. Here, the ESSs aim to alleviate energy management difficulties by charging/discharging energy to complete energy shifting, and the DLC contributes to maintaining a power generation and consumption balance under uncertainties of renewable power generation.

7.2.2 First stage – hourly energy storage operation

It is assumed that ESSs are invested and installed at the same locations with RESs such as WTs and PVs. The ESSs operate to charge power from the RESs, when the power generation exceeds microgrid loads. On the other hand, the stored energy is discharged from the ESSs to the microgrid when necessary, e.g. when the renewable power generation is lower than the loads or when the electricity price is high. As a result, the microgrid can decrease energy purchase from the main grid, which in turn increases the overall operating profit. Thus, in microgrid operation, the ESSs aim to cooperate with RES uncertain output power to maximise the operating profit for microgrid operators. Since frequent changes of ESS charging/discharging states lead to reduced lifetimes and increased O&M costs, ESSs should operate to change the states between charging and discharging at a relatively long time interval to minimise the O&M costs [3]. Considering this characteristic of the ESSs, in the proposed coordination method, the states of the ESSs are planned in a relatively long timescale which is set as 1 h in this chapter, and the ESSs act hourly with predicted renewable power generation. Within each operation hour, the states of the ESSs are fixed. This ESS operation is regarded as the first-stage operation. It is emphasised that the predicted RES output power according to weather forecasts can be in a reasonably accurate range, but it still varies randomly from expectation. Thus, once the uncertain renewable output power deviates from the predicted value heavily within an hour, the economic benefits may be deteriorated and microgrid operating constraints may not be guaranteed.

7.2.3 Second stage – direct load control

To achieve robust microgrid operation against the uncertainty realisation of renewable power generation, a second-stage operation is designed in this chapter,

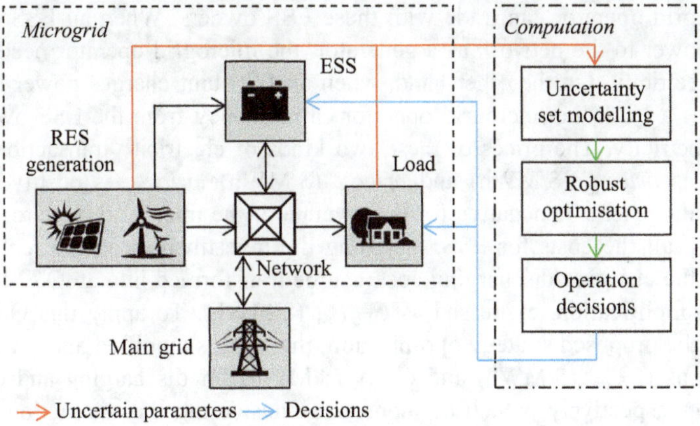

Figure 7.1 An hourly two-stage coordination method for microgrid

in which the DLC is implemented to complement the ESS operation. As the DLC is relatively faster, it can act in a shorter timescale which is set as one quarter hour, i.e. 15 min in this chapter. The DLC aims to modify the loads when the RES output power has significant deviations from the hour-ahead expected values. A 15-min prediction is applied for the renewable power generation in this stage as well. However, unlike the first-stage issue, this prediction accuracy can be significantly higher, as the prediction leading time is much shorter. This indicates that the uncertainties are much lower, and they can be regarded as realised. As a result, the power supply of the ESSs and the RESs can fulfil power consumption requirement of the responded loads, resulting in more operating profits.

Figure 7.1 shows the structure of the proposed two-stage coordination method. In the microgrid, RES units, ESSs, and responsive loads can coordinate to maximise the operator's economic and technical benefits. The RES output power, ESS charging/discharging state, and load data can be collected by measurement equipment and smart meters, and they are applied in a computation procedure as optimisation parameters. The proposed mathematical modelling and optimisation method are introduced in Sections 7.3 and 7.4, respectively. With operation decisions optimised, the ESS operates to change charging/discharging states in the first stage and the DLC works in the second stage.

7.3 Mathematical formulation

7.3.1 Energy storage models

The ESSs have an economic model and an operational model. The former is used in the overall operating cost calculation for the microgrid operator, while the latter is used as microgrid operating constraints.

For the sake of modelling, it is assumed that each ESS unit has a dummy owner which can be the microgrid operator itself or a private entity in practice. As a result, the microgrid operator can trade with these ESS owners. When an ESS unit discharges power to the network as a generator, the microgrid operator needs to pay for the electricity. On the other hand, when an ESS unit charges power from the network as a load, the microgrid operator earns money from the ESS owners for selling electricity. The prices of these two kinds of electricity transaction can be designed as $price_{dis}$ (\$/MWh) and $price_{ch}$ (\$/MWh). $price_{dis}$ is positive for the payment and $price_{ch}$ is negative for the revenue of the microgrid operator.

In [3], all the costs for ESS, including the investment cost, the replacement cost, and the charging/discharging cost, can be transformed into an O&M cost for analysis simplification, expressed as $C_{ES,OM}$ (\$/MWh). To apply this O&M cost model in the proposed strategy optimisation, this cost is modified and divided into two parts as $C_{ES,dis}$ (\$/MWh) and $C_{ES,ch}$ (\$/MWh) for discharging and charging electricity, respectively, which are constant. Considering the transaction concept described above, they can be $price_{dis}$ and $price_{ch}$, respectively. The following relationship should be satisfied for cost equality:

$$C_{ES,dis}E_{dis} + C_{ES,ch}E_{ch} = C_{ES,OM}E_{stored} \tag{7.1}$$

Considering the characteristics of ESS, the following equations need to be satisfied,

$$E_{stored} = \eta_{dis}E_{dis} = \eta_{ch}E_{ch},\ \eta_{dis} > 1, \eta_{ch} < 1 \tag{7.2}$$

where E_{stored}, E_{dis}, and E_{ch} represent the energy stored, the energy discharged, and the energy charged by ESS for a cycle, respectively. η_{ch} and η_{dis} are ESS efficiency factors during charging/discharging. The cycle means the period starting from some certain energy being charged and ending in this certain energy being discharged.

Substituting (7.2) into (7.1), we have

$$\frac{C_{ES,dis}}{\eta_{dis}} + \frac{C_{ES,ch}}{\eta_{ch}} = C_{ES,OM} \tag{7.3}$$

Thus, $C_{ES,dis}$ and $C_{ES,ch}$ can be designed by using (7.3) and the proposed ESS transaction model is developed as above. This ESS economic model is applied in the proposed microgrid operation optimisation.

After the economic model of the ESS cost is developed, a complementary operational model of the ESS is developed here. Considering that the operation states of the ESS units are fixed during the planned hour in the first stage, a discrete charging/discharging model is developed in this chapter.

In this model, the maximum output power an ESS unit can generate at a node, i.e. the maximum discharging power, is divided into several levels (10 levels in this chapter simulation). Each level represents a percentage of the maximum discharging power – this is aligned with industry-grade ESS discharging controller which needs a set point to control the discharging level. For the sake of the optimisation purpose, a binary decision variable is used to denote each level: 1 for the ESS unit discharging the corresponding power of the level, and 0 for no operating. Thus, for a specified ESS unit at the node m, the planned discharging power, $P_{\text{ES,dis},m}$ is calculated as follows:

$$P_{\text{ES,dis},m} = P_{\text{dis},m}^{\max} \sum_{j \in J_{\text{dis}}} \alpha_{\text{dis},m,j} L_{\text{dis},m,j} \tag{7.4}$$

Similarly, several levels (e.g. 10) for charging power of each ESS unit are derived and their corresponding binary decision variables are defined. The planned charging power of the ESS unit at the node m, $P_{\text{ES,ch},m}$ is derived as

$$P_{\text{ES,ch},m} = P_{\text{ch},m}^{\max} \sum_{j \in J_{\text{ch}}} \alpha_{\text{ch},m,j} L_{\text{ch},m,j} \tag{7.5}$$

In (7.4) and (7.5), $\alpha_{\text{ch},m,j}$ and $\alpha_{\text{dis},m,j}$ are, respectively, binary charging and discharging decisions of ESS unit at node m on charging and discharging level j; $P_{\text{ch},m}^{\max}$ and $P_{\text{dis},m}^{\max}$ are, respectively, rated charging and discharging power of ESS unit at node m; and $L_{\text{ch},m,j}$ and $L_{\text{dis},m,j}$ are, respectively, charging and discharging power rates (% of rated power) of ESS unit at node m on level j.

With these binary variables and the pre-designed charging/discharging levels, this operational model can be optimised to decide the ESS charging/discharging power to maximise the profit with the proposed ESS economic model.

7.3.2 Direct load control model

A DLC model is derived by [20]. In this DLC model, the total loads can be divided into two groups, controllable loads, i.e. DLC loads, and uncontrollable loads.

The ratio of the controllable loads to the total loads is calculated as

$$K_{\text{L,con}} = \frac{P_{\text{L,con}}}{P_{\text{L}}} \tag{7.6}$$

This ratio also represents the maximum percentage of the loads which can be controlled. During a certain period when the DLC is implemented, the controllable loads can be cut off by a load cutting rate which is the control variable in the DLC model. This load cutting rate is limited by

$$P_{\text{L,unc}} = \left(1 - K_{\text{L,con}}\right) P_{\text{L}} \tag{7.7}$$

In the DLC model, the load cutting rate can be a continuous and adjustable variable which is expected to be optimised by microgrid operator. However, this variable can also be discrete when loads are clustered (say, loads can be clustered into several groups, e.g. several houses can have only one DLC switch).

The real and reactive power loads in the DLC model can be calculated as (assuming the power factor is fixed)

$$P_{\text{DLC},i,q} = \left(1 - K_{\text{DLC},q}\right) P_{\text{L},i}, \ \forall i, q \tag{7.8}$$

$$Q_{\text{DLC},i,q} = \left(1 - K_{\text{DLC},q}\right) Q_{\text{L},i}, \ \forall i, q \tag{7.9}$$

The controllable loads can be cut off by remotely controlled switches. During the DLC programme, these customers with the cut-off loads can use limited power, even no power. Normally, to make up the loss for the controllable load customers, the microgrid operator offers lower electricity prices for these controllable loads than the normal prices for the uncontrollable loads. The microgrid operator can reset the controllable load prices to modify the number of the customers who agree to join in the DLC group [20]. For example, the microgrid operator can reduce the controllable load prices to involve more loads as controllable loads with the customers' agreement, so that more loads can be cut off for the modifying load demand purpose. However, considerably surplus controllable loads mean a much lower overall revenue from the customers, since the controllable load prices are much lower. Thus, the microgrid operator should design the controllable load prices based on the cautious expectation of the DLC capacity.

7.3.3 Coordination optimisation model

The two-stage coordination optimisation model is formulated as follows:

$$\min C_{\mathrm{ES}} + C_{\mathrm{WT}} + C_{\mathrm{PV}} + C_{\mathrm{grid}} - C_{\mathrm{rev}} \tag{7.10}$$

$$\begin{aligned}\text{s.t.}\quad C_{\mathrm{ES}} = {} & C_{\mathrm{ES,ch}} \sum_{m\in N_{\mathrm{ES}}} P^{\max}_{\mathrm{ch},m} \sum_{j\in J_{\mathrm{ch}}} \alpha_{\mathrm{ch},m,j} L_{\mathrm{ch},m,j} \\ & + C_{\mathrm{ES,dis}} \sum_{m\in N_{\mathrm{ES}}} P^{\max}_{\mathrm{dis},m} \sum_{j\in J_{\mathrm{dis}}} \alpha_{\mathrm{dis},m,j} L_{\mathrm{dis},m,j}\end{aligned} \tag{7.11}$$

$$C_{\mathrm{WT}} = C_{\mathrm{WT,OM}} \sum_{n\in N_{\mathrm{WT}}} \sum_{t\in T} P_{\mathrm{WT},n,t}\pi \tag{7.12}$$

$$C_{\mathrm{PV}} = C_{\mathrm{PV,OM}} \sum_{n\in N_{\mathrm{PV}}} \sum_{t\in T} P_{\mathrm{PV},n,t}\pi \tag{7.13}$$

$$C_{\mathrm{grid}} = C_{\mathrm{buy}} \sum_{t\in T} P_{\mathrm{def},t}\pi - C_{\mathrm{sell}} \sum_{t\in T} P_{\mathrm{sur},t}\pi \tag{7.14}$$

$$C_{\mathrm{rev}} = C_{\mathrm{L,unc}}\left(1 - K_{\mathrm{L,con}}\right) \sum_{i\in N_{\mathrm{L}}} P_{\mathrm{L},i} + C_{\mathrm{L,con}} \sum_{q\in HQ} \left(K_{\mathrm{L,con}} - K_{\mathrm{DLC},q}\right) \sum_{i\in N_{\mathrm{L}}} P_{\mathrm{L},i}\pi \tag{7.15}$$

$$\alpha_{\mathrm{ch},m,j} \in \{0,\ 1\},\ \alpha_{\mathrm{dis},m,j} \in \{0,\ 1\},\ \forall m,j \tag{7.16}$$

$$\sum_{j\in J_{\mathrm{ch}}\cup J_{\mathrm{dis}}} \alpha_{\mathrm{ch},m,j} + \alpha_{\mathrm{dis},m,j} \le 1,\ \forall m \tag{7.17}$$

$$\begin{aligned}-E_{0,m} \le {} & \eta_{\mathrm{ch}} P^{\max}_{\mathrm{ch},m} \sum_{j\in J_{\mathrm{ch}}} \alpha_{\mathrm{ch},m,j} L_{\mathrm{ch},m,j} - \eta_{\mathrm{dis}} P^{\max}_{\mathrm{dis},m} \sum_{j\in J_{\mathrm{dis}}} \alpha_{\mathrm{dis},m,j} L_{\mathrm{dis},m,j} \\ & \le E_{r,m} - E_{0,m},\ \forall m\end{aligned} \tag{7.18}$$

$$0 \le K_{\mathrm{DLC},q} \le K_{\mathrm{L,con}},\ \forall q \tag{7.19}$$

$$P_{\mathrm{DLC},i,q} = (1 - K_{\mathrm{DLC},q})P_{\mathrm{L},i},\ \forall i,q \tag{7.20}$$

$$Q_{\mathrm{DLC},i,q} = (1 - K_{\mathrm{DLC},q})Q_{\mathrm{L},i},\ \forall i,q \tag{7.21}$$

$$\begin{aligned} P_{b+1,t} = {} & P_{b,t} - P_{0,b+1,t} - P^{\max}_{\mathrm{ch},m}\sum_{j\in J_{\mathrm{ch}}}\alpha_{\mathrm{ch},m,j}L_{\mathrm{ch},m,j} + P^{\max}_{\mathrm{dis},m}\sum_{j\in J_{\mathrm{dis}}}\alpha_{\mathrm{dis},m,j}L_{\mathrm{dis},m,j} \\ & - P_{\mathrm{DLC},i,q} + P_{\mathrm{WT},n,t} + P_{\mathrm{PV},n,t},\ b \in Br(i),\ \forall i,t,q \end{aligned} \tag{7.22}$$

$$Q_{b+1,t} = Q_{b,t} - Q_{0,b+1,t} - Q_{\mathrm{DLC},i,q},\ b \in Br(i),\ \forall i,t,q \tag{7.23}$$

$$V_{i+1,t} = V_{i,t} - \frac{R_bP_{b,t} + X_bQ_{b,t}}{V_0},\ b \in Br(i,i+1),\ \forall i,t \tag{7.24}$$

$$1 - \Delta V^{\max} \le V_{i,t} \le 1 + \Delta V^{\max},\ \forall i,t \tag{7.25}$$

$$P_{1,t} = P_{\mathrm{def},t} - P_{\mathrm{sur},t},\ P_{\mathrm{def},t} \ge 0,\ P_{\mathrm{sur},t} \ge 0,\ \forall t \tag{7.26}$$

$$P^{\min}_{\mathrm{WT},n,t} \le P_{\mathrm{WT},n,t} \le P^{\max}_{\mathrm{WT},n,t},\ \forall n,t \tag{7.27}$$

$$P^{\min}_{\mathrm{PV},n,t} \le P_{\mathrm{PV},n,t} \le P^{\max}_{\mathrm{PV},n,t},\ \forall n,t \tag{7.28}$$

The objective function (7.10) considers all operating costs and revenue during the microgrid operation. Equations (7.11–7.15) are the calculation functions of these costs and revenue, i.e. the O&M costs of the ESSs, WTs, and PVs, respectively, the transaction cost with the main grid including the electricity payment and the electricity revenue, and the revenue of selling electricity to the customers. The objective is to maximise the total operating profit, which is equivalent to minimise all the operating costs minus the revenue.

Constraint (7.16) describes the first-stage decision variables for the ESS operation are binary for all the charging/discharging levels. Constraint (7.17) guarantees that only one charging/discharging level is planned to be operated for each ESS. Constraint (7.18) limits the maximum charging/discharging power of each ESS with the consideration of the battery efficiency.

Constraint (7.19) expresses the allowed range of the load cutting rate during the DLC. Equations (7.20) and (7.21) represent the active and reactive power loads during each quarter of the hour with the consideration of the planned DLC.

Constraints (7.22–7.24) are the linearised Dist-Flow model. Constraint (7.25) guarantees the voltage at each bus is kept within the allowed operating range.

Constraint (7.26) denotes the relationship between the power flow from the main grid and transacted electricity, where $P_{1,t}$ means the power transformed from the main grid to the microgrid. It is noted that $P_{\mathrm{def},t} > 0$ means the microgrid buys electricity from the main grid and $P_{\mathrm{sur},t} > 0$ means the microgrid sells electricity to the main grid.

Constraints (7.27) and (7.28) describe that due to the uncertain nature of the wind and solar power, the power generation of the WTs and PVs varies randomly in the predicted ranges.

Note that the first-stage ESS operation acts hourly and the operation time is set to 1 h. Thus, the first-stage time summation is 1 hand it can be hidden in (7.11), (7.15), and (7.18).

Objective (7.10) and constraints (7.11–7.28) make up a mixed-integer linear programming with uncertainty variables. Herein, the ESS charging/discharging operation levels $\alpha_{\text{ch},m,j}$, $\alpha_{\text{dis},m,j}$ are the first-stage decisions and the DLC load cutting rates $K_{\text{DLC},q}$ are the second-stage decisions in this coordination optimisation problem. The uncertainty variables include renewable power generation $P_{\text{WT},n,t}$ and $P_{\text{PV},n,t}$.

7.4 Two-stage robust optimisation method

7.4.1 Principle of two-stage robust optimisation

A TSRO method aims to solve optimisation problems under the worst case of possible uncertainty realisation, to obtain robust solutions against any uncertainty realisation.

First, uncertainties are modelled into uncertainty sets as constraints used in a TSRO model. Second, the TSRO method searches for the worst case of possible uncertainty realisation which can lead to infeasibility or the maximal objective for the second-stage optimisation problem. Then, the TSRO method optimises the first-stage decisions to minimise the objective under the worst case. Since it takes the worst case into account, the TSRO method can guarantee full solution robustness against any uncertainty realisation within the uncertainty sets.

Note that the objective would be further minimised with actual uncertainty realisation in the second stage. Thus, the actual objective in the second stage is less than or equal to that under the worst case. In other words, the optimised solutions based on the worst case are regarded as robust against all the uncertainties, since the second-stage problem is always feasible and the actual objective cannot be worse than the optimised objective.

Compared to the conventional SO methods, the RO method has two major advantages summarised in [26]: (1) the RO method does not need a probability density function or scenario-based data to model uncertainties; (2) the RO method achieves a robust solution under the worst case of uncertainty realisation instead of a solution based on the optimal expectation. Besides, the RO method can achieve high computational efficiency, since it utilises uncertainty sets to provide the worst case and solves problems only under the worst case rather than with a large number of scenarios.

In order to make the solutions fully robust against the uncertain RES output power, a TSRO method is developed to coordinate the ESS operation and the DLC in the two operation stages.

7.4.2 Uncertainty set modelling

For an RO problem, the uncertainty variables are modelled into uncertainty sets formed by bounds of each uncertainty variable and budgets for an overall uncertainty degree.

For the uncertainty variables, WT output power $P_{\mathrm{WT},n,t}$ and PV output power $P_{\mathrm{PV},n,t}$ limited by (7.27) and (7.28), respectively, two polyhedral uncertainty sets are modelled as follows:

$$U_{\mathrm{WT}} = \left\{ P_{\mathrm{WT},n,t} \in \mathbb{R}^{n_{\mathrm{wt}}}: \right.$$

$$\left. \mu_{\mathrm{WT},l} \leq \frac{1}{n_n n_t} \sum_{n \in N_{\mathrm{WT}}} \sum_{t \in T} \frac{P_{\mathrm{WT},n,t}}{P^{\mathrm{fc}}_{\mathrm{WT},n,t}} \leq \mu_{\mathrm{WT},u}, \; P^{\min}_{\mathrm{WT},n,t} \leq P_{\mathrm{WT},n,t} \leq P^{\max}_{\mathrm{WT},n,t}, \; \forall n, t \right\} \tag{7.29}$$

$$U_{\mathrm{PV}} = \left\{ P_{\mathrm{PV},n,t} \in \mathbb{R}^{n_{\mathrm{pv}}}: \right.$$

$$\left. \mu_{\mathrm{PV},l} \leq \frac{1}{n_n n_t} \sum_{n \in N_{\mathrm{PV}}} \sum_{t \in T} \frac{P_{\mathrm{PV},n,t}}{P^{\mathrm{fc}}_{\mathrm{PV},n,t}} \leq \mu_{\mathrm{PV},u}, \; P^{\min}_{\mathrm{PV},n,t} \leq P_{\mathrm{PV},n,t} \leq P^{\max}_{\mathrm{PV},n,t}, \; \forall n, t \right\} \tag{7.30}$$

$P^{\min}_{\mathrm{WT},n,t}$, $P^{\max}_{\mathrm{WT},n,t}$, $P^{\min}_{\mathrm{PV},n,t}$ and $P^{\max}_{\mathrm{PV},n,t}$ are predicted lower and upper bounds of the WT and PV outputs at a bus n during a period t. They can be obtained from interval forecasting tools [21]. In practice, they are obtained from the hour-ahead prediction.

Furthermore, a pair of uncertainty budgets are two user-defined parameters as μ_l and μ_u to limit overall uncertainty degree for each type of uncertainty. They are lower and upper bounds of a summation of the ratios between the possibly realised renewable power generation and the predicted one, averaged by the total numbers of periods n_t and RESs n_n. These budgets are the indicators for measuring the uncertainty degree caused by prediction errors of WT and PV power generation. For example, if both μ_l and μ_u are set as 100%, it means that the overall hourly prediction is fully accurate. Otherwise, it means the overall hourly prediction has an average accuracy rate between μ_l and μ_u.

The uncertainty budgets should be designed considering a few aspects such as the historical data (e.g. the accuracy statistics of the prediction tools) and the expectation of the solution robustness. For example, the budgets should have a larger range to improve the robustness if heavy uncertainties are predicted to occur. Besides, if the budget range is not set up large enough, realisation of the uncertainties may be out of the uncertainty sets and the robustness may not be guaranteed, which may lead to an increased realised objective and/or operating constraint violations. It is emphasised that the budget range is enlarged when μ_l falls and μ_u rises, so that the final optimisation solutions are robust against more uncertainty realisation cases, which means the solution robustness degree is improved. However, a large budget range may lead to a conservative solution (i.e. a large objective result) under uncertainty realisation. Therefore, the uncertainty budgets should be designed by the TSRO user with consideration of a trade-off between solution robustness and conservativeness.

7.4.3 Two-stage robust optimisation model

The proposed microgrid coordination optimisation model is converted to a TSRO model, which can be formulated in the following compact matrix form:

$$\min_{x} \boldsymbol{c}^T\boldsymbol{x} + \max_{u} \min_{y} \boldsymbol{d}^T\boldsymbol{y} + \boldsymbol{e}^T\boldsymbol{u} \tag{7.31}$$

$$\text{s.t.} \quad \boldsymbol{Ax} \geq \boldsymbol{b} \tag{7.32}$$

$$\boldsymbol{y} \in O(\boldsymbol{x}, \boldsymbol{u}) = \{\boldsymbol{Fx} + \boldsymbol{Gy} \leq \boldsymbol{v},\ \boldsymbol{Hx} + \boldsymbol{Iy} + \boldsymbol{Ju} = \boldsymbol{w}\} \tag{7.33}$$

$$\boldsymbol{u} \in \boldsymbol{U} \tag{7.34}$$

The TSRO model is a '*min-max-min*' tri-level optimisation problem with corresponding constraints for different operation stages. The first '*min*' is to minimise the first-stage objective by optimising the first-stage variable vector $\boldsymbol{x}$. The '*max*' is to find the worst uncertainty case $\boldsymbol{u}$ in given uncertainty sets $\boldsymbol{U}$. The second '*min*' is to minimise the second-stage objective by optimising the second-stage variable vector $\boldsymbol{y}$. It can be seen that the worst case from the uncertainty set is obtained by maximising the minimal second-stage objective, which guarantees the solution robustness against any uncertainty realisation.

The objective described in (7.31) has the '*min-max-min*' objective form. According to the characteristics of the '*min-max-min*' form, the proposed coordination optimisation objective (7.10) can be rewritten as the following:

$$\min_{\alpha_{\text{dis}},\alpha_{\text{ch}}} C_{\text{ES}} + \max_{P_{\text{WT}},P_{\text{PV}}} \min_{K_{\text{DLC}},V,P,Q\in O} C_{\text{WT}} + C_{\text{PV}} + C_{\text{grid}} - C_{\text{rev}} \tag{7.35}$$

Here, C_{ES} is the first-stage objective, while $C_{\text{WT}} + C_{\text{PV}} + C_{\text{grid}} - C_{\text{rev}}$ presents the second-stage objective.

The variables at the different stages can be classified into three types as follows.

$\boldsymbol{x}$ represents a vector of decision variables in the first operation stage which are not subject to the uncertainty variables or the decision variables in the second operation stage. Considering the unadjustable characteristic of these decision variables, $\boldsymbol{x}$ can be regarded as the 'here-and-now' decisions. In the proposed coordination model formulated in Section 7.3.3, $\boldsymbol{x}$ indicates the vector of the binary variables for all the ES operation levels, i.e. α_{ch} and α_{dis}. Besides, their constraints (7.16–7.18) are converted into (7.32).

$\boldsymbol{y}$ stands for a vector of decision variables in the second operation stage including the DLC decision variables, K_{DLC}, and dependent variables, P, Q, V. Herein, K_{DLC} for each 15-min period can be optimised in the second operation stage after the realisation of the uncertainty variables, and they are referred as the 'wait-and-see' decisions. Constraint (7.33) indicates that $\boldsymbol{y}$ must be adjusted in the feasible set, i.e. $O(x,u)$ which is based on the first-stage decision variable vector $\boldsymbol{x}$ and any single case of uncertainty variables $\boldsymbol{u}$. $O(x,u)$ is formed by constraints (7.19–7.26).

$\boldsymbol{u}$ expresses a vector for uncertainty variables for WT and PV power generation, i.e. P_{WT} and P_{PV}. Constraint (7.34) means that with the prediction and designed uncertainty budgets, the uncertainty variables can vary within the uncertainty sets $\boldsymbol{U}$ such as (7.29) and (7.30) which support the worst case of possible uncertainty realisation. In other words, with (7.34) as the constraint, the minimal second-stage objective is maximised.

The other bold uppercase symbols mean matrices of parameters while the other bold lowercase symbols the parameter vectors.

It should be noticed that the 'wait-and-see' decisions which can be adjusted in the second operation stage can improve the objective value when the worst case does not occur. In other words, the objective optimised under the worst case may be further reduced by the second-stage decisions according to the uncertainty realisation so that the solution conservativeness can be reduced.

7.4.4 Solution algorithm

To solve a '*min-max-min*' TSRO problem, two decomposition methodologies are widely applied. One is a Benders decomposition algorithm with dual-cutting [15] and the other one is a C&CG algorithm which can be regarded as a primal cutting plane algorithm [22]. It is concluded in [22] that the convergence speed of the C&CG algorithm is much faster than that of Benders decomposition. Therefore, the C&CG algorithm is applied in this chapter.

In the C&CG algorithm, the TSRO problem is divided into a master problem and a sub-problem which can formulate the first and the second operation stages, respectively. The master problem and the sub-problem are solved iteratively until they converge with a gap not greater than the preset threshold ε.

The master problem is formulated as follows:

$$\min_{\boldsymbol{x}} \boldsymbol{c}^T\boldsymbol{x} + \lambda \tag{7.36}$$

$$\text{s.t.} \quad \boldsymbol{Ax} \geq \boldsymbol{b} \tag{7.37}$$

$$\lambda \geq \boldsymbol{d}^T\boldsymbol{y}_l + \boldsymbol{e}^T\boldsymbol{u}_l^*, \ \forall \boldsymbol{u}_l^* \in \boldsymbol{S} \tag{7.38}$$

$$\boldsymbol{Fx} + \boldsymbol{Gy}_l \leq \boldsymbol{v} \tag{7.39}$$

$$\boldsymbol{Hx} + \boldsymbol{Iy}_l + \boldsymbol{Ju}_l^* = \boldsymbol{w}, \ \forall \boldsymbol{u}_l^* \in \boldsymbol{S} \tag{7.40}$$

Since in the proposed coordination optimisation model, $\boldsymbol{x}$ is a vector of binary decision variables, the master problem is still a mixed-integer linear programming problem. In this problem, an optimal solution can be derived as $(\boldsymbol{x}^*, \lambda^*)$ with a set of given uncertainty variables $\boldsymbol{u}_l^*$ obtained from the sub-problem. It is denoted that the solution, $\boldsymbol{x}^*$ is the current optimal solution for the first operation stage and it is used for solving the sub-problem. In addition, $\boldsymbol{y}_l$ corresponding to each uncertainty variable $\boldsymbol{u}_l^*$ is optimised as well. However, it is emphasised that only the final $\boldsymbol{x}^*$ which is the 'here-and-now' decision is obtained and used as the final operation solution, since the second-stage variables $\boldsymbol{y}_l$ treated as 'wait-and-see' decisions can be modified further during the second operation stage.

On the other hand, a sub-problem is modelled as follows:

$$S(\boldsymbol{u},\boldsymbol{x}^*) = \max_{\boldsymbol{u}} \min_{\boldsymbol{y}} \boldsymbol{d}^T\boldsymbol{y} + \boldsymbol{e}^T\boldsymbol{u} \tag{7.41}$$

$$\text{s.t.} \quad \boldsymbol{F}\boldsymbol{x}^* + \boldsymbol{G}\boldsymbol{y} \leq \boldsymbol{v} \tag{7.42}$$

$$\boldsymbol{H}\boldsymbol{x}^* + \boldsymbol{I}\boldsymbol{y} + \boldsymbol{J}\boldsymbol{u} = \boldsymbol{w} \tag{7.43}$$

$$\boldsymbol{u} \in \boldsymbol{U} \tag{7.44}$$

It is noticed that exactly solving the '*max-min*' sub-problem is significantly challenging with a polyhedral uncertainty set [22]. To solve the sub-problem, in this chapter, the classic Lagrange dual is proposed to make strong duality. Thus, the '*max-min*' sub-problem is converted into a bilinear maximisation problem as follows:

$$\max_{u,\varphi,\rho} (\boldsymbol{F}\boldsymbol{x}^* - \boldsymbol{v})^T \varphi + (\boldsymbol{w} - \boldsymbol{H}\boldsymbol{x}^* - \boldsymbol{J}\boldsymbol{u})^T \rho + \boldsymbol{e}^T\boldsymbol{u} \tag{7.45}$$

$$\text{s.t.} \quad \boldsymbol{G}^T \varphi + \boldsymbol{I}^T \rho + \boldsymbol{d} = \boldsymbol{0} \tag{7.46}$$

$$\varphi \geq \boldsymbol{0}, \rho \text{ free} \tag{7.47}$$

$$\boldsymbol{u} \in \boldsymbol{U} \tag{7.48}$$

Here, φ and ρ are the dual variables of the sub-problem constraints (7.42) and (7.43), respectively. This dual problem can be solved by non-linear and bilinear solvers such as SCIP [23]. By solving the dual problem of the sub-problem, the worst case is searched with a corresponding uncertainty variable $\boldsymbol{u}^*$ as the sub-problem solution. Besides, $\boldsymbol{u}^*$ is added into an uncertainty solution set $\boldsymbol{S}$, for the master problem in future iterations. After solving the sub-problem, new second-stage variables are generated with their constraints (7.38–7.40) and added in the master problem.

The C&CG algorithm can be implemented as demonstrated in Figure 7.2. ε is a predefined termination threshold of the bound gap.

7.5 Case study

7.5.1 Test system description

In this chapter, a 33-bus radial distribution system as a microgrid is used to demonstrate the proposed hourly coordination approach. The system topology is shown in Figure 7.3 and its data can be obtained from [24]. The allowed maximum voltage deviation is set as 0.05 p.u. Location, power/energy capacity, and charging/discharging parameters of the ESSs are shown in Table 7.1. In addition, location and rated power data of the WTs and the PVs are shown in Table 7.2. Moreover, Table 7.3 lists economic parameters for the microgrid operation.

Simulations are conducted on a 64-bit PC with 3.30-GHz CPU and 8 GB RAM using YALMIP toolbox [25] in the MATLAB® platform. Both the bilinear sub-problem and the master problem are solved by SCIP solver [23]. The termination threshold of the TSRO method is set as 0.01.

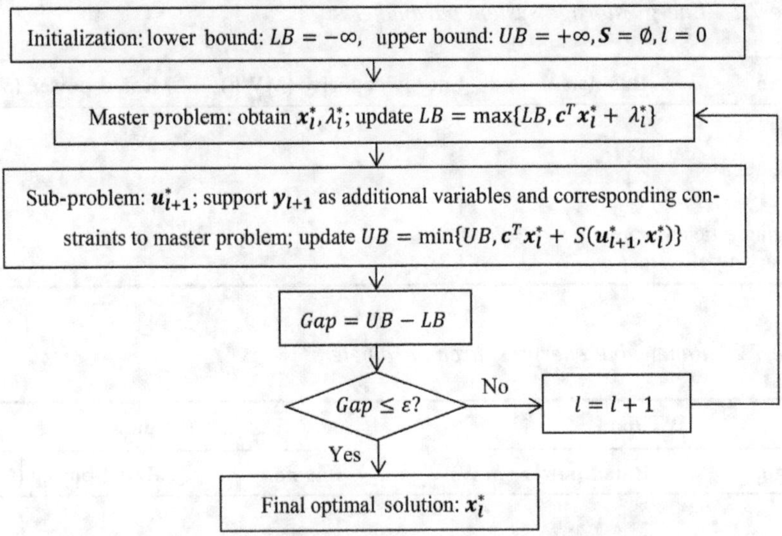

Figure 7.2 Column-and-constraint generation algorithm

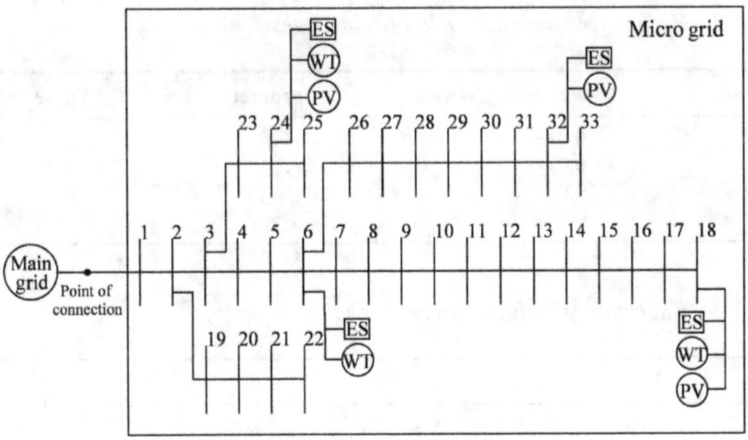

Figure 7.3 Test microgrid topology

7.5.2 Initial tests

A base system state is considered for the initial tests. The base state has relatively low energy of 1.4 MWh stored in total in the ESSs, 35% of the WT rated power, 90% of the PV rated power, and 100% of the loads indicating a peak-load condition. For this system state, six groups of different uncertainty budgets are considered, which are listed in Table 7.4. For the six groups of uncertainty sets, six corresponding tests are carried out by using the C&CG algorithm to solve the TSRO model.

Table 7.1 Energy storage system parameters

ESS no.	Bus no.	Energy capacity (MWh)	Rated power (MW)
1	6	2	1.5
2	18	1	1
3	24	1	1
4	32	1	1
Charging efficiency η_{ch}		0.9	
Discharging efficiency η_{dis}		1.2	

Table 7.2 Renewable energy source parameters

WT data		PV data	
Bus no.	Rated power (MW)	Bus no.	Rated power (MW)
6	2	18	0.5
18	1	24	0.5
24	1	32	0.5

Table 7.3 Parameters for microgrid operation tests

Parameter	Value (\$/kWh)	Parameter	Value (\$/kWh)
C_{sell}	0.015	$C_{ES,dis}$	0.051
C_{buy}	0.07	$C_{ES,ch}$	−0.02
$C_{L,con}$	0.06	$C_{WT,OM}$	0.01
$C_{L,unc}$	0.1	$C_{PV,OM}$	0.01

Table 7.4 Uncertainty budget settings for tests

Test no.	1	2	3	4	5	6
$\mu_{WT,l}$	95%	90%	85%	80%	75%	70%
$\mu_{WT,u}$	105%	110%	115%	120%	125%	130%
$\mu_{PV,l}$	97.5%	95%	92.5%	90%	87.5%	85%
$\mu_{PV,u}$	102.5%	105%	107.5%	110%	112.5%	115%

Taking Test 1 as an example, the second column of Table 7.4 shows that the lower and upper uncertainty budgets for the WT output power, $\mu_{WT,l}$ and $\mu_{WT,u}$, are set 95% and 105%, respectively. This means that the average prediction accuracy ratio of the actual wind power to the predicted one can vary between 95% and 105%.

The optimisation results of these six tests are given in Table 7.5, which shows the ESS discharging states at the first stage, the DLC results under the worst case,

Table 7.5 Solution results for base case under different uncertainty sets

Test no.		1	2	3	4	5	6
ESS Discharging, % of rated power	ESS 1	0%	10%	0%	10%	10%	10%
	ESS 2	0%	0%	10%	0%	0%	0%
	ESS 3	20%	20%	40%	20%	40%	40%
	ESS 4	30%	20%	20%	30%	30%	30%
DLC under the worst case	0–15 min	0%	0%	0%	0%	0%	0%
	15–30 min	46%	0%	43%	39%	38%	38%
	30–45 min	0%	0%	0%	0%	31%	0%
	45–60 min	3%	6%	2%	2%	0%	0%
Operating profit under the worst case ($)		192.39	187.94	184.45	179.86	177.30	174.29
Iteration number		5	5	3	3	3	2
Solver time (s)		61.39	15.96	12.04	13.84	18.34	7.01

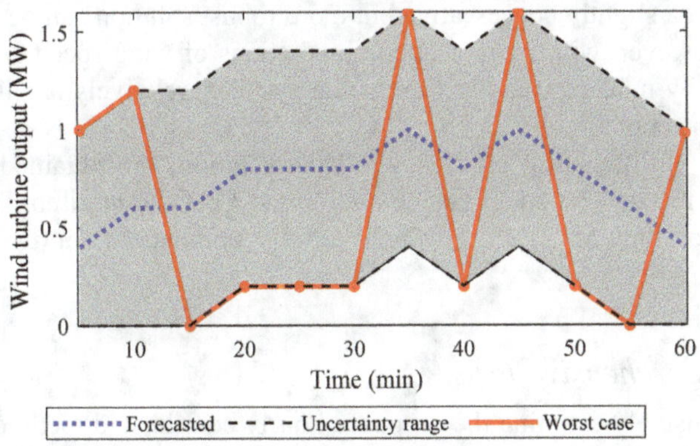

Figure 7.4 Illustration of the worst case in WT uncertainty set

the operating profit under the worst case as well as the iteration number of the C&CG algorithm, and the solver time.

For an illustration purpose, with the uncertainty sets for Test 1, the predicted wind output power, the uncertainty range, and the worst case found by the TSRO method for the WT at bus 6 are shown in Figure 7.4. The uncertainty range of the WT output power is depicted as the grey area with the blue dotted line as the predicted value, and the red curve presents the worst case found in the optimisation process. The worst case corresponds to maximising the minimisation of the second-stage objective, i.e. '*max-min*'. It can be seen that the uncertainty variable of each period reaches the boundaries of the range for this uncertainty set. However, note that the uncertainty variables may not always lay at the boundaries if the uncertainty budgets change.

For Test 1, the ESS operation states, i.e. discharging rates, are shown in the second column of Table 7.5 and they are the hour-ahead decisions of the coordination method. The ESSs are dispatched to act these operation states in the following hour. Besides, under the worst case, the 15-min DLC decisions, i.e. the load cutting rates for the controllable loads, are also given. It is emphasised that the actual DLC decisions are optimised 15 min ahead with the uncertainty realisation, and then they are implemented for the corresponding 15-min interval. The operating profit under the worst case is calculated as well, but note that the actual operating profit would be calculated according to the DLC implementation and the realisation of the uncertainties.

According to the solution results for the other five groups of the uncertainty sets, it can be seen that as the uncertainty budget ranges increase, the optimal solutions are more conservative. In Table 7.4, from Test 1 to Test 6, the uncertainty budget ranges are enlarged, and as a result, the overall operating profit decreases. For a solution with larger budgets, although the expected operating profit is relatively low, the solution can fit more uncertainty conditions, which means more robust. Thus, for a solution robustness improvement purpose, large uncertainty budgets can be selected by the microgrid operators to dispatch the ESS hour-ahead operation, and a slightly conservative but more robust solution can be prepared for operation. However, with more accurate predictions of wind speed and solar irradiance given by modern weather forecasting systems, relatively small uncertainty budgets can be enough.

In terms of computing efficiency, all the solutions are obtained within five iterations by the developed Lagrange-dual-based C&CG algorithm. Besides, the average solver time is 21.43 s, which is fully compatible for the hour-ahead optimisation.

7.5.3 Comprehensive tests

To comprehensively examine the proposed hourly coordinated method, a range of different load/ESS/RES operating conditions are used for testing. These operating conditions make up 16 tests and they are shown in Table 7.6. For the uncertainty sets of the TSRO model, 90% and 110% are, respectively, set as the lower and upper uncertainty budgets for the WT output power, while 95% and 105% are for the PV one.

Table 7.7 demonstrates the TSRO solutions of the proposed hourly coordination method for all the 16 tests. For the ESS operation states, negative values stand for discharging power and positive values for charging power during the hour.

From this table, for the high WT and PV output power tests (Tests 1, 5, 9, and 13), some ESSs are dispatched to be charged since the renewable power generation is greater than the loads in the microgrid. Furthermore, from the Tests 9 and 13 indicating the off-peak periods, more energy is charged by the ESSs with the adequate renewable power generation. It is noted from Tests 11 and 15 that although the PV output power is low, the ESSs still operate charging energy due to the sufficient power generated by the WTs supplying to the half loads. Besides,

Table 7.6 Operating conditions for comprehensive tests

Test no.	Operating conditions			
	Loads (%)	**ESS stored energy (MWh)**	**WT output power rate (%)**	**PV output power rate (%)**
1	100	2.7	95	90
2	100	2.7	25	90
3	100	2.7	95	10
4	100	2.7	25	10
5	100	0.8	95	90
6	100	0.8	25	90
7	100	0.8	95	10
8	100	0.8	25	10
9	50	2.7	95	90
10	50	2.7	25	90
11	50	2.7	95	10
12	50	2.7	25	10
13	50	0.8	95	90
14	50	0.8	25	90
15	50	0.8	95	10
16	50	0.8	25	10

Table 7.7 Energy storage operation planning results

Test no.	Operation states, % of rated power				Iteration	Solver time (s)
	ES 1	**ES 2**	**ES 3**	**ES 4**		
1	0	20	0	0	4	21.82
2	−50	−10	−50	0	3	11.72
3	0	0	0	0	3	10.22
4	−50	−30	−50	−50	3	13.34
5	0	20	0	0	4	26.13
6	−20	−20	0	0	2	23.47
7	0	0	0	0	3	10.35
8	−20	−20	0	0	2	1.87
9	60	60	40	10	3	3.54
10	−10	20	0	0	3	3.39
11	10	40	40	0	4	8.87
12	−10	−10	−50	−10	4	8.47
13	0	60	40	100	3	9.01
14	−10	20	0	0	3	3.49
15	10	30	0	50	5	24.22
16	−20	−10	0	0	3	2.97

when the WT output power is low (Tests 2, 4, 6, 8, 12, and 16), some ESSs are dispatched to discharge power, for the WTs are the main DG in the microgrid and the overall renewable power generation is insufficient. For Tests 10 and 14, No.1

ESS discharges but No. 2 ESS charges, because the power imbalance conditions in the areas around buses 6 and 18 are quite different. To keep all the bus voltages within the allowed range, these two ESSs are operated differently. Last but not least, it is noted that for Tests 3 and 7, the ESSs are not operated, since the power is almost balanced in the microgrid and possible operating constraint violations can be addressed by the DLC.

Besides, the iteration number and the solver time are shown in Table 7.7. The average solver time is 11.43 s, which again demonstrates the high computational efficiency of the proposed Lagrange-dual-based C&CG algorithm.

7.6 Conclusion

This chapter introduces an hourly two-stage coordination method of ESSs and DLC in microgrids and its corresponding optimisation problem is solved by a TSRO to deal with uncertain renewable power generation. In the proposed coordination method, operating states of the ESSs are optimised to charge or discharge power an hour ahead in the first operation stage, and an adjustable 15-min DLC method is implemented in the second operation stage to make power balanced and operating profits maximum. The proposed TSRO problem is solved by a developed C&CG algorithm to obtain robust solutions under the worst case of uncertainty realisation, such that the proposed coordination method fully takes renewable power generation uncertainties into account.

The proposed hourly coordination method is verified on a 33-bus testing microgrid with comprehensive specific tests and characteristics of uncertainty budgets are analysed. The test results indicate high solution efficiency and robustness of the hourly two-stage coordinated method in coordinating the ESS operation and the DLC, to achieve maximum operating profits under the uncertainties of renewable power generation.

References

[1] P. Denholm, E. Ela, B. Kirby, and M. Milligan, 'The role of energy storage with renewable electricity generation', Available: www.osti.gov, accessed March 2018.

[2] A. Nasiri, 'Integrating energy storage with renewable energy systems', *Proc. 34th Annual Conf. of IEEE Indus. Electro.*, pp. 17–18, Orlando, FL, USA, 2008.

[3] P. Poonpun and W.T. Jewell, 'Analysis of the cost per kilowatt hour to store electricity', *IEEE Trans. Energy Conv.*, vol. 23, no. 2, pp. 529–534, 2008.

[4] S.H. Jang, J.B. Park, J.H. Roh, S.Y. Son, and K.Y. Lee, 'Short-term resource scheduling for power systems with energy storage systems', *Proc. IEEE PES General Meeting*, pp. 1–7, San Diego, CA, USA, 2012.

[5] S. Lakshminarayana, T.Q.S. Quek, and H.V. Poor, 'Combining cooperation and storage for the integration of renewable energy in smart grids', *Proc.*

IEEE Conf. on Comput. Commun. Workshops, pp. 622–627, Toronto, ON, Canada, 2014.

[6] M. Ross, R. Hidalgo, C. Abbey, and G. Joos, 'Energy storage system scheduling for an isolated microgrid', *IET Renew. Power Gen.*, vol. 5, no. 2, pp. 117–123, 2011.

[7] J. Kumano and A. Yokoyama, 'Optimal weekly operation scheduling on pumped storage hydro power plant and storage battery considering reserve margin with a large penetration of renewable energy', *Proc. 2014 Inter. Conf. on Power Syst. Tech.*, pp. 1120–1126, Chengdu, China, 2014.

[8] P. Harsha and M. Dahleh, 'Optimal management and sizing of energy storage under dynamic pricing for the efficient integration of renewable energy', *IEEE Trans. Power Syst.*, vol. 30, no. 3, pp. 1164–1181, 2015.

[9] P.P. Zeng, Z. Wu, X.P. Zhang, C. Liang, and Y. Zhang, 'Model predictive control for energy storage systems in a network with high penetration of renewable energy and limited export capacity', *Proc. 18th Power Syst. Comput. Conf. (PSCC)*, pp. 1–7, Wroclaw, Poland, 2014.

[10] F. García and C. Bordons, 'Optimal economic dispatch for renewable energy microgrids with hybrid storage using model predictive control', *Proc. 39th Annual Conf. of the IEEE Indus. Electro. Society*, pp. 7932–7937, Vienna, Austria, 2013.

[11] S. Dutta and T.J. Overbye, 'Optimal storage scheduling for minimising schedule deviations considering variability of generated wind power', *Proc. IEEE PES Power Syst. Conf. and Expo.*, pp. 1–7, Phoenix, AZ, USA, 2011.

[12] N. Li and K.W. Hedman, 'Economic assessment of energy storage in systems with high levels of renewable resources', *IEEE Trans. Sustain. Energy*, vol. 6, no. 3, pp. 1103–1111, 2015.

[13] K. Rahbar, J. Xu, and R. Zhang, 'Real-time energy storage management for renewable integration in microgrid: an off-line optimisation approach', *IEEE Trans. Smart Grid*, vol. 6, no. 1, pp. 124–134, 2015.

[14] M.H. Albadi and E.F. El-Saadany, 'A summary of demand response in electricity markets', *Electric Power Syst. Research*, vol. 78, no. 11, pp. 1989–1996, 2008.

[15] D. Bertsimas, E. Litvinov, X.A. Sun, J. Zhao, and T. Zheng, 'Adaptive robust optimisation for the security constrained unit commitment problem', *IEEE Trans. Power Syst.*, vol. 28, no. 1, pp. 52–63, 2013.

[16] R. Jiang, J. Wang, and Y. Guan, 'Robust unit commitment with wind power and pumped storage hydro', *IEEE Trans. Power Syst.*, vol. 27, no. 2, pp. 800–810, 2012.

[17] L. Zhao and B. Zeng, 'Robust unit commitment problem with demand response and wind energy', *Proc. IEEE PES General Meeting*, pp. 1–8, San Diego, CA, USA, 2012.

[18] Z. Wang, B. Chen, J. Wang, J. Kim, and M.M. Begovic, 'Robust optimisation based optimal DG placement in microgrids', *IEEE Trans. Smart Grid*, vol. 5, no. 5, pp. 2173–2182, 2014.

[19] R.A. Jabr, I. Džafić, and B.C. Pal, 'Robust optimisation of storage investment on transmission networks', *IEEE Trans. Power Syst.*, vol. 30, no. 1, pp. 531–539, 2015.

[20] K.-H. Ng and G.B. Sheble, 'Direct load control – A profit-based load management using linear programming', *IEEE Trans. Power Systems*, vol. 13, pp. 688–694, 1998.

[21] C. Wan, Z. Xu, P. Pinson, Z.Y. Dong, and K.P. Wong, 'Probabilistic forecasting of wind power generation using extreme learning machine', *IEEE Trans. Power Syst.*, vol. 29, no. 3, pp. 1033–1044, 2014.

[22] B. Zeng and L. Zhao, 'Solving two-stage robust optimisation problems using a column-and-constraint generation method', *Operations Research Letters*, vol. 41, no. 5, pp. 457–461, 2013.

[23] T. Achterberg, 'SCIP: solving constraint integer programs', *Math. Prog. Comput.*, vol. 1, no. 1, pp. 1–41, 2009.

[24] M. Baran and F.F. Wu, 'Optimal sizing of capacitors placed on a radial distribution system', *IEEE Trans. Power Del.*, vol. 4, no. 1, pp. 735–743, 1989.

[25] J. Lofberg, 'YALMIP: a toolbox for modelling and optimisation in MATLAB', *Proc. IEEE Inter. Conf. on Robotics and Automation*, pp. 284–289, Taipei, Taiwan, 2004.

[26] Y. An and B. Zeng, 'Exploring the modelling capacity of two-stage robust optimisation: variants of robust unit commitment model', *IEEE Trans. Power Syst.*, vol. 30, no. 1, pp. 109–122, 2015.

Chapter 8

Daily coordination of microturbines and demand response

Microgrid is an effective means to integrate distributed generation (DG) resources. However, uncertain renewable DG such as wind turbine (WT) and photovoltaic (PV) power outputs and uncertain loads can bring in tremendous difficulties for energy management in microgrids. In this chapter, a two-stage robust microgrid coordination strategy is proposed: a price-based demand response (PBDR) programme is scheduled a day ahead and microturbine power outputs are modified hourly. A two-stage robust optimisation (RO) model is applied to address the coordinated operation problem with guaranteed robustness against the uncertainties of renewable DG and loads. Simulation results show the PBDR programme and multiple DG units can coordinate effectively to accommodate the renewable and load uncertainties while maximising microgrid benefits.

Nomenclature

A Acronyms

C&CG	Column and constraint generation
DG	Distributed generation
MILP	Mixed-integer linear programming
O&M	Operation and maintenance
PBDR	Price-based demand response
PV	Photovoltaic
RO	Robust optimisation
SO	Stochastic optimisation
TSRO	Two-stage robust optimisation
WT	Wind turbine

B Sets and indices

$Br(i)$	Set of all branches that connect to bus i
$Br(i,j)$	Branch between buses i and j
b	Index of branches

J	Set of all PBDR price levels
N_{L}, N_{WT}, N_{PV}, N_{MT}	Set of buses that have loads, WTs, PVs and microturbines, respectively
T	Set of all operation periods
U_{WT}, U_{PV}, U_{L}	Uncertainty set for WT output power/PV output power/loads

C Parameters

A	Constant used in the demand response equation
$C_{\mathrm{MT,OM}}$, $C_{\mathrm{WT,OM}}$, $C_{\mathrm{PV,OM}}$	Operation and maintenance cost of microturbines/WTs/PVs (\$/MWh)
C_{sell}, C_{buy}	Price for selling/buying electricity to/from the main grid (\$/MWh)
$L_{j,t}$	Load demand response rate on price level j during period t (%)
n_i, n_t	Total number of the system buses and the operation periods
$P_{b,\mathrm{cap}}$	Capacity of branch b (MW)
$P_{\mathrm{L},i,t}^{\mathrm{pr}}$, $Q_{\mathrm{L},i,t}^{\mathrm{pr}}$	Predicted active and reactive power loads at bus i during period t (MW/MVar)
$P_{\mathrm{MT},m}^{\mathrm{rated}}$	Rated output power of the microturbines at bus m (MW)
P_{tie}	Capacity of the tie line between the microgrid and the main grid (MW)
$P_{\mathrm{WT},n}^{\mathrm{rated}}$, $P_{\mathrm{PV},n}^{\mathrm{rated}}$	Rated output power of the WTs/PVs at bus n (MW)
$Pr_{0,t}$	Original electricity price for customers without PBDR during period t (\$/MWh)
$Pr_{j,t}$	Electricity price for customers on price level j during period t (\$/MWh)
R_b, X_b	Resistance/reactance of branch b
$R_{\mathrm{L},i,j,t}^{\mathrm{low}}$, $R_{\mathrm{L},i,j,t}^{\mathrm{up}}$	Lower/upper bound of predicted load rate at bus i on price level j during period t (%)
$R_{\mathrm{WT},n,t}^{\mathrm{low}}$, $R_{\mathrm{WT},n,t}^{\mathrm{up}}$, $R_{\mathrm{PV},n,t}^{\mathrm{low}}$, $R_{\mathrm{PV},n,t}^{\mathrm{up}}$	Lower/upper bound of predicted WT/PV output power rate (% of rated power) at bus n during period t
$R_{\mathrm{WT},n,t}^{\mathrm{pr}}$, $R_{\mathrm{PV},n,t}^{\mathrm{pr}}$	Predicted output power rate (% of rated power) of the WTs/PVs at bus n during period t
V_0	Voltage at substation, normally 1 p.u.
$\Delta V^{\max}$	Allowed maximal voltage deviation (p.u.)
ε	Price elasticity of loads
$\mu_{\mathrm{WT},l}$, $\mu_{\mathrm{WT},u}$, $\mu_{\mathrm{PV},l}$, $\mu_{\mathrm{PV},u}$, $\mu_{\mathrm{L},l}$, $\mu_{\mathrm{L},u}$	Lower/upper budget of WT output power/PV output power/loads
τ	Time length of one operation period (h)

D Variables

C_{grid}	Total transaction cost between the main grid and the microgrid ($)
$C_{\text{MT}}, C_{\text{WT}}, C_{\text{PV}}$	Total O&M cost of microturbines/WTs/PVs ($)
C_{rev}	Total revenue from customer loads ($)
$P_{0,b,t}, Q_{0,b,t}$	Active/reactive power through the lateral branch of branch b during period t (MW/MVar)
$P_{b,t}, Q_{b,t}$	Active/reactive power through branch b during period t (MW/MVar)
$P_{\text{def},t}, P_{\text{sur},t}$	Power deficiency/surplus of the microgrid during period t (MW)
$P_{\text{L},i,t}, Q_{\text{L},i,t}$	Active/reactive power loads at bus i during period t with PBDR (MW/MVar)
$P_{\text{MT},m,t}$	Output power of the microturbines at bus m during period t (MW)
$R^{\text{unc}}_{\text{L},i,j,t}$	Uncertain variation rate of loads at bus i on price level j during period t (%)
$R_{\text{WT},n,t}, R_{\text{PV},n,t}$	Uncertain output power rate (% of rated power) of the WTs/PVs at bus n during period t
$V_{i,t}$	Voltage at bus i during period t (p.u.)
$\alpha_{j,t}$	Binary on/off decision of PBDR on price level j during period t

8.1 Introduction

Modern power systems are penetrated with growing DG, including utility owned middle-scale DG units and individually owned units. DG has a high degree of flexibility and provides a number of technical and economic benefits such as reduction of network energy losses, enhancement of power quality and system stability, and reduction of network maintenance costs [1,2]. In general, DG can be classified into two groups according to controllability of its generation resources [3], i.e. dispatchable and non-dispatchable. The dispatchable DG, such as microturbines and fuel cells, has cost-effective operation and good reliability, due to its sufficient and certain energy resources. On the other hand, the non-dispatchable DG including WTs and PVs generally has eco-friendly advantages and sustainable development incentives. In practice, these two types of DG can work together to maximise the overall benefits. With these DG units sufficiently installed, microgrids are developed as clusters of DG, energy storage systems, and flexible loads. Microgrids can efficiently supply power to loads by local power generation systems and interact with the main grid with electricity trading.

Considering their high efficiency and wide-area availability, WTs and PVs are two types of dominant non-dispatchable DG, and they have remarkably proliferated in microgrids. However, their intermittent and uncertain power generation poses a challenge for energy management [4]: when uncertain output power is excessively high, it may be curtailed; when it is low, power from a microgrid itself may not fulfil loads.

To alleviate this problem, demand response is an effective approach. Demand response is defined as payments designed to decrease electricity demand when wholesale market prices are high or when system reliability is at a high risk [5]. For energy management problems, demand response has been applied to reshape load profiles. Reference [6] proposes a day-ahead PBDR schedule to curtail or shift price-responsive loads to solve an energy management problem, indicating the day-ahead schedule can maximise welfare efficiently. The authors of [7,8] focus on optimising a day-ahead demand response market clearing procedure when wind power uncertainty is involved. In [9], the authors apply optimal power flow to optimise the day-ahead demand response in distribution networks considering renewable power uncertainties. All these works have indicated that day-ahead demand response can solve energy management problems efficiently, improving economic benefits for operators and customers.

However, a day-ahead demand response method alone is not always adequate to achieve optimal operating profits, especially when non-dispatchable DG output power heavily deviates from predicted values. Besides, the price-responsive loads are uncertain as well [10], which has an adverse impact on optimisation results. In addition, bus voltage constraints may not be guaranteed because of power imbalance caused by uncertainties of renewable power generation and loads. In order to tackle these problems, an output power adjustment of dispatchable DG is essential to deal with the possible power imbalance. Meanwhile, this output power adjustment can also help make more operating profits for the microgrid.

This chapter aims to coordinate the PBDR programme and the dispatchable DG for robust and optimal operation of a microgrid. Specifically, a daily two-stage robust coordination method consisting of the day-ahead PBDR and the hourly DG output power adjustments is proposed. In the first operation stage, day-ahead PBDR programme is optimally scheduled for 24 h with day-ahead predictions of renewable power generation and loads. Then, the dispatchable DG output power is adjusted hourly as the second-stage operation to track the uncertain variations of renewable power generation and loads. In addition, in the second operation stage, more accurate hourly predictions of renewable power generation and responded loads are applied to improve hourly operating efficiency. In this way, the second stage complements the first-stage decisions for any imperfect information. In other words, the hourly dispatchable DG can coordinate with the day-ahead PBDR efficiently for maximising operating profits while satisfying microgrid operating constraints in presence of the uncertain renewable power generation and loads.

To solve an optimisation problem under uncertainties, there are two major methodologies. The first one is conventional stochastic optimisation (SO) methods [11–13], which require scenario generation techniques, such as Monte Carlo sampling, to obtain a large number of random scenarios to present uncertainty realisations. However, SO methods can only provide probabilistic guarantees for constraint satisfaction. Moreover, it is usually difficult to accurately characterise probability distribution of uncertainty variables, and due to the requirement for a large number of scenarios, computational burdens of SO methods are always heavy. On the other hand, RO methods have gained increased attention in power system planning and operation research. RO methods are applied to solve unit commitment problems in [14–16], a DG planning problem in [17], an energy storage system planning problem in [18], and a plug-in hybrid electric vehicle transition planning problem in [19]. RO methods search for the worst case within predefined uncertainty sets first, and then objectives are optimised under the worst case. Since RO solutions are obtained under the worst case, the solutions are robust for any realisation of uncertainties. Compared to SO methods, RO methods have two major advantages [15]: (1) they utilise uncertainty sets instead of probability distribution data to achieve high computing performance; (2) they utilise the worst case instead of the optimal expectation to achieve robust solutions. Considering these advantages of RO methods, this chapter proposes a two-stage robust optimisation (TSRO) model for the coordinated operation method.

The major technical contents in this chapter to be introduced are summarised as follows.

A new two-stage coordination method for a microgrid is proposed, which coordinates a day-ahead PBDR programme and hourly output power dispatch of dispatchable DG to maximise operating profits under uncertainties.

A new PBDR model is developed, which takes into consideration load uncertainty and is suitable for the two-stage coordination framework.

A TSRO method is developed to optimise the proposed two-stage coordination method, which maximises the total operating profit with operating constraints satisfied under any uncertainty realisation.

8.2 Two-stage coordination of day-ahead demand response and microturbine dispatch

Operation of a microgrid aims to maximise operating profits and meanwhile satisfy operating constraints. To address uncertainties arisen by renewable power generation and loads, a two-stage coordinated operation method is developed for a microgrid. This method contains two operation stages which are day-ahead PBDR scheduling and hourly dispatchable DG output power adjustment, to increase solution robustness against the uncertainties on a coordinated optimisation basis. In this chapter, WTs and PVs are considered as the non-dispatchable DG and microturbines as the dispatchable DG. It is worth indicating that other DG systems such as fuel cells and biomass generators can also be considered as the second-stage power adjustment devices when necessary.

The proposed coordinated operation framework for the microgrid is shown in Figure 8.1.

8.2.1 First stage – day-ahead PBDR

In the first operation stage, the PBDR programme is scheduled one day ahead for 24 h of the next day. In the PBDR model, price-responsive loads respond to electricity price offers given to customers in advance. The mechanism of the PBDR is

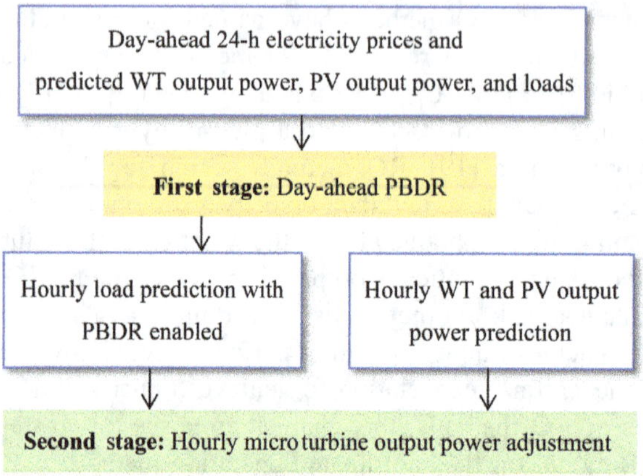

Figure 8.1 Two-stage coordination framework for microgrid

that the price-responsive loads are expected to increase when the lower electricity price is set, and on the other hand, the price-responsive loads are expected to decrease when the higher electricity price is offered. The day-ahead 24-hour PBDR programme aims to achieve the maximal total operating profit by optimising 24-hour electricity prices.

This PBDR scheduling must be performed based on predictions of renewable output power and loads which are uncertainty variables in the optimisation model. Despite the best efforts, the predictions can only support reasonable intervals for possible uncertainty realisations of renewable power generation and loads. As a result, these uncertainties can impact negatively on the operating profit maximisation and the microgrid operating constraints.

8.2.2 Second stage – hourly microturbine dispatch

In order to mitigate the negative impacts of the uncertainties of renewable power generation and loads, in the second operation stage, the output power of microturbines is adjusted after realisations of the uncertainties. In this stage, hourly renewable output power is more accurately predicted by modern weather-based power generation prediction systems, and more accurate hourly predictions of the loads responding to the first-stage PBDR programme are also completed with smart meter measurements and data analytics approaches. Thus, by employing the hourly predicted data, the microturbine power generation dispatch is optimised to complement the first-stage PBDR decisions. With the coordination of the day-ahead PBDR scheduling and the hourly microturbine dispatch, the overall operating profit can be maximised while ensuring the microgrid operating constraints.

Note that in practice the second-stage operation period should be designed carefully with several concerns, such as dispatchable DG ramping characteristics, uncertainty levels, and accuracy of predictions, as well as market bid/offer periods if any. In this chapter, an hour is set as the second-stage operation period, given that most of power generation dispatch is on the hourly basis. However, any other time lengths, such as 15 min, can also be considered if necessary, and it does not affect the effectiveness of the proposed coordination method.

8.3 Mathematical formulation

8.3.1 Price-based demand response model

This chapter applies a PBDR model which follows the demand response equation as

$$P_{\mathrm{L},t} = \mathrm{A}Pr_t^{\varepsilon} \tag{8.1}$$

where ε is price elasticity of load demand which is defined as the ratio of the change in the load demand to the change in the electricity price, and A is a constant. The calculation methods of ε and A can be found in [20].

The price elasticity depends on the price changing range and the load demand responding capability. For example, according to the electricity price and the load data given by [21], a price elasticity of load demand can be estimated as −0.2122.

In order to apply the PBDR model efficiently with the stepwise price-elastic demand curve concept proposed by [22], N^{Pr} price level rates are designed first. Then, by using A with the price elasticity of demand, N^{Pr} corresponding load demand response rates are computed. For example, given that $N^{\mathrm{Pr}} = 10$, ten stepwise PBDR levels are generated and shown in Table 8.1. It is important to indicate that other PBDR levels can also be considered according to practical data.

With these price and response levels, the real and reactive power loads in the stepwise PBDR model can be calculated as (assuming the power factor is fixed)

$$P_{\mathrm{L},i,t} = P_{\mathrm{L},i,t}^{\mathrm{pr}} \sum_{j \in J} \alpha_{j,t} L_{j,t} \left(1 + R_{\mathrm{L},i,j,t}^{\mathrm{unc}}\right) \tag{8.2}$$

$$Q_{\mathrm{L},i,t} = Q_{\mathrm{L},i,t}^{\mathrm{pr}} \sum_{j \in J} \alpha_{j,t} L_{j,t} \left(1 + R_{\mathrm{L},i,j,t}^{\mathrm{unc}}\right) \tag{8.3}$$

8.3.2 Coordination optimisation model

The two-stage coordination optimisation model is formulated as follows:

$$\min C_{\mathrm{MT}} + C_{\mathrm{WT}} + C_{\mathrm{PV}} + C_{\mathrm{grid}} - C_{\mathrm{rev}} \tag{8.4}$$

$$\text{s.t. } C_{\mathrm{MT}} = \sum_{t \in T} \sum_{m \in N_{\mathrm{MT}}} C_{\mathrm{MT,OM}} P_{\mathrm{MT},m,t} \tau \tag{8.5}$$

$$C_{\mathrm{WT}} = \sum_{t \in T} \sum_{n \in N_{\mathrm{WT}}} C_{\mathrm{WT,OM}} R_{\mathrm{WT},n,t} P_{\mathrm{WT},n}^{\mathrm{rated}} \tau \tag{8.6}$$

Table 8.1 Price-based demand response level

Price level	Price rate (%)	Expected response rate (%)
1	70	107.9
2	80	104.8
3	90	102.3
4	100	100.0
5	110	98.0
6	120	96.2
7	130	94.6
8	140	93.1
9	150	91.8
10	160	90.5

$$C_{\mathrm{PV}} = \sum_{t\in T}\sum_{n\in N_{\mathrm{PV}}} C_{\mathrm{PV,OM}} R_{\mathrm{PV},n,t} P_{\mathrm{PV},n}^{\mathrm{rated}} \tau \tag{8.7}$$

$$C_{\mathrm{grid}} = \sum_{t\in T}\left(C_{\mathrm{buy}} P_{\mathrm{def},t}\tau - C_{\mathrm{sell}} P_{\mathrm{sur},t}\tau\right) \tag{8.8}$$

$$C_{\mathrm{rev}} = C_{\mathrm{rev}}^{\mathrm{pr}} + C_{\mathrm{rev}}^{\mathrm{unc}} \tag{8.9}$$

$$C_{\mathrm{rev}}^{\mathrm{pr}} = \sum_{t\in T}\sum_{i\in N_{\mathrm{L}}} P_{\mathrm{L},i,t}^{\mathrm{pr}}\tau \sum_{j\in J} \alpha_{j,t} L_{j,t} Pr_j \tag{8.10}$$

$$C_{\mathrm{rev}}^{\mathrm{unc}} = \sum_{t\in T}\sum_{i\in N_{\mathrm{L}}} P_{\mathrm{L},i,t}^{\mathrm{pr}}\tau \sum_{j\in J} \alpha_{j,t} L_{j,t} Pr_j R_{\mathrm{L},i,j,t}^{\mathrm{unc}} \tag{8.11}$$

$$\alpha_{j,t} \in \{0,\ 1\},\ \forall j,t \tag{8.12}$$

$$\sum_{j\in J} \alpha_{j,t} = 1,\ \forall t \tag{8.13}$$

$$\sum_{t\in T}\sum_{i\in N_{\mathrm{L}}} P_{\mathrm{L},i,t}^{\mathrm{pr}} \sum_{j\in J} \alpha_{j,t} L_{j,t} Pr_{j,t} \le \sum_{t\in T}\sum_{i\in N_{\mathrm{L}}} P_{\mathrm{L},i,t}^{\mathrm{pr}} Pr_{0,t} \tag{8.14}$$

$$\sum_{t\in T}\sum_{i\in N_{\mathrm{L}}} P_{\mathrm{L},i,t}^{\mathrm{pr}} \sum_{j\in J} \alpha_{j,t} L_{j,t} \ge \sum_{t\in T}\sum_{i\in N_{\mathrm{L}}} P_{\mathrm{L},i,t}^{\mathrm{pr}} \tag{8.15}$$

$$P_{\mathrm{L},i,t} = P_{\mathrm{L},i,t}^{\mathrm{pr}} \sum_{j\in J} \alpha_{j,t} L_{j,t}\left(1 + R_{\mathrm{L},i,j,t}^{\mathrm{unc}}\right),\ \forall i,t \tag{8.16}$$

$$Q_{\mathrm{L},i,t} = Q_{\mathrm{L},i,t}^{\mathrm{pr}} \sum_{j\in J} \alpha_{j,t} L_{j,t}\left(1 + R_{\mathrm{L},i,j,t}^{\mathrm{unc}}\right),\ \forall i,t \tag{8.17}$$

$$0 \le P_{\mathrm{MT},m,t} \le P_{\mathrm{MT},m}^{\max},\ \forall m,t \tag{8.18}$$

$$\begin{aligned} P_{b+1,t} = {} & P_{b,t} - P_{0,b+1,t} + P_{\mathrm{MT},m,t} - P_{\mathrm{L},i,t} + R_{\mathrm{WT},n,t} P_{\mathrm{WT},n}^{\mathrm{rated}} \\ & + R_{\mathrm{PV},n,t} P_{\mathrm{PV},n}^{\mathrm{rated}},\ b \in Br(i),\ \forall i,t \end{aligned} \tag{8.19}$$

$$Q_{b+1,t} = Q_{b,t} - Q_{0,b+1,t} - Q_{L,i,t},\ b \in Br(i),\ \forall i,t \tag{8.20}$$

$$V_{i+1,t} = V_{i,t} - \frac{R_b P_{b,t} + X_b Q_{b,t}}{V_0},\ b \in Br(i,i+1),\ \forall i,t \tag{8.21}$$

$$1 - \Delta V^{\max} \leq V_{i,t} \leq 1 + \Delta V^{\max}, \ \forall i, t \tag{8.22}$$

$$-P_{b,\text{cap}} \leq P_{b,t} \leq P_{b,\text{cap}}, \ b \in Br(i), \ \forall i, t \tag{8.23}$$

$$-P_{\text{tie}} \leq P_{1,t} \leq P_{\text{tie}}, \ \forall t \tag{8.24}$$

$$P_{1,t} = P_{\text{def},t} - P_{\text{sur},t}, \ P_{\text{def},t} \geq 0, \ P_{\text{sur},t} \geq 0, \ \forall t \tag{8.25}$$

$$R_{\text{WT},n,t}^{\text{low}} \leq R_{\text{WT},n,t} \leq R_{\text{WT},n,t}^{\text{up}}, \ \forall n, t \tag{8.26}$$

$$R_{\text{PV},n,t}^{\text{low}} \leq R_{\text{PV},n,t} \leq R_{\text{PV},n,t}^{\text{up}}, \ \forall n, t \tag{8.27}$$

$$R_{\text{L},i,j,t}^{\text{low}} \leq 1 + R_{\text{L},i,j,t}^{\text{unc}} \leq R_{\text{L},i,j,t}^{\text{up}}, \ \forall i, j, t \tag{8.28}$$

The objective function (8.4) includes all the costs and the revenues during the whole day microgrid operation. Equations (8.5)–(8.9) are the computation formulae of these costs and revenues, i.e. the operation and maintenance (O&M) costs of the microturbines, the WTs, and the PVs, respectively, the transactions with the main grid, and the revenue of selling electricity to customers. Herein, the O&M costs of the WTs and the PVs are modelled considering their ageing along with the operation, regular maintenance, component replacement, and refurbishment [23]. In renewable system planning and operation, the O&M costs of the WTs and the PVs in (8.6) and (8.7) are considered as an important economic concern and they are normally modelled based on energy generated by the units, i.e. \$/kWh [24]. In addition, (8.8) computes the microgrid's transactions with the main grid including the buying electricity payment and the selling electricity revenue, and the transactions represent an important feature of the microgrid grid-connected mode. Considering the characteristics of the uncertain loads, in (8.9), the revenue from the customers is split into two parts, i.e. the predicted revenue based on the predicted loads and the uncertain revenue difference from the predicted one. Constraints (8.10) and (8.11) support the calculation functions of these two revenues, respectively.

Constraint (8.12) denotes the decision variable for each PBDR level is binary. Constraint (8.13) guarantees that only one PBDR level decision can be carried out for each hour. Constraints (8.14) and (8.15) guarantee the bills for the customers cannot increase and the energy which the customers can use cannot decrease. These mean that the proposed PBDR model does not reduce the customer economic benefits. Equations (8.16) and (8.17) compute the load active and reactive power consumption, respectively, after the PBDR applied with the load uncertainties.

Constraint (8.18) expresses the allowed range of the adjustable microturbine outputs.

Constraints (8.19)–(8.21) are the linearised distribution system power flow (Dist-Flow) model [25]. Constraint (8.22) ensures the voltage at each bus is kept within the allowed maximum deviation from the nominal value. Constraint (8.23) represents the branch power flow limits. Constraint (8.24) indicates the active power flow between the main grid and the microgrid cannot be over the tie line capacity.

Constraint (8.25) states the relationship between the power flow from the main grid to the microgrid and the electricity transactions, where $P_{\mathrm{def},t} > 0$ means the microgrid buys electricity from the main grid and $P_{\mathrm{sur},t} > 0$ means the microgrid sells electricity to the main grid.

Constraints (8.26) and (8.27) limit WT and PV output power in predicted intervals. Constraint (8.28) presents a possible variation interval for each uncertain load.

The objective (8.4) subject to constraints (8.5)–(8.28) presents a mixed-integer linear programming (MILP) with uncertainty variables. Herein, the day-ahead PBDR levels $\alpha_{j,t}$ are the first-stage decisions and the hourly microturbine power dispatch decisions $P_{\mathrm{MT},m,t}$ are the second-stage decisions in this coordination optimisation problem. The uncertainty variables include $R_{\mathrm{WT},n,t}$, $R_{\mathrm{PV},n,t}$, and $R^{\mathrm{unc}}_{\mathrm{L},i,j,t}$.

8.4 Two-stage robust optimisation method

To achieve robust coordination under uncertainties of renewable power generation and loads, the TSRO method is applied to optimise the coordination model of the day-ahead PBDR programme and the hourly microturbine power dispatch in the two operation stages.

The principle of the TSRO method is introduced in Section 7.4.1.

8.4.1 Uncertainty set modelling

For the uncertainty variables, uncertain renewable output power rates, $R_{\mathrm{WT},n,t}$, $R_{\mathrm{PV},n,t}$, and uncertain load variation rate $R^{\mathrm{unc}}_{\mathrm{L},i,j,t}$, according to their constraints (8.26)–(8.28), three polyhedral uncertainty sets are modelled as follows:

$$U_{\mathrm{WT}} = \{R_{\mathrm{WT},n,t} \in \mathbb{R}^{n_{\mathrm{wt}}}: \mu_{\mathrm{WT},l} \leq \frac{\sum\limits_{n\in N_{\mathrm{WT}}}\sum\limits_{t\in T} R_{\mathrm{WT},n,t}}{\sum\limits_{n\in N_{\mathrm{WT}}}\sum\limits_{t\in T} R^{\mathrm{pr}}_{\mathrm{WT},n,t}} \leq \mu_{\mathrm{WT},u},\ R^{\mathrm{low}}_{\mathrm{WT},n,t} \leq R_{\mathrm{WT},n,t} \leq R^{\mathrm{up}}_{\mathrm{WT},n,t},\ \forall n,t\} \tag{8.29}$$

$$U_{\mathrm{PV}} = \{R_{\mathrm{PV},n,t} \in \mathbb{R}^{n_{\mathrm{pv}}}: \mu_{\mathrm{PV},l} \leq \frac{\sum\limits_{n\in N_{\mathrm{PV}}}\sum\limits_{t\in T} R_{\mathrm{PV},n,t}}{\sum\limits_{n\in N_{\mathrm{PV}}}\sum\limits_{t\in T} R^{\mathrm{pr}}_{\mathrm{PV},n,t}} \leq \mu_{\mathrm{PV},u},\ R^{\mathrm{low}}_{\mathrm{PV},n,t} \leq R_{\mathrm{PV},n,t} \leq R^{\mathrm{up}}_{\mathrm{PV},n,t},\ \forall n,t\} \tag{8.30}$$

$$U_{\mathrm{L}} = \{P^{\mathrm{unc}}_{\mathrm{L}} \in \mathbb{R}^{n_l}: \mu_{\mathrm{L},l} \leq 1 + \frac{1}{n_i n_t}\sum_{i\in N_{\mathrm{L}}}\sum_{t\in T}\sum_{j\in J} \alpha_{j,t} R^{\mathrm{unc}}_{\mathrm{L},i,j,t} \leq \mu_{\mathrm{L},u},\ R^{\mathrm{low}}_{\mathrm{L},i,j,t} \leq 1 + R^{\mathrm{unc}}_{\mathrm{L},i,j,t} \leq R^{\mathrm{up}}_{\mathrm{L},i,j,t},\ \forall i,j,t\} \tag{8.31}$$

In these sets, the lower and upper bounds, $R^{\text{low}}_{\text{WT/PV}}, R^{\text{low}}_{\text{L}}$ and $R^{\text{up}}_{\text{WT/PV}}, R^{\text{up}}_{\text{L}}$ can be obtained from the prediction intervals of the variations of WT output power, PV output power, and loads, respectively. Several interval prediction techniques have been proposed in [26]. As a result, these bounds indicate the possible intervals of the different uncertainties for each single hour.

Furthermore, for each group of the uncertainty variables, uncertainty budgets defined in pair of μ_l and μ_u are used to limit the overall uncertainty degree. These budgets indicate the possible intervals of the different uncertainties for the whole day. For example, it is expressed by (8.29) that the actual total WT output of the whole day is assumed to be in an interval formed by the lower and upper budgets with the predicted total output as reference. Note that the size of the uncertainty set is enlarged when μ_l falls and μ_u rises. As a result, the final optimisation solution would be more robust, but more conservative. Therefore, the uncertainty budgets play a key role in the TSRO method and the budget design depends on a few aspects including the accuracy of prediction and the expected solution robustness.

8.4.2 Two-stage robust optimisation model

The proposed microgrid coordination optimisation model is converted to a TSRO model, which can be formulated in the following compact matrix form:

$$\min_{x} \boldsymbol{c}^T\boldsymbol{x} + \max_{u} \min_{y} \boldsymbol{d}^T\boldsymbol{y} + \boldsymbol{e}^T\boldsymbol{u} \tag{8.32}$$

$$\text{s.t.} \quad \boldsymbol{Ax} \geq \boldsymbol{b} \tag{8.33}$$

$$\boldsymbol{Fx} + \boldsymbol{Gy} + \boldsymbol{Hu} \leq \boldsymbol{v} \tag{8.34}$$

$$\boldsymbol{I}(\boldsymbol{u})\boldsymbol{x} + \boldsymbol{Jy} + \boldsymbol{Ku} = \boldsymbol{w} \tag{8.35}$$

$$\boldsymbol{u} \in \boldsymbol{U} \tag{8.36}$$

The TSRO model is a '*min-max-min*' tri-level optimisation problem with corresponding constraints for different operation stages. The detailed characteristics of the TSRO model are given in Section 7.4.3.

The objective described in (8.32) has the '*min-max-min*' objective form. According to the characteristics of the '*min-max-min*' form, the proposed coordination optimisation objective (8.4) can be rewritten as the following:

$$\min_{\alpha} -C^{\text{pr}}_{\text{rev}} + \max_{R_{\text{WT}},R_{\text{PV}},R^{\text{unc}}_{\text{L}}} \min_{P_{\text{MT}},V,P,Q} C_{\text{MT}} + C_{\text{WT}} + C_{\text{PV}} + C_{\text{grid}} - C^{\text{unc}}_{\text{rev}} \tag{8.37}$$

It is indicated from (8.37) that the predicted revenue from the loads $-C^{\text{pr}}_{\text{rev}}$ is the first-stage objective and the summation of the other costs and revenues $C_{\text{MT}} + C_{\text{WT}} + C_{\text{PV}} + C_{\text{grid}} - C^{\text{unc}}_{\text{rev}}$ is the second-stage objective.

There are three types of the variables at the different stages.

$\boldsymbol{x}$ represents a vector of decision variables in the first operation stage, regarded as the 'here-and-now' decisions. In the proposed coordination model formulated in Section 8.3.2, $\boldsymbol{x}$ indicates the vector of the binary variables for all the PBDR levels, i.e. α. Accordingly, the PBDR constraints (8.12)–(8.15) are converted into (8.33).

$\boldsymbol{y}$ stands for a vector of decision variables and dependent variables in the second operation stage, as the 'wait-and-see' decisions. The dispatchable microturbine output power P_{MT} and the power flow variables P, Q, V are included in $\boldsymbol{y}$. The TSRO constraint (8.34) is made up by (8.18) and (8.22)–(8.25), while the constraint (8.35) is formed by (8.19)–(8.21). Additionally, $\boldsymbol{I}(\boldsymbol{u})$ is the coefficient matrix of $\boldsymbol{x}$ and it includes the uncertainty variables $\boldsymbol{u}$ related to (8.16) and (8.17).

$\boldsymbol{u}$ expresses a vector for the uncertainty variables which include the output power rates of the WTs and the PVs, i.e. R_{WT} and R_{PV} and the uncertain load proportions to the predicted values, i.e. $R_{\mathrm{L}}^{\mathrm{unc}}$. The uncertainty sets (8.29)–(8.31) are used as the TSRO constraint (8.36).

The other bold uppercase symbols mean matrices of parameters while the other bold lowercase symbols for the parameter vectors.

In this TSRO modelling, the uncertainty variables are searched first within the uncertainty sets to form the worst case. Then these variables are fixed and applied in optimisation as parameters. Thus, the solution obtained with the worst case of possible uncertainty realisation is robustly optimal for all uncertainty realisation provided by the uncertainty sets.

8.4.3 Solution algorithm

Since the column-and-constraint generation (C&CG) algorithm has a faster convergence speed than the Benders decomposition algorithm, which is concluded in [27], this chapter applies the C&CG algorithm to solve the proposed TSRO problem.

In the C&CG algorithm, the TSRO problem is divided into a master problem and a sub-problem which can formulate the first and the second operation stages, respectively. The master problem and the sub-problem are solved iteratively until they converge with a gap not greater than the preset threshold γ.

The master problem is formulated as follows:

$$\min_{\boldsymbol{x}} \boldsymbol{c}^T\boldsymbol{x} + \lambda \tag{8.38}$$

$$\text{s.t.} \quad \boldsymbol{A}\boldsymbol{x} \geq \boldsymbol{b} \tag{8.39}$$

$$\lambda \geq \boldsymbol{d}^T\boldsymbol{y}_l + \boldsymbol{e}^T\boldsymbol{u}_l^* \tag{8.40}$$

$$\boldsymbol{F}\boldsymbol{x} + \boldsymbol{G}\boldsymbol{y}_l + \boldsymbol{H}\boldsymbol{u}_l^* \leq \boldsymbol{v} \tag{8.41}$$

$$\boldsymbol{I}(\boldsymbol{u}_l^*)\boldsymbol{x} + \boldsymbol{J}\boldsymbol{y}_l + \boldsymbol{K}\boldsymbol{u}_l^* = \boldsymbol{w}, \ \forall \boldsymbol{u}_l^* \in \boldsymbol{S} \tag{8.42}$$

Since in the proposed coordination optimisation model, $\boldsymbol{x}$ is a vector of binary decision variables, the master problem is an MILP. In this problem, with a vector of the fixed uncertainty variables $\boldsymbol{u}_l^*$ which are generated from the sub-problem, an optimal solution can be obtained as $(\boldsymbol{x}^*, \lambda^*)$. It is denoted that the solution $\boldsymbol{x}^*$ is the current optimal solution for the first-stage optimisation and it is fixed temporarily and used for solving the sub-problem further.

On the other hand, a sub-problem is modelled as follows:

$$S(\boldsymbol{u},\boldsymbol{x}^*) = \max_{\boldsymbol{u}} \min_{\boldsymbol{y}} \boldsymbol{d}^T\boldsymbol{y} + \boldsymbol{e}^T\boldsymbol{u} \tag{8.43}$$

$$\text{s.t. } \boldsymbol{Fx}^* + \boldsymbol{Gy} + \boldsymbol{Hu} \le \boldsymbol{v} \tag{8.44}$$

$$\boldsymbol{I}(\boldsymbol{u})\boldsymbol{x}^* + \boldsymbol{Jy} + \boldsymbol{Ku} = \boldsymbol{w} \tag{8.45}$$

$$\boldsymbol{u} \in \boldsymbol{U} \tag{8.46}$$

It is worth noting that exactly solving the '*max-min*' sub-problem with polyhedral uncertainty sets is significantly challenging [27]. To solve the sub-problem, this chapter uses Karush–Kuhn–Tucker conditions [28] and a Big-M constraint approach to further convert the '*max-min*' form optimisation problem into an MILP model.

First, the '*max-min*' optimisation problem can be converted into a bilinear constrained maximisation problem by Karush–Kuhn–Tucker conditions as follows:

$$S(\boldsymbol{u},\boldsymbol{x}^*) = \max_{\boldsymbol{u}} \min_{\boldsymbol{y}} \boldsymbol{d}^T\boldsymbol{y} + \boldsymbol{e}^T\boldsymbol{u} \tag{8.47}$$

$$\text{s.t. } \boldsymbol{Fx}^* + \boldsymbol{Gy} + \boldsymbol{Hu} \le \boldsymbol{v} \tag{8.48}$$

$$\boldsymbol{I}(\boldsymbol{u})\boldsymbol{x}^* + \boldsymbol{Jy} + \boldsymbol{Ku} = \boldsymbol{w} \tag{8.49}$$

$$\boldsymbol{d} + \boldsymbol{G}^T\boldsymbol{\rho} + \boldsymbol{J}^T\boldsymbol{\varphi} = \boldsymbol{0} \tag{8.50}$$

$$(\boldsymbol{Fx}^* + \boldsymbol{Gy} + \boldsymbol{Hu} - \boldsymbol{v})_i\rho_i = 0, \ \forall \boldsymbol{i} \tag{8.51}$$

$$\boldsymbol{\rho} \ge \boldsymbol{0}, \boldsymbol{\varphi} \text{ free} \tag{8.52}$$

$$\boldsymbol{u} \in \boldsymbol{U} \tag{8.53}$$

Here, $\boldsymbol{\rho}$ and $\boldsymbol{\varphi}$ are the dual variables of the sub-problem constraints (8.44) and (8.45), respectively.

Then, the bilinear constraint (8.51) can be further converted through the Big-M approach as follows:

$$\rho_i \le Mz_i, \ \forall i \tag{8.54}$$

$$(\boldsymbol{v} - \boldsymbol{Fx}^* - \boldsymbol{Gy} - \boldsymbol{Hu})_i \le M(1 - z_i), \ \forall i \tag{8.55}$$

$$z_i \in \{0, 1\}, \ \forall i \tag{8.56}$$

Herein, z_i is a binary additional variable and M is a large constant number. The selection of M is normally based on the variable ranges. To improve the Big-M method efficiency, it should be the maximal limit of the left sides of '$\le$'. However, in this chapter, the left sides of '$\le$' in (8.54) and (8.55) do not have a maximal limit. In this case, M is set larger than the possibly maximal value of the optimised left sides of (8.54) and (8.55). On the other hand, if M is set too large, the computation efficiency would be reduced. Thus, it can be suggested that a very large number is selected as initial M to do the TSRO method for a new system, and then M is reduced to a number which is larger than all the optimised left sides results but not much larger, to improve computation efficiency of the TSRO method for this system.

Therefore, the objective (8.47) and the constraints (8.48)–(8.50) and (8.52)–(8.56) make up an MILP sub-problem. The optimal solution of this sub-problem $\boldsymbol{u}_l^*$ can stand for one worst case of possible uncertainty realisation, and it is added in an uncertainty solution set $\boldsymbol{S}$ and passed to the master problem. Besides, additional constraints including $\boldsymbol{u}_l^*$ as a vector of parameters and $\boldsymbol{y}_l$ as a vector of additional variables are generated and added in the master problem of the next iteration.

The C&CG process continues to solve the master problem and the sub-problem iteratively until the gap between the lower bound given by the master problem and the upper bound given by the sub-problem is reduced to an allowed termination value. Finally, the first-stage decision variable vector, $\boldsymbol{x}$, is robustly optimised under the worst case of uncertainty realisation.

The complete C&CG algorithm can be summarised by the following steps. γ is the predefined allowed maximal gap of the bounds.

Step 1. Initialise the lower bound LB as $-\infty$, the upper bound UB as $+\infty$, the sub-problem solution set $\boldsymbol{S}$ as $\varnothing$, and the iteration number l as 0.

Step 2. Solve the master problem to obtain the optimal solutions $\boldsymbol{x}_l^*, \lambda_l^*$ and update LB by $\max\{LB, \boldsymbol{c}^T\boldsymbol{x}_l^* + \lambda_{l+1}^*\}$.

Step 3. Solve the sub-problem to compute and add the optimal solution $\boldsymbol{u}_{l+1}^*$ to $\boldsymbol{S}$, provide y_{l+1} as additional variables and its corresponding constraints for the master problem and update UB by $\min\{UB, \boldsymbol{c}^T\boldsymbol{x}_l^* + S(\boldsymbol{u}_{l+1}^*, \boldsymbol{x}_l^*)\}$.

Step 4. Compute the bound gap, $Gap = |(UB - LB)/LB|$.

Step 5. If $Gap \leq \gamma$, obtain the final optimal solution $\boldsymbol{x}_l^*$ and terminate the procedure. Otherwise, increase l by 1, go to Step 2, and continue the procedure.

Note that only the final $\boldsymbol{x}^*$ is regarded as the final optimisation solution, since the other variables can be re-optimised during the second operation stage with actual uncertainty realisation.

8.5 Case study

8.5.1 Test system description

The proposed daily coordination method is tested on a 33-bus radial distribution system which can be modified to a microgrid and its data can be obtained in [29]. All loads in this microgrid are assumed to be price-responsive loads. The data of WTs, PVs, and microturbines, including locations, rated power, and O&M costs, are given in Table 8.2. The day-ahead market electricity price for each hour is the average price of NSW 2014 winter (July–September) prices from AEMO [30] and it is shown in Figure 8.2. This market price can be used as the price for buying energy from the main grid, i.e. C_{buy}, and the price for selling energy to the main grid, C_{sell}, is 30% of the buying price. In addition, the original electricity price for

Table 8.2 Wind turbine, photovoltaic, and microturbine data

Location	**Rated power (MW)**		
	Wind turbine	**Photovoltaic**	**Microturbine**
6	1	0.7	0.8
18	0.6	0.5	0.6
24	0.5	0.3	0.6
32	0.4	0.5	0.5
O&M cost ($/kWh)	0.010	0.008	0.023

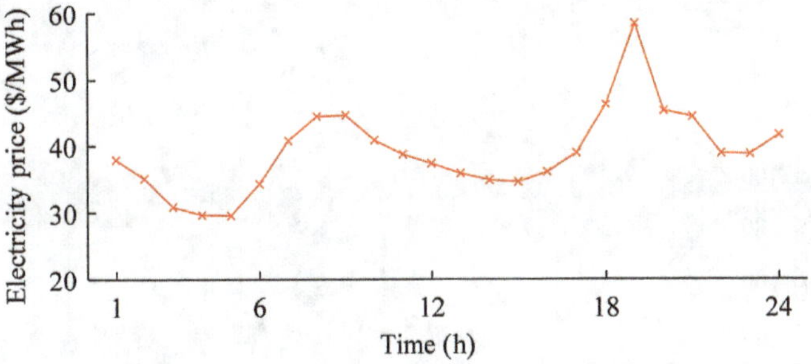

Figure 8.2 Day-ahead market electricity price

selling energy to customers, Pr_0, is set as $60/MWh in this chapter. It must be noted that any practical price profile can be used, and it does not affect the effectiveness of the proposed coordination method.

Simulations are implemented on a 64-bit PC with 3.30-GHz CPU and 8 GB RAM with YALMIP toolbox [31] in MATLAB®. The optimisation problems are solved by GORUBI [32].

8.5.2 Uncertainty sets

The uncertainty sets are constructed based on the interval predictions for each hour. The predicted data used in this chapter are shown and explained as follows.

Figure 8.3 shows a specified day-ahead prediction of 24-hour loads (percentages of expected peak loads). The predicted mean values (M_{L}) in this figure are modified from the IEEE RTS-96 winter weekday hourly load data and they are multiplied by the peak loads to obtain the predicted mean loads $P_{\mathrm{L}}^{\mathrm{pr}}$. The lower and upper bounds of the interval are set as $R_{\mathrm{L}}^{\mathrm{low}}$ and $R_{\mathrm{L}}^{\mathrm{up}}$ for each hour in (8.31).

Figure 8.4 shows a day-ahead prediction of 24-hour output power of the WTs and the PVs (percentages of corresponding rated power). The mean values of the WT output power rate (M_{WT}) are generated based on a practical high wind speed

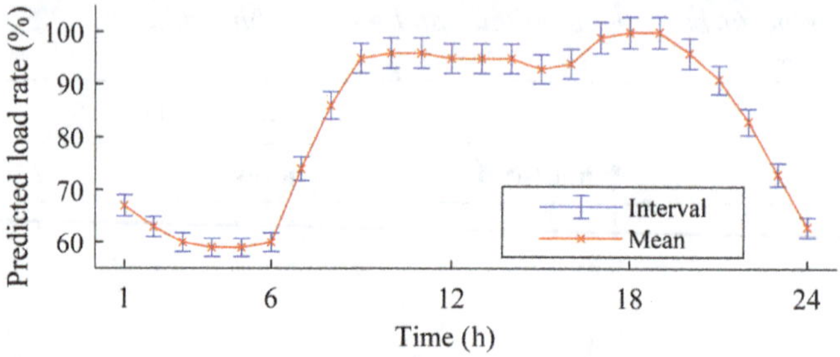

Figure 8.3 Day-ahead load interval prediction

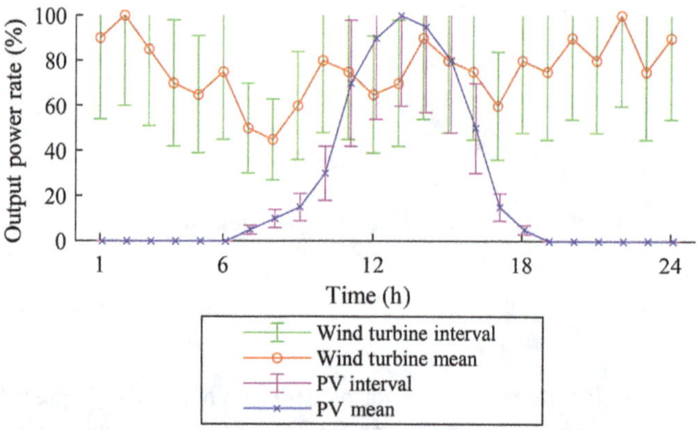

Figure 8.4 Day-ahead WT and PV output power interval predictions

case and they are applied as $R_{\mathrm{WT}}^{\mathrm{pr}}$ in (8.29) to calculate the predicted total output power of the whole day. The lower and upper bounds of the WT output power interval are set as $R_{\mathrm{WT}}^{\mathrm{low}}$ and $R_{\mathrm{WT}}^{\mathrm{up}}$ for each hour in (8.29). Similarly, the mean values of the PV output power rate (M_{PV}) are generated based on a standard winter sunny-day solar case and they are applied as $R_{\mathrm{PV}}^{\mathrm{pr}}$ in (8.30). The lower and upper bounds of the PV output power interval are set as $R_{\mathrm{PV}}^{\mathrm{low}}$ and $R_{\mathrm{PV}}^{\mathrm{up}}$ for each hour in (8.30). In the case study, it is assumed that in this microgrid, all the WTs have the same output power rates for each hour and so do all the PVs. Note that in practice the output power rates for different WTs or PVs can be different if detailed local renewable power generation predictions are available.

Table 8.3 Uncertainty budgets

Budgets	WT output power	PV output power	Loads
Lower μ_l	90%	90%	97.5%
Upper μ_u	110%	110%	102.5%

In the test microgrid, all simulations are implemented with these load, WT output power, and PV output power profiles. It is worth noting that other profiles can also be used when necessary and they do not affect the effectiveness of the proposed coordination method.

The uncertainty budgets of these uncertainty sets are shown in Table 8.3. In this table, the lower and upper budgets are defined for WT output power, PV output power, and loads, respectively. Taking the WT output power as an example, 90% is set as the lower budget $\mu_{\mathrm{WT},l}$ in (8.29) and 110% is set as the upper budget $\mu_{\mathrm{WT},u}$ in (8.29). These uncertainty budgets state that the actual 24-hour total WT output power is predicted to vary between 90% and 110% of the predicted total output power which is calculated according to the mean values. In addition, since the renewable power generation is far more uncertain than the loads, the WT and PV output power has much larger uncertainty set sizes than the loads.

The above uncertainty sets are applied in daily coordination method simulations in Sections 8.5.3 and 8.5.4, as well as in feasibility check in Section 8.5.5.

8.5.3 First-stage price-based demand response results

With the uncertainty sets constructed in Section 8.5.2, the proposed daily coordination method is tested with the TSRO termination threshold γ as 0.5%. Solution results including the lower and upper bounds of the objective and their gaps during the C&CG algorithm process are listed in Table 8.4.

Since the objective (8.37) is to minimise the overall operating cost, the negative signs in Table 8.4 indicate that the microgrid operator can make operating profits. This table also shows that the gap of the bounds declines for each iteration, which implies that the C&CG algorithm converges effectively. Moreover, the final bounds represent a reasonable range of the total operating profit under the worst case, i.e. between \$3,279.32 and \$3,288.18. Note that in practice the final actual total operating profit can be larger than this range, since the microturbine adjustable output power will compensate hourly to further maximise the profit under uncertainty realisations. The final actual profit analysis is demonstrated in Section 8.5.4.

In addition, Figure 8.5 shows the 24-hour PBDR scheduling results, i.e. the first-stage decisions. In the figure, the expected loads in megawatts without PBDR and with the first-stage PBDR scheduling are displayed, respectively. It is reasonable to see that the day-ahead PBDR decisions are optimised to enlarge the loads between 13:00 and 15:00 when the PV power generation is sufficient. Moreover, the loads are expected to be higher between 2:00 and 5:00, because the WT power

Table 8.4 Column and constraint generation algorithm results

Iteration	Upper bound	Lower bound	Gap (%)
1	−3,263.22	−3,492.15	6.56
2	−3,263.22	−3,342.71	2.38
3	−3,263.22	−3,309.14	1.39
4	−3,268.68	−3,295.83	0.82
5	−3,268.94	−3,292.37	0.71
6	−3,273.37	−3,290.52	0.52
7	−3,279.32	−3,288.18	0.27

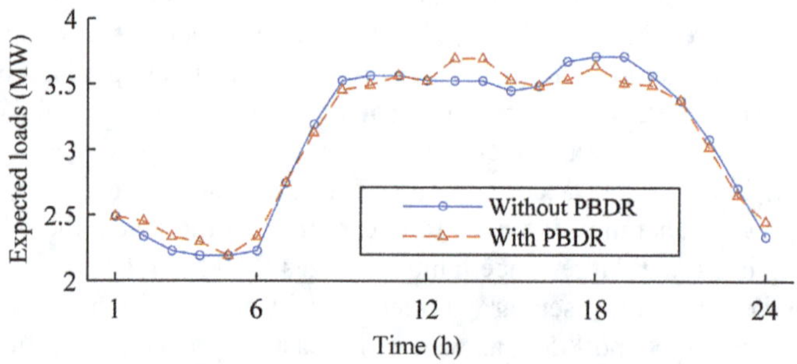

Figure 8.5 24-hour price-based demand response results

generation is sufficient. On the other hand, the PBDR levels are set up to reduce the loads, when the PV and WT output power is not enough and the electricity price is high between 17:00 and 20:00.

8.5.4 Second-stage implementation results

Followed by the first-stage day-ahead PBDR scheduling, the second-stage microturbine dispatch is implemented. The accurate hourly predictions in the second stage can realise the uncertainties of renewable power generation and loads. Thus, the microturbine power generation adjustments are optimised and implemented according to the accurate hourly predictions. For each hour in the second stage, the objective is

$$\min_{P_{\mathrm{MT}},V,P,Q} (C_{\mathrm{MT}} + C_{\mathrm{grid}}) + [C_{\mathrm{WT}} + C_{\mathrm{PV}} - C_{\mathrm{rev}}] \tag{8.57}$$

Since the uncertainties of the WT output power, the PV output power, and the loads are realised by the accurate hourly predictions, the O&M costs of the WTs and the PVs as well as the revenue from the customers $[C_{WT} + C_{PV} - C_{rev}]$ are fixed. Thus, the second stage is actually to minimise the summation of the O&M cost of the microturbines and the transaction cost with the main grid, i.e. $(C_{MT} + C_{grid})$. The microturbine power outputs are the controllable variables in the second stage. With constraints (8.5)–(8.11) and (8.19)–(8.28), the second-stage microturbine dispatch optimisation problem is of linear programming.

The day-ahead PBDR scheduling results obtained in Section 8.5.3 are used as parameters in the second-stage optimisation problem. Accurate hourly predictions of the load rate, the WT output power rate, and the PV output power rate are given as the black curves in Figures 8.6, 8.7, and 8.8, respectively. It is worth indicating that other hourly prediction scenarios can also be applied to verify the second-stage operation if necessary.

The 24-hour total operating profit is maximised as \$3,489.89 with the second stage implemented. The microgrid sells only 1.694 MWh to the main grid and it does not buy energy during the 24 h.

In addition, the hourly microturbine energy generation and the hourly energy sold to the main grid are shown in Figure 8.9. The microturbine energy generation

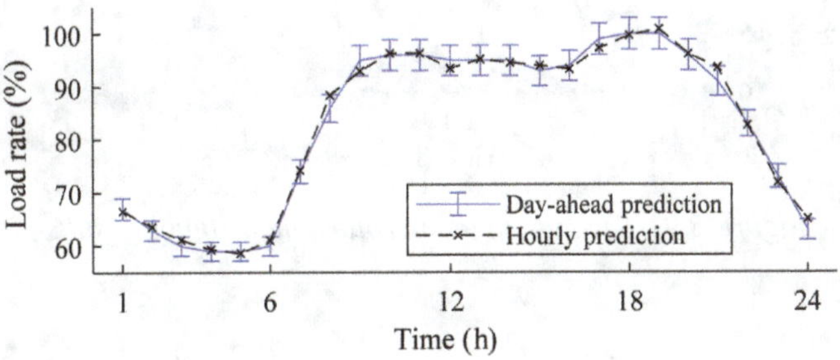

Figure 8.6 Predictions of loads for second-stage implementation

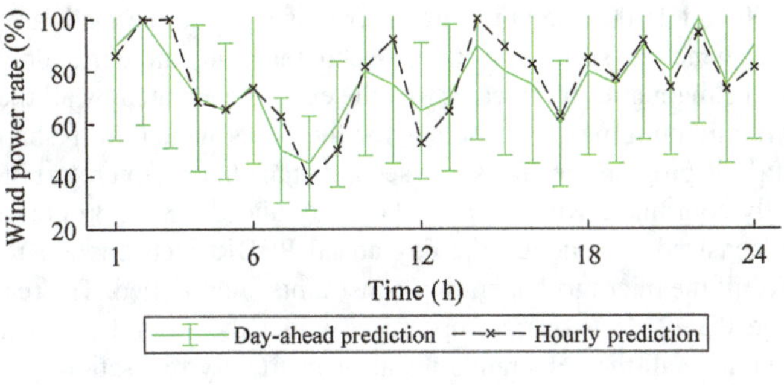

Figure 8.7 Second-stage WT output power predictions

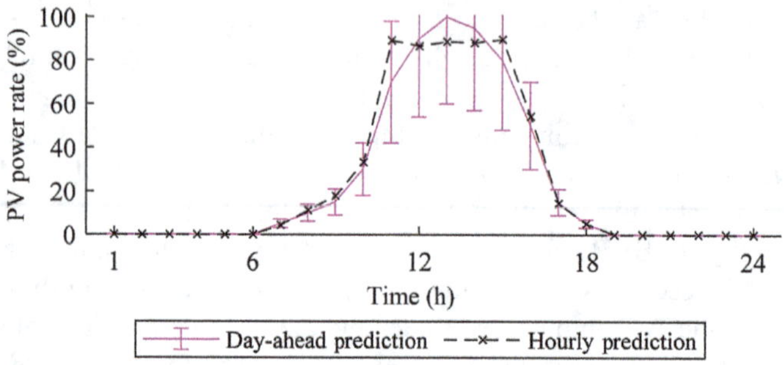

Figure 8.8 Second-stage PV output power predictions

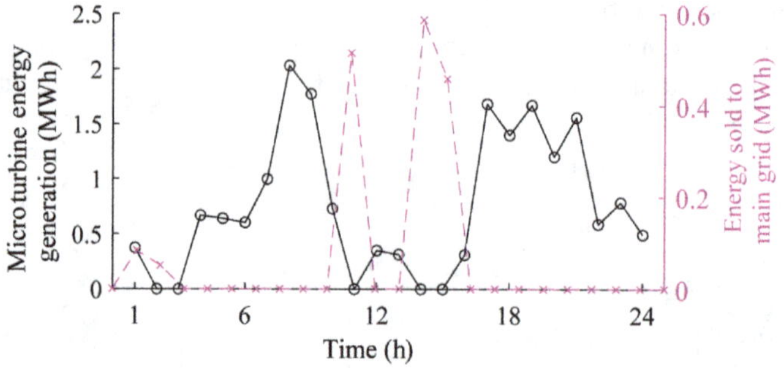

Figure 8.9 Second-stage microturbine dispatch results

is the total energy generation of the four units in the microgrid. During the hours of 2:00, 3:00, 11:00, 14:00, and 15:00, the energy generated by the renewable DG is sufficient, so that the microgrid sells energy to the main grid and the microturbines does not generate energy. On the other hand, the microturbines generate high energy between 8:00 and 9:00 when the wind and solar energy is low, as well as between 17:00 and 21:00 when the solar energy is low and the loads are high. The microturbines generate sufficient energy such that the microgrid does not buy energy from the main grid. It indicates that the energy generated by all the DG units including the microturbines is sufficient for the loads which are reshaped by the day-ahead PBDR programme. Thus, the second-stage hourly microturbine dispatch can efficiently coordinate with the first-stage day-ahead PBDR programme.

It is emphasised that though the day-ahead PBDR decisions are fixed in the second stage, all the microgrid operating constraints are satisfied. The reason is that the first-stage PBDR decisions are optimised under the worst case, which means they can guarantee all the constraints for any uncertainty realisation.

8.5.5 Monte Carlo simulation check

Monte Carlo simulations are carried out to check performance of the proposed daily two-stage coordination method under uncertainty realisations. In these simulations, the second-stage operation is implemented with multiple 24-hour scenarios which are generated by Monte Carlo sampling. Moreover, the results are compared with a conventional single-stage operation method. In this single-stage method, the PBDR levels and the microturbine output power dispatch are scheduled simultaneously one day ahead for 24 h without the uncertainties considered, and the PBDR and the microturbine output power decisions are fixed during the 24-hour operation period. Figure 8.10 shows the PBDR scheduling results of the proposed two-stage coordination method, i.e. the expected loads with the PBDR decisions, and the conventional single-stage operation method results including both the expected loads with the PBDR decisions and the microturbine output power.

Normal distribution-based Monte Carlo sampling is performed to generate hourly predictions of the WT output power, the PV output power, and loads. The mean values shown in Figures 8.3 and 8.4 are used as the expectations ($M_{\mathrm{WT}}, M_{\mathrm{PV}}, M_{\mathrm{D}}$) of the normal distribution for the hourly WT output power, PV output power, and loads. In terms of standard deviations ($\sigma_{\mathrm{WT}}, \sigma_{\mathrm{PV}}, \sigma_{\mathrm{D}}$), three different scenario groups with different standard deviations are generated and shown in Table 8.5. Taking Deviation group 1 as an example, for each hour, the standard deviations $\sigma_{\mathrm{WT}} = 5\%M_{\mathrm{WT}}$, $\sigma_{\mathrm{PV}} = 5\%M_{\mathrm{PV}}$, and $\sigma_{\mathrm{D}} = 1\%M_{\mathrm{D}}$. Since the renewable power generation has a higher uncertainty degree, it has larger standard deviations than the loads. From Deviation group 1 to 3, the standard deviations increase, which means the uncertainty level increases. For each group, 1,000 hourly prediction scenarios are randomly generated with the normal distribution. These

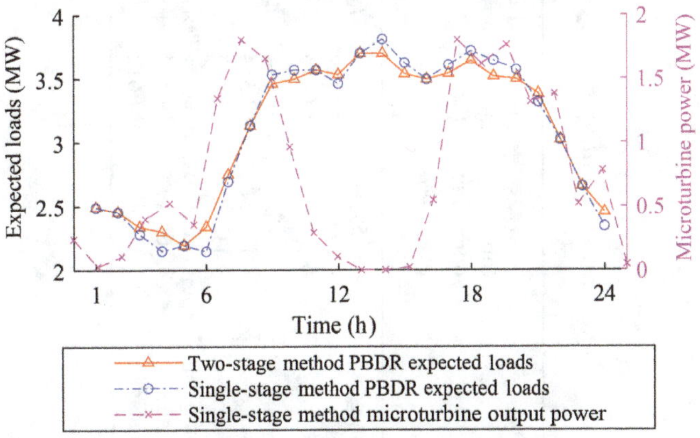

Figure 8.10 Day-ahead scheduling results of different methods

Table 8.5 Monte Carlo simulation results

Operation method	Average total operating profit ($)	Microturbine generation (MWh)	Energy bought from main grid (MWh)	Energy sold to main grid (MWh)	Average maximal voltage deviation p.u.	Two-stage beat single-stage in profit (%)	Two-stage beat single-stage in voltage deviation (%)
Deviation group 1: $\sigma_{\mathrm{WT}} = 5\%M_{\mathrm{WT}}$; $\sigma_{\mathrm{PV}} = 5\%M_{\mathrm{PV}}$; $\sigma_{\mathrm{D}} = 1\%M_D$							
Two-stage	3,484.56	18.148	0.000	0.704	0.0151	100%	85.2%
Single-stage	3,465.65	17.612	0.970	1.225	0.0152		
Deviation group 2: $\sigma_{\mathrm{WT}} = 10\%M_{\mathrm{WT}}$; $\sigma_{\mathrm{PV}} = 10\%M_{\mathrm{PV}}$; $\sigma_{\mathrm{D}} = 2\%M_{\mathrm{D}}$							
Two-stage	3,479.21	18.514	0.000	0.796	0.0152	100%	93.6%
Single-stage	3,437.12	17.612	1.989	1.970	0.0160		
Deviation group 3: $\sigma_{\mathrm{WT}} = 20\%M_{\mathrm{WT}}$; $\sigma_{\mathrm{PV}} = 20\%M_{\mathrm{PV}}$; $\sigma_{\mathrm{D}} = 4\%M_{\mathrm{D}}$							
Two-stage	3,464.49	19.564	0.009	1.055	0.0154	100%	99.5%
Single-stage	3,378.48	17.612	4.064	3.246	0.0183		

scenarios are applied to check performances of the proposed two-stage coordination method and the conventional single-stage method. The Monte Carlo simulation results are shown in Table 8.5.

For each deviation group, the first row lists the results of the proposed two-stage method while the second row lists those of the conventional single-stage method. The results include the average total operating profit, the average total microturbine energy generation, the average total energy bought/sold from/to the main grid, and the average maximal bus voltage deviation of the 1,000 scenarios. In addition, considering comparison of the results for each scenario, two performance analysis rates are calculated and shown in Table 8.5 as well. In the last two columns, the numbers indicate the percentages of the scenarios where the two-stage method outperforms the single-stage one in terms of the total operating profit and the bus voltage deviation, respectively. These comparison results are explained as follows.

First, for the average total operating profit of the 1,000 scenarios, the two-stage coordination method outperforms the single-stage one for all the three deviation groups. In comparison among the different groups, as the uncertainty degree increases from Deviation group 1 to 3, both the operating profits of the two-stage and single-stage methods decrease. However, it is noted that the operating profit of the single-stage method decreases significantly more than that of the two-stage coordination one. Specifically, for Deviation group 1, the two-stage method has the average total operating profit of \$3,484.56 slightly larger than \$3,465.65 which the single-stage method obtains by \$18.92. But for Deviation group 3, the two-stage operating profit is \$3,464.49 which is much larger than the single-stage one (\$3,378.48) by \$86.01. The major reason is that as the deviations increase, the energy exchange (the 4th and 5th columns) with the main grid increases almost proportionally for the single-stage method, which is not economical and decreases the operating profits. However, through the adjustment of the microturbine power generation (the 3rd column), the two-stage method can reduce the energy exchange efficiently to remain the operating profits as much as possible when the heavy uncertainties occur. It is concluded that the proposed two-stage coordination method can maximise the total operating profit effectively, especially for the high uncertainties.

Second, in terms of bus voltage performance (the 6th column), the two-stage coordination method reports smaller average maximal bus voltage deviations in all the three deviation groups. In addition, as the uncertainty level increases, the average maximal bus voltage deviations of the two-stage method increase slightly while those of the single-stage one increase obviously.

Third, in comparison for each scenario, for the operating profit (the 7th column), the two-stage coordination method outperforms the single-stage one for all the 1,000 scenarios in all the three deviation groups; while for the bus voltage deviation (the 8th column), the two-stage coordination method beats the single-stage one again with the percentages of 85.2%, 93.6%, and 99.5 % for the three deviation groups, respectively. This performance analysis indicates higher effectiveness of the proposed two-stage coordination method for the scenarios where the heavy uncertainties are likely to occur.

In summary, compared to the single-stage method, the proposed two-stage coordination method can achieve higher operating profits and lower bus voltage deviations under the uncertainties.

8.6 Conclusion

This chapter develops a daily two-stage coordination method for a microgrid where PBDR decisions are optimised one day ahead as the first-stage operation and hourly adjustments of microturbine output power are carried out to track uncertain renewable power generation and loads in the second stage. In the proposed two-stage coordination method, the total operating profit is maximised by optimising the day-ahead 24-hour PBDR price levels and the hourly microturbine output power dispatch. To ensure the optimisation results are robust against uncertainty realisations of WT output power, PV output power, and loads, a TSRO method is developed to optimise the proposed two-stage optimisation problem with full consideration of the uncertainties.

The numerical simulations show satisfying results that the operating profits are maximised under the uncertainty realisations. In addition, the Monte Carlo simulation check shows that when the uncertainties realise, the proposed two-stage coordination method can achieve higher operating profits and lower bus voltage deviations than a conventional single-stage method.

References

[1] J.A.P. Lopes, N. Hatziargyriou, J. Mutale, P. Djapic, and N. Jenkins, 'Integrating distributed generation into electric power systems: a review of drivers, challenges and opportunities', *Electric Power Syst. Research*, vol. 77, no. 9, pp. 1189–1203, 2008.

[2] P. Chiradeja and R. Ramakumar, 'An approach to quantify the technical benefits of distributed generation', *IEEE Trans. Energy Conv.*, vol. 19, no. 4, pp. 764–773, 2004.

[3] K. Zou, A.P. Agalgaonkar, K.M. Muttaqi, and S. Perera, 'An analytical approach for reliability evaluation of distribution systems containing dispatchable and nondispatchable renewable DG units', *IEEE Trans. Smart Grid*, vol. 5, no. 6, pp. 2657–2665, 2014.

[4] P. Denholm, E. Ela, B. Kirby, and M. Milligan, 'The role of energy storage with renewable electricity generation', Available: www.osti.gov, accessed March 2018.

[5] M.H. Albadi and E.F. El-Saadany, 'Demand response in electricity markets: an overview', *Proc. IEEE PES General Meeting*, pp. 1–5, Tampa, FL, USA, 2007.

[6] H. Wu, M. Shahidehpour, and A. Al-Abdulwahab, 'Hourly demand response in day-ahead scheduling for managing the variability of renewable energy', *IET Gen. Transm. Distrib.*, vol. 7, no. 3, pp. 226–234, 2013.

[7] M. Parvania and M. Fotuhi-Firuzabad, 'Integrating load reduction into wholesale energy market with application to wind power integration', *IEEE Systems Journal*, vol. 6, no. 1, pp. 35–45, 2012.

[8] A. Radaideh and V. Ajjarapu, 'Demand response planning in day-ahead market for improving power system flexibility with high wind penetration levels', *Proc. North American Power Sympo. (NAPS) 2015*, pp. 1–6, Charlotte, NC, USA, Oct. 2015.

[9] M. Khanabadi and S. Kamalasadan, 'Day ahead scheduling of distribution system with distributed energy resources considering demand response and energy storage', *Proc. North American Power Sympo. (NAPS) 2013*, pp. 1–6, Manhattan, KS, USA, Sept. 2013.

[10] C. Zhao, J. Wang, J.P. Watson, and Y. Guan, 'Multi-stage robust unit commitment considering wind and demand response uncertainties', *IEEE Trans. Power Syst.*, vol. 28, no. 3, pp. 2708–2717, 2013.

[11] P.A. Ruiz, C.R. Philbrick, E. Zak, K.W. Cheung, and P.W. Sauer, 'Uncertainty management in the unit commitment problem', *IEEE Trans. Power Syst.*, vol. 24, no. 2, pp. 642–651, 2009.

[12] J. Wang, M. Shahidehpour, and Z. Li, 'Security-constrained unit commitment with volatile wind power generation', *IEEE Trans. Power Syst.*, vol. 23, no. 3, pp. 1319–1327, 2008.

[13] Q. Wang, Y. Guan, and J. Wang, 'A chance-constrained two-stage stochastic program for unit commitment with uncertain wind power output', *IEEE Trans. Power Syst.*, vol. 27, no. 1, pp. 206–215, 2012.

[14] Q. Wang, X. Wang, K. Cheung, Y. Guan, and F.S.S. Bresler, 'A two-stage robust optimisation for PJM look-ahead unit commitment', *Proc. Power Tech. 2013*, pp. 1–6, Grenoble, France, 2013.

[15] Y. An and B. Zeng, 'Exploring the modelling capacity of two-stage robust optimisation: variants of robust unit commitment model', *IEEE Trans. Power Syst.*, vol. 30, no. 1, pp. 109–122, 2015.

[16] H. Dashti, A.J. Conejo, R. Jiang, and J. Wang, 'Weekly two-stage robust generation scheduling for hydrothermal power systems', *IEEE Trans. Power Syst.*, vol. 31, no. 6, pp. 4554–4564, 2016.

[17] Z. Wang, B. Chen, J. Wang, J. Kim, and M.M. Begovic, 'Robust optimisation based optimal DG placement in microgrids', *IEEE Trans. Smart Grid*, vol. 5, no. 5, pp. 2173–2182, 2014.

[18] R.A. Jabr, I. Džafić, and B.C. Pal, 'Robust optimisation of storage investment on transmission networks', *IEEE Trans. Power Syst.*, vol. 30, no. 1, pp. 531–539, 2015.

[19] A.H. Hajimiragha, C.A. Canizares, M.W. Fowler, S. Moazeni, and A. Elkamel, 'A robust optimisation approach for planning the transition to plug-in hybrid electric vehicles', *IEEE Trans. Power Syst.*, vol. 26, no. 4, pp. 2264–2274, 2011.

[20] P.R. Thimmapuram, J. Kim, A. Botterud, and Y. Nam, 'Modeling and simulation of price elasticity of demand using an agent-based model', *Proc. IEEE Conf. Innovative Smart Grid Technologies (ISGT)*, pp. 1–8, 2010.

[21] A. Faruqui and S. Sergici, 'Household response to dynamic pricing of electricity: a survey of 15 experiments', *J. Regulatory Economics*, vol. 38, no. 2, pp. 193–225, 2010.

[22] A. Khodaei, M. Shahidehpour, and S. Bahramirad, 'SCUC with hourly demand response considering intertemporal load characteristics', *IEEE Trans. Smart Grid*, vol. 2, no. 3, pp. 564–571, 2011.

[23] A. Evans, V. Strezov, and T.J. Evans, 'Assessment of sustainability indicators for renewable energy technologies', *Renew. Sustain. Energy Rev.*, vol. 13, no. 5, pp. 1082–1088, 2009.

[24] K. Zou, A.P. Agalgaonkar, K.M. Muttaqi, and S. Perera, 'Distribution system planning with incorporating DG reactive capability and system uncertainties', *IEEE Trans. Sustain. Energy*, vol. 3, no. 1, pp. 112–123, 2012.

[25] H.G. Yeh, D.F. Gayme, and S.H. Low, 'Adaptive VAR control for distribution circuits with photovoltaic generators', *IEEE Trans. Power Syst.*, vol. 27, no. 3, pp. 1656–1663, 2012.

[26] C. Wan, Z. Xu, P. Pinson, Z.Y. Dong, and K.P. Wong, 'Probabilistic forecasting of wind power generation using extreme learning machine', *IEEE Trans. Power Syst.*, vol. 29, no. 3, pp. 1033–1044, 2014.

[27] B. Zeng and L. Zhao, 'Solving two-stage robust optimisation problems using a column-and-constraint generation method', *Operations Research Letters*, vol. 41, no. 5, pp. 457–461, 2013.

[28] S. Boyd and L. Vandenberghe, *Convex Optimisation*, Cambridge University Press, UK, 2004.

[29] M.E. Baran and F.F. Wu, 'Network reconfiguration in distribution systems for loss reduction and load balancing', *IEEE Trans. Power Del.*, vol. 4, no. 2, pp. 1401–1407, 1989.

[30] Australian Energy Market Operator (AEMO), Available: www.aemo.com.au, accessed March 2018.

[31] J. Lofberg, 'YALMIP: a toolbox for modelling and optimisation in MATLAB', *Proc. IEEE Inter. Conf. on Robotics and Automation*, pp. 284–289, Taipei, Taiwan, 2004.

[32] GUROBI, Available: www.gurobi.com, accessed March 2018.

Chapter 9

Optimal dispatch of MEMGs

This chapter proposes a system-wide coordinated optimal energy dispatch method for multi-energy microgrids in both the grid-connected and islanded modes. The studied microgrid consists of multiple energy carriers covering controllable generation (CG) units (fuel cell (FC), electric boiler (EB), combined cooling, heat and power (CCHP) plant, and electric chiller (EC)), uncontrollable generation units (wind turbine (WT) and photovoltaic cell (PV)), and energy storage (ES) devices (battery storage, heat storage tank (HST), and ice storage tank (IST)). The proposed energy dispatch method aims to minimise the microgrid net operation cost and enhance the dispatch flexibility of CCHP plant in supplying power, heat, and cooling in the day-ahead energy market. The problems are formulated as the mixed-integer linear programming (MILP) models for both grid-connected and islanded modes, which can be efficiently solved by commercial solvers. Comprehensive case studies are performed to examine the effectiveness of the proposed models and compared with traditional dispatch methods which separately supply electricity and heat/cooling energies. Simulation results have shown that the proposed method has much higher energy efficiency.

Nomenclature

A Sets

j	Set of CG, running from 1 to N_{CG}
k	Set of the blocks, running from 1 to N_{L}

B Variables

$\eta^t_{\mathrm{MT}}/\eta^t_{\mathrm{FC}}$	Electrical efficiency of MT/FC
$P^t_{\mathrm{MT}}/P^t_{\mathrm{FC}}$	Power outputs of MT/FC
$Q^t_{\mathrm{MTH}}/Q^t_{\mathrm{MTC}}$	Heat/cooling output of MT
$P^t_{\mathrm{ESC}}/P^t_{\mathrm{ESD}}$	Charging/discharging power of ES
E^t_{ES}	Energy stored in ES
$P^t_{\mathrm{EB}}/P^t_{\mathrm{EC}}$	Power inputs of EB/EC
$Q^t_{\mathrm{EB}}/Q^t_{\mathrm{EC}}$	Heat/cooling outputs of EB/EC
$F_{\mathrm{G}}/F_{\mathrm{S}}$	Net operation cost of grid-connected/islanded microgrid

$C^t_{\mathrm{F,FC}}/C^t_{\mathrm{F,MT}}$	Fuel cost of FC/MT
$C^t_{\mathrm{U,FC}}/C^t_{\mathrm{U,PG}}$	Unit power generation cost of FC/MT for only power generation
$C^t_{\mathrm{U,S}}/C^t_{\mathrm{U,W}}$	Unit power generation cost of MT in summer/winter
$C^t_{\mathrm{F}}/C^t_{\mathrm{OM}}$	Fuel/maintenance cost
$C^t_{\mathrm{ST}}/C^t_{\mathrm{SD}}$	Start-up/shutdown cost
$C^t_{\mathrm{EX}}/C^t_{\mathrm{PH}}$	Power exchange/energy deviation cost
$C^t_{\mathrm{HC}}/C^t_{\mathrm{SHC}}$	Revenue of supplying heat/cooling in grid-connected/islanded microgrid
$C^t_{\mathrm{H}}/C^t_{\mathrm{C}}$	Revenue of supplying heat/cooling in grid-connected microgrid
$P^t_{\mathrm{BSC}}/P^t_{\mathrm{BSD}}$	Discharging/charging power of BS
$P^t_{\mathrm{ISTC}}/P^t_{\mathrm{ISTD}}$	Discharging/charging power of IST
$P^t_{\mathrm{HSTC}}/P^t_{\mathrm{HSTD}}$	Discharging/charging power of HST
$C^t_{\mathrm{SH}}/C^t_{\mathrm{SC}}$	Revenue of supplying heat/cooling in islanded microgrid
U^t_j	Binary status of unit j, $U^t_j = 1$ is on, 0 is off
$P^t_{\mathrm{B}}/P^t_{\mathrm{S}}$	Purchasing/selling power of microgrid
$P^t_{\mathrm{WT}}/P^t_{\mathrm{PV}}$	Power outputs of WT/PV
P^t_{L}	Power loads
$P^t_{\mathrm{CG},j}$	Power of controllable unit j
$\gamma^t_{\mathrm{B}}/\gamma^t_{\mathrm{S}}$	Binary state variables with $\gamma^t_{\mathrm{B}} = 1/\gamma^t_{\mathrm{S}} = 1$, microgrid purchases/sells power
$\gamma^t_{\mathrm{ESC}}/\gamma^t_{\mathrm{ESD}}$	Binary state variables with $\gamma^t_{\mathrm{ESC}} = 1/\gamma^t_{\mathrm{ESD}} = 1$, ES is charging/discharging
$Q^t_{\mathrm{TS}}/Q^t_{\mathrm{IS}}/P^t_{\mathrm{DS}}$	Heat/cooling/power shortage
$Q^t_{\mathrm{TE}}/Q^t_{\mathrm{IE}}/P^t_{\mathrm{DE}}$	Heat/cooling/power surplus
$\gamma^t_{\mathrm{TS}}/\gamma^t_{\mathrm{IS}}/\gamma^t_{\mathrm{DS}}$	Binary indicators of heat/cooling/power shortage, 1 means that there is energy shortage
$\gamma^t_{\mathrm{TE}}/\gamma^t_{\mathrm{IE}}/\gamma^t_{\mathrm{DE}}$	Binary indicators of the heat/cooling/power surplus, 1 means that there is energy surplus
B^t_k	Binary state of block k

C Parameters

$\eta_{\mathrm{H}}/\eta_{\mathrm{C}}$	Heat recovery ratio of heat recovery unit/absorption chiller
$COP_{\mathrm{H}}/COP_{\mathrm{C}}$	Coefficient of performance of heat recovery unit/absorption chiller
η_{L}	Heat loss ratio of MT
τ	Decay rate of ES
$COP_{\mathrm{EB}}/COP_{\mathrm{EC}}$	Heat/cooling coefficient of performance of EB/EC
$\eta_{\mathrm{ESC}}/\eta_{\mathrm{ESD}}$	Charging/discharging rates of ES
$c_{\mathrm{G}}/L_{\mathrm{NG}}$	Unit price/lower heating value of natural gas
$K^{\mathrm{om}}_{\mathrm{WT}}/K^{\mathrm{om}}_{\mathrm{PV}}$	Unit maintenance cost of WT/PV

$K_{\mathrm{EC}}^{\mathrm{om}}/K_{\mathrm{EB}}^{\mathrm{om}}/K_{\mathrm{MT}}^{\mathrm{om}}$	Unit maintenance cost of EC/EB/MT
$K_{\mathrm{FC}}^{\mathrm{om}}/K_{\mathrm{ES}}^{\mathrm{om}}$	Unit maintenance cost of FC/ES
$C_{\mathrm{ST},j}^{\mathrm{U}}/C_{\mathrm{SD},j}^{\mathrm{D}}$	Unit start-up/shutdown cost of unit j
$c_{\mathrm{H}}/c_{\mathrm{C}}$	Unit price of supplying heat/cooling to customers
$P_{\mathrm{CG},j}^{\min}/P_{\mathrm{CG},j}^{\max}$	Minimum/maximum power of unit j
$R_{\mathrm{CG},j}^{\mathrm{up}}/R_{\mathrm{CG},j}^{\mathrm{down}}$	Maximum ramp up/down rate of unit j
$P_{\mathrm{ESC}}^{\max}/P_{\mathrm{ESD}}^{\max}$	Maximum charge/discharge power of ES
$\xi_{\mathrm{ES}}^{\min}/\xi_{\mathrm{ES}}^{\max}$	Minimum/maximum state of charge of ES
Cap_{ES}	Capacity of ES
$E_{\mathrm{ES}}^{0}/E_{\mathrm{ES}}^{N\Delta t}$	Starting/ending energy of ES
$c_{\mathrm{TE}}/c_{\mathrm{IE}}/c_{\mathrm{DE}}$	Unit cost for heat/cooling/power surplus
$c_{\mathrm{TS}}/c_{\mathrm{IS}}/c_{\mathrm{DS}}$	Unit cost for heat/cooling/power shortage
α_k/β_k	Slope/intercept of linear function in block k
P_k	Starting point of kth block
P_1/P_{N+1}	Minimum/maximum power outputs of N_{L} blocks
Δt	Unit dispatch interval, 1 h

9.1 Introduction

Multi-energy systems aim to integrate diverse energy carriers such as electricity, heat, and cooling, etc., to achieve higher energy utilisation efficiency [1]. In practice, multi-energy can be simultaneously generated by highly efficient co- or tri-generation units such as CCHP plant [2]. Once deployed in the distribution network, the CCHP plant, ES, and renewable energy-based distributed generators (DGs) such as solar and wind power can form a multi-energy microgrid [2,3]. Therefore, one key research problem is to optimally dispatch the multiple energy resources for maximum operation efficiency.

In the literature, most of the research works focus on the operation of individual CCHP plants, which can generally fall into two operation strategies, i.e. following power load (FPL) and following heat load (FHL) [4–8]. Ref. [4] worked on the FPL/FHL switching strategies to enhance the integrated performance of energy consumption, operation cost, and CO_2 emissions of CCHP plant. In [5] and [6], operation strategies for CCHP plant were designed in terms of IF-THEN conditions and corresponding evaluation criteria. With a ground source heat pump, Ref. [7] configured a CCHP-organic Rankine cycle system and analysed its performance in the cost savings, gas emission reduction, and system efficiency improvement. In [8], matrix modelling approach was introduced to optimise the input and output power of CCHP plant as well as its capacity. However, the coordinated operation of CCHP plant with other DGs at a system level was not considered in the above works.

At the microgrid level, traditional research works are mostly on the single-energy microgrid [9,10], which do not integrate multiple energy resources. The coordination of CCHP plant with other DGs in supplying multiple energies is mainly based on the grid-connected microgrids [11,12]. Reference [13] investigated the coordination of CCHP plant and PV prosumers to minimise the operation cost in a grid-connected microgrid. In [14], the robust optimisation method was employed to coordinate CCHP plant with DGs of uncertain output to handle their intermittency. Reference [15] proposed an integrated multi-objective dispatch model for microgrid to minimise its net operation cost and gas emissions. In [16], an optimisation model for the operation and design of microgrid satisfying power and heat demands was established considering the uncertainty of renewable energies. In our recent research work [17], dynamic dispatch of the grid-connected multi-energy microgrids considering the opportunity profit from the electrical ES was studied.

It can be observed in the literature that very few research works consider the multi-energy dispatch of the islanded microgrids. Yet, most of the islanded microgrids research still focus on the single energy dispatch, i.e. power dispatch [10,18–20]. In practice, multi-energy management of the islanded microgrids is quite necessary especially for remote areas like the Semakau Island microgrid in Singapore [3]. Without the grid as the backup, operation objectives for islanded microgrids can be rather different from the grid-connected ones. Reference [21] is one of the few studies which proposed a multi-objective optimisation model for the minimisation of the fuel cost and gas emission.

Besides, all the above works dispatch the individual CCHP plant via FPL/FHL or hybrid of the two strategies, while at the system level, CCHP plant typically operates under FHL strategy in grid-connected microgrids, since the power imbalance can be compensated by the grid. For islanded microgrids, FPL strategy for CCHP plant is preferred in that power system requires a precise power balance all the time, while the imbalance of heat/cooling for short periods is acceptable [11]. However, under all the above strategies, power, heat/cooling outputs of CCHP plant are highly coupled and dependent on one another, which makes the operation of microgrid not flexible enough, and potentials of CCHP plant in the cost reduction are not fulfilled.

Moreover, given the non-linear nature of the problem, the solution methods are mostly the evolutionary algorithms like the Genetic Algorithm or the Particle Swarm Optimisation [10,11,13–15,18,21]. However, these algorithms are computationally demanding especially when the problem size is high. Further, they often suffer from issues such as inconsistent solutions and difficulty to rigorously converge. Since the new competitive environment in the power system requires more accurate and efficient tools to support decisions for the resource dispatch, more efficient solution process and solution quality is needed [22].

This chapter studies a comprehensive multi-energy microgrid which consists of a CCHP plant, a battery storage, an EC, an HST, a WT, a PV, an IST, an EB, and an FC. For higher dispatch flexibility and energy efficiency, system-wide coordinated optimal dispatching models under both the grid-connected and islanded modes are proposed. The heat/cooling storage tanks are utilised to decouple the inflexible operation strategies of CCHP plant, and then its heat/cooling and power outputs can participate in the day-ahead energy market flexible. The conversion units, i.e. EB and EC, which transfer power to heat/cooling, can further enhance the coordination of different energies. The objectives are to minimise net operation cost in both modes subject to operational limits. Besides, the non-linear models are linearised to MILP ones for high solution efficiency. Finally, typical scenarios for winter and summer periods are used to fully test the validity of the proposed dispatch models.

Compared with existing research works, the major contributions of this chapter are summarised as

1. The optimal day-ahead energy dispatch of multi-energy microgrids in both the grid-connected and islanded modes is proposed.
2. The conventional inflexible operation strategies (FPL or FHL) of CCHP plant are decoupled, and different DGs and energy resources are coordinated at the microgrid level.
3. Detailed and practical models of the multi-energy microgrid components are considered, and the non-linear problems are linearised to MILP models which can be efficiently solved by commercial solvers.

The rest of this chapter is organised as follows. System description and modelling are given in Section 9.2. In Section 9.3, the coordinated dispatch models of a microgrid in both modes are introduced. Case studies are presented and analysed in Section 9.4. Then we end the chapter with some conclusions in Section 9.5.

9.2 Multi-energy microgrid modelling

In this study, the multi-energy microgrid consists of a WT, a PV, a battery storage, a CCHP plant, an FC, an EB, an HST, an EC, and an IST as shown in Figure 9.1. Besides, there is a switch between the grid and microgrid: when the switch is on, microgrid operates in grid-connected mode, and while off, microgrid operates in islanded mode [18–20].

The components' modelling mainly focuses on their ON/OFF statuses and generation output [12], and the detailed modelling is given next in this section. It should be noted that the microturbine (or CCHP plant), EB, FC, and EC are all of the small sizes and flexible control fast start-up/shoutdown rates. Thus, they can be regarded as the CG units. The WT and PV whose power outputs are uncertain and volatile are categorised as the uncontrollable generation units.

9.2.1 Combined cooling, heat and power (CCHP) plant

Typically, the CCHP plant consists of three parts: a microturbine (MT), an absorption chiller, and a heat recovery unit. MT consumes natural gas to generate power, and at the same time, the waste heat can be utilised by heat recovery unit/absorption chiller to produce heat/cooling energy [3,16]. Thus, the energy utilisation efficiency can be greatly improved. In this chapter, the Capstone's MT C65

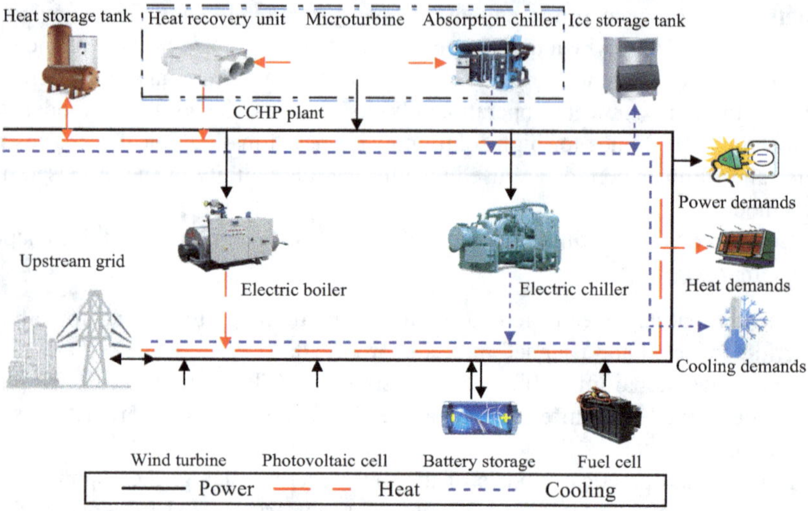

Figure 9.1 Structure of a multi-energy microgrid

[24], which has a high international market share, is used as the reference for the CCHP plant modelling. The relation between the power outputs and electrical efficiency can be modelled as

$$\eta_{\mathrm{MT}}^{t} = 0.076\left(\frac{P_{\mathrm{MT}}^{t}}{65}\right)^{3} - 0.310\left(\frac{P_{\mathrm{MT}}^{t}}{65}\right)^{2} + 0.415\left(\frac{P_{\mathrm{MT}}^{t}}{65}\right) + 0.107 \tag{9.1}$$

The corresponding heat and cooling outputs are

$$Q_{\mathrm{MTH}}^{t} = \eta_{\mathrm{H}} COP_{\mathrm{H}} P_{\mathrm{MT}}^{t}(1 - \eta_{\mathrm{MT}}^{t} - \eta_{\mathrm{L}})/\eta_{\mathrm{MT}}^{t} \tag{9.2}$$

$$Q_{\mathrm{MTC}}^{t} = \eta_{\mathrm{C}} COP_{\mathrm{C}} P_{\mathrm{MT}}^{t}(1 - \eta_{\mathrm{MT}}^{t} - \eta_{\mathrm{L}})/\eta_{\mathrm{MT}}^{t} \tag{9.3}$$

Both the heat and cooling come from the waste heat. Thus, they share similar properties. If the waste heat is not recovered, MT operates only for the power generation [10,24,25].

9.2.2 Energy storage

ES can decouple the energy generation from consumption and reduce the imbalance between energy supply and demand [26]. In this chapter, ES mainly consists of three types: electrical ES, heat storage, and cooling storage.

1. Electrical ES: Currently, the electrical ES including the power-type like supercapacitors and the energy-type like electrochemical batteries are intensively investigated. The power-type ES is mostly utilised to alleviate the effects of fluctuating outputs of renewable energy generators on power system

operation, while the energy-type ES is mainly used for the peak saving in the energy management field [27]. For the day-ahead dispatch scheme in this chapter, the battery storage (BS), which belongs to the energy-type electrical ES is applied [17,28].

2. Heat/cooling storage: Under FHL/FPL strategies, the power or heat/cooling outputs of CCHP plant are limited by each other, which restricts the operation of CCHP plant in the overall multi-energy dispatch. To decouple the inflexible operation strategies of CCHP plant, the HST and IST, whose operation principles are similar to the electrical ES are deployed. Nowadays, HST and IST have already been widely employed in European nations like Denmark [29].

The power and energy relationship of the electrical ES and heat/cooling storage share the same model as follows:

$$E_{\mathrm{ES}}^{t} = (1-\tau)E_{\mathrm{ES}}^{t-1} + (P_{\mathrm{ESC}}^{t}\eta_{\mathrm{ESC}} + P_{\mathrm{ESD}}^{t}/\eta_{\mathrm{ESD}})\Delta t \tag{9.4}$$

It should be noted that ES is a generalised abbreviation, in this study, ES should be BS, HST, or IST.

9.2.3 Electric boiler (EB) and Electric chiller (EC)

To enhance the interaction of different energy systems, EBs and ECs are utilised to transfer power into heat/cooling with the following models [14]:

$$Q_{\mathrm{EB}}^{t} = P_{\mathrm{EB}}^{t}COP_{\mathrm{EB}} \tag{9.5}$$

$$Q_{\mathrm{EC}}^{t} = P_{\mathrm{EC}}^{t}COP_{\mathrm{EC}} \tag{9.6}$$

9.2.4 Fuel cells

FCs convert the energy stored in O_2 and H_2 into pollution-free electrical energy by an electrochemical process. Due to the high electrical efficiency, waste heat of FC is not recovered [10,17,30]. Proton-exchange membrane fuel cell IFC PC-29 powered by natural gas [31] is chosen as a reference in this study, the electrical efficiency can be given by

$$\eta_{\mathrm{FC}}^{t} = -0.0023P_{\mathrm{FC}}^{t} + 0.674 \tag{9.7}$$

9.3 Coordinated optimal dispatch

With regard to different microgrid operation modes, system-wide coordinated dispatch models are proposed, respectively.

9.3.1 Grid-connected mode

In grid-connected microgrids, the power imbalance can be compensated by the grid. Thus, the objective is to minimise the net operation cost:

1. Objective function:

$$\text{MIN } F_{\mathrm{G}} = \sum_{t=1}^{N} (C_{\mathrm{F}}^{t} + C_{\mathrm{EX}}^{t} + C_{\mathrm{OM}}^{t} + C_{\mathrm{ST}}^{t} + C_{\mathrm{SD}}^{t} - C_{\mathrm{HC}}^{t})\Delta t \tag{9.8}$$

   (a) Fuel cost:

$$C_{\mathrm{F}}^{t} = C_{\mathrm{F,FC}}^{t} + C_{\mathrm{F,MT}}^{t} = [c_{\mathrm{G}}(P_{\mathrm{MT}}^{t}/\eta_{\mathrm{MT}}^{t} + P_{\mathrm{FC}}^{t}/\eta_{\mathrm{FC}}^{t})]/L_{\mathrm{NG}} \tag{9.9}$$

   (b) Maintenance cost:

$$C_{\mathrm{OM}}^{t} = \begin{bmatrix} P_{\mathrm{EC}}^{t} \cdot K_{\mathrm{EC}}^{\mathrm{om}} + P_{\mathrm{WT}}^{t} \cdot K_{\mathrm{WT}}^{\mathrm{om}} + P_{\mathrm{PV}}^{t} \cdot K_{\mathrm{PV}}^{\mathrm{om}} + P_{\mathrm{EB}}^{t} \cdot K_{\mathrm{EB}}^{\mathrm{om}} \\ + P_{\mathrm{MT}}^{t} \cdot K_{\mathrm{MT}}^{\mathrm{om}} + P_{\mathrm{FC}}^{t} \cdot K_{\mathrm{FC}}^{\mathrm{om}} + (P_{\mathrm{ESC}}^{t} + P_{\mathrm{ESD}}^{t}) \cdot K_{\mathrm{ES}}^{\mathrm{om}} \end{bmatrix} \tag{9.10}$$

   (c) Start-up cost and shutdown cost:

$$C_{\mathrm{ST}}^{t} = \sum_{j=1}^{N_{\mathrm{CG}}} \max\left\{0, U_{j}^{t} - U_{j}^{t-1}\right\} C_{\mathrm{ST},j}^{\mathrm{U}} \tag{9.11}$$

$$C_{\mathrm{SD}}^{t} = \sum_{j=1}^{N_{\mathrm{CG}}} \max\{0, U_{j}^{t-1} - U_{j}^{t}\} C_{\mathrm{SD},j}^{\mathrm{D}} \tag{9.12}$$

   (d) Power exchange cost:

$$C_{\mathrm{EX}}^{t} = C_{\mathrm{BU}}^{t} - C_{\mathrm{SE}}^{t} = c_{\mathrm{B}}^{t} P_{\mathrm{B}}^{t} - c_{\mathrm{S}}^{t} P_{\mathrm{S}}^{t} \tag{9.13}$$

   (e) Revenue of supplying heat/cooling:

$$C_{\mathrm{HC}}^{t} = C_{\mathrm{H}}^{t} + C_{\mathrm{C}}^{t} = c_{\mathrm{H}}^{t} Q_{\mathrm{H}}^{t} + c_{\mathrm{C}}^{t} Q_{\mathrm{C}}^{t} \tag{9.14}$$

2. Constraints:
   The operation constraints include the following:
   (a) Energy balance:

$$P_{\mathrm{WT}}^{t} + P_{\mathrm{PV}}^{t} + P_{\mathrm{MT}}^{t} + P_{\mathrm{FC}}^{t} + P_{\mathrm{BSD}}^{t} - P_{\mathrm{BSC}}^{t} + P_{\mathrm{B}}^{t} - P_{\mathrm{S}}^{t} = P_{\mathrm{L}}^{t} + P_{\mathrm{EB}}^{t} + P_{\mathrm{EC}}^{t} \tag{9.15}$$

$$Q_{\mathrm{EC}}^{t} + Q_{\mathrm{MTC}}^{t} - P_{\mathrm{ISTC}}^{t} + P_{\mathrm{ISTD}}^{t} = Q_{\mathrm{C}}^{t} \tag{9.16}$$

$$Q_{\mathrm{EB}}^{t} + Q_{\mathrm{MTH}}^{t} - P_{\mathrm{HSTC}}^{t} + P_{\mathrm{HSTD}}^{t} = Q_{\mathrm{H}}^{t} \tag{9.17}$$

   (b) Controllable units:

$$\text{Power outputs limits: } U_{j}^{t} \cdot P_{\mathrm{CG},j}^{\min} \le P_{\mathrm{CG},j}^{t} \le U_{j}^{t} \cdot P_{\mathrm{CG},j}^{\max} \tag{9.18}$$

$$\text{Ramp-up/down limits: } R_{\mathrm{CG},j}^{\mathrm{down}} \Delta t \le P_{\mathrm{CG},j}^{t} - P_{\mathrm{CG},j}^{t-1} \le R_{\mathrm{CG},j}^{\mathrm{up}} \Delta t \tag{9.19}$$

(c) Power exchanges:

$$\text{Power outputs limits: } 0 \le P_B^t \le \gamma_B^t P_B^{\max}, 0 \le P_S^t \le \gamma_S^t P_S^{\max} \tag{9.20}$$

$$\text{State limits : } \gamma_B^t + \gamma_S^t \le 1 \tag{9.21}$$

(d) Energy storage:

$$\text{Charging power limits: } 0 \le P_{ESC}^t \le \gamma_{ESC}^t P_{ESC}^{\max} \tag{9.22}$$

$$\text{Discharging power limits: } 0 \le P_{ESD}^t \le \gamma_{ESD}^t P_{ESD}^{\max} \tag{9.23}$$

$$\text{Energy limits: } \xi_{ES}^{\min} Cap_{ES} \le E_{ES}^t \le \xi_{ES}^{\max} Cap_{ES} \tag{9.24}$$

$$\text{Starting/ending energy limits: } E_{ES}^0 = E_{ES}^{N\Delta t} \tag{9.25}$$

$$\text{State limits: } \gamma_{ESC}^t + \gamma_{ESD}^t \le 1 \tag{9.26}$$

Note that CG is a generalised abbreviation which includes MT, EB, EC, or FC.

9.3.2 Islanded mode

Without the grid as the backup, the energy generation and consumption have to be balanced within the islanded microgrid. In this condition, demand response should be incurred. Thus, compared with the grid-connected microgrids, energy deviation cost (demand response cost) should be considered in the net operation cost.

1. Objective function:

$$\text{MIN } F_S = \sum_{t=1}^{N} (C_F^t + C_{OM}^t + C_{ST}^t - C_{SHC}^t + C_{PH}^t)\Delta t \tag{9.27}$$

(a) Energy deviation cost:

$$C_{PH}^t = c_{TS} Q_{TS}^t + c_{TE} Q_{TE}^t + c_{IS} Q_{IS}^t + c_{IE} Q_{IE}^t + c_{DS} P_{DS}^t + c_{DE} P_{DE}^t \tag{9.28}$$

(b) Revenue of supplying heat/cooling to customers:

$$C_{SHC}^t = C_{SH}^t + C_{SC}^t = c_H^t (Q_H^t - Q_{TS}^t) + c_C^t (Q_C^t - Q_{IS}^t) \tag{9.29}$$

The cost items with the same symbols are the same with the grid-connected microgrid.

2. Constraints:

The operation constraints for the islanded microgrid are the following:

(a) Energy balance:

$$\begin{aligned} &P_{WT}^t + P_{PV}^t + P_{MT}^t + P_{FC}^t + P_{BSD}^t - P_{BSC}^t \\ &= P_L^t - P_{DS}^t + P_{DE}^t + P_{EB}^t + P_{EC}^t \end{aligned} \tag{9.30}$$

$$Q^t_{\mathrm{EC}} + Q^t_{\mathrm{MTC}} - P^t_{\mathrm{ISTC}} + P^t_{\mathrm{ISTD}} = Q^t_{\mathrm{C}} - Q^t_{\mathrm{IS}} + Q^t_{\mathrm{IE}} \tag{9.31}$$

$$Q^t_{\mathrm{EB}} + Q^t_{\mathrm{MTH}} - P^t_{\mathrm{TSTC}} + P^t_{\mathrm{TSTD}} = Q^t_{\mathrm{H}} - Q^t_{\mathrm{TS}} + Q^t_{\mathrm{TE}} \tag{9.32}$$

(b) Energy deviations:

$$\text{Energy deviation limits:} \begin{cases} 0 \le P^t_{\mathrm{DS}} \le \gamma^t_{\mathrm{DS}} P^t_{\mathrm{L}}, 0 \le P^t_{\mathrm{DE}} \le \gamma^t_{\mathrm{DE}} P^t_{\mathrm{L}} \\ 0 \le Q^t_{\mathrm{TS}} \le \gamma^t_{\mathrm{TS}} Q^t_{\mathrm{H}}, 0 \le Q^t_{\mathrm{TE}} \le \gamma^t_{\mathrm{TE}} Q^t_{\mathrm{H}} \\ 0 \le Q^t_{\mathrm{IS}} \le \gamma^t_{\mathrm{IS}} Q^t_{\mathrm{C}}, 0 \le Q^t_{\mathrm{IE}} \le \gamma^t_{\mathrm{IE}} Q^t_{\mathrm{C}} \end{cases} \tag{9.33}$$

$$\text{State limits:} \quad \gamma^t_{\mathrm{DS}} + \gamma^t_{\mathrm{DE}} \le 1, \gamma^t_{\mathrm{TS}} + \gamma^t_{\mathrm{TE}} \le 1, \gamma^t_{\mathrm{IS}} + \gamma^t_{\mathrm{IE}} \le 1 \tag{9.34}$$

Other constraints are in line with the grid-connected microgrid.

9.3.3 Model linearisation

The linear constant electrical efficiencies of CCHP plant and FC of the overall operation problems are utilised in the literature [4–8,16,17]. However, it is not so accurate to represent the realistic conditions during their operation. In practice, with the influence of ambient temperature and elevation, the electrical efficiencies are non-linear functions of power outputs [23,31], hence those items with the efficiencies, i.e. the fuel cost function (9.9) and heat/cooling outputs functions (9.2) and (9.3), are non-linear. Besides, the objective function and start-up/shutdown cost functions (9.11) and (9.12) are also non-linear. Thus, the proposed models are mixed integer non-linear programming problems.

In order to reduce the computation burdens and improve the solution quality, the non-linear problems are linearised into MILP models in this chapter.

1. Linearisation of fuel cost

   The fuel cost curves of CCHP plant and FC are plotted as the solid line in Figure 9.2(a), and the heat/cooling outputs curves of CCHP plant are plotted as the solid line in Figure 9.2(b). N_{L} blocks of linear functions in dash lines are

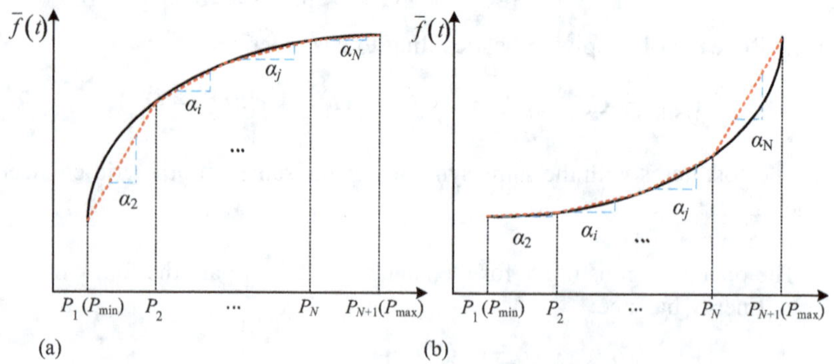

Figure 9.2 Piecewise linearisation of the non-linear items: (a) fuel cost of CCHP plant or FC; (b) heat or cooling outputs of CCHP plant

used to approximate the origin non-linear functions by introducing a binary variable and a continuous variable in each block.

Piecewise-linear functions for the fuel cost functions and heat/cooling outputs of CCHP plant are the following:

$$\bar{f}(t) = \sum_{k=1}^{N_L} (\alpha_k P_k^t + \beta_k B_k^t) \tag{9.35}$$

$$\alpha_k = [f(P_{k+1}) - f(P_k)/(P_{k+1} - P_k)] \tag{9.36}$$

$$\beta_k = f(P_k) - P_k \alpha_k \tag{9.37}$$

$$P_k B_k^t \leq P_k^t \leq B_k^t P_{k+1} \tag{9.38}$$

$$U_j^t = \sum_{k=1}^{N_L} B_k^t \leq 1 \tag{9.39}$$

2. The linearisation of the start-up/shutdown cost is achieved by separating of the max function into two individual parts as

$$\begin{cases} C_{ST}^t \geq 0 \\ C_{ST}^t \geq \sum_{j=1}^{N_{CG}} (U_j^t - U_j^{t-1}) C_{ST,j}^U \end{cases} \tag{9.40}$$

$$\begin{cases} C_{SD}^t \geq 0 \\ C_{SD}^t \geq \sum_{j=1}^{N_{CG}} (U_j^{t-1} - U_j^t) C_{SD,j}^D \end{cases} \tag{9.41}$$

Then, after the linearisation, commercial solvers, which have high performance in solving MILP problems, can be employed to solve the problems.

9.4 Case studies

9.4.1 Data input

In this section, the proposed dispatch models are tested in a hypothetical multi-energy microgrid. The hourly power outputs of WT and PV for the typical winter and summer days can be obtained by forecasting [12]. It is assumed that there is no heat load in summer and no cooling load in winter [15]. Power outputs of WT, PV, and system load profiles are shown in Figure 9.3. The data are obtained from [10,17,25] with minor changes.

For case studies, simulation is conducted for a 24-h dispatch horizon with unit dispatch interval as 1 h. The real-time power pricing is shown in Figure 9.4 [25]. For supplying heat/cooling, fixed pricing scheme is adopted [16]. The parameters for case studies are listed in Tables 9.1 and 9.2. All the units are assumed to be off before the dispatch.

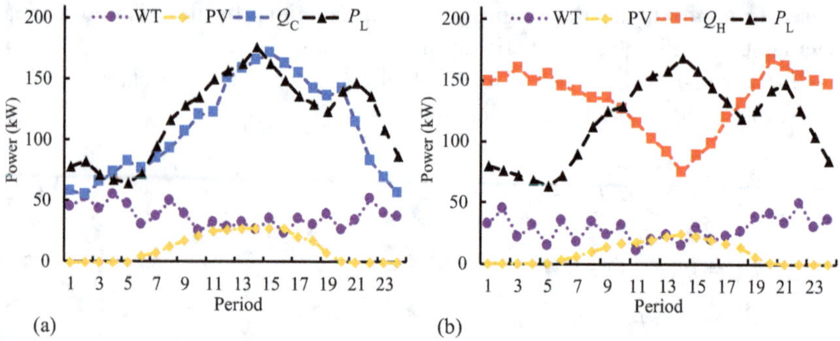

Figure 9.3 Forecast loads, PV, and WT power outputs: (a) summer and (b) winter

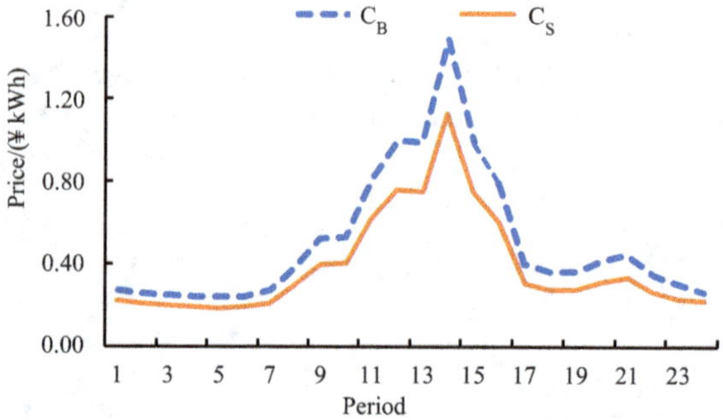

Figure 9.4 The real-time price structure

Table 9.1 Parameters of ES (¥/kWh) [4–6]

Type	η_C/η_D	Decay	P_{ESC}^{max}	P_{ESD}^{max}	ξ_{min}	ξ_{max}	K_{ES}^{om} (¥/kWh)	E_{ES}^{0}	Cap_{ES}
BS	0.9	0.001	37.5	37.5	0.2	0.8	0.0073	30	150
HST	0.87	0.01	45	45	0	0.9	0.0012	0	160
IST	0.85	0.01	40	40	0	0.9	0.0014	0	120

Two cases based on the conventional models for supplying multiple energies in microgrids are compared in this chapter.

Case 1: Separate generation: Heat and cooling loads are met by the gas boiler (GB) with heat efficiency of 85% and absorption chiller, and power loads are all satisfied by power generation units [8,16].

Case 2: Typical tri-generation: CCHP plant operates under FHL strategy in the grid-connected microgrids and under FPL strategy in the islanded microgrids with GB working as a supplement [12–16,21].

Case 3: The system-wide coordinated generation proposed in this chapter.

Table 9.2 Values of the parameters [8,9,23,25]

Parameters	Value	Parameters	Value	Parameters	Value
η_{H}, η_{C}	85%	P_{MT}^{min}	15 kW	K_{EC}^{om}	¥0.0104/kWh
η_{L}	15%	P_{MT}^{max}	65 kW	K_{WT}^{om}	¥0.0145/kWh
COP_{H}	1.20	P_{EB}^{min}	0 kW	K_{PV}^{om}	¥0.0133/kWh
COP_{C}	2.00	P_{EB}^{max}	50 kW	K_{EB}^{om}	¥0.0089/kWh
COP_{EB}	0.85	P_{FC}^{min}	5 kW	K_{MT}^{om}	¥0.0990/kWh
COP_{EC}	3.5	P_{FC}^{max}	40 kW	K_{FC}^{om}	¥0.0841/kWh
$C_{ST,MT}^{U}$	¥1.94	P_{EC}^{min}	0 kW	C_{IS}	¥0.12/kWh
$C_{ST,EB}^{U}$	¥1.23	P_{EC}^{max}	40 kW	C_{IE}	¥0.10/kWh
$C_{ST,FC}^{U}$	¥2.21	P_{B}^{max}	60 kW	C_{DS}	¥1.50/kWh
$C_{ST,EC}^{U}$	¥1.32	P_{S}^{max}	60 kW	C_{DE}	¥0.90/kWh
$C_{SD,MT}^{D}$	¥1.82	C_{TS}	¥0.20/kWh	R_{MT}^{up}	10 kW/min
$C_{SD,EB}^{D}$	¥1.14	C_{TE}	¥0.15/kWh	R_{MT}^{down}	5 kW/min
$C_{SD,FC}^{D}$	¥2.05	C_{H}	¥0.30/kWh	R_{FC}^{up}	2 kW/min
$C_{SD,EC}^{D}$	¥1.15	C_{C}	¥0.20/kWh	R_{FC}^{down}	2 kW/min
R_{EB}^{down}	2 kW/min	R_{EC}^{up}	4 kW/min	C_{G}	¥2.70/m^3
R_{EB}^{up}	3 kW/min	R_{EC}^{down}	3 kW/min	L_{NG}	9.78 kWh/m^3

Table 9.3 Combination of all the units in three cases

No.	Common units				Summer			Winter		
	PV/WT	FC	BS	CCHP	GB	EC	IST	GB	EB	HST
Case 1	√	√	√	√	√			√		
Case 2	√	√	√	√	√			√		
Case 3	√	√	√	√		√	√		√	√

The detailed combination of the units in the three cases is summarised in Table 9.3.

The mathematical models are implemented in the General Algebraic Modelling System via the CPLEX solver. The program is run on an Intel(R) Xeon(R) E5-1630 3.70 GHz personal computer with 16 G memory.

9.4.2 Remarks

For the waste heat is not recovered in Case 1, MT works only for the power generation. In Cases 2 and 3, MT generates power and cooling in summer and power and heat in winter. To better illustrate the simulation results for the grid-connected microgrids, the unit cost of FC for power generation can be calculated as

$$C_{U,FC}^{t} = \frac{C_{F,FC}^{t}}{P_{FC}^{t}} + K_{FC}^{om} \tag{9.43}$$

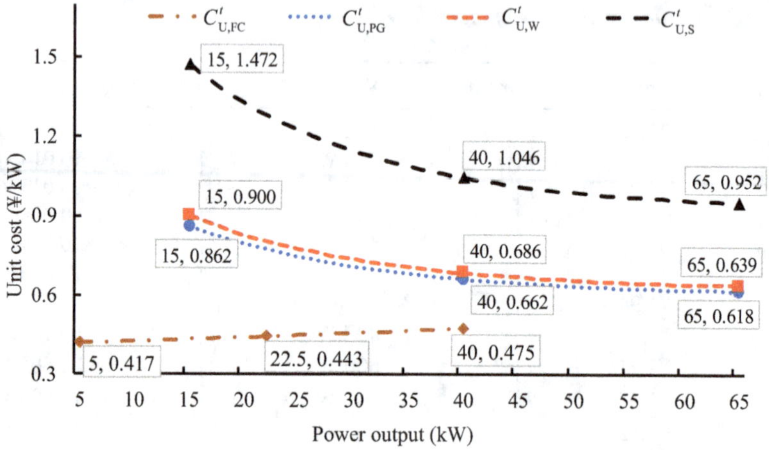

Figure 9.5 Unit power generation cost of FC and MT in different modes

Similarly, for the power generation, the unit cost of MT in the only power generation mode is

$$C^t_{U,PG} = \frac{C^t_{F,MT}}{P^t_{MT}} + K^{om}_{MT} \tag{9.44}$$

The unit cost of MT (CCHP pant) operating in summer and winter is as follows:

$$C^t_{U,W} = \frac{C^t_{F,MT} - C^t_{H,m}}{P^t_{MT}} + K^{om}_{MT} \tag{9.45}$$

$$C^t_{U,S} = \frac{C^t_{F,MT} - C^t_{C,m}}{P^t_{MT}} + K^{om}_{MT} \tag{9.46}$$

Figure 9.5 presents the unit cost curves of FC and MT in different conditions.

As we can see, for the non-linear electrical efficiencies, the unit cost of FC is increasing, but the unit cost of MT in different conditions is decreasing. Since the coefficient of performance of the absorption chiller is higher than that of the heat recovery unit, the unit cost of MT (CCHP plant) operating in winter is greater than that in summer, but with the recovery of waste heat, they are both lower than the unit cost of MT in only power generation condition. In addition, Figure 9.5 shows that within FC's maximum capacity, its unit cost is always lower than that of MT.

9.4.3 Grid-connected mode

Figures 9.6 and 9.7 show the simulation results of the grid-connected microgrid in all three cases. The net operation cost is listed in Table 9.4. In those figures, P_N denotes the total power loads minus the summation of WT and PV power outputs;

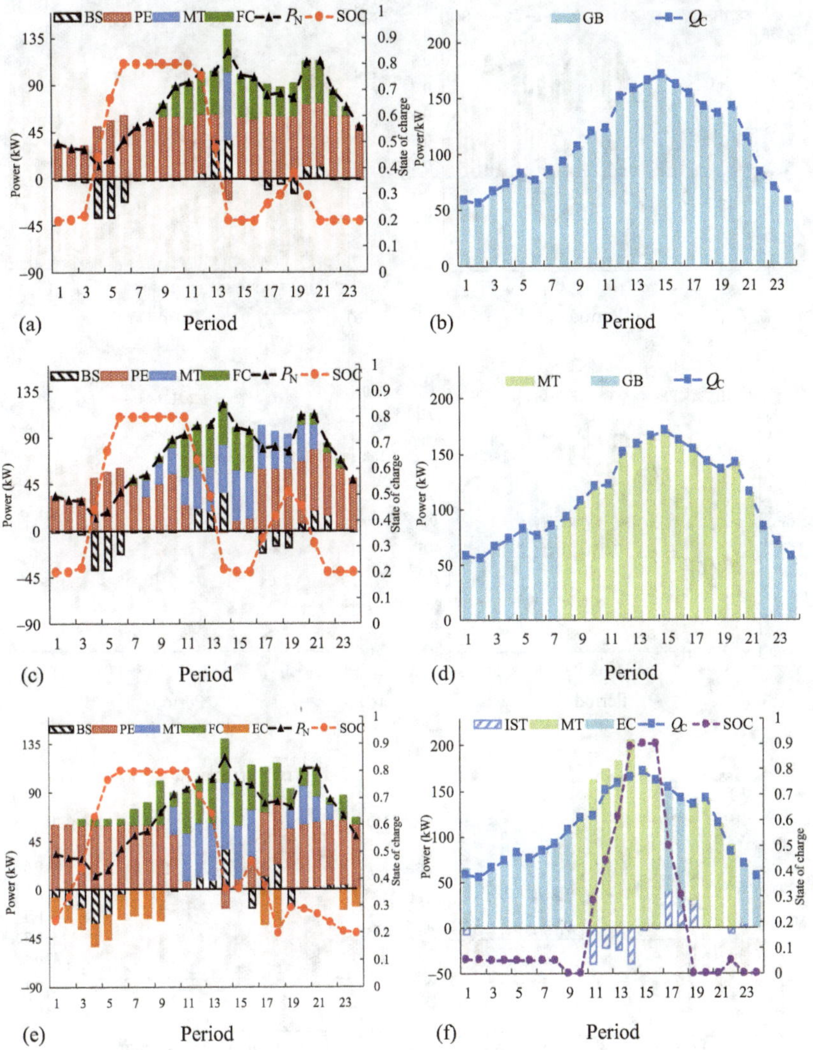

Figure 9.6 Power and heat balance conditions of the three cases in summer: Case 1 (a), (b); Case 2 (c), (d); Case 3 (e), (f)

PE denotes the subtraction of purchasing power with selling power; and BS is the discharging power minus the charging power. HST and IST denote the releasing heat/cooling energy minus the absorbing, respectively.

It is indicated in Figures 9.6 and 9.7 that all the units in the three cases are scheduled based on their unit cost as well as operation constraints. Since the unit power generation cost of FC is lower than MT (Figure 9.5), FC outputs in priority to MT when there is power shortage. Except for satisfying the internal power loads,

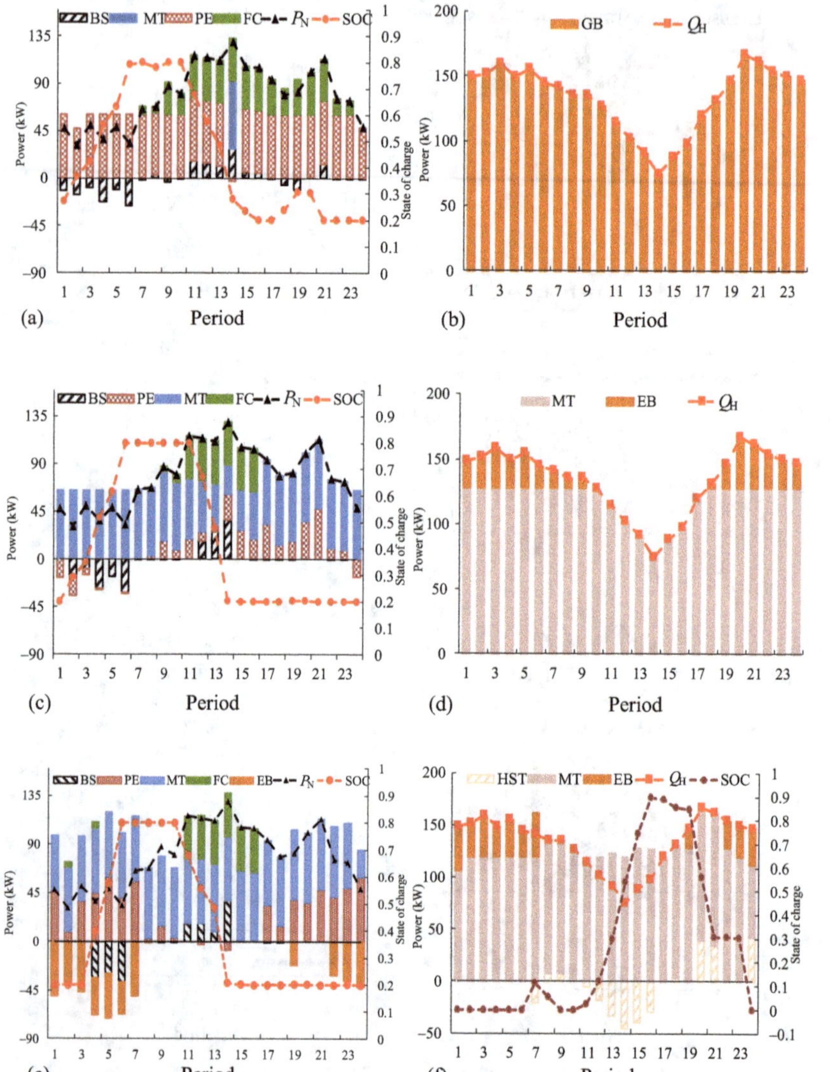

Figure 9.7 Power and heat balance conditions of the three cases in winter: Case 1 (a), (b); Case 2 (c), (d); Case 3 (e), (f)

FC and MT will output extra power when their unit cost is less than the selling prices in order to obtain more power exchange revenue. Besides, the charging or discharging decisions of BS is made with respect to real-time pricing to shift loads and lower the net operation cost.

In Case 1, the heat/cooling loads are met by GB (Figures 9.6(b) and 9.7(b)), the power loads are supplied by FC, MT, BS, and power exchange (Figures 9.6(a) and

Table 9.4 Net operation cost of grid-connected microgrid (¥)

Cost/revenue item	**Summer**			**Winter**		
	Case 1	**Case 2**	**Case 3**	**Case 1**	**Case 2**	**Case 3**
Fuel	822.7	808.2	685.6	1342	1564.3	1507.5
Maintenance	77.4	103.6	94.5	88.9	195	175.5
Start-up and shutdown	9.4	16.1	6.11	9.4	9.8	7.4
Power exchange	537.3	350.5	377.6	629.9	177.6	187.8
Heat or cooling revenue	528.4	528.4	528.4	960.1	960.1	960.1
Net operation cost	918.6	750.1	635.7	1110.3	986.6	918.3

9.7(a)) based on their individual cost functions. The simulation results show that the heat and power are scheduled without any interaction, and the net operation cost is highest since the efficiency of GB is low, and heat generated by MT is all wasted (Figures 9.6(a) and 9.7(a), period 14.00, Table 9.4).

In Case 2, MT supplies heat/cooling in priority under FHL strategy and GB works as a supplement (Figures 9.6(d) and 9.7(d)). The fixed power outputs of MT coordinate with FC, BS, and the power exchange to satisfy the total power loads (Figures 9.6(c) and 9.7(c)). For the waste heat is fully utilised, the net operation cost is thus lower than Case 1. However, under FHL strategy, MT's power outputs are highly limited by its cooling/heat outputs. Hence MT cannot join in the energy market flexible. But as the waste heat is fully utilised, the net operation cost is lower than Case 1 (Table 9.4).

By marked contrast, it can be seen that the system-wide coordinated optimal dispatch of power, heat, and cooling is realised in Case 3. When the purchasing prices are lower than the unit cost of MT operating in summer or in winter, power participates in the heat/cooling dispatch, i.e. EB/EC outputs in priority to MT to transfer power to heat/cooling (Figure 9.6(e) and (f): periods 1.00–9.00, 17.00–18.00, 23.00–24.00; Figure 9.7(e) and (f): periods 1.00–7.00, 22.00–24.00). When the purchasing prices are higher than the unit cost of MT operating in summer or winter, HST/IST stores the surplus heat/cooling to decouple inflexible strategies of CCHP plant, and then more power can be generated by MT to supply the power loads or sold to grid for revenue (Figure 9.6(e) and (f): period 11.00–14.00; Figure 9.7(e) and (f): period 12.00–17.00). When the purchasing prices fall in between the above two price intervals, heat/cooling participate in the power dispatch, i.e. HST or IST releases heat/cooling in order to decrease the MT power outputs (Figure 6(e) and (f): period 17.00–19.00; Figure 9.7(e) and (f): periods 20.00–21.00, 24.00). Further, from Figure 9.5, with the increasing power outputs of MT, its unit cost is decreasing. Thus, in this way, the HST/IST helps MT to fully utilise this advantage to further reduce the net operation cost. Hence, the operation cost of Case 3 is the lowest of all (Table 9.4).

In summary, the proposed models can effectively achieve the system-wide coordination for the grid-connected microgrid and lower its net operation cost.

In terms of the solution speed, all the solutions are obtained within 3.39 s with the minimal solver time of 1.24 s, which is computationally efficient.

9.4.4 *Islanded mode*

Simulation results of the islanded microgrid are shown in Figures 9.8 and 9.9. The net operation cost is listed in Table 9.5. Similarly, P_D, Q_T, and Q_I denote the subtraction of power, heat, and cooling shortage with their respective surplus.

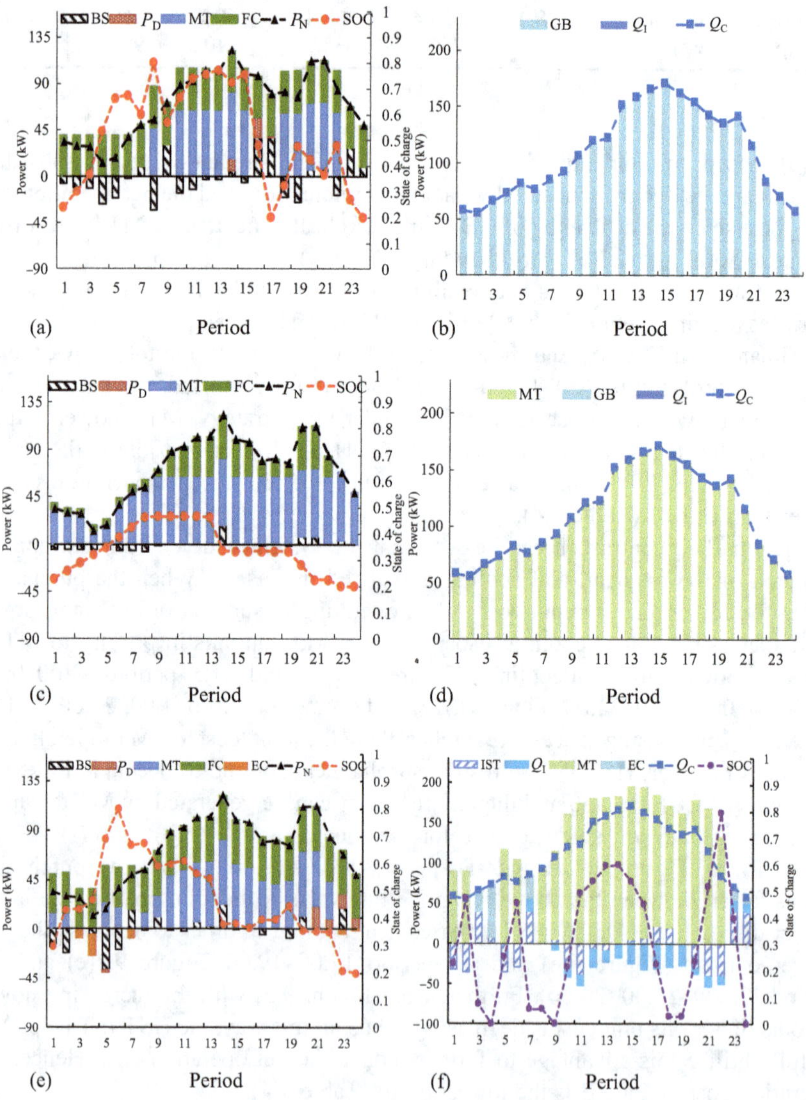

Figure 9.8 Power and heat balance condition of the three cases in summer: Case 1 (a), (b); Case 2 (c), (d); Case 3 (e), (f)

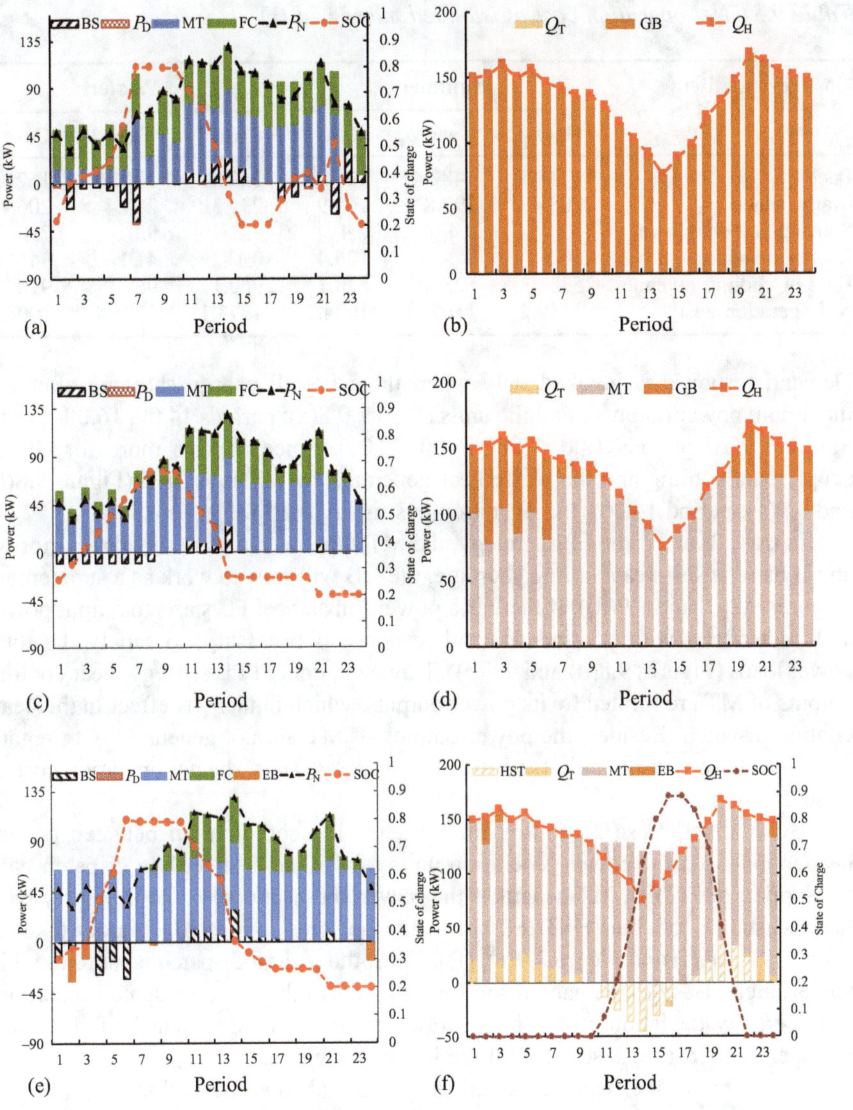

Figure 9.9 Power and heat balance conditions of the three cases in winter: Case 1 (a), (b); Case 2 (c), (d); Case 3 (e), (f)

It is shown in Figures 9.8 and 9.9 that since there is no power support from the grid, demand response will be invoked with the energy deviation cost. As shown in the Table 9.2, the unit power deviation cost is much higher than the unit cost of DGs, and then all the DGs and ES are scheduled to reduce power deviations cost.

In Case 1, the heat/cooling loads are all met by GB (Figure 9.8(b) and 9.9(b)), and the power loads are supplied by FC, MT, and BS (Figures 9.8(a) and 9.9(a)).

Table 9.5 Net operation cost of islanded microgrid (¥)

Cost/revenue item	Summer			Winter		
	Case 1	Case 2	Case 3	Case 1	Case 2	Case 3
Fuel	1713.8	1482.4	1254.6	2494.6	1796.1	1671.5
Maintenance	208.4	197.8	165.9	231.3	218.4	206.7
Start-up and shutdown	15	6.4	7.4	7.2	9.8	7.4
Energy deviation	50.3	9	126.4	0.1	41.1	34
Heat or cooling revenue	528.4	528.4	520.1	960.1	960.1	910.7
Net operation cost	1459.2	1167.2	1034.2	1773.1	1105.4	1008.9

Demand response is invoked only when there is still energy shortage after the maximum power outputs of all the units (Figure 9.8(a): periods 14.00, 16.00–17.00; Figure 9.9(a): periods 1.00, 7.00, 16.00–20.00). Since it costs more for GB to supply heat/cooling and the waste heat generated by MT is wasted (Figure 9.8(a) and 9.9(a): period 14.00), the operation cost is highest (Table 9.5).

In Case 2, under the FPL strategy, the MT supplies the power loads in priority, and if there is still heat/cooling shortage, the GB will start to work as a supplement (Figures 9.8(d) and 9.9(d)). If there is a power imbalance, FC starts to output power and coordinates with BS and demand response if necessary, to satisfy the total power loads (Figures 9.8(d) and 9.9(b)). However, under FPL strategy, heat/cooling outputs of MT are limited by its power outputs, which hinders its effect in the heat/cooling dispatch. Besides, the power outputs of MT are not generated with regard to its operation cost but only the inflexible strategy. Thus, the net operation cost of Case 2 is high.

By contrast, Case 3 realises the system-wide coordination between power, heat/cooling, i.e. the power generation units join in the heat/cooling dispatch with the support of EC/EB, MT generates the power and heat/cooling coordinately with the assistance of the HST/IST just as the grid-connected microgrid does (Figure 9.8(e), (f) and Figure 9.9(e), (f)). Thus, the overall dispatch is more flexible and efficient. Besides, demand response also joins in the overall dispatch according to the energy deviation cost, which further enhances the dispatch flexibility. The net operation cost of Case 3 is still the lowest (Table 9.5).

Furthermore, the maximal solution time for all the islanded cases is 4.51 s, which again shows the high solution efficiency. In addition, the solution time is enough for the day-ahead dispatch scheme.

9.5 Conclusions

This chapter proposes comprehensive system-wide coordinated optimal dispatch models for multi-energy microgrid in both the grid-connected and islanded mode. In the proposed models, the HST/IST could decouple the inflexible operation strategies of CCHP plant and make it join in the power and heat/cooling dispatch with higher freedom. Besides, EB/EC can significantly enhance the interactions between the dispatch of power and heat/cooling. The comprehensive case studies based on the separate generation model, tri-generation model, and coordinated optimal dispatch model proposed in this chapter fully demonstrate that the proposed models can reduce the net operation cost, and achieve more efficient and flexible operation of multi-energy microgrid. It is concluded that the system-wide coordinated optimal dispatch models can provide beneficial references to the optimal design, planning, and operation of multi-energy supplying networks in different modes and have great potential in further reducing the net operation cost.

References

[1] P. Mancarella, 'MES (multi-energy systems): an overview of concepts and evaluation models', *Energy*, vol. 65, pp. 1–17, 2014.

[2] M. Liu, Y. Shi, and F. Fang, 'Combined cooling, heating and power systems: a survey', *Renewable and Sustainable Energy Reviews*, vol. 35, pp. 1–22, 2014.

[3] 'NTU to build an offshore integrated system of renewable energy sources', 2016, http://media.ntu.edu.sg/NewsReleases/Pages/newsdetail.aspx?news=830aa91e-aa88-4c7b-8f95-00c171c17560&from=timeline&isappinstalled=0.

[4] F. Fang, Q.H. Wang, and Y. Shi, 'A novel optimal operational strategy for the CCHP system based on two operating modes', *IEEE Transactions on Power Systems*, vol. 27, no. 2, pp. 1032–1041, 2012.

[5] N. Fumo and L.M. Chamra, 'Analysis of combined cooling, heating, and power systems based on source primary energy consumption', *Applied Energy*, vol. 87, no. 6, pp. 2023–2030, 2010.

[6] N. Fumo, P.J. Mago, and L.M. Chamra, 'Emission operational strategy for combined cooling, heating, and power systems', *Applied Energy*, vol. 86, no. 11, pp. 2344–2350, 2009.

[7] L. Kang, J. Yang, Q. An, *et al.*, 'Complementary configuration and performance comparison of CCHP-ORC system with a ground source heat pump under three energy management modes', *Energy Conversion and Management*, vol. 135, pp. 244–255, 2017.

[8] M. Liu, Y. Shi, and F. Fang, 'Optimal power flow and PGU capacity of CCHP systems using a matrix modeling approach', *Applied Energy*, vol. 102, pp. 794–802, 2013.

[9] Q. Jiang, M. Xue, and G. Geng, 'Energy management of microgrid in grid-connected and stand-alone modes', *IEEE Transactions on Power Systems*, vol. 28, no. 3, pp. 3380–3389, 2013.

[10] H. Kanchev, F. Colas, V. Lazarov, *et al.*, 'Emission reduction and economical optimization of an urban microgrid operation including dispatched PV-based active generators', *IEEE Transactions on Sustainable Energy*, vol. 5, no. 4, pp. 1397–1405, 2014.

[11] M.H. Moradi, M. Eskandari, and S.M. Hosseinian, 'Operational strategy optimization in an optimal sized smart microgrid', *IEEE Transactions on Smart Grid*, vol. 6, no. 3, pp. 1087–1095, 2015.

[12] W. Gu, Z. Wu, R. Bo, *et al.*, 'Modeling, planning and optimal energy management of combined cooling, heating and power microgrid: a review', *International Journal of Electrical Power & Energy Systems*, vol. 54, pp. 26–37, 2014.
[13] L. Ma, N. Liu, J. Zhang, *et al.*, 'Energy management for joint operation of CHP and PV prosumers inside a grid-connected microgrid: a game theoretic approach', *IEEE Transactions on Industrial Informatics*, vol. 12, no. 5, pp. 1930–1942, 2016.
[14] L. Wang, Q. Li, M. Sun, *et al.*, 'Robust optimisation scheduling of CCHP systems with multi-energy based on minimax regret criterion', *IET Generation, Transmission & Distribution*, vol. 10, no. 9, pp. 2194–2201, 2016.
[15] L. Wang, Q. Li, R. Ding, *et al.*, 'Integrated scheduling of energy supply and demand in microgrids under uncertainty: a robust multi-objective optimization approach', *Energy*, vol. 130, no. 1, pp. 1–14, 2017.
[16] B. Yan, P.B. Luh, G. Warner, *et al.*, 'Operation and design optimization of microgrids with renewables', *IEEE Transactions on Automation Science and Engineering*, vol. 14, no. 2, pp. 573–585, 2017.
[17] Z. Li and Y. Xu, 'Dynamic dispatch of grid-connected multi-energy microgrids considering opportunity profit', *2017 IEEE Power & Energy Society General Meeting*, pp. 1–5, 2017.
[18] M. Marzband, M. Ghadimi, A. Sumper, *et al.*, 'Experimental validation of a real-time energy management system using multi-period gravitational search algorithm for microgrids in islanded mode', *Applied Energy*, vol. 128, pp. 164–174, 2014.
[19] B. Belvedere, M. Bianchi, A. Borghetti, *et al.*, 'A microcontroller-based power management system for standalone microgrids with hybrid power supply', *IEEE Transactions on Sustainable Energy*, vol. 3, no. 3, pp. 422–431, 2012.
[20] M. Marzband, A. Sumper, A. Ruiz-Álvarez, *et al.*, 'Experimental evaluation of a real time energy management system for stand-alone microgrids in day-ahead markets', *Applied Energy*, vol. 106, pp. 365–376, 2013.
[21] A.K. Basu, A. Bhattacharya, S. Chowdhury, *et al.*, 'Planned scheduling for economic power sharing in a CHP-based micro-grid', *IEEE Transactions on Power Systems*, vol. 27, no. 1, pp. 30–38, 2012.
[22] M. Carrión and J.M. Arroyo, 'A computationally efficient mixed-integer linear formulation for the thermal unit commitment problem', *IEEE Transactions on Power Systems*, vol. 21, no. 3, pp. 1371–1378, 2006.
[23] S. Campanari and E. Macchi, 'Technical and tariff scenarios effect on microturbine trigenerative applications', *ASME Turbo Expo 2003*, collocated with the *2003 International Joint Power Generation Conference*, American Society of Mechanical Engineers, pp. 747–757, 2003.
[24] M.G. Rasul, C. Ault, and M. Sajjad, 'Bio-gas mixed fuel micro gas turbine co-generation for meeting power demand in Australian remote areas', *Energy Procedia*, vol. 75, pp. 1065–1071, 2015.

[25] M.Q. Wang and H.B. Gooi, 'Spinning reserve estimation in microgrids', *IEEE Transactions on Power Systems*, vol. 26, no. 3, pp. 1164–1174, 2011.

[26] D.J. Swider, 'Compressed air energy storage in an electricity system with significant wind power generation', *IEEE Transactions on Energy Conversion*, vol. 22, no. 1, pp. 95–102, 2007.

[27] X. Luo, J. Wang, M. Dooner, *et al.*, 'Overview of current development in electrical energy storage technologies and the application potential in power system operation', *Applied Energy*, vol. 137, no. 6, pp. 511–536, 2015.

[28] C. Zhang, Y. Xu, Z.Y. Dong, *et al.*, 'Robust operation of microgrids via two-stage coordinated energy storage and direct load control', *IEEE Transactions on Power Systems*, vol. 5, no. 99, pp. 1–10, 2016.

[29] 'State of Green Company', 2017, https://stateofgreen.com/en/profiles/ramboll/solutions/heat-accumulator-for-chp-plants.

[30] Z. Zhao, W.C. Lee, Y. Shin, *et al.*, 'An optimal power scheduling method for demand response in home energy management system', *IEEE Transactions on Smart Grid*, vol. 4, no. 3, pp. 1391–1400, 2013.

[31] F. Barbir and T. Gomez, 'Efficiency and economics of proton exchange membrane (PEM) fuel cells', *International Journal of Hydrogen Energy*, vol. 21, no. 10, pp. 891–901, 1996.

Chapter 10

Temporally coordinated dispatch of MEMGs under diverse uncertainties

This chapter proposes a temporally coordinated operation method for a multi-energy microgrid under diverse uncertainties from renewable energy sources (RESs), power loads, and electricity transaction prices between the microgrid and the utility grid. The method aims to coordinate multiple energies in different timescales considering distinct properties of thermal and power energy: the combined cooling heat, and power (CCHP) plants, power-to-thermal conversion units, and thermal storage tanks (TSTs) are dispatched hourly in the day-ahead operation stage; in the intraday online operation stage, the battery storage units (BSs) are dispatched to supplement the day-ahead operation decisions every 5 min after the uncertainties are realised. Based on constraints linearisation and uncertain scenario generation/reduction, the problem is converted to a deterministic two-stage mixed-integer linear programming (MILP) equivalent model which can be solved efficiently. Finally, the proposed method is verified on an IEEE 33-bus distribution network-based multi-energy microgrid. Compared with the existing methods, the simulation results indicate that the proposed method can better coordinate multiple energies with low operating cost and high robustness.

Nomenclature

A Sets and indices

t/i	Index of dispatch periods/buses
b/k	Index of branches/linear pieces
N_T	Set of dispatch periods
$N_M/N_W/N_P$	Set of buses that install CCHP plants/WTs/PVs
$N_E/N_B/N_H$	Set of buses that install PTCs/BSs/TSTs
$Br(i)$	Set of the branches that connect to bus i
$Br(i, l)$	Set of the branch between bus i and l

B Constants and parameters

η_H	Efficiency of heat recovery unit/absorption chiller
η_{TSTC}/η_{TSTD}	Charging/discharging efficiency of TSTs

τ_{TST}/τ_{BS}	Decay rate of TSTs/BSs
η_{BSC}/η_{BSD}	Charging/discharging efficiency of TSTs
θ/ψ	Scale parameters for Beta distribution
N_F	Normalised factor for Beta distribution
μ/σ	Mean/standard deviation for normal distribution
Δt	Unit dispatch interval
γ_G/γ_{HR}	Unit gas/thermal energy selling price
$\gamma_{WT}/\gamma_{PV}/\gamma_{MT}$	Unit maintenance cost of WTs/PVs/MTs
$\gamma_{PT}/\gamma_{BS}/\gamma_{TST}$	Unit maintenance cost of PTCs/BSs/TSTs
γ_B/γ_S	Purchasing/selling price
C_{CG}^{U}/C_{CG}^{D}	CG's unit start-up/shutdown cost
$P_{CG}^{max,i}/P_{CG}^{min,i}$	Maximal/minimal CG's power outputs
$R_{CG}^{up,i}/R_{CG}^{down,i}$	CG's ramp up/down rates
$P_{ESC}^{max,i}/R_{ESD}^{max,i}$	ES's maximal charging/discharging power
$\xi_{ESC}^{min,i}/\xi_{ESC}^{max,i}$	ES's maximal/minimal state of charge (SOC)
Cap_{ES}^{i}	ES's capacity
R^b/X^b	Resistance/reactance of branch b
α_k/β_k	Slope/intercept of linear function in kth piece
P_k	Starting point of kth block
η_{PC}	Efficiency of power to thermal conversion (PTC)
ΔV_{BUS}^{max}	Maximal voltage deviation

C Variables

f/f	Original non-linear function/approximated linearised function
$P_{MT}^{t,i}/H_{MT}^{t,i}$	Power or thermal outputs of MTs
$\eta_{MT}^{t,i}$	Electrical efficiency of MTs (CCHP plant)
$P_{PT}^{t,i}/H_{PT}^{t,i}$	Power or thermal energy outputs of PTCs
$E_{TST}^{t,i}/E_{BS}^{t,i}$	Energy stored in TSTs/BSs
$P_{TSTC}^{t,i}/P_{TSTD}^{t,i}$	Absorbing/releasing power of TSTs
$P_{BSC}^{t,i}/P_{BSD}^{t,i}$	Charging/discharging power of BSs
$P_{RES}^{t,i}$	Power outputs of RES
$P_{L}^{t,i}/H_{L}^{t,i}$	System power/thermal loads
F_G	Net operating cost
$C_{FC}/C_{OM}/C_{EX}$	Fuel/maintenance/power transaction cost
$C_{ST}/C_{SD}/C_{HR}$	Start-up/shutdown cost/thermal energy selling revenue
$P_{BUY}^{t,i}/P_{SELL}^{t,i}$	Purchasing/selling power
$U_{CG}^{t,i}$	Binary status of controllable generators (CG)
$P_{CG}^{t,i}$	CG's power outputs

$P_{\mathrm{ESC}}^{t,i}/P_{\mathrm{ESD}}^{t,i}$	ES's charging and discharging power
$\varepsilon_{\mathrm{ESC}}^{t,i}/\varepsilon_{\mathrm{ESD}}^{t,i}$	ES's charging/discharging state
$E_{\mathrm{ES}}^{t,i}$	Energy stored in ESs
$P_{\mathrm{PF}}^{t,b}/Q_{\mathrm{PF}}^{t,b}$	Active/reactive power that flows on branch b
$P_{\mathrm{L}}^{t,i}/Q_{\mathrm{L}}^{t,i}$	Active/reactive power loads at bus i
$V_{\mathrm{BUS}}^{t,i}$	Voltage at bus i
$P_{\mathrm{PF}}^{t,0,b}/Q_{\mathrm{PF}}^{t,0,b}$	Active/reactive power that flows on the lateral branch of branch b
B_k^t	Binary state of piece k
P_k^t	Power output of the MT in kth piece

10.1 Introduction

10.1.1 Background and motivation

Compared with traditional energy systems in which different energy sectors are treated separately, multi-energy systems can integrate different energy carriers such as gas, power, and thermal (heat and cooling) energy for simultaneous energy supply with a higher energy utilisation efficiency [1]. When structured at the distribution network level, they can form a multi-energy microgrid [2]. Due to the different properties of multiple energies and diverse uncertainties from intermittent distributed generators, power transaction prices, energy demands, etc., how to optimally coordinate multiple energy carriers is a key yet challenging problem.

First, in terms of differences in energy properties, compared with electric power, thermal energy is of slow response speed due to the thermal inertia. Besides, slow transmission speed and large transfer rate of thermal energy will cause a high time delay and great thermal losses [3]. Second, distributed generators can be categorised into the controllable and uncontrollable ones: the former can achieve a reliable and cost-effective dispatch like the CCHP plants and power-to-thermal conversion units; the latter, specifically RES-based units, e.g. wind turbines (WTs) and photovoltaics (PVs), are inherently uncertain and intermittent [4]. Third, given the random human and market behaviours [5], power loads and electricity transaction prices are highly stochastic; together with the uncontrollable units, the diverse uncertainties will pose considerable technical and economic challenges to the operation of multi-energy microgrids [6].

In this light, this chapter aims to coordinate different energies and address the diverse uncertainties to achieve an optimal and robust operation of a multi-energy microgrid.

10.1.2 Literature survey

To supply multiple energies, traditional research works focus on the dispatch of a single CCHP plant, which generally operates under the following-power-load or following-thermal-load strategy [7–9]. Smith *et al.* [7] proposes a hybrid strategy to minimise the excess power and thermal energy produced by CCHP plants. In [8], an improved operational strategy for CCHP plants based on the following-power-load strategy is introduced to minimise the energy consumption for a building. Li *et al.* [9] model a CCHP plant to investigate its benefits in reducing the operating cost, CO_2 emissions, and energy consumption, and then based on the conventional strategies, five operational strategies are employed for the dispatching. In the above works, only the operation of the single plant has been studied, but the multi-energy coordination of CCHP plants with other units like the renewable energy-driven generators and energy storage at a system level (e.g. microgrid) is not systematically investigated.

At microgrid level, CCHP plants typically operate under the following-thermal-load strategy in grid-tied microgrids since the main grid can compensate power imbalance [10–12]. In [10], an optimisation method for the operation of a grid-tied microgrid with power and thermal demands is proposed considering RES. Reference [11] studies the coordination of a CCHP plant with PV prosumers to minimise the net operating cost. In our recent work [12], a multi-period dispatch approach for grid-tied microgrids considering the opportunity profit from BS is proposed. For islanded microgrids, given that power system needs to be real-time balanced while the thermal imbalance can be tolerable for short periods [3,13], the following-power-load strategy is preferred for CCHP plants. In [14], a multi-objective optimisation method is proposed to minimise fuel cost and gas emission.

However, for the above research works either at a single plant or system level, CCHP power and thermal outputs are highly coupled, which makes the multiple energies supply not flexible enough. In practice, TSTs (heat storage tanks or ice storage tanks) and power-to-thermal conversion units (electric boiler or electric chiller) can be introduced to enhance the dispatch flexibility and coordination degree of multiple energies [15]. At the single plant level, with the installation of the both, a so-called variable electric cooling ratio is introduced to dispatch the CCHP plants [16], and then the annual benefit is greatly improved with respect to traditional methods. At the microgrid level, Zheng *et al.* [15] propose a TST-based operational strategy for the CCHP plants. Compared with the traditional operation strategies, simulation results show that TSTs can enhance the CCHP operational flexibility and economy as well as reduce CO_2 emissions and energy consumption. In our recent work [17], a system-wide optimal coordinated energy dispatch method for a microgrid in both the grid-tied and islanded modes is proposed. The TSTs and power-to-thermal conversion units enhance the operational flexibility, and the operating cost of the microgrid in both modes is greatly reduced.

In the above literature [10–17], the multi-energy coordination is based on the deterministic day-ahead operation, but the uncertainties of the RES, power loads, and transaction prices are ignored, which is not practical for application. To handle uncertainties from RES outputs and power loads, Guan *et al.* [18] employ the stochastic optimisation method to minimise the power and natural gas cost at the building level. In [19] and [20], a robust optimisation-based scheduling method is proposed to attenuate the disturbance by uncertainties and minimise the operating cost and gas emissions. The research work in [21] conducts a two-stage operation to smooth out the RES fluctuations and follow the load variations; at the first stage, the deterministic operation is employed to set the reference for the second-stage rolling process. However, in the above existing works, different properties of power and thermal energy are not fully considered and coordinated in the system level. Further, Refs. [10–21] ignore the power network's constraints, so the operation decisions may lead to security violations in some cases.

To guarantee the system operating security, the online optimisation of the energy networks, like the real-time power flow, is essential. In this regard, Ref. [22] presents an overview of the recent studies on real-time power flow of the power system under wind energy penetration and provides an essential review of the major advancements in this area. Excellent works have been done in the real-time power flow area and provides insightful references for the secure system operation. On top of this, other energies can be further integrated for a coordinated and comprehensive system operation.

In terms of solution algorithms, a number of references have employed the evolutionary algorithms such as the Particle Swarm Optimisation and Genetic Algorithm like Refs. [8,9,11–14,19–21]. However, these algorithms usually suffer from issues such as excessive computational burden, inconsistent solutions, and difficulty to rigorously converge.

Besides, apart from offline dispatch methods which are based on classic algorithms, real-time optimal power flow (RT-OPF) methods have also been proposed in the literature [22]. The study in [23] develops a real-time active – reactive optimal power flow method to simultaneously optimise the wind active power curtailment and reactive power dispatch as well as discrete reference values of slack bus voltage. Besides, a novel reconciliation algorithm is proposed to converge the real-time mixed-integer non-linear programming problem. Authors in [24] propose a two-phase, i.e. prediction phase and realisation phase, solution approach to solve the large-scale complex real-time power flow optimisation problem under the penetration of renewable energies. Reference [25] proposes a novel sensitivity-based approach for simulating local controllers in the distribution network. This approach improves the computational performance of the power flow model and is of practical value for real-time applications. While such RT-OPF techniques are promising, they have not been applied in multi-energy microgrid dispatch problems.

10.1.3 Contribution of this chapter

This chapter proposes a temporally coordinated stochastic operation method for a grid-tied multi-energy microgrid which consists of WTs, CCHP plants, PVs, power-to-thermal conversion units, TSTs, and BSs. To accommodate different properties of the energy carriers, thermal-related units including CCHP plants, power-to-thermal conversion units, and thermal storages are dispatched to manage the slow variations in a longer timescale at the day-ahead operation stage. The power-related units BSs act in a shorter timescale at the intraday online operation stage and respond to the fast variations of uncertain resources. The two timescales are coordinated via the stochastic optimisation model, where the day-ahead operation makes initial decisions irrespective of uncertainties, and online operation replies on a recourse action to compensate day-ahead decisions after the system uncertainties are revealed.

Compared with the existing works and practices, the main contributions of this chapter are summarised as follows:

1. A temporally coordinated optimal operation method for a multi-energy microgrid considering different energy properties is proposed. The different energy properties (including electricity and heat) are fully utilised optimal dispatch of the microgrid.
2. Diverse uncertainties from RESs, power, and heat loads as well as power transaction prices are integrated into the system operation model and addressed by the multi-energy coordination through the scenario-based multi-stage stochastic programming method.
3. The power network constraints including the power flow and bus voltage limits for both day-ahead and online dispatch stage are incorporated in the system operation to guarantee the operating security.
4. The original non-linear problem is linearised to an MILP one, which can be efficiently solved by some commercial solvers. Compared with the traditional evolutionary algorithms, the solution quality and speed are greatly improved.

The rest of this chapter is organised as follows. The multi-energy microgrid is modelled in Section 10.2. The proposed operation method is detailed in Section 10.3. The mathematical formulations for the system operation are presented in Section 10.4. Then, the solution method is given in Section 10.5. In Section 10.6, comprehensive case studies are done. Finally, the conclusions are drawn in Section 10.7.

10.2 Multi-energy microgrid modelling

10.2.1 Structure of the microgrid

The typical structure of a multi-energy microgrid is shown in Figure 10.1. In the microgrid, the thermal loads (heat or cooling loads) are supplied by CCHP plants, power-to-thermal conversion units, and TSTs, while the power loads are supplied by CCHP plants, BSs, WTs, and PVs [26]. The microgrid is connected to the main grid for power transaction [15,17].

10.2.2 Components modelling

Since this study focuses on the system operation, the component modelling mainly focuses on the external characteristic, i.e. the ON/OFF statuses and energy outputs of each unit.

1. *CCHP plants*: Microturbine (MT) is the core unit of a CCHP plant, and it consumes natural gas to generate power. At the same time, the waste heat can be utilised by heat recovery unit or absorption chiller to produce heat or

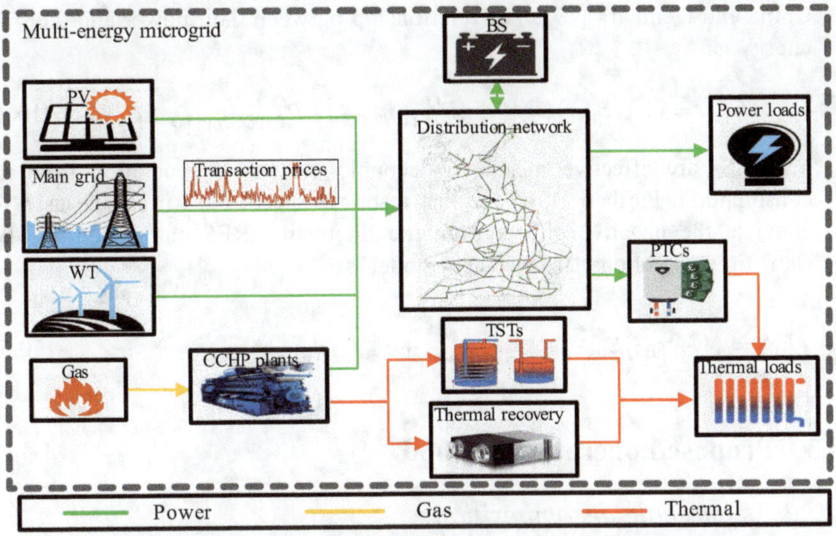

Figure 10.1 Typical structure of a multi-energy microgrid

cooling [4,12]. In this chapter, Capstone's MT C200, which has a high international market share, is employed in this chapter [27]. The relation between its power outputs and electrical efficiency is

$$\eta_{\mathrm{MT}}^{t,i} = 2.499 \times 10^{-6} (P_{\mathrm{MT}}^{t,i})^3 - 1.544 \times 10^{-3} (P_{\mathrm{MT}}^{t,i})^2 + 0.2910 P_{\mathrm{MT}}^{t,i} + 9.092 \tag{10.1}$$

The corresponding thermal outputs can be expressed as

$$H_{\mathrm{MT}}^{t,i} = \eta_{\mathrm{H}} [1.222 \times 10^{-4} (P_{\mathrm{MT}}^{t,i})^3 + 2.062 \times 10^{-4} (P_{\mathrm{MT}}^{t,i})^2 + 0.621 P_{\mathrm{MT}}^{t,i} - 370.4] \tag{10.2}$$

Note that multiple C200 MTs can be integrated as a single larger unit named a *MultiPac*, and when the power loads are larger than the maximal power outputs of a single unit, the loads are shared evenly among all the C200 MTs [21].

2. *Power-to-thermal energy conversion units*: To enhance the flexibility of supplying multiple energies, the power-to-thermal conversion units can be utilised to convert electric power to thermal power with the following model:

$$H_{\mathrm{PT}}^{t,i} = P_{\mathrm{PT}}^{t,i} \eta_{\mathrm{PT}} \tag{10.3}$$

3. *TSTs*: To decouple the inflexible CCHP operational strategies, TSTs are deployed. In practice, they are already widely employed in countries like

China and Denmark [17]. The relationship between their power and thermal energy is

$$E_{\mathrm{TST}}^{t,i} = (1 - \tau_{\mathrm{TST}})E_{\mathrm{TST}}^{t-1,i} + (P_{\mathrm{TSTC}}^{t,i}\eta_{\mathrm{TSTC}} + P_{\mathrm{TSTD}}^{t,i}/\eta_{\mathrm{TSTD}})\Delta t \tag{10.4}$$

4. *BSs*: BSs are effective means to decouple the power generation from its consumption. Further, BSs have fast response speed; thus, it can be used to alleviate the negative effects from the fluctuating RES outputs during the short-time operation [10,28]. Their model is

$$E_{\mathrm{BS}}^{t,i} = (1 - \tau_{\mathrm{BS}})E_{\mathrm{BS}}^{t-1,i} + (P_{\mathrm{BSC}}^{t,i}\eta_{\mathrm{BSC}} + P_{\mathrm{BSD}}^{t,i}/\eta_{\mathrm{BSD}})\Delta t \tag{10.5}$$

10.3 Proposed operation method

10.3.1 Timescale decomposition

Given the large thermal inertia [3], thermal energy changing rate is rather low. Besides, thermal loads are usually much steadier than power loads, so the short-term thermal load forecasting (i.e. hourly-ahead) can be considered accurate enough. In this chapter, the operating states of the CCHP plants, power-to-thermal conversion units, and TSTs are planned in a relatively long timescale, which is set as 1 h. Their operation decisions are made in the day-ahead time window (first stage). Then, to counteract the adverse effects from the diverse uncertainties, during the intraday online operation phase (second stage), fast-responding devices BSs are dispatched to complement the day-ahead operation decisions after the uncertainty is realised. As BSs can respond faster, it acts in a shorter timescale (5 min), which can be used in the online operation.

This timescale separation acknowledges the distinct properties of different energy carriers and units' respond speeds. Note that other timescale lengths (i.e. half-hour, 15 min, etc.) can also be considered if necessary, which does not affect the effectiveness of the proposed method.

10.3.2 Temporally coordinated operation framework

To address the uncertainties and make full use of different energy carriers, a temporally coordinated operation framework is developed in Figure 10.2: the day-ahead operation (long timescale) aims to dispatch the status of CCHP plants, TSTs, and power-to-thermal conversion units every hour based on the forecasts of power loads, transaction prices, and RES outputs; within each hour, intraday online operation (short timescale) dispatches batteries every 5 min in response to the faster variations of the uncertain sources.

For implementation, at the day-ahead decision cycle, CCHP plants, TSTs, and power-to-thermal conversion units are dispatched a day ahead for next day's 24 h based on the hourly uncertainty predictions. Then, the intraday online operation

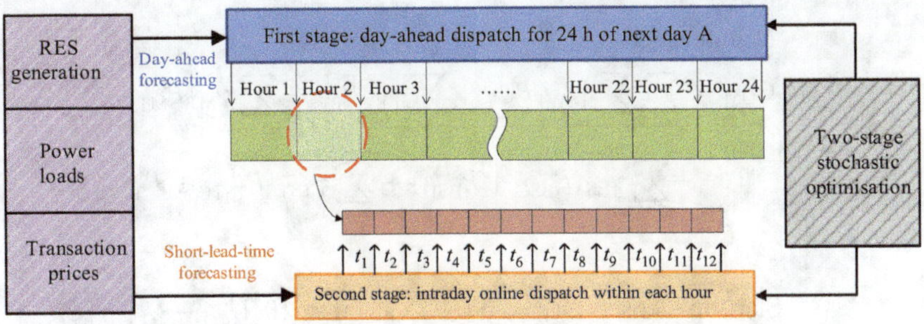

Figure 10.2 Temporally coordinated stochastic operation framework

compensates the day-ahead decisions for the negative effects from the fast and stochastic variations. At each online dispatch interval, the random variations of the next 5-min's power loads, transaction prices, and RES outputs are predicted and used as the inputs to optimise the BSs' states. Since the battery's charging/discharging power speed is very fast (within seconds and minutes), 5-min dispatch interval is feasible. However, other dispatch intervals (e.g. half-hour) can also be considered without affecting the effectiveness of the proposed method. It is worth noting that many forecasting techniques have been developed in the literature, showing that the shorter lead time forecasting tends to be more accurate than longer lead time one, hence uncertainties can be better grasped in the online short-term operation time window [29].

10.4 Mathematical formulation

10.4.1 Multi-energy microgrid coordination model

The operation of multi-energy microgrids aims to minimise the net operating cost while satisfying network operational limits over different timescales in presence of diverse uncertainties; the mathematical models for the two stages are as follows:

1. Objective function: Minimise the system net operating cost:

$$\text{Net operation cost:}\quad \text{MIN } F_{\text{G}} = C_{\text{FC}} + C_{\text{OM}} + C_{\text{EX}} + C_{\text{ST}} + C_{\text{SD}} - C_{\text{HR}} \tag{10.6}$$

where

$$\text{Fuel cost:}\quad C_{\text{FC}} = \sum_{t\in N_{\text{T}}}\sum_{i\in N_{\text{M}}}(\gamma_{\text{G}}P_{\text{MT}}^{t,i}/\eta_{\text{MT}}^{t,i})\Delta t \tag{10.7}$$

$$\text{Power transaction cost:}\quad C_{\text{EX}} = \sum_{t\in N_{\text{T}}}(\gamma_{\text{B}}P_{\text{BUY}}^{t,1} - \gamma_{\text{S}}P_{\text{SELL}}^{t,1})\Delta t \tag{10.8}$$

Maintenance cost:

$$C_{\mathrm{OM}} = \sum_{t\in N_{\mathrm{T}}}\sum_{i\in N_{\mathrm{W}}}[\gamma_{\mathrm{WT}}P_{\mathrm{WT}}^{t,i} + \sum_{i\in N_{\mathrm{P}}}\gamma_{\mathrm{PV}}P_{\mathrm{PV}}^{t,i} + \sum_{i\in N_{\mathrm{M}}}\gamma_{\mathrm{MT}}P_{\mathrm{MT}}^{t,i} + \sum_{i\in N_{\mathrm{E}}}\gamma_{\mathrm{PT}}P_{\mathrm{PT}}^{t,i}$$
$$+\sum_{i\in N_{\mathrm{B}}}\gamma_{\mathrm{BS}}(P_{\mathrm{BSC}}^{t,i} + P_{\mathrm{BSD}}^{t,i}) + \sum_{i\in N_{\mathrm{H}}}\gamma_{\mathrm{TST}}(P_{\mathrm{TSTC}}^{t,i} + P_{\mathrm{TSTD}}^{t,i})]\Delta t \tag{10.9}$$

Start-up cost: $$C_{\mathrm{ST}} = \sum_{t\in N_{\mathrm{T}}}\sum_{i\in(N_{\mathrm{M}}\cup N_{\mathrm{E}})}\max\{0, U_{\mathrm{CG}}^{t,i} - U_{\mathrm{CG}}^{t-1,i}\}C_{\mathrm{CG}}^{\mathrm{U}} \tag{10.10}$$

Shutdown cost: $$C_{\mathrm{SD}} = \sum_{t\in N_{\mathrm{T}}}\sum_{i\in(N_{\mathrm{M}}\cup N_{\mathrm{E}})}\max\{0, U_{\mathrm{CG}}^{t-1,i} - U_{\mathrm{CG}}^{t,i}\}C_{\mathrm{CG}}^{\mathrm{D}} \tag{10.11}$$

Heat selling revenue: $$C_{\mathrm{HR}} = \sum_{t\in N_{\mathrm{T}}}\sum_{i\in N_{\mathrm{M}}}\gamma_{\mathrm{HR}}H_{\mathrm{L}}^{t,i}\Delta t \tag{10.12}$$

2. Constraints:

(a) *CCHP plants and power-to-thermal conversion units:*

As mentioned before, CCHP plants and power-to-thermal conversion units are controllable (denoted by subscript CG) and they are subject to their power output (13) and ramp up/down limits (14).

$$U_{\mathrm{CG}}^{t,i}\cdot P_{\mathrm{CG}}^{\min,i} \le P_{\mathrm{CG}}^{t,i} \le U_{\mathrm{CG}}^{t,i}\cdot P_{\mathrm{CG}}^{\max,i} \tag{10.13}$$

$$R_{\mathrm{CG}}^{\mathrm{down},i}\Delta t \le P_{\mathrm{CG}}^{t,i} - P_{\mathrm{CG}}^{t-1,i} \le R_{\mathrm{CG}}^{\mathrm{up},i}\Delta t \tag{10.14}$$

(b) *TSTs and BSs:*

The TST and BS are energy storage (ES) units. Their operation models can be formulated as (10.15)–(10.19), where the subscript ES represents either TST or BS. The charging and discharging power of the energy storage is constrained by the corresponding minimal and maximal power as (10.15)–(10.16); (10.17) denotes that energy stored in the energy storage should be within the available capacity; (10.18) denotes that for the energy storage at each dispatching cycle, the starting energy is the same as the ending, in order to keep the same dispatch flexibility at different dispatching cycles; (10.19) means that in each dispatch interval, the summation of the binary charging and discharging states of energy storage should be less than or equal to 1, and in this way, the charging and discharging states cannot happen simultaneously [18].

$$0 \le P_{\mathrm{ESC}}^{t,i} \le \varepsilon_{\mathrm{ESC}}^{t,i}P_{\mathrm{ESC}}^{\max,i} \tag{10.15}$$

$$0 \le P_{\mathrm{ESD}}^{t,i} \le \varepsilon_{\mathrm{ESD}}^{t,i}P_{\mathrm{ESD}}^{\max,i} \tag{10.16}$$

$$\xi_{\mathrm{ES}}^{\min,i}Cap_{\mathrm{ES}}^{i} \le E_{\mathrm{ES}}^{t,i} \le \xi_{\mathrm{ES}}^{\max,i}Cap_{\mathrm{ES}}^{i} \tag{10.17}$$

$$E_{\mathrm{ES}}^{0,i} = E_{\mathrm{ES}}^{N_T \Delta t,i} \tag{10.18}$$

$$\varepsilon_{\mathrm{ESC}}^{t,i} + \varepsilon_{\mathrm{ESD}}^{t,i} \leq 1 \tag{10.19}$$

(c) *Linearised distribution load flow:*

The power distribution network's power flow in the multi-energy microgrid can be modelled as linearised *Dist-Flow* [7] as (10.20)–(10.22) with the assumption that all the DGs are of the unity power factor as Ref. [2]. Besides, the cost from power loss is relatively low – for simplicity, we ignore it without loss of the solution accuracy as Refs. [2] and [5].

$$\text{Active power: } P_{\mathrm{PF}}^{t,b+1} = P_{\mathrm{PF}}^{t,b} - P_{\mathrm{PF}}^{t,0,b+1} - P_{\mathrm{L}}^{t,i} + P_{\mathrm{WT}}^{t,i} + P_{\mathrm{PV}}^{t,i}$$

$$+ P_{\mathrm{BSD}}^{t,i} - P_{\mathrm{BSC}}^{t,i} + P_{\mathrm{MT}}^{t,i} - P_{\mathrm{PT}}^{t,i}, b \in Br(i), \forall i, t \tag{10.20}$$

$$\text{Reactive power: } Q_{\mathrm{PF}}^{t,b+1} = Q_{\mathrm{PF}}^{t,b} - Q_{\mathrm{PF}}^{t,0,b+1} - Q_{L}^{t,i}, b \in Br(i), \forall i, t \tag{10.21}$$

$$\text{Voltage: } V_{\mathrm{BUS}}^{t,i+1} = V_{\mathrm{BUS}}^{t,i} - (R^b P_{\mathrm{PF}}^{t,b} + X^b Q_{\mathrm{PF}}^{t,b})/V_0, b \in Br(i, i+1), \forall i, t \tag{10.22}$$

(d) *Thermal energy balance, i.e. the thermal energy generation should be equal to the consumption.*

$$H_{\mathrm{L}}^{t,i} = H_{\mathrm{MT}}^{t,i} + H_{\mathrm{PT}}^{t,i} + P_{\mathrm{TSTD}}^{t,i} - P_{\mathrm{TSTC}}^{t,i} \tag{10.23}$$

(e) *Voltage limits, i.e. the voltage should be maintained in a secure range.*

$$1 - \Delta V_{\mathrm{BUS}}^{\max} \leq V_{\mathrm{BUS}}^{t,i} \leq 1 + \Delta V_{\mathrm{BUS}}^{\max} \tag{10.24}$$

10.4.2 Model linearisation

It can be seen that the coordination model is a mixed-integer non-linear problem due to the non-linear fuel cost in (10.7), thermal outputs in (10.2), and start-up/shutdown cost in (10.10)–(10.11). Such a non-linear problem is difficult and time-consuming to solve. To reduce the computation burdens and improve solution quality, in this chapter, the original problem is linearised into an MILP one.

1. *Linearisation of fuel cost and thermal outputs*: N_R pieces of linear functions can approximate the non-linear functions by introducing a binary variable and a continuous variable in each piece. The corresponding linear functions are the following:

$$\bar{f} = \sum_{k \in N_R} (\alpha_k P_k^t + \beta_k B_k^t) \tag{10.25}$$

$$\alpha_k = [(f(P_{k+1}) - f(P_k))/(P_{k+1} - P_k)] \tag{10.26}$$

$$\beta_k = f(P_k) - P_k \alpha_k \tag{10.27}$$

$$P_k B_k^t \le P_k^t \le B_k^t P_{k+1} \tag{10.28}$$

$$U_{\text{CG}}^t = \sum_{k \in N_R} B_k^t \le 1 \tag{10.29}$$

2. *Linearisation of start-up/shutdown cost*: separating the max function in (10.10) and (10.11) into two individual parts as follows:

$$C_{\text{ST}} \ge 0; \quad C_{\text{ST}} \ge \sum_{t \in T} \sum_{i \in (N_{\text{M}} \cup N_{\text{E}})} (U_{\text{CG}}^{t,i} - U_{\text{CG}}^{t-1,i}) C_{\text{CG}}^{\text{U}} \tag{10.30}$$

$$C_{\text{SD}} \ge 0; \quad C_{\text{SD}} \ge \sum_{t \in T} \sum_{i \in (N_{\text{M}} \cup N_{\text{E}})} (U_{\text{CG}}^{t-1,i} - U_{\text{CG}}^{t,i}) C_{\text{CG}}^{\text{D}} \tag{10.31}$$

In this way, the coordination model becomes an MILP problem that can be effectively and efficiently solved by powerful solvers such as CPLEX and GUROBI.

10.5 Solution method

Due to the diverse uncertain variables, the mathematical model of (10.6)–(10.24) is a stochastic programming problem. This section will convert the model into a tractable deterministic equivalent model based on scenario generation and reduction techniques.

10.5.1 Uncertainty modelling

For RES units like WTs and PVs, their stochastic variations or forecasting errors usually follow the Beta distribution [6,25]:

$$f_{P_{\text{RES}}^{t,i}} = (P_{\text{RES}}^{t,i})^{\theta-1} \cdot (1 - P_{\text{RES}}^{t,i})^{\psi-1} \cdot N_{\text{F}} \tag{10.32}$$

Since power loads and transaction prices are mainly affected by human behaviours, the normal distribution is employed to represent their forecasting errors [4]:

$$f_{P_{\text{L}}^{t,i}} = 1/(\sqrt{2\pi}\sigma) \cdot \exp[-(P_{\text{L}}^{t,i} - \mu)^2/(2\sigma^2)] \tag{10.33}$$

10.5.2 Stochastic programming model

The stochastic optimisation method constitutes an effective tool for making decisions under uncertainties and reveals the fact when the uncertainties become known as time evolves. The coordination model of (10.6)–(10.24) is formulated into a two-stage stochastic optimisation model as

$$\min_{z \in F_{\mathrm{A}}} \{F(z) + E[G(z, \omega)]\} \tag{10.34}$$

Here, $F(z)$ is the day-ahead operation objective, where z is the day-ahead decision vector; $E[G(z,\omega)]$ is the expectation of the intraday online operation problem; and $G(z, \omega)$ is the intraday online operation objective, which is equal to $\min_{y \in \Omega(z,\omega)} L(y)$, where y is the online decision vector and ω is the random vector.

10.5.3 Deterministic equivalence

Assuming that ω has a finite number of possible realisations, called scenarios, denoted as $\omega_1, \omega_2, \ldots, \omega_n$ with respective possibilities of $\chi_1, \chi_2, \ldots, \chi_n$, then the expectation term in (10.34) can be written as

$$E[G(z, \omega)] = \sum_{n \in N_{\mathrm{S}}} \chi_n G(z, \omega_n) \tag{10.35}$$

where N_S is the number of scenarios and n is the scenario index.

In this way, the original stochastic optimisation model can be reformulated into the *deterministic equivalence* as

$$\min_{z, y_1, y_2, \ldots, y_n} \quad F(z) + \sum_{n \in N_{\mathrm{S}}} \chi_n L(y_n) \tag{10.36}$$

$$\begin{aligned} \text{s.t.} \quad & z \in F_{\mathrm{A}} \\ & y_n \in \Omega(z, \omega_n), \forall n \end{aligned} \tag{10.37}$$

From the *deterministic equivalence*, a '*here-and-now*' decision z is made in the day-ahead operation before the realisation of uncertainties; after a realisation of the uncertain scenario ω in the intraday online operation, a '*wait-and-see*' decision y is made to compensate the difference between the prediction and reality in the day-ahead operation (y is also named as the '*recourse actions*').

Based on objective function in (10.36), the two stages are coordinated since day-ahead decisions are obtained with effects from online operation stage; when optimising the online operation decisions, day-ahead decisions are passed down as state variables.

Specifically, for day-ahead operation, $F(z)$ and z correspond to the operating cost and power outputs of day-ahead operation units, respectively; set F_A corresponds to the constraints of (10.13)–(10.24) for each hour. For intraday online operation, $L(y_n)$ corresponds to the summation of the operating cost of WTs, PVs, BSs, and power transaction cost; y corresponds to battery's power outputs in every 5-min interval; ω corresponds to the stochastic variations of the RES outputs, power loads and transaction prices; and $\Omega(z, \omega)$ corresponds to constraints of (10.15)–(10.24) for each 5-min interval.

10.5.4 Scenario generation and reduction

Based on the probability distributions of diverse uncertain variables in (10.32) and (10.33), Latin hypercube sampling method is used to generate a near-random sample of parameter values. It partitions each input distribution into several intervals of equal probability and selects one sample from each interval. Compared with the Monte Carlo method, Latin hypercube can spread the sample points more evenly across all possible values.

Then, to alleviate the computational burden from the large number of scenarios, Simultaneous backward reduction method [7] is implemented to choose a much smaller but rather representative set of scenarios for practical applications.

After the scenario reduction, the problem is converted to a small-scale deterministic MILP equivalent model. Commercial solvers like CPLEX or GUROBI can be utilised to efficiently solve the model.

10.6 Simulation results

10.6.1 Test system

The proposed method is tested on the grid-tied multi-energy microgrid shown in Figure 10.3, which is structured on the IEEE 33-bus radial distribution system [2]. A winter case is tested here. Since long-distance thermal transmission will lead to high time delay and great thermal loss [3,11], in this chapter, the thermal energy is supplied locally within a geographical region (denoted as heat supply region 1–3) and no heat network is modelled. The substation voltage V_0 is 1.0 p.u. and voltage limits $\Delta V_{\mathrm{BUS}}^{\max}$ are set to be ±5% of the nominal level (12.66 kV) [2,5]. The ramp-up

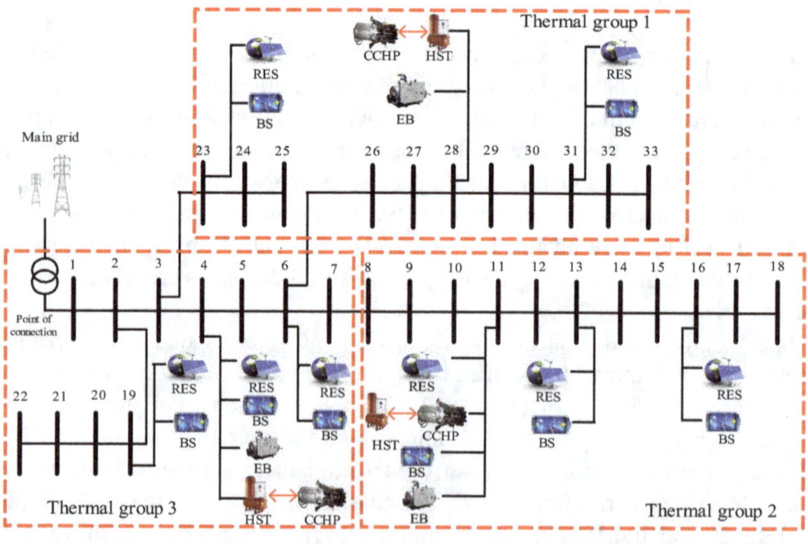

Figure 10.3 A 33-bus radial distribution system-based multi-energy microgrid

and down rates for the CCHP plants are, respectively, set as 30 and 15 kW/min [17,27], while those rates for electric boilers are 24 and 20 kW/min [17].

The location and size of all the units are given in Table 10.1.

Within each heat supply region, CCHP plants, power-to-thermal conversion units, and thermal storages are assumed to be installed at the same bus. The heat ratio of each region to the base heat capacity (3 MW) is set as 30%, 37%, and 33%. The total capacity of all distributed generators (WTs, PVs, and CCHP plants) is 6.6 MW, which is sufficient for the microgrid to support itself under islanding condition [2,5]. In addition, the capacity of the point of connection linking the microgrid and main grid is set as 1.5 times of microgrid's peak power loads [5]. Stochastic variations for RES generation, power loads, and transaction prices are set as 25%, 7.5%, and 15% deviation to their forecasted values, respectively.

Based on the average data in Tianjin, China, 2017 winter (Dec.–Feb.), Figure 10.4 shows a day-ahead prediction of 24-h output rates for RES generation and loads. The day-ahead prediction for transaction prices is shown in Figure 10.5, which is from [30]. For the heat energy supplied to the customers, the fixed pricing scheme is considered [12,17].

It should be indicated that any other real-world data can also be applied, and they won't influence the effectiveness of the proposed method. The simulations are

Table 10.1 Capacity of MT, WTs, PVs, PTCs, and TSTs [2,5,6,17]

Bus	**Rated power (kW)/rated energy (kWh)**					
	WT	**PV**	**PTC**	**MT**	**TST**	**BS**
4	200	100	600	1,000	500/1,500	250/800
6	250	150	–	–	–	250/800
11	300	100	500	1,000	600/1,800	150/480
13	200	100	–	–	–	200/640
16	350	200	–	–	–	100/320
19	300	250	–	–	–	150/480
23	350	250	–	–	–	200/640
28	–	–	400	1,000	600/1,800	–
31	350	150	–	–	–	100/320

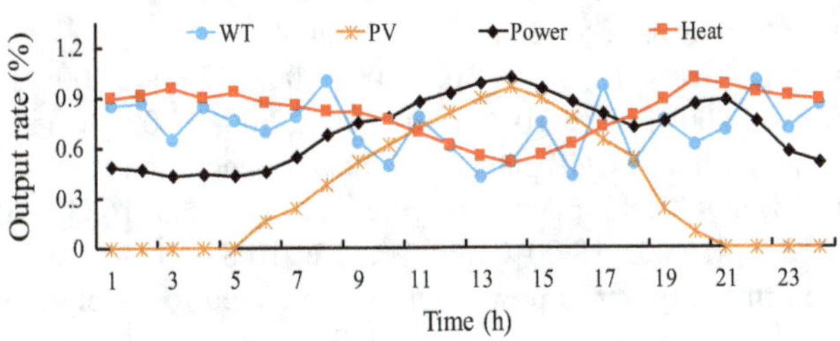

Figure 10.4 Day-ahead forecasted rates for RES outputs and loads

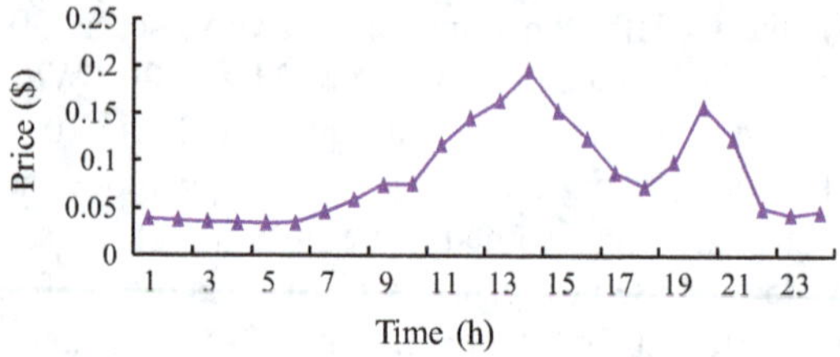

Figure 10.5 Day-ahead forecasted transaction prices

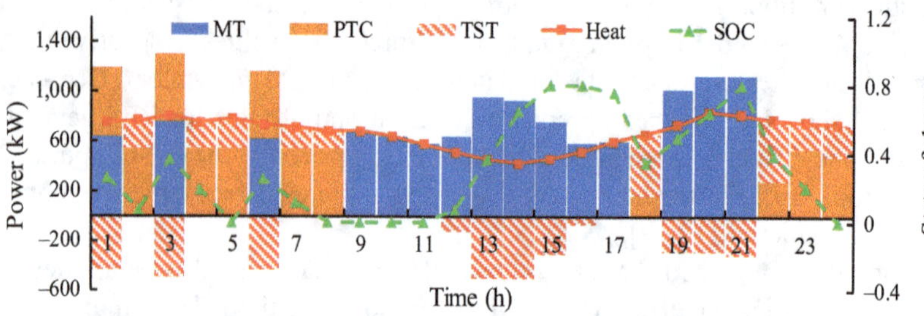

Figure 10.6 Heat dispatch of Thermal group 1

run on an Intel(R) Xeon(R) E5-1630 3.70GHz 64-bit PC of 16G RAM. Commercial MILP solver CPLEX is used to solve the mathematical model on the General Algebraic Modelling System (GAMS) platform (Version 25.1.1).

10.6.2 Day-ahead operation results and discussion

In the day-ahead operation, 2,000 random scenarios are initially generated based on day-ahead predictions. Then, they are reduced to ten representative ones with corresponding possibilities. The heat dispatch, i.e. the day-ahead operation decisions for each heat supply region, is shown in Figures 10.6–10.8, respectively. The total expected operation cost is $2,118.69.

It can be seen that through CCHP plants, the power-to-thermal conversion units, and TSTs, multiple energies are coordinated and CCHP operation is more flexible: when the transaction prices are lower than the CCHP unit power generation cost, power-to-thermal conversion units output in priority to CCHP plants to transfer power to heat (Figure 10.6, 1.00–8.00, 22.00–24.00; Figure 10.7, 1.00–6.00, 23.00–24.00; Figure 10.8, 1.00–7.00, 22.00–24.00); when the transaction prices are higher than CCHP plants' unit power generation cost, TSTs store the surplus heat from CCHP plants (Figure 10.6, 12.00–16.00, 19.00–21.00; Figure 10.7, 12.00–16.00, 18.00–21.00; Figure 10.8, 8.00–9.00, 13.00–16.00, 19.00–21.00), in this way, more power can be generated to supply power loads or sold to grid for revenue; when the prices fall in between the above two price

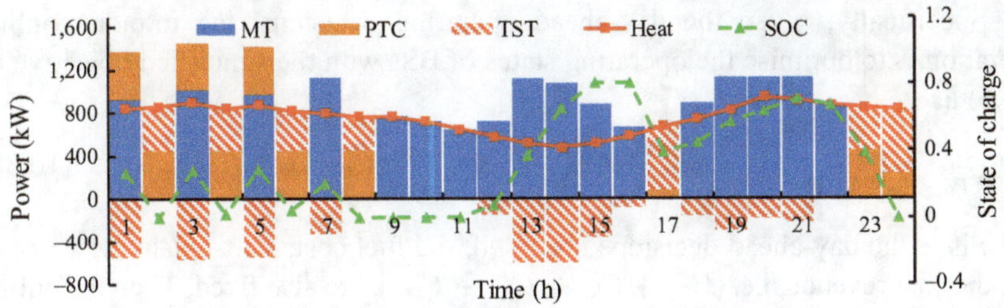

Figure 10.7 Heat dispatch of Thermal group 2

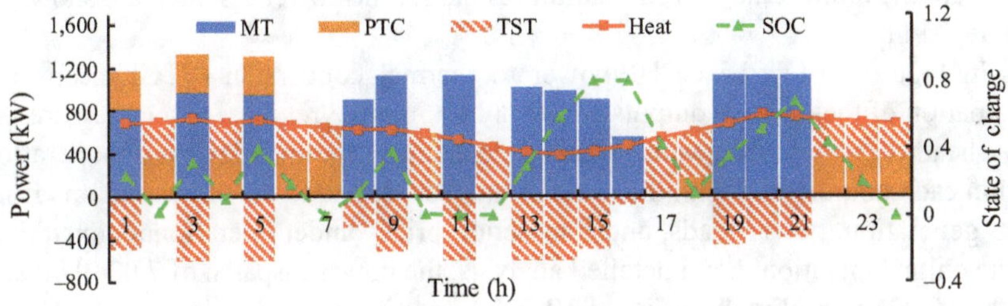

Figure 10.8 Heat dispatch of Thermal group 3

intervals, TSTs releases heat to decrease energy outputs of CCHP plants (Figure 10.6, 18.00; Figure 10.7, 17.00; Figure 10.8, 17.00, 18.00). Thus, the proposed method achieves a flexible coordination of different energies.

Besides, the heat dispatch in the three different regions are similar but not the same – this is because the heat supply regions are located at different buses – the power flow constraints can influence the heat supply via the CCHP plants and power-to-thermal conversion units. In this regard, the heat and power energy are also well coordinated to satisfy the network security constraints.

The number of the state and decision variables of the day-ahead dispatch model is 33,816 and 37,392, respectively. Using the CPLEX solver, the solution is obtained after 94 iterations, which only takes 3.1 min. Therefore, it is very computationally efficient for the practical day-ahead implementation.

10.6.3 Intraday online operation results and discussion

For the online operation stage, the RES outputs, power loads, and transaction prices are predicted in a 5-min interval in the 3-h ahead rolling horizon [8] (note that other time horizons like 6 or 12 h can also be used based on different forecasting lead times [15]). Then, the operating decisions are optimised for the whole 3-h time horizon, but only the first 5-min operating decision is practically adopted, and the remaining are discarded. The whole process is rolling every 5 min.

Specifically, given the day-ahead operating decisions, the intraday online operation is to optimise the operating states of BSs with the simplified objective of (10.6) as

$$\underset{P_{\text{BSC}},P_{\text{BSD}},P,V,Q}{\text{MIN}} \quad F_{\text{S}} = C_{\text{OM}} + C_{\text{EX}} + (C_{\text{FC}} + C_{\text{ST}} + C_{\text{SD}} - C_{\text{HR}}) \tag{10.38}$$

Since the day-ahead decisions are fixed, the fuel cost, start-up/shutdown cost, and thermal revenue, i.e. $(C_{\text{FC}} + C_{\text{ST}} + C_{\text{SD}} - C_{\text{HR}})$, are also fixed. Then the online operation is to minimise the maintenance cost of BSs and real-time electricity transaction cost, i.e. $(C_{\text{OM}} + C_{\text{EX}})$ under constraints (10.15)–(10.24).

The specified 5-min predictions of the load rates, RES generation rates, and transaction prices are given in Figures 10.9 and 10.10.

With the online data, the simulation results for the power dispatch are shown in Figure 10.11.

In Figure 10.11, the CCHP/power to thermal conversion (PTC) means the summation of their power outputs/consumption in three regions. Since they are the day-ahead decisions, their values remain constant in the intraday online operation within each hour. Along with the main grid, BSs respond to the fast variations from RES generation, power loads, and transaction prices under operational constraints in the online operation. For a detailed analysis, the power dispatch of 7.00–9.00 and 13.00–15.00 are replotted in Figure 10.12(a) and (b).

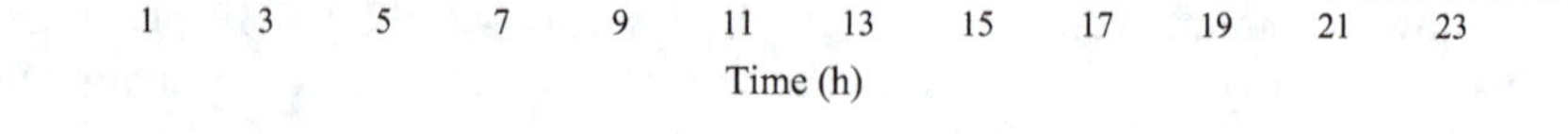

Figure 10.9 Online values of WT, PV, and power loads

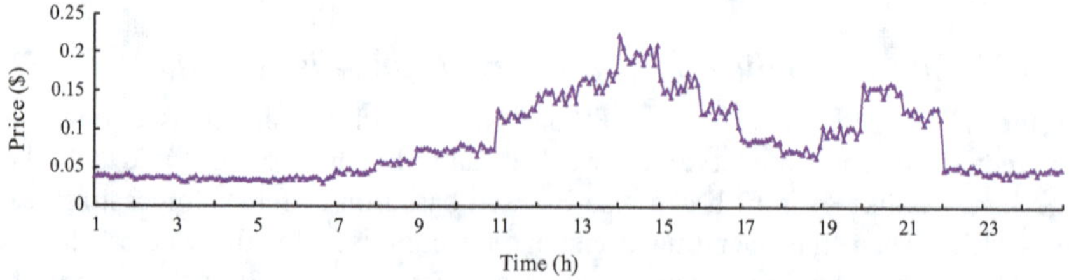

Figure 10.10 Online transaction prices

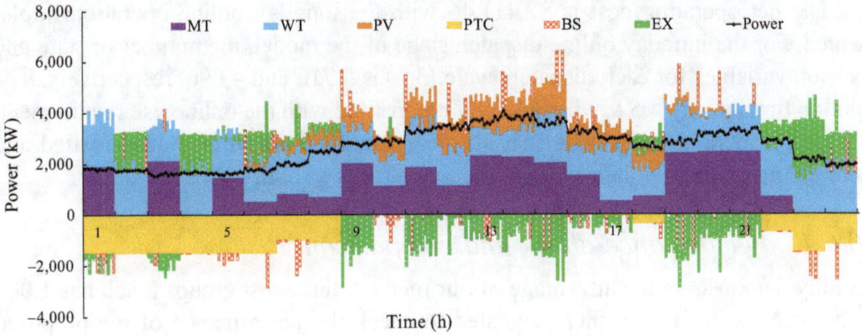

Figure 10.11 The 24-h online power dispatch

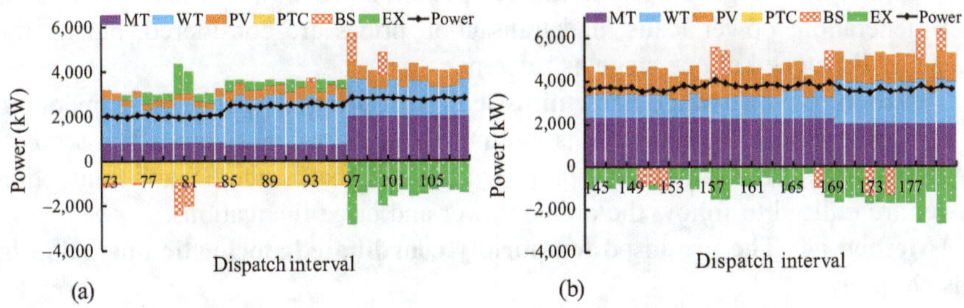

Figure 10.12 Intraday online power dispatch. (a) Dispatch interval 73–108 (7.00–9.00). (b) Dispatch interval 145–180 (13.00–15.00)

In Figure 10.12(a), it can be seen that due to the low prices at 7.00–8.00 (5-min interval 73–96), CCHP power outputs are kept at a low level, microgrid purchases power from the main grid to charge BSs, and generates heat via power-to-thermal conversion units; when the price is higher at 9.00 (5-min interval 97–108), CCHP plants generate more power for selling to the main grid for more transaction revenue. Meanwhile, the extra heat is used to charge TSTs. Besides, we can see that the energy stored in BSs at 7.00–8.00 which is of low prices is then discharged at 9.00 with a higher price, and at the same time, the microgrid purchases power at 7.00–8.00 and sell it to the main grid at 9.00. In this way, BSs cooperate with the grid to make full use of price differences to reduce the operating cost.

Similar analysis can be done for 13.00–15.00 (5-min interval 145–180) in Figure 10.12(b): since the transaction prices of those hours are high, no power is converted to heat via power-to-thermal conversion units, while the CCHP power outputs remain at a high level to satisfy the system heat and power loads (see also in Figures 10.6–10.8); microgrid sells power to the main grid all the time with high prices; BSs can still take the advantage of price difference to charge/discharge to reduce the operating cost.

The net operating cost is \$2,411.46 with the intraday online operation implemented. For the intraday online dispatch stage of the model, the number of state and decision variables for each dispatch cycle (3 h) is 4,716 and 4,896, respectively. The solution time is only 4.5 s, which is fully compatible with the online use requirement. It is also important to indicate that all the operational constraints are satisfied and there are no voltage violations and branch overflow at both operating stages.

10.6.4 Performance check and comparison

To fully demonstrate the advantage of our method, three test groups (each has 1,000 random scenarios) are further generated to check the performance of the proposed method and compared with three traditional operational methods in the literature.

Method #1: Single-stage deterministic operation [15–17]. All the controllable resources are dispatched at the same stage (day-ahead stage) and no uncertainties are considered.

Method #2: Single-stage stochastic operation [18–20]. The uncertainties for RES generation, power loads, and transaction prices are considered, but all the units are dispatched at one stage as reference.

Method #3: Two-stage deterministic optimisation. The first-stage operation is the same with Method #1; that is, no uncertainties are considered. The second-stage operation is the same as our proposed method; that is, the BS and transaction power are utilised to follow the online power and price fluctuations.

Method #4: The proposed temporally coordinated stochastic operation in this chapter

The three test groups have different levels of uncertainty as shown in Table 10.2, where the percentage value means the deviation percentage to their respective forecasted values.

These random scenarios are substituted into the optimisation results of the four methods. For the 1,000 scenarios in each group, the average operation cost and average voltage violation rates are compared in Table 10.3.

According to Table 10.3, it can be found the following:

1. In terms of the average costs, the stochastic methods are lower than the corresponding deterministic methods; that is, the cost of Method #2 is lower than Method #1, and the cost of Method #4 is lower than Method #3. This is because stochastic scenarios are addressed in their day-ahead operation, and so the operation decision are immunised to the uncertainties [31]. This justifies the effectiveness of stochastic optimisation method.

Table 10.2 Three uncertainty levels

Uncertainty	**Level 1**	**Level 2**	**Level 3**
RES generation	25%	35%	45%
Power loads	7.5%	10%	12.5%
Transaction prices	15%	20%	25%

Table 10.3 Results of different uncertainty levels

Item	Method #1	Method #2	Method #3	Method #4
Uncertainty level 1 (lower uncertainty)				
Average cost ($)	2,183.46	2,149.65	2,468.20	2,440.22
Average voltage violation (%)	30.40	16.50	0	0
Uncertainty level 2 (medium uncertainty)				
Average cost ($)	2,218.89	2,188.97	2,483.19	2,450.78
Average voltage violation (%)	74.70	49.80	0	0
Uncertainty level 3 (high uncertainty)				
Average cost ($)	2,341.64	2,282.66	2,556.04	2,508.65
Voltage violation (%)	97.20	77.90	0	0

Table 10.4 The computation parameters for different methods

Item	First stage			Second stage		
	State variables	Decision variables	Computation time (min)	State variables	Decision variables	Computation time
Method #1	5,520	8,016	1.3	-	-	-
Method #2	26,688	8,880	1.8	-	-	-
Method #3	5,520	8,016	1.3	3,528	144	1.7 s
Method #4	33,816	37,392	3.1	4,716	4,896	4.5 s

2. In terms of voltage violation, the single-stage methods have considerable violation rates while the two-stage methods have no violation at all, which justifies the effectiveness of the recourse effect of the two-stage operation method. Comparing the single-stage methods, the stochastic method has a much lower voltage violation rate since the uncertainties are considered. For the same reason, comparing the two-stage methods, while both can ensure the voltage security, the proposed method (Method #4) has the lower operating cost.
3. Moreover, when uncertainty level increases, the average costs of all the methods will increase and the voltage violation rate for single stage methods will increase accordingly. This clearly implies that uncertainty has a negative impact on the grid operating performance and should be well addressed in the two-stage coordinated stochastic optimisation framework.

The number of the state and decision variables as well as the computing time for the four methods is listed as Table 10.4.

It can be seen from Table 10.4 that the proposed method has most state and decision variables due to the coordination modelling of two operating stages and the uncertain variables. However, since the whole model is linearised into an MILP problem, its computation speed is still highly fast which is fully compatible for online use.

In summary, the comparison results verify the general superiority of the proposed method in terms of achieving reasonable operating cost with satisfying security constraints under uncertainty. Especially, with increased uncertainty, the superiority of our method becomes more manifest.

10.7 Conclusion and future work

Based on different properties of multiple energies, a temporally coordinated operation method is proposed for a grid-tied multi-energy microgrid under diverse uncertainties. Simulation results on the IEEE 33-bus distribution network illustrate that the proposed method can achieve the multi-energy coordination with higher flexibility while satisfying system security constraints. Compared with three traditional methods, the proposed method is generally superior in terms of achieving reasonable operating cost with the guarantee of system security. Besides, with the increasing uncertainty level, the superiority of our method becomes more prominent. For the real application, the proposed method can be embedded in the microgrid energy management system to provide reference power output set points to distributed generators, energy storage, etc., to minimise the total operating cost.

In this study, since the thermal energy is locally supplied within a small area, heat and gas network is not considered, and the future work can consider large-area multi-energy supply with the thermal and gas network integrated. Besides, although this chapter does not meet any numerical solution difficulties, in the future, more efficient and effective algorithms to enhance the convergence and feasibility can be used. Moreover, the proposed method can be integrated with RT-OPF methods for real-time optimising the multi-energy microgrid operation.

References

[1] P. Mancarella, 'MES (multi-energy systems): an overview of concepts and evaluation models', *Energy*, vol. 65, pp. 1–17, 2014.

[2] C. Zhang, Y. Xu, Z.Y. Dong, and K.P. Wong, 'Robust coordination of distributed generation and price-based demand response in microgrids', *IEEE Trans. Smart Grid*, vol. 9, no. 5, pp. 4236–4247, 2018.

[3] W. Gu, J. Wang, S. Lu, Z. Luo, and C. Wu, 'Optimal operation for integrated energy system considering thermal inertia of district heating network and buildings', *Appl. Energy*, vol. 199, no. 1, pp. 234–246, 2017.

[4] M. Alipour, B. Mohammadi-Ivatloo, and K. Zare, 'Stochastic scheduling of renewable and CHP-based microgrids', *IEEE Trans. Ind. Informat.*, vol. 11, no. 5, pp. 1049–1058, 2015.

[5] C. Zhang, Y. Xu, Z. Li, and Z. Y. Dong, 'Robustly coordinated operation of a multi-energy microgrid with flexible electric and thermal loads', *IEEE Trans. Smart Grid*, vol. 10, no. 3, pp. 2765–2775, 2019.

[6] Y. Xu, Z.Y. Dong, R. Zhang, and D.J. Hill, 'Multi-timescale coordinated voltage/var control of high renewable-penetrated distribution systems', *IEEE Trans. Power Syst.*, vol. 32, no. 6, pp. 4398–4408, 2017.

[7] A.D. Smith and P.J. Mago, 'Effects of load-following operational methods on combined heat and power system efficiency', *Appl. Energy*, vol. 115, no. 15, pp. 337–351, 2014.

[8] J. Wang, J. Sui, and H. Jin, 'An improved operation strategy of combined cooling heating and power system following electrical load', *Energy*, vol. 85, no. 1, pp. 654–666, 2015.

[9] L. Li, H. Mu, N. Li, and M. Li, 'Analysis of the integrated performance and redundant energy of CCHP systems under different operation strategies', *Energy Build.*, vol. 99, no. 15, pp. 231–242, 2015.

[10] B. Yan, P.B. Luh, G. Warner, and P. Zhang, 'Operation and design optimization of microgrids with renewables', *IEEE Trans. Autom. Sci. Eng.*, vol. 14, no. 2, pp. 573–585, 2017.

[11] L. Ma, N. Liu, J. Zhang, W. Tushar, and C. Yuen, 'Energy management for joint operation of CHP and PV prosumers inside a grid-connected microgrid: a game theoretic approach', *IEEE Trans. Ind. Informat.*, vol. 12, no. 5, pp. 1930–1942, 2016.

[12] Z.M. Li and Y. Xu, 'Dynamic dispatch of grid-connected multi-energy microgrids considering opportunity profit', *IEEE PES General Meeting/ Conference & Exposition*, 2017, IEEE, 2017.

[13] M.H. Moradi, M. Eskandari, and S.M. Hosseinian, 'Operational strategy optimization in an optimal sized smart microgrid', *IEEE Trans. Smart Grid*, vol. 6, no. 3, pp. 1087–1095, 2015.

[14] A.K. Basu, A. Bhattacharya, S. Chowdhury, and S.P. Chowdhury, 'Planned scheduling for economic power sharing in a CHP-based micro-grid', *IEEE Trans. Power Syst.*, vol. 27, no. 1, pp. 30–38, 2012.

[15] C.Y. Zheng, J.Y. Wu, X.Q. Zhai, and R.Z. Wang, 'A novel thermal storage strategy for CCHP system based on energy demands and state of storage tank', *Int. J. Elec. Power Energ. Syst.*, vol. 85, no. 1, pp. 117–129, 2017.

[16] H. Hajabdollahi, A. Ganjehkaviri, and M.N.M. Jaafar, 'Assessment of new operational strategy in optimization of CCHP plant for different climates using evolutionary algorithms', *Appl. Therm. Eng.*, vol. 75, no. 22, pp. 468–480, 2015.

[17] Z. Li and Y. Xu, 'Optimal coordinated energy dispatch of a multi-energy microgrid in grid-connected and islanded modes', *Appl. Energy*, vol. 210, no. 15, pp. 974–986, 2018.

[18] X. Guan, Z. Xu, and Q.S. Jia, 'Energy-efficient buildings facilitated by microgrid', *IEEE Trans. Smart Grid*, vol. 1, no. 3, pp. 243–252, 2010.

[19] L. Wang, Q. Li, M. Sun, and G. Wang, 'Robust optimisation scheduling of CCHP systems with multi-energy based on minimax regret criterion', *IET Gener. Transm. Dis.*, vol. 10, no. 9, pp. 2194–2201, 2016.

[20] L. Wang, Q. Li, R. Ding, M. Sun, and G. Wang, 'Integrated scheduling of energy supply and demand in microgrids under uncertainty: a robust multi-objective optimization approach', *Energy*, vol. 130, no. 1, pp. 1–14, 2017.

[21] Z. Bao, Q. Zhou, Z. Yang, Q. Yang, L. Xu, and T. Wu, 'A multi time-scale and multi energy-type coordinated microgrid scheduling solution – Part I: Model and methodology', *IEEE Trans. Power Syst.*, vol. 30, no. 5, pp. 2257–2266, 2015.

[22] E. Mohagheghi, M. Alramlawi, A. Gabash, and P. Li, 'A survey of real-time optimal power flow', *Energies*, vol. 11, no. 11, pp. 3142–3142, 2018.

[23] E. Mohagheghi, A. Gabash, M. Alramlawi, and P. Li, 'Real-time optimal power flow with reactive power dispatch of wind stations using a reconciliation algorithm', *Renew. Energy*, vol. 126, pp. 509–523, 2018.

[24] E. Mohagheghi, A. Gabash, and P. Li, 'A framework for real-time optimal power flow under wind energy penetration', *Energies*, vol. 10, no. 4, pp. 1–28, 2017.

[25] R.A. Jabr, I. Dzafic, and S. Karaki, 'Tracking transformer tap position in real-time distribution network power flow applications', *IEEE Trans. Smart Grid*, vol. 9, no. 4, pp. 2442–2452, 2016.

[26] W. Gu, S. Lu, Z. Wu, *et al.*, 'Residential CCHP microgrid with load aggregator: operation mode, pricing strategy, and optimal dispatch', *Appl. Energy*, vol. 205, pp. 173–186, 2017.

[27] Capstone Turbine Corporation, 'Capstone C200 microturbine technical reference', Chatsworth, CA, USA, *Tech. Rep. 410066, Rev C*, Jun. 2009.

[28] B. Zhang, Q. Li, L. Wang, and W. Feng, 'Robust optimization for energy transactions in multi-microgrids under uncertainty', *Appl. Energy*, vol. 217, pp. 346–360, 2018.

[29] C. Wan, Z. Xu, P. Pinson, Z.Y. Dong, and K.P. Wong, 'Probabilistic forecasting of wind power generation using extreme learning machine', *IEEE Trans. Power Syst.*, vol. 29, no. 3, pp. 1033–1044, 2014.

[30] 'Power prices for the Tianjin Grid', 2019, http://www.tjdpc.gov.cn/gzcx/syjgcx/gd/201307/t20130705_30044.shtml.

[31] A. Shapiro and A.P. Ruszczynski, *Stochastic Programming*, Amsterdam, The Netherlands: Elsevier, 2003.

Chapter 11

Robustly optimal dispatch of MEMGs with flexible loads

A multi-energy microgrid (MEMG) can simultaneously supply electric and thermal energy to customers to improve overall energy utilisation efficiency. However, intermittency and uncertainty from renewable power generation such as wind turbines (WTs) and solar photovoltaics (PVs), as well as electric and temperature-dependent thermal loads, can significantly challenge and complicate the operation of an MEMG. To conquer the challenges, this chapter introduces the utilisation of price-based demand response (PBDR) and indoor temperature control to flexibilise the electric and thermal loads, respectively. Then, a two-stage coordinated operation method is proposed to optimally coordinate the combined cooling, heat and power (CCHP) plants, flexible electric and thermal loads, and thermal storage under multiple uncertainties. The mathematical problem is modelled as a two-stage robust optimisation (TSRO) model and solved by column-and-constraint generation algorithm. Simulation results verify high energy utilisation efficiency and operating robustness of the proposed method.

Nomenclature

A Acronyms

CCHP	Combined cooling, heat and power
C&CG	Column and constraint generation
DG	Distributed generation
MEMG	Multi-energy microgrid
PBDR	Price-based demand response
PGU	Power generation unit
PV	Photovoltaic
RO	Robust optimisation
TES	Thermal energy storage
TSRO	Two-stage robust optimisation
WT	Wind turbine

B Sets and indices

I, i	Set/index of buses in an MEMG
J, j	Set/index of PBDR price levels
K, k	Set/index of thermal zones
N, n	Set/index of buses with renewable generators
M, m	Set/index of buses with CCHP plants and thermal loads
T, t	Set/index of operation periods
$U_{(\cdot)}$	Uncertainty set

C Parameters

A	Constant used in the demand response equation
$C^{(\cdot)}$	Operation and maintenance cost (\$/kWh)
$C^{\text{buy}}, C^{\text{sell}}$	Price for selling/buying electricity to/from the main grid (\$/kWh)
C^{fuel}	Fuel cost for CCHP plants (\$/m^3)
$c^{\text{air}}_{m,k}$	Air specific heat capacity (kWh/°C)
$L_{j,t}$	Electric load demand response rate (%)
P^{cap}_i	Capacity of the branch from bus i (kW)
$Pr^{\text{CE}}, Pr^{\text{HE}}$	Cooling/heat energy price for customers (\$/kWh)
$Pr^{\text{EL}}_{0,t}$	Original electricity price for customers without PBDR (\$/kWh)
$Pr^{\text{EL}}_{j,t}$	Electricity price for customers with PBDR (\$/kWh)
$q^{\text{TS}}_{m,0}$	Initial energy stored in thermal energy storage units (kWh)
R_i, X_i	Resistance/reactance of branch from bus i
$R^{\text{T}}_{m,k}$	Thermal resistance of building envelope (°C/kW)
r^{CCHP}_m	Ramping rate of CCHP electric power output
V_0	Voltage at substation, normally 1 p.u.
ε	Price elasticity of load
η	Power transfer efficiency
$\underline{\mu}, \bar{\mu}$	Lower/upper budget of uncertainty set
σ	Energy converting factor of 1 m^3 fuel to kWh
τ	Time length of one operation period (h)
$(\cdot)^{min/max}$	Minimal/maximal limits
$\hat{(\cdot)}$	Predicted value
$\underline{(\cdot)}, \overline{(\cdot)}$	Lower and upper bounds of interval predictions

D Variables

$E^{\text{PGU}}_{m,t}$	Electric energy output of PGU (kWh)
$F^{\text{PGU}}_{m,t}$	Fuel energy for PGU (kWh)
$P^{(\cdot)}_{m/n,t}$	Electric power output of DG (kW)

$P_{i,t}, Q_{i,t}$	Active/reactive power through the main branch from bus i (kW/kVar). $P_{1,t}$ means active power through the branch from substation
$P^{\text{EL}}_{0,i,t}, Q^{\text{EL}}_{0,i}$	Uncertain active and reactive electric loads (kW)
$P^{\text{EL}}_{i,t}, Q^{\text{EL}}_{i,t}$	Active/reactive electric loads with PBDR (kW/kVar)
$P^{\text{def}}_{t}, P^{\text{sur}}_{t}$	Power deficiency/surplus of MEMG (kW)
$P^{\text{lat}}_{i,t}, Q^{\text{lat}}_{i,t}$	Active/reactive power through the lateral branch from bus i (kW/kVar)
$q^{(\cdot)}_{m,t}$	Thermal energy generation/consumption at bus m (kWh)
$q^{\text{CE}}_{m,k,t}, q^{\text{HE}}_{m,k,t}$	Cooling/heat energy supported to a thermal zone (kWh)
$q^{\text{CE,bill}}_{m,k,t}, q^{\text{HE,bill}}_{m,k,t}$	Cooling/heat energy of customer thermal usage bill (kWh)
$V_{i,t}$	Voltage at bus i (p.u.)
$V^{\text{fuel}}_{m,t}$	Fuel consumption of CCHP plants (m^3)
$\alpha_{j,t}$	Binary on/off decision of PBDR level j
$\beta^{\text{TSC/D}}_{m,t}$	Binary decision of thermal energy storage charging/discharging state
$\theta^{\text{in}}_{m,k,t}$	Controlled indoor temperature (°C)
$\theta^{\text{am}}_{m,t}$	Uncertain ambient temperature (°C)

11.1 Introduction

Multiple energy carriers such as electric and thermal energy can be integrated together to enhance overall energy utilisation efficiency [1]. In practice, an MEMG consists of renewable distributed generation (DG) units such as WTs and PVs as well as multi-energy generation units like CCHP plants, to simultaneously supply electricity and thermal energy. In practice, two typical operation modes of CCHP plants are used, i.e. following-electric-load and following-thermal-load modes [2]. In the former mode, electric output of CCHP plants is required to match local electric loads as a priority, while in the later mode, thermal energy generated by CCHP plants is required to match local thermal loads as a priority. However, due to coupling nature of electric and thermal outputs of CCHP plants, local electric and thermal loads may not be matched simultaneously by the CCHP plants. Imbalance between energy generation and consumption can lead to waste of energy, which impairs overall energy utilisation efficiency.

In order to decouple electric and thermal generation of CCHP plants, Refs. [3] and [4] utilise thermal energy storage (TES) units to store excessive thermal energy and then release it when thermal loads are high. As a result, operating profits, i.e. revenues minus operating costs, are improved significantly. On the other hand, from electric side, electric load demand response can be used to allow electric loads to match electric power generation. In [5], efficiency of different types of demand response is studied and multi-energy systems are operated through model predictive control. In terms of thermal loads, temperature-dependent thermal loads are modelled and controlled in [6] and [7]. These two works coordinate electric load shifting/curtailment, thermal load control, and TES to improve the CCHP plant generation efficiency and operating profits. However, the electric load shifting/curtailment may hamper customer benefits. Besides, these methods are limited to shiftable electric loads which may not be ample in practice. Moreover, Refs. [5–7] only consider small systems like residential energy hubs to realise the electric load shifting, but distribution network operating constraints are not considered.

In the meantime, uncertainties, such as renewable power generation as well as electric and thermal loads, add a significantly challenging dimension to operation of MEMGs. In the above literature, Refs. [3–7] all assume accurate forecast of renewable power generation and loads without uncertain variations. However, these uncertainties have non-negligible impacts on system operating robustness. Reference [8] reports that the risk level of operation is increased with forecast errors of uncertain loads. In order to handle uncertainties, Ref. [9] applies prediction errors in optimising reserve capacity of an MEMG. The prediction errors are generated randomly to model uncertainty variables. A stochastic game approach which randomly generates uncertainty samples is proposed in [10] to optimise MEMG operation. However, if the generated prediction errors and samples differ from actual values heavily, performance of these methods would be deteriorated. In [11], a regret-based robust optimisation (RO) method is proposed for single-stage operation of CCHP systems. The RO method can guarantee robustness of operation decisions. However, such a single-stage operation framework does not consider different response speeds of different energy supplies and loads, and it tends to be considerably conservative if the worst case given by the RO method does not occur. Reference [12] proposes a two-stage operation method, where CCHP plants, batteries, and air conditioners are dispatched day-ahead and real-time dispatch is used to compensate prediction errors of uncertain wind and solar power. Reference [13] applies rolling optimisation for dispatch of CCHP plants, gas boilers, batteries, and TES tanks in the first stage. Then it applies a feedback correction in 15-min time intervals of these components in the second stage to compensate prediction errors of uncertain renewable energy and electric loads. References [12] and [13] can improve operation solution robustness of MEMGs effectively. However, these works do not consider flexibility of electric and thermal demand response, such that the overall energy utilisation efficiency of the MEMGs may be restricted. Besides, due to large thermal inertia which slows down the thermal device response speed, efficiency of frequent dispatch of the CCHP plants and the TES units can be restricted. Importantly, [9–13] all neglect thermal load uncertainty so that the operation decisions may still not be robust. Moreover, all the works [9–13] do not consider the distribution networks and associated operating constraints.

This chapter presents a new coordinated operation method for an MEMG, which aims to optimally and robustly coordinate supply side (CCHP plants and renewable power generators), load side (flexible electric and thermal loads), and TES in the presence of multiple uncertainties (renewable power generation, electric loads, ambient temperature). This coordinated operation method has two operation stages, i.e. day-ahead and intraday. In the day-ahead scheduling stage, the PBDR levels and the TES units are scheduled to shift next-day electric loads and thermal energy. Then in the intraday dispatch stage, to counteract multiple uncertainties, the CCHP plants are re-dispatched and the thermal loads are controlled through indoor temperature adjustment on an hourly basis, which acts as a recourse to complement the day-ahead scheduling decisions. The uncertainties are modelled in uncertainty sets based on interval forecasting. The MEMG operation is optimised by a TSRO method.

Compared with the existing works, this chapter presents the following new technical contents.

Multiple (electric and thermal) energy supplies and loads are coordinated in multiple operation stages with full consideration of their different response characteristics.

Multi-energy load side is flexibilised through indoor temperature control and a PBDR method.

Multiple uncertainties (renewable power generation, electric loads, and ambient temperature) are robustly addressed, such that network operating constraints and customer comfort can be fully satisfied.

Distribution networks and associated operating constraints (including a power flow model and bus voltage constraints) are fully modelled and applied in optimisation of MEMG operation.

11.2 Two-stage coordinated operation of multi-energy microgrid

11.2.1 Multi-energy microgrid with flexible thermal and electric loads

An MEMG consists of electric and thermal loads, renewable power generators, and CCHP plants with TES units, and it can exchange electric energy with the main grid. The MEMG operator sells electric and thermal energy to customers, aiming to maximise the overall operating profit, i.e. revenues of selling energy minus operating costs, while fulfilling customer benefits and operating constraints. The electric and thermal energy balances in the MEMG can be modelled as

$$E^{\mathrm{MG}} + E^{\mathrm{RG}} + E^{\mathrm{PGU}} = E^{\mathrm{EL}} \tag{11.1}$$

$$q^{\mathrm{HR}} + q^{\mathrm{TSD}} = q^{\mathrm{TSC}} + q^{\mathrm{AC}} + q^{\mathrm{HC}} \tag{11.2}$$

where $E^{(\cdot)}$ and $q^{(\cdot)}$ express electric and thermal energy, respectively. The superscripts MG, RG, PGU, EL, HR, TSD, TSC, AC, and HC represent main grid, renewable generation, power generation unit (PGU), electric load, heat recovery, thermal storage discharging, thermal storage charging, absorption chiller, and heating coil, respectively.

However, due to the coupling nature of the CCHP electric and thermal outputs and the limited capacity of the TES units, the electric and thermal energy flows may complicate and impact each other. To solve this problem, this chapter uses a

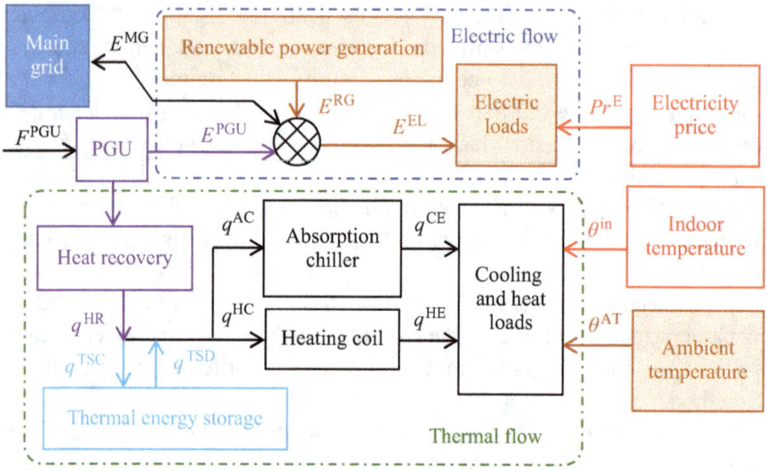

Figure 11.1 Multiple energy flows in a multi-energy microgrid

PBDR method and indoor temperature control to flexibilise the electric and thermal loads, respectively. The electric and thermal energy flows in a grid-connected MEMG with the flexible loads are illustrated in Figure 11.1, where the shadowed parts are uncertainties including renewable power generation, electric loads, and ambient temperature that affects the thermal loads.

At the generation side, the CCHP plants generate electric energy by PGUs and thermal energy by heat recovery systems. Meanwhile, the TES units charge/discharge thermal energy to shift thermal energy generation from low-load periods to high-load periods. Thus, the thermal energy supply to the customers can be decoupled from the CCHP plant electric energy generation.

At the load side, based on the PBDR method, by setting high and low electricity prices, the electric loads can be reduced in peak periods and increased in off-peak periods, respectively. An advantage of the PBDR method is that it can improve usage rates of low-cost energy from the main grid and clean energy from the local renewable generators. Meanwhile, the thermal loads (cooling and heat loads) can be controlled through indoor temperature adjustment of air conditioning systems. Within a thermal comfort temperature range, the energy required by the customers, i.e. the thermal loads, can vary in a corresponding range. Therefore, the thermal outputs of the CCHP plants can be dispatched in a flexible range.

11.2.2 First stage – day-ahead PBDR and thermal energy storage scheduling

This chapter aims to maximise the overall operating profit of the MEMG by optimising the multiple energy flows among the generation, storage, and load sides, while satisfying customer benefits and comfort as well as network operating constraints. A two-stage coordinated operation method is proposed to coordinate

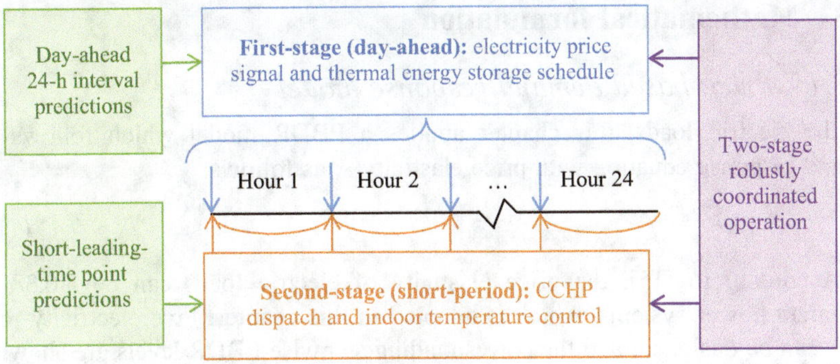

Figure 11.2 Two-stage coordinated operation framework

different controllable resources considering their different response characteristics at different operation stages. The proposed coordinated operation framework is given in Figure 11.2.

In the day-ahead scheduling time window, i.e. the first stage, 24-h PBDR and TES charging/discharging decisions are optimised based on day-ahead interval predictions of renewable power generation, electric loads, and ambient temperature, which affects thermal loads. This aims to shift electric loads and thermal energy generation in a large time horizon to utilise more low-cost and renewable energy. Besides, the PBDR decisions must be delivered to customers one day ahead to allow sufficient time for the electric loads to respond. The optimisation decisions are to be implemented in each hour of the following day. Unlike conventional point predictions, the interval predictions provide lower and upper bounds of uncertainties and they are applied to more efficiently reflect the uncertainties [14].

11.2.3 Second stage – hourly CCHP dispatch

During the intraday dispatch time window, i.e., the second stage, CCHP electric and thermal outputs and indoor temperature control decisions are dispatched hourly based on hour-ahead predictions. The second-stage dispatch aims to compensate the first-stage scheduling when the uncertainties realise. In the second stage, thermal loads are controlled through indoor temperature adjustment. Thus, the CCHP plants can re-optimise the electric flow by re-dispatching their electric power generation, while the thermal energy generation matches the controlled thermal loads. In other words, the second-stage dispatch is a short-period recourse to deal with prediction errors from the first stage and track the realisation of the uncertainties. Besides, it can maintain bus voltages in an allowed range and the indoor temperature in a comfortable range, e.g. ±2°C from preferred temperature. Note that the time length of this operation stage cannot be too short since the thermal dispatching with strong thermal inertia needs relatively long time to respond. In this chapter, it is set as 1 h.

11.3 Mathematical formulation

11.3.1 Price-based demand response model

For the electric loads, this chapter applies a PBDR model which follows the demand response equation with price elasticity ε, as follows:

$$P_t^{\mathrm{EL}} = \mathrm{A}Pr_t^{\,\varepsilon} \tag{11.3}$$

According to [15], the price elasticity of electric loads can be −0.38 for Australian power systems, which is adopted in this chapter. Five electricity price levels can be designed, and the corresponding stepwise PBDR levels are shown in Table 11.1.

With these PBDR levels ($L_{j,t}$), electric loads can be expected to respond with PBDR level selection decisions ($\alpha_{j,t}$) as follows (assuming power factors are fixed). With uncertain electric loads ($P_{0,i,t}^{\mathrm{EL}}, Q_{0,i,t}^{\mathrm{EL}}$), responsive electric loads with PBDR for each bus can be calculated as

$$P_{i,t}^{\mathrm{EL}} = \sum_{j\in J} \alpha_{j,t} L_{j,t} P_{0,i,t}^{\mathrm{EL}},\ Q_{i,t}^{\mathrm{EL}} = \sum_{j\in J} \alpha_{j,t} L_{j,t} Q_{0,i,t}^{\mathrm{EL}} \tag{11.4}$$

The detailed information of the PBDR model can be found in Section 8.3.1.

11.3.2 Indoor temperature control model

In this chapter, temperature-dependent thermal loads are considered. The thermal loads are dependent on the required indoor temperature set point and the outdoor ambient temperature. The ambient temperature impacts the indoor temperature by thermal conduction through building envelope. The thermal transfer rate p^T is calculated based on Fourier's law as follows:

$$p^{\mathrm{T}} = \left(\theta^{\mathrm{am}} - \theta^{\mathrm{in}}\right)/R^{\mathrm{T}} \tag{11.5}$$

In (11.5), the thermal power transferred from ambient environment to indoor space is calculated by the temperature difference over the building envelope divided by the thermal resistance of the building envelope. Considering thermal power supplied from air conditioning systems, thermal loads (p^{HE} expresses heat power and p^{CE} expresses cooling power) are modelled through a thermodynamic equation as

Table 11.1 Price-based demand response level

Level	Electricity price ($/kWh)	Expected electric load demand response rate (%)
1	0.1331	121.42
2	0.1775	108.85
3	0.2219	100
4	0.2774	91.87
5	0.3550	83.64

$$p^{\mathrm{HE}} - p^{\mathrm{CE}} = c^{\mathrm{air}}\left(\mathrm{d}\theta^{\mathrm{in}}/\mathrm{d}t\right) - p^{\mathrm{T}} \tag{11.6}$$

Equation (11.6) means that the temperature-dependent thermal loads are equal to the indoor thermal power variation minus the thermal conduction from the ambient environment. Thus, the thermal loads depend on the indoor temperature set point of the air conditioning systems and the thermal conduction from the ambient environment. Due to the large inertia of the thermal energy, the indoor temperature can only vary slowly, so the variation in each short-term period can be neglected. Thus, the indoor temperature can be assumed as constant within each intraday dispatch interval. Then, (11.6) can be converted into a discrete state model, where the temperature and thermal loads are modelled in hourly states, as

$$\left(q_t^{\mathrm{HE}} - q_t^{\mathrm{CE}}\right)/\tau = c^{\mathrm{air}}\left(\theta_t^{\mathrm{in}} - \theta_{t-1}^{\mathrm{in}}\right)/\tau - \left(\theta_t^{\mathrm{am}} - \theta_t^{\mathrm{in}}\right)/R^{\mathrm{T}} \tag{11.7}$$

Based on (11.7), when the indoor temperature is set to a different set point, the thermal energy required by the air conditioning systems, i.e. the heat load q_t^{HE} or cooling load q_t^{CE}, is adjusted accordingly. Since the thermal comfort temperature usually has a certain range, the thermal loads are flexible in a corresponding range. Therefore, the thermal loads can be adjusted through the indoor temperature control.

Without impairing the customer benefits, the customer thermal load bill payment is based on minimal cooling/heat energy required to keep the comfortable indoor temperature. The limits of the comfortable indoor temperature are applied to calculate this thermal energy as

$$q^{\mathrm{bill}} = \begin{cases} \left(\theta_t^{\mathrm{in,min}} - \theta_t^{\mathrm{am}}\right)\tau/R^{\mathrm{T}}, \mathit{if}\,\theta_t^{\mathrm{am}} < \theta_t^{\mathrm{in,min}} \\ 0, \mathit{if}\,\theta_t^{\mathrm{in,min}} \le \theta_t^{\mathrm{am}} \le \theta_t^{\mathrm{in,max}} \\ \left(\theta_t^{\mathrm{am}} - \theta_t^{\mathrm{in,max}}\right)\tau/R^{\mathrm{T}}, \mathit{if}\,\theta_t^{\mathrm{am}} > \theta_t^{\mathrm{in,max}} \end{cases} \tag{11.8}$$

This minimal thermal energy usage depends on only the varying ambient temperature rather than the controlled indoor temperature. As a result, the thermal usage bills are not affected by the indoor temperature control and the customer benefits are protected.

Note that human activities, electric/gas appliances, and solar radiation also impact the indoor temperature, but these concerns are usually minor and can be ignored. A full thermal load model can be referred in [7] and considered if necessary.

11.3.3 Coordination optimisation model

The two-stage coordination optimisation model is formulated as follows:

$$\min C_{\mathrm{CCHP}} + C_{\mathrm{OM}} + C_{\mathrm{grid}} - C_{\mathrm{rev}}^{\mathrm{elec}} - C_{\mathrm{rev}}^{\mathrm{thm}} \tag{11.9}$$

$$\text{s.t. } C_{\mathrm{CCHP}} = \sum_{t \in T} \sum_{m \in M} C^{\mathrm{fuel}} V_{m,t}^{\mathrm{fuel}} \tag{11.10}$$

$$C_{\mathrm{OM}} = \sum_{t \in T} \left(\sum_{m \in M} C^{\mathrm{CCHP}} P_{m,t}^{\mathrm{CCHP}} \tau + \sum_{n \in N_{\mathrm{WT}}} C^{\mathrm{WT}} P_{n,t}^{\mathrm{WT}} \tau + \sum_{n \in N_{\mathrm{PV}}} C^{\mathrm{PV}} P_{n,t}^{\mathrm{PV}} \tau \right) \tag{11.11}$$

$$C_{\mathrm{grid}} = \sum_{t \in T} C^{\mathrm{buy}} P_t^{\mathrm{def}} \tau - C^{\mathrm{sell}} P_t^{\mathrm{sur}} \tau \tag{11.12}$$

$$C_{\mathrm{rev}}^{\mathrm{elec}} = \sum_{t \in T} \sum_{i \in I} \sum_{j \in J} \alpha_{j,t} L_{j,t} Pr_{j,t}^{\mathrm{EL}} P_{0,i,t}^{\mathrm{EL}} \tau \tag{11.13}$$

$$C_{\mathrm{rev}}^{\mathrm{thm}} = \sum_{t \in T} \sum_{m \in M} \left(Pr^{\mathrm{CE}} \sum_{k \in K} q_{m,k,t}^{\mathrm{CE,bill}} + Pr^{\mathrm{HE}} \sum_{k \in K} q_{m,k,t}^{\mathrm{HE,bill}} \right) \tag{11.14}$$

$$\sum_{j \in J} \alpha_{j,t} = 1, \forall t \tag{11.15}$$

$$P_{i,t}^{\mathrm{EL}} = \sum_{j \in J} \alpha_{j,t} L_j P_{0,i,t}^{\mathrm{EL}}, \forall i, t \tag{11.16}$$

$$Q_{i,t}^{\mathrm{EL}} = \sum_{j \in J} \alpha_{j,t} L_j \widehat{Q}_{i,t}^{\mathrm{EL}} P_{0,i,t}^{\mathrm{EL}} / \widehat{P}_{i,t}^{\mathrm{EL}}, \forall i, t \tag{11.17}$$

$$\sum_{t \in T} \sum_{i \in I} \sum_{j \in J} \alpha_{j,t} L_{j,t} Pr_{j,t}^{\mathrm{EL}} \widehat{P}_{i,t}^{\mathrm{EL}} \tau \leq \sum_{t \in T} \sum_{i \in I} Pr_{0,t}^{\mathrm{EL}} \widehat{P}_{i,t}^{\mathrm{EL}} \tau \tag{11.18}$$

$$\sum_{t \in T} \sum_{i \in I} \sum_{j \in J} \alpha_{j,t} L_{j,t} \widehat{P}_{i,t}^{\mathrm{EL}} \tau \geq \sum_{t \in T} \sum_{i \in I} \widehat{P}_{i,t}^{\mathrm{EL}} \tau \tag{11.19}$$

$$\beta_{m,t}^{\mathrm{TSC}} + \beta_{m,t}^{\mathrm{TSD}} \leq 1, \forall m, t \tag{11.20}$$

$$0 \leq q_{m,t}^{\mathrm{TSC/D}} \leq \beta_{m,t}^{\mathrm{TSC/D}} q_m^{\mathrm{TSC/D,max}}, \forall m, t \tag{11.21}$$

$$q_{m,t}^{\mathrm{TS}} = \eta^{\mathrm{TS}} q_{m,t-1}^{\mathrm{TS}} + \eta^{\mathrm{TSC}} q_{m,t}^{\mathrm{TSC}} - q_{m,t}^{\mathrm{TSD}} / \eta^{\mathrm{TSD}}, \forall m, t \tag{11.22}$$

$$q_{m,t}^{\mathrm{TS,min}} \leq q_{m,t}^{\mathrm{TS}} \leq q_{m,t}^{\mathrm{TS,max}}, \forall m, t \tag{11.23}$$

$$q_{m,24}^{\mathrm{TS}} = q_{m,0}^{\mathrm{TS}}, \forall m \tag{11.24}$$

$$E_{m,t}^{\mathrm{PGU}} = P_{m,t}^{\mathrm{CCHP}} \tau, \forall m, t \tag{11.25}$$

$$E_{m,t}^{\mathrm{PGU}} = \eta^{\mathrm{PGU}} F_{m,t}^{\mathrm{PGU}}, \forall m, t \tag{11.26}$$

$$q_{m,t}^{\mathrm{HR}} = \eta^{\mathrm{HR}} \left(F_{m,t}^{\mathrm{PGU}} - E_{m,t}^{\mathrm{PGU}} \right), \forall m, t \tag{11.27}$$

$$q_{m,t}^{\mathrm{HR}} + q_{m,t}^{\mathrm{TSD}} = q_{m,t}^{\mathrm{TSC}} + q_{m,t}^{\mathrm{AC}} + q_{m,t}^{\mathrm{HC}}, \forall m, t \tag{11.28}$$

$$q_{m,t}^{\mathrm{CE/HE}} = \eta^{\mathrm{AC/HC}} q_{m,t}^{\mathrm{AC/HC}}, \forall m, t \tag{11.29}$$

$$F_{m,t}^{\mathrm{PGU}} = \sigma V_{m,t}^{\mathrm{fuel}}, \forall m, t \tag{11.30}$$

$$P_m^{\mathrm{CCHP,min}} \le P_{m,t}^{\mathrm{CCHP}} \le P_m^{\mathrm{CCHP,max}}, \forall m, t \tag{11.31}$$

$$-r_m^{\mathrm{CCHP}} P_m^{\mathrm{CCHP,max}} \le P_{m,t}^{\mathrm{CCHP}} - P_{m,t-1}^{\mathrm{CCHP}} \le r_m^{\mathrm{CCHP}} P_m^{\mathrm{CCHP,max}}, \forall m, t \tag{11.32}$$

$$0 \le q_{m,t}^{\mathrm{HR/AC/HC}} \le q_m^{\mathrm{HR/AC/HC,max}}, \forall m, t \tag{11.33}$$

$$c^{\mathrm{air}} \frac{\theta_{m,k,t}^{\mathrm{in}} - \theta_{m,k,t-1}^{\mathrm{in}}}{\tau} = \frac{q_{m,k,t}^{\mathrm{HE}} - q_{m,k,t}^{\mathrm{CE}}}{\tau} + \frac{\theta_{m,k,t}^{\mathrm{am}} - \theta_{m,k,t}^{\mathrm{in}}}{R_{m,k}^{T}}, \forall m, k, t \tag{11.34}$$

$$0 \le q_{m,k,t}^{\mathrm{CE/HE}} \le q_{m,k}^{\mathrm{CE/HE,max}}, \forall m, k, t \tag{11.35}$$

$$\theta_{m,k,t}^{\mathrm{in,min}} \le \theta_{m,k,t}^{\mathrm{in}} \le \theta_{m,k,t}^{\mathrm{in,max}}, \forall m, k, t \tag{11.36}$$

$$\eta^{\mathrm{TD}} q_{m,t}^{\mathrm{CE/HE}} = \sum_{k \in K} q_{m,k,t}^{\mathrm{CE/HE}}, \forall m, t \tag{11.37}$$

$$q_{m,k,t}^{\mathrm{HE,bill}} = \begin{cases} \dfrac{\left(\theta_{m,k,t}^{\mathrm{in,min}} - \theta_{m,k,t}^{\mathrm{am}}\right)\tau}{R_{m,k}^{\mathrm{T}}}, \text{if } \theta_{m,k,t}^{\mathrm{am}} < \theta_{m,k,t}^{\mathrm{in,min}}, \forall m, k, t \\ 0, if\, \theta_{m,k,t}^{\mathrm{am}} \ge \theta_{m,k,t}^{\mathrm{in,min}} \end{cases} \tag{11.38}$$

$$q_{m,k,t}^{\mathrm{CE,bill}} = \begin{cases} \dfrac{\left(\theta_{m,k,t}^{\mathrm{am}} - \theta_{m,k,t}^{\mathrm{in,max}}\right)\tau}{R_{m,k}^{\mathrm{T}}}, \text{if } \theta_{m,k,t}^{\mathrm{am}} > \theta_{m,k,t}^{\mathrm{in,max}}, \forall m, k, t \\ 0, if\, \theta_{m,k,t}^{\mathrm{am}} \le \theta_{m,k,t}^{\mathrm{in,max}} \end{cases} \tag{11.39}$$

$$P_{i+1,t} = P_{i,t} - P_{i+1,t}^{\mathrm{lat}} + P_{i+1,t}^{\mathrm{DG}} - P_{i+1,t}^{\mathrm{EL}}, \forall i, t \tag{11.40}$$

$$Q_{i+1,t} = Q_{i,t} - Q_{i+1,t}^{\mathrm{lat}} - Q_{i+1,t}^{\mathrm{EL}}, \forall i, t \tag{11.41}$$

$$V_{i+1,t} = V_{i,t} - \frac{R_i P_{i,t} + X_i Q_{i,t}}{V_0}, \forall i, t \tag{11.42}$$

$$P_{i,t}^{\mathrm{DG}} = P_{m,t}^{\mathrm{CCHP}} + P_{n,t}^{\mathrm{WT}} + P_{n,t}^{\mathrm{PV}}, \forall i \in M \cup N_{\mathrm{WT}} \cup N_{\mathrm{PV}}, t \tag{11.43}$$

$$V_{i,t}^{\min} \le V_{i,t} \le V_{i,t}^{\max}, \forall i, t \tag{11.44}$$

$$-P_i^{\mathrm{cap}} \le P_{i,t} \le P_i^{\mathrm{cap}}, \forall i, t \tag{11.45}$$

$$P_{1,t} = P_t^{\mathrm{def}} - P_t^{\mathrm{sur}}, P_t^{\mathrm{def}} \ge 0, P_t^{\mathrm{sur}} \ge 0, \forall t \tag{11.46}$$

$$\overline{P}_{n,t}^{\mathrm{WT}} \le P_{n,t}^{\mathrm{WT}} \le \bar{P}_{n,t}^{\mathrm{WT}}, \forall n, t \tag{11.47}$$

$$\overline{P}_{n,t}^{\mathrm{PV}} \le P_{n,t}^{\mathrm{PV}} \le \bar{P}_{n,t}^{\mathrm{PV}}, \forall n, t \tag{11.48}$$

$$\overline{P}_{0,i,t}^{\mathrm{EL}} \le P_{0,i,t}^{\mathrm{EL}} \le \bar{P}_{0,i,t}^{\mathrm{EL}}, \forall i, t \tag{11.49}$$

$$\overline{\theta}_{m,k,t}^{\mathrm{am}} \le \theta_{m,k,t}^{\mathrm{am}} \le \bar{\theta}_{m,k,t}^{\mathrm{am}}, \forall m, k, t \tag{11.50}$$

Equation (11.9) is to minimise the total operating cost during the whole day operation of the MEMG. Equations (11.10)–(11.14) are the detailed functions for these cost and revenue items, i.e. the fuel cost of the CCHP plants, the operation and maintenance costs of the CCHP plants, WTs and PVs, the electric energy transaction with the main grid, as well as the revenues of selling electric and thermal energy to the customers.

Equation (11.15) means that only one PBDR level can be delivered for each operating period. The calculation of the electric loads with PBDR is expressed in (11.16) and (11.17). Here, the power factor of each bus is assumed as a constant. Constraints (11.18) and (11.19) ensure the customers have access to sufficient electric energy without paying more electric usage bills after the PBDR is implemented. This would protect the customer economic benefits.

Constraint (11.20) means that the TES is either charged or discharged for each period. The limits of thermal energy charging and discharging are denoted as (11.21). Equation (11.22) calculates the current TES state considering the previous storage state, energy loss, and energy which is charged/discharged in the current period. The limits of the thermal energy which can be stored during a period are denoted by (11.23). Equation (11.24) ensures that the starting and the ending storage states of a day are equal.

Constraints (11.25)–(11.29) represent the energy conversion and flows in the CCHP systems. The fuel volume used for the PGU is computed by (11.30). The CCHP electric power output is limited with the capacity in (11.31) and the ramping rate in (11.32). The thermal energy generation of the heat recovery unit, absorption chiller, and heating coil is limited by their capacity in (11.33).

A story of a commercial building or a residential house can be defined as a single thermal zone. The indoor temperature control is implemented for each thermal zone. It is assumed that the air conditioner of each thermal zone utilises the cooling/heat energy from the CCHP plants to control the indoor temperature. Equation (11.34) denotes that the thermal energy state of a thermal zone varies with cooling and heat supported by the CCHP plants and the thermal conduction through the building envelope. A maximal limit is used for the cooling/heat energy that a thermal zone can absorb through the air conditioning system in (11.35). Constraint (11.36) means that the indoor temperature must be kept within the human comfort range. The energy loss during the cooling and heat distribution from the CCHP plants to the thermal loads is modelled in (11.37). Equations (11.38) and (11.39) calculate the customer bills for thermal energy usage, where the customer economic benefits are protected.

The linearised Dist-Flow equations are presented as (11.40)–(11.42), which are explained in detail in [16]. Equation (11.43) computes power injection from the DG, i.e. the summation of the CCHP plant, WT, and PV output power at a bus. Constraint (11.44) ensures that the voltage at each bus is kept within an allowed range. The branch power flow limits are modelled in (11.45).

The electric power exchange between the MEMG and the main grid is modelled in (11.46), where $P_t^{\text{def}} > 0$ means the MEMG buys energy from the main grid and $P_t^{\text{sur}} > 0$ means the MEMG sells energy to the main grid.

Constraints (11.47) and (11.48) provide the predicted bounds for the output power of the WTs and PVs, respectively. The electric load is constrained within the predicted range as (11.49). Moreover, the ambient temperature is also considered as uncertainty and it is limited by (11.50) within the predicted range as well.

The objective (11.9) with the constraints (11.10)–(11.50) forms a mixed-integer linear programming with uncertainty variables. The day-ahead PBDR levels $\alpha_{j,t}$ and TES charging/discharging outputs $\beta_{m,t}^{\mathrm{TSC/D}}, q_{m,t}^{\mathrm{TSC/D}}$ are the first-stage decisions, while the hourly CCHP outputs $P_{m,t}^{\mathrm{CCHP}}$ and indoor temperature set points $\theta_{m,k,t}^{\mathrm{in}}$ are the second-stage decisions in this coordination optimisation problem. The uncertainty variables include $P_{n,t}^{\mathrm{WT}}, P_{n,t}^{\mathrm{PV}}, P_{0,i,t}^{\mathrm{EL}}$, and $\theta_{m,k,t}^{\mathrm{am}}$.

11.4 Two-stage robust optimisation method

The proposed coordination optimisation problem of the MEMG has multiple uncertainties. To fully address these uncertainties and obtain a robust solution, a TSRO method is applied for the MEMG coordination optimisation problem.

The principle of the TSRO method is introduced in Section 7.4.1.

11.4.1 Uncertainty set modelling

The uncertain renewable power generation $P_{n,t}^{\mathrm{WT}}, P_{n,t}^{\mathrm{PV}}$, uncertain electric loads $P_{0,i,t}^{\mathrm{EL}}$, and uncertain ambient temperature $\theta_{m,k,t}^{\mathrm{am}}$ with their constraints (11.47)–(11.50) are modelled into four polyhedral uncertainty sets, respectively, as follows:

$$U_{\mathrm{WT}} = \left\{ P_{n,t}^{\mathrm{WT}} \in \mathbb{R}^{N_{\mathrm{WT}}} : \overline{P}_{n,t}^{\mathrm{WT}} \le P_{n,t}^{\mathrm{WT}} \le \bar{P}_{n,t}^{\mathrm{WT}}, \forall n, t, \overline{\mu}^{\mathrm{WT}} \le \frac{\sum_{n\in N_{\mathrm{WT}}} \sum_{t\in T} P_{n,t}^{\mathrm{WT}}}{\sum_{n\in N_{\mathrm{WT}}} \sum_{t\in T} \widehat{P}_{n,t}^{\mathrm{WT}}} \le \bar{\mu}^{\mathrm{WT}} \right\} \tag{11.51}$$

$$U_{\mathrm{PV}} = \left\{ P_{n,t}^{\mathrm{PV}} \in \mathbb{R}^{N_{\mathrm{PV}}} : \overline{P}_{n,t}^{\mathrm{PV}} \le P_{n,t}^{\mathrm{PV}} \le \bar{P}_{n,t}^{\mathrm{PV}}, \forall n, t, \overline{\mu}^{\mathrm{PV}} \le \frac{\sum_{n\in N_{\mathrm{PV}}} \sum_{t\in T} P_{n,t}^{\mathrm{PV}}}{\sum_{n\in N_{\mathrm{PV}}} \sum_{t\in T} \widehat{P}_{n,t}^{\mathrm{PV}}} \le \bar{\mu}^{\mathrm{PV}} \right\} \tag{11.52}$$

$$U_{\mathrm{EL}} = \left\{ P_{0,i,t}^{\mathrm{EL}} \in \mathbb{R}^{N} : \overline{P}_{0,i,t}^{\mathrm{EL}} \le P_{0,i,t}^{\mathrm{EL}} \le \bar{P}_{0,i,t}^{\mathrm{EL}}, \forall i, t, \overline{\mu}^{\mathrm{EL}} \le \frac{\sum_{i\in N} \sum_{t\in T} P_{0,i,t}^{\mathrm{EL}}}{\sum_{i\in N} \sum_{t\in T} \widehat{P}_{0,i,t}^{\mathrm{EL}}} \le \bar{\mu}^{\mathrm{EL}} \right\} \tag{11.53}$$

$$U_{\mathrm{am}} = \left\{ \theta_{m,k,t}^{\mathrm{am}} \in \mathbb{R}^{M} : \overline{\theta}_{m,k,t}^{\mathrm{am}} \le \theta_{m,k,t}^{\mathrm{am}} \le \bar{\theta}_{m,k,t}^{\mathrm{am}}, \forall m, k, t, \overline{\mu}^{\mathrm{am}} \le \frac{\sum_{m\in M} \sum_{k\in K} \sum_{t\in T} \theta_{m,k,t}^{\mathrm{am}}}{\sum_{m\in M} \sum_{k\in K} \sum_{t\in T} \widehat{\theta}_{m,k,t}^{\mathrm{am}}} \le \bar{\mu}^{\mathrm{am}} \right\} \tag{11.54}$$

In each uncertainty set, the uncertainty variable can vary randomly between the lower and upper bounds. Determining the bounds can be accomplished by using available historical data [17] or interval prediction tools which can provide uncertainty variation ranges [14]. The bounds of the predicted uncertainty variation range are set as the lower and upper bounds in the uncertainty sets.

Besides, for each set, a pair of uncertainty budgets $\overline{\mu}$ and $\underline{\mu}$ are set to quantify the overall uncertainty degree. The design of the uncertainty budgets is based on accuracy of prediction and expectation of solution robustness. When the lower budget falls and the upper budget rises, it means that the uncertainty set can cover a higher uncertainty degree to achieve higher solution robustness. However, as the solution robustness is improved, the solution tends to be more conservative, i.e. a larger objective. Thus, there should be a trade-off between solution robustness and conservativeness. Detailed description of the budget design can be referred in Section 7.4.2.

11.4.2 Two-stage robust optimisation model

The proposed MEMG coordination optimisation model is converted to the TSRO model which is given in Section 7.4.3.

The TSRO objective is formulated as follows:

$$\min_{x} \max_{u} \min_{y} C_{\text{CCHP}} + C_{\text{OM}} + C_{\text{grid}} - C_{\text{rev}}^{\text{elec}} - C_{\text{rev}}^{\text{thm}} \tag{11.55}$$

Here, $C_{\text{CCHP}} + C_{\text{OM}} + C_{\text{grid}} - C_{\text{rev}}^{\text{elec}} - C_{\text{rev}}^{\text{thm}}$ expresses the second-stage objective. Note that there is no objective for the first operation stage, but the first-stage variables impact on the second-stage decisions and the corresponding objective.

This model consists of three types of variables: (1) $\boldsymbol{x}$ is the vector of the first-stage decision variables, including the PBDR levels $\alpha_{j,t}$ as well as the TES operation states and outputs $\beta_{m,t}^{\text{TSC/D}}$ and $q_{m,t}^{\text{TSC/D}}$; (2) $\boldsymbol{y}$ stands for the vector of the second-stage decision variables such as the CCHP electric power outputs $P_{m,t}^{\text{CCHP}}$ and the indoor temperature set points $\theta_{m,k,t}^{\text{in}}$, as well as the dependent power and energy flow variables; (3) $\boldsymbol{u}$ is the vector of the uncertainty variables which include the renewable power generation $P_{n,t}^{\text{WT}}$, $P_{n,t}^{\text{PV}}$, the electric loads $P_{0,i,t}^{\text{EL}}$ and the ambient temperature $\theta_{m,k,t}^{\text{am}}$.

The TSRO objective is subject to the constraints (11.10)–(11.46) and $\boldsymbol{u}$ is constrained by (11.51)–(11.54).

Within the uncertainty sets, the worst case of uncertainty realisation is obtained through the '*max*'. Simultaneously, the minimisation process obtains the optimal intraday dispatch decisions by the second '*min*' under the worst case. Then, the day-ahead scheduling decisions are optimised through the first '*min*' with the worst case. Note that only the first-stage decisions are implemented in the day-ahead time window, while the second-stage decisions would be re-optimised and implemented with realised uncertainties.

Detailed descriptions of the TSRO model can be found in Section 7.4.3.

11.4.3 Solution algorithm

To solve the '*min-max-min*' TSRO problem, the column-and-constraint generation (C&CG) algorithm which has a convergence speed [18] is applied in this chapter.

In the C&CG algorithm, the '*min-max-min*' TSRO problem is divided into a '*min*' master problem and a '*max-min*' sub-problem which can formulate the two operation stages, respectively. The C&CG algorithm solves the master problem and the sub-problem iteratively until the gap between the lower bound given by the master problem and the upper bound given by the sub-problem is reduced to an allowed termination threshold. Finally, the first-stage decision variable vector, $\boldsymbol{x}$, is robustly optimised. The complete procedure of the C&CG algorithm is provided in Section 7.4.4.

It is worth noting that exactly solving the '*max-min*' sub-problem with the polyhedral uncertainty sets is significantly challenging [18]. This '*max-min*' sub-problem can be converted into a bilinear problem by strong duality, which is described in Section 7.4.4. To improve computing efficiency, the chapter applies an outer approximation (OA) algorithm [19] to solve the bilinear problem. In this algorithm, the bilinear problem is decomposed into two linear programming problems. The OA algorithm can linearise the bilinear items around intermediate solution points and pass the linear approximation to the OA formulation. Then, the linear programming problems are solved iteratively at a fast computing speed even though variables are numerous.

The complete procedure of the OA algorithm to solve the bilinear problem can be found in [19].

The master problem as well as the linearised problems of the subproblem can be efficiently solved by mixed integer linear programming solvers such as GUROBI [20].

11.5 Case study

11.5.1 Test system description

In this chapter, a 33-bus radial distribution network as a microgrid [21] is used to test the proposed two-stage coordinated operation method of the MEMG. Table 11.2 shows system parameters, where type '*rh*' means a residential house and type '*cbs*' means a commercial building story space. Table 11.3 shows parameters of CCHP plants and numbers of thermal zones which are supplied by these CCHP plants. Herein, the energy limits are for the period of an hour. Table 11.4 shows locations and capacities of WTs and PVs. In this chapter, the period of the intraday

Table 11.2 System parameters

Name	Value	Name	Value	Name	Value
C^{CCHP}	\$0.008/kWh	Pr^{HE}	\$0.04/kWh	η^{HR}	0.7
C^{WT}	\$0.010/kWh	Pr_0^{EL}	\$0.2219/kWh	η^{TS}	0.95
C^{PV}	\$0.012/kWh	r^{CCHP}	0.5	$\eta^{\mathrm{TSC/D}}$	0.97
C^{fuel}	\$0.5/m^3	$R_{\mathrm{rh}}^{\mathrm{T}}$	2.8°C/kW	$\eta^{\mathrm{AC/HC}}$	0.7/0.8
$c_{\mathrm{rh}}^{\mathrm{air}}$	0.53 kWh/°C	$R_{\mathrm{cbs}}^{\mathrm{T}}$	1.3°C/kW	η^{TD}	0.95
$c_{\mathrm{cbs}}^{\mathrm{air}}$	1.85 kWh/°C	σ	9.78 kWh/m^3	$\theta^{\mathrm{in,min}}$	18°C
Pr^{CE}	\$0.05/kWh	η^{PGU}	0.33	$\theta^{\mathrm{in,max}}$	22°C

Table 11.3 Combined cooling, heat and power plant and thermal load parameters

Bus, *m*	**8**	**14**	**24**	**25**	**30**	**32**
Thermal zone type	*cbs*	*cbs*	*rh*	*rh*	*rh*	*cbs*
No. of thermal zones, K	13	15	75	60	30	18
$P_m^{\mathrm{CCHP,min}}$(kW)	20	20	40	40	20	20
$P_m^{\mathrm{CCHP,max}}$ (kW)	200	200	400	400	200	200
$Q_{m,t}^{\mathrm{CE,max}}$(kWh)	200	300	600	500	300	400
$Q_{m,t}^{\mathrm{HE,max}}$(kWh)	200	300	600	500	300	400
$Q_{m,t}^{\mathrm{HR,max}}$(kWh)	285	285	570	570	285	285
$Q^{\mathrm{TS,max}}$ (kWh)	600	800	1,200	400	800	1,000
$Q_{m,t}^{\mathrm{TSC/D,max}}$ (kWh)	75	100	150	50	100	125

Table 11.4 WT and PV location and capacity

Bus	**Rated power (kW)**		**Bus**	**Rated power (kW)**		**Bus**	**Rated power (kW)**	
	WT	**PV**		**WT**	**PV**		**WT**	**PV**
3	200	0	16	200	50	25	0	250
4	100	0	18	100	100	29	100	100
7	200	100	20	100	0	30	100	200
10	0	100	21	200	0	31	200	150
12	100	50	24	0	300	32	200	50

dispatch stage τ is set as 1 h. Note that other systems or parameters can also be applied, and they do not affect the effectiveness of the proposed method.

The day-ahead interval predictions of WT and PV power generation, electric loads, and ambient temperature are shown in Figure 11.3. The expected WT output power is sampled randomly with summer medium wind speeds. The expected PV output power is sampled with clear-sky characteristics. The electric load profile of 30 January 2017 in Sydney is provided by AEMO [22] as the expected value. The ambient temperature (Temp.) profile of 30 January 2017 is extracted as well. It is assumed that the ambient temperature is identical for the whole MEMG territory. The intervals are set as $\pm$30% of the expected values for the renewable power generation, $\pm$3% for the electric loads, and $\pm$5% for the ambient temperature according to typical prediction accuracy. These intervals are set as the lower and upper bounds of the uncertainty variables. The lower and upper budgets are, respectively, set as 90% and 110% for the renewable power generation, 98% and 102% for the electric loads, 97.5% and 102.5% for the ambient temperature. These bounds and budgets are used to construct the uncertainty sets in (11.51)–(11.54).

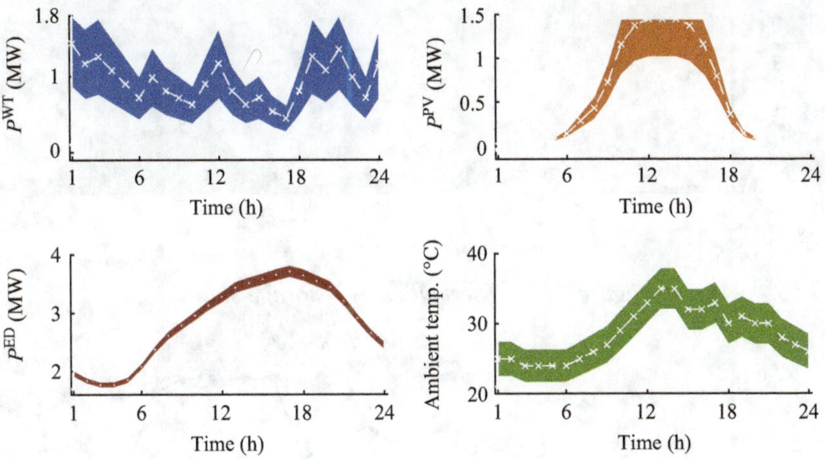

Figure 11.3 Day-ahead interval predictions

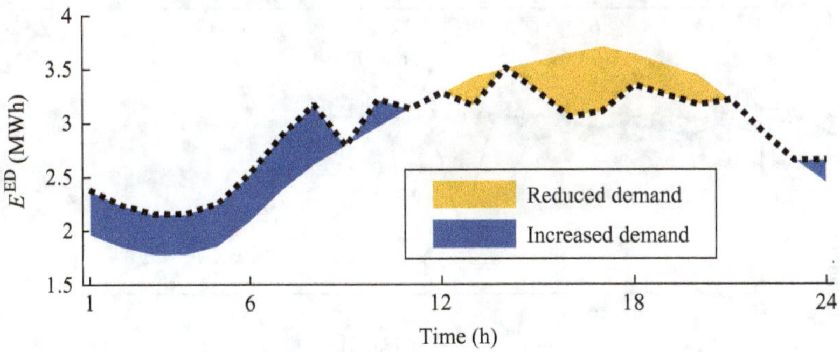

Figure 11.4 Electric load demand response decisions

The electric power transaction price of 30 January 2017 provided by AEMO [22] is used in this case study.

11.5.2 Day-ahead scheduling results

The day-ahead scheduling decisions are obtained by solving the TSRO model. The computation speed is considerably high, and the average solver time using GUROBI is only 4 s.

The overall operating profit under the worst case is \$9,914 for the whole day, meaning that at least \$9,914 would be obtained. Note that this value may increase after the intraday dispatch, as the worst case may not occur, which would be illustrated in Section 11.5.3.

The electric loads with the PBDR decisions are shown in Figure 11.4. It shows that the electric loads respond to increase between 0.00 and 8.00 and decrease between 14.00 and 20.00.

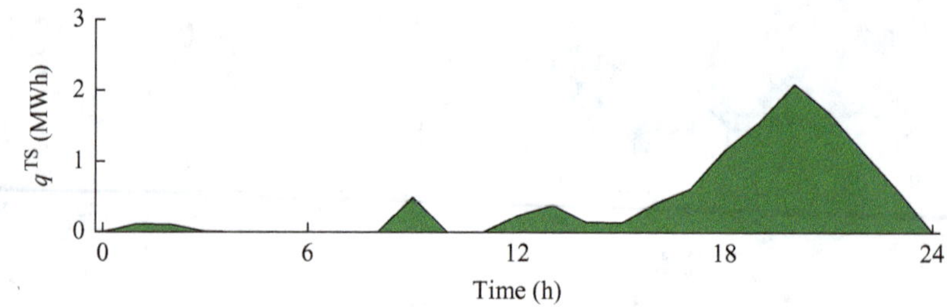

Figure 11.5 Thermal energy storage states

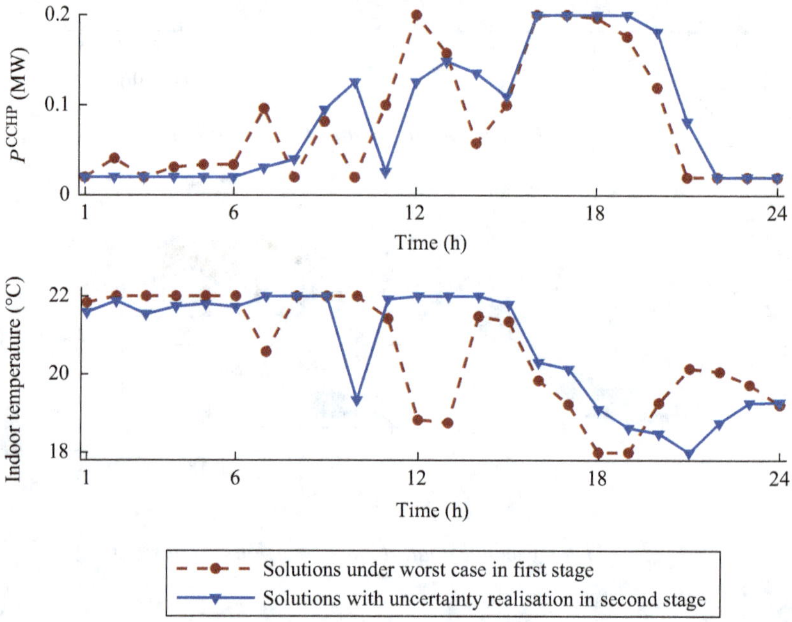

Figure 11.6 Intraday dispatch decisions

The TES states are shown in Figure 11.5. The TES units charge thermal energy during the periods 0.00–1.00, 8.00–9.00, 11.00–13.00, and 15.00–20.00 when thermal energy generation is excessive and discharge thermal energy during the periods 2.00–3.00, 9.00–10.00, 13.00–14.00, and 20.00–24.00 when the thermal loads are high. It can be seen that the thermal energy is stored and released for several times during the day to decouple the electric and thermal outputs of the CCHP plants.

The CCHP electric power output and the indoor temperature obtained under the worst case are shown in Figure 11.6. Between 15.00 and 20.00, since PV output power drops, the CCHP plants increase the electric power generation. As a by-product, the corresponding thermal output also increases. However, the thermal loads reduce since the ambient temperature drops during this period. Therefore, the excessive thermal energy is stored as shown in Figure 11.5.

11.5.3 Intraday dispatch results

During the intraday dispatch stage, the CCHP outputs and indoor temperature set points are re-optimised based on hour-ahead predictions of the renewable power generation, electric loads, and ambient temperature which are regarded as uncertainty realisations. With the implemented day-ahead decisions and the realised uncertainties, the intraday dispatch optimisation problem is modelled as follows:

$$-C_{\mathrm{rev}}^{\mathrm{elec}} + \min_{y}\left(C_{\mathrm{CCHP}} + C_{\mathrm{OM}} + C_{\mathrm{grid}} - C_{\mathrm{rev}}^{\mathrm{thm}}\right) \tag{11.56}$$

s.t. (11.10)–(11.14), (11.25)–(11.46)

In this optimisation problem, the electric and thermal outputs of the CCHP plants are dispatched and the indoor temperature is controlled within the customer comfort range. The obtained decisions are implemented for each hour to coordinate with the renewable power generation, the PBDR decisions, and the TES scheduling, while fulfilling the customer benefits and the network operating constraints.

The intraday dispatch is simulated by solving the problem (11.56) based on the hour-ahead predictions of the renewable output power, electric loads, and ambient temperature.

The intraday electric power output of the CCHP plant and the indoor temperature of the thermal zones at bus 8 are shown in Figure 11.6. For comparison, this figure also shows the day-ahead optimisation results obtained under the worst case. Between 11.00 and 12.00, the realised renewable output power is higher than that under the worst case, and the CCHP electric power output becomes less than that in the day-ahead optimisation. Meanwhile, the coupled thermal energy output is reduced, but the indoor temperature is kept within the customer comfort range. This means the intraday dispatch can compensate the day-ahead scheduling effectively against the uncertainty realisation to improve the overall energy utilisation efficiency while satisfying the customer comfort.

The total flexible thermal loads in the system are shown in Figure 11.7. As the realised ambient temperature differs from the worst case and the indoor temperature is controlled, the actual thermal loads are adjusted from those under the worst case. The intraday flexible thermal loads allow the CCHP plants to dispatch their electric and thermal outputs in a flexible range.

For this intraday dispatch result, the operating profit becomes \$10,763 for the whole day which is higher than that obtained under the worst case.

11.5.4 Operation robustness check

This chapter compares the proposed robust two-stage coordination method (Method A) and other two methods, a single-stage operation method with flexible loads (Method B) and a single-stage operation method without flexible loads (Method C). Both Methods B and C are optimised by a deterministic optimisation

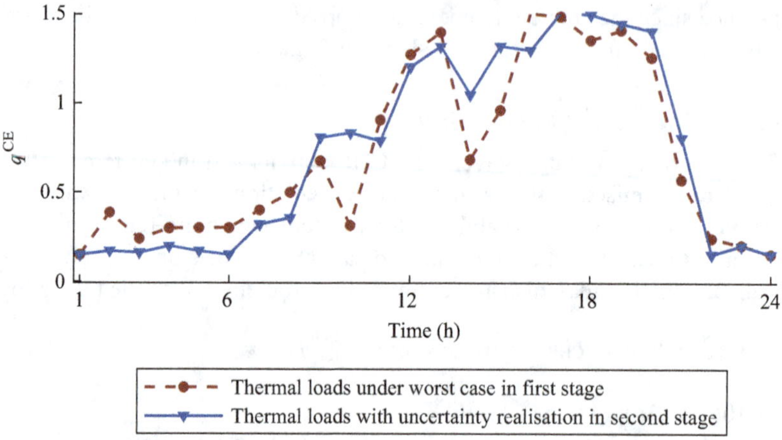

Figure 11.7 Thermal loads with indoor temperature control

Table 11.5 Comparison among different methods

Method	A (proposed)	B	C
Operating profit ($)	10,900	10,033	9,424
CCHP fuel cost ($)	2,824	2,760	3,272
Temperature absolute deviation (°C)	1.49	1.75	N/A
Temperature comfort rate (%)	100.00	0.00	N/A
Voltage satisfaction rate (%)	100.00	51.6	98.9

approach which uses the predicted mean values of the uncertainties (cross signs in Figure 11.3). In Method B, all decisions are optimised through (11.9)–(11.46). In Method C, the thermal loads referred to indoor temperature of 20°C are calculated before optimisation.

In total, 5,000 scenarios are randomly generated by Monte Carlo sampling for the solution robustness check and these scenarios are regarded as realisations of the uncertainties. Methods A, B, and C are tested by these uncertainty realisations. The comparison results are shown in Table 11.5. All the profit, cost, and indoor temperature values are average values of all the scenarios. The temperature absolute deviation is calculated according to 20°C, which is the most preferred indoor temperature. Temperature comfort and voltage satisfaction rates are percentages of the scenarios where the indoor temperature satisfies the customer comfort range and the bus voltages satisfy the allowed range in all the scenarios, which indicates operation robustness.

Comparing the operation methods with the flexible loads (Methods A and B) and the one without the flexible loads (Method C), the CCHP fuel costs of the former are much less than that of the latter, which contributes in increasing the overall operating profits of the MEMG. It indicates that the flexible loads can improve the energy utilisation efficiency and operating profits effectively. Besides, the proposed robustly coordinated operation method achieves the highest operating profit.

In terms of indoor temperature, the temperature absolute deviation of the proposed robust two-stage coordination method is least. In addition, the indoor temperature of the proposed method in all the scenarios fulfils the customer comfort, but that achieved by the single-stage method (Method B) does not at all.

For the bus voltages, the voltage satisfaction rate of the proposed robust two-stage coordination method is 100%, while that of the single-stage one with the flexible loads is only 51.62% for the 5,000 scenario checks.

The results of the indoor temperature and the bus voltages indicate that the proposed robust two-stage coordination method has significantly high robustness in keeping the indoor temperature and the bus voltages in the allowed ranges against the uncertainty realisations.

11.6 Conclusion

In this chapter, the MEMG is improved with the flexible electric and thermal loads which are achieved by the PBDR method and the indoor temperature control. A two-stage coordinated operation method of the MEMG is proposed to achieve the maximal overall operating profit while fulfilling customer economic benefits and thermal comfort as well as network operating constraints. A TSRO method is applied with interval predictions of multiple uncertainties to optimise the two-stage coordinated operation, resulting in robust operation of the MEMG against uncertainty realisation.

The proposed two-stage coordinated operation of the MEMG is tested with sufficient uncertainty scenarios and compared with conventional single-stage methods. The simulation results indicate that the operating profit is robustly maximised while satisfying the customer benefits, the thermal comfort, and the network operating constraints. Thus, it indicates that the proposed two-stage coordination of the MEMG can achieve high energy utilisation efficiency and guaranteed solution robustness.

References

[1] P. Mancarella, 'MES (multi-energy systems): an overview of concepts and evaluation models', *Energy*, vol. 65, pp. 1–17, 2014.

[2] F. Fang, Q.H. Wang, and Y. Shi, 'A novel optimal operational strategy for the CCHP system based on two operating modes', *IEEE Trans. Power Syst.*, vol. 27, no. 2, pp. 1032–1041, 2012.

[3] R. Hashemi, 'A developed offline model for optimal operation of combined heating and cooling and power systems', *IEEE Trans. Energy Conv.*, vol. 24, no. 1, pp. 222–229, 2009.

[4] G.K.H. Larsen, N.D. van Foreest, and J.M.A. Scherpen, 'Distributed MPC applied to a network of households with micro-CHP and heat storage', *IEEE Trans. Smart Grid*, vol. 5, no. 4, pp. 2106–2114, 2014.

[5] M. Houwing, R.R. Negenborn, and B. De Schutter, 'Demand response with micro-CHP systems', *Proc. of the IEEE*, vol. 99, no. 1, pp. 200–213, 2011.
[6] F. Brahman, M. Honarmand, and S. Jadid, 'Optimal electric and thermal energy management of a residential energy hub, integrating demand response and energy storage system', *Energy and Buildings*, vol. 90, pp. 65–75, 2015.
[7] A. Anvari-Moghaddam, H. Monsef, and A. Rahimi-Kian, 'Optimal smart home energy management considering energy saving and a comfortable lifestyle', *IEEE Trans. Smart Grid*, vol. 6, no. 1, pp. 324–332, 2015.
[8] M. Tasdighi, H. Ghasemi, and A. Rahimi-Kian, 'Residential microgrid scheduling based on smart meters data and temperature dependent thermal load modelling', *IEEE Trans. Smart Grid*, vol. 5, no. 1, pp. 349–357, 2014.
[9] J. Wang, H. Zhong, Q. Xia, C. Kang, and E. Du, 'Optimal joint-dispatch of energy and reserve for CCHP-based microgrids', *IET Gen. Transm. Distrib.*, vol. 11, no. 3, pp. 785–794, 2017.
[10] C. Li, Y. Xu, X. Yu, C. Ryan, and T. Huang, 'Risk-averse energy trading in multienergy microgrids: a two-stage stochastic game approach', *IEEE Trans. on Indus. Info.*, vol. 13, no. 5, pp. 2620–2630, 2017.
[11] L. Wang, Q. Li, M. Sun, and G. Wang, 'Robust optimisation scheduling of CCHP systems with multi-energy based on minimax regret criterion', *IET Gen. Transm. Distrib.*, vol. 10, no. 9, pp. 2194–2201, 2016.
[12] Z. Bao, Q. Zhou, Z. Yang, Q. Yang, L. Xu, and T. Wu, 'A multi time-scale and multi energy-type coordinated microgrid scheduling solution – Part I: Model and methodology', *IEEE Trans. Power Syst.*, vol. 30, no. 5, pp. 2257–2266, 2015.
[13] W. Gu, Z. Wang, Z. Wu, Z. Luo, Y. Tang, and J. Wang, 'An online optimal dispatch schedule for CCHP microgrids based on model predictive control', *IEEE Trans. Smart Grid*, vol. 8, no. 5, pp. 2332–2342, 2017.
[14] C. Zhang, Y. Xu, Z.Y. Dong, and K.P. Wong, 'Robust coordination of distributed generation and price-based demand response in microgrids', *IEEE Trans. Smart Grid*, vol. 9, no. 5, pp. 4236–4247, 2018.
[15] A. Faruqui and S. Sergici, 'Household response to dynamic pricing of electricity: a survey of 15 experiments', *J. Regulatory Economics*, vol. 38, no. 2, pp. 193–225, 2010.
[16] C. Zhang, Y. Xu, Z.Y. Dong, and J. Ma, 'Robust operation of microgrids via two-stage coordinated energy storage and direct load control', *IEEE Trans. Power Syst.*, vol. 32, no. 4, pp. 2858–2868, 2017.
[17] Y. Guan and J. Wang, 'Uncertainty sets for robust unit commitment', *IEEE Trans. Power Syst.*, vol. 29, no. 3, pp. 1439–1440, 2014.
[18] B. Zeng and L. Zhao, 'Solving two-stage robust optimisation problems using a column-and-constraint generation method', *Operations Research Letters*, vol. 41, no. 5, pp. 457–461, 2013.
[19] D. Bertsimas, E. Litvinov, X. A. Sun, J. Zhao, and T. Zheng, 'Adaptive robust optimisation for the security constrained unit commitment problem', *IEEE Trans. Power Syst.*, vol. 28, no. 1, pp. 52–63, 2013.

[20] GUROBI, Available: www.gurobi.com, accessed March 2018.
[21] M.E. Baran and F.F. Wu, 'Network reconfiguration in distribution systems for loss reduction and load balancing', *IEEE Trans. Power Del.*, vol. 4, no. 2, pp. 1401–1407, 1989.
[22] Australian Energy Market Operator (AEMO), Available: www.aemo.com.au, accessed March 2018.

Chapter 12
Multi-timescale coordinated voltage/var control optimisation

This chapter presents a multi-timescale coordinated voltage/var control (VVC) optimisation approach for the grid-connected microgrid with high-level solar photovoltaic (PV) units. It aims to utilise multiple devices to counteract uncertain voltage fluctuation and deviation caused by stochastic solar PV power output and load demand. In the hourly timescale (first stage), mechanical devices including capacitor banks (CBs) and transformer tap changers are scheduled. In the 15-min timescale (second stage), PV inverters are dispatched to provide reactive power support to supplement the first-stage decision after stochastic PV power output and load variations are realised. The coordination problem is formulated as a two-stage stochastic programming model and converted to a deterministic mixed-integer quadratic programming (MIQP) model, and then can be efficiently solved by off-the-shelf solvers. Compared with existing methods, the proposed approach can achieve lower expected energy loss and can sustain a secure voltage level under random PV power output and load variation.

12.1 Introduction

For low-voltage distribution networks, the R/X ratio of distribution feeder is usually very high; therefore, the active power is also coupled with the bus voltage. This can significantly challenge the VVC of microgrids: First, high-level penetration of renewable energy source (RES), such as rooftop solar PV, can inject considerable active power into the distribution grid and reduce the net power load or even cause reverse power flow, thereby raising the bus voltage magnitude. Second, the RES power output is highly uncertain and intermittent; for example, the power output of a PV unit can ramp on the order of 15% of its capacity per minute under transient cloud movements [1], which can cause stochastic and fast voltage fluctuations and deviations in the distribution network.

Traditionally, the VVC is scheduled on a centralised optimisation basis in the form of the reactive power optimisation problem, which usually minimises the network power loss while satisfying voltage constraints [2,3]. Conventional VVC equipment including on-load tap changers (OLTCs) and CBs are mechanical devices which have a slow response speed (e.g. from several seconds to minutes); besides, their lifetime can dramatically reduce with excessive number of switching operations. As an alternative, power-electronic equipment such as static VAR compensator (SVC) and static synchronous compensator (STATCOM) have a much faster response speed. To regulate the voltage under fast RES power output fluctuations, Ref. [4] presents a multi-objective VVC optimisation method to schedule the next day's VVC equipment and RES power outputs for simultaneously minimising power loss, voltage deviations, and pollution emission. The load demand and RES uncertainties are modelled through Monte Carlo sampling, and the optimisation model is solved by an evolutionary algorithm (EA). However, the day-ahead VVC schedule cannot be able to accommodate the intraday stochastic power output variations of RESs. Besides, the EA usually suffers from a slow convergence speed and inconsistent solutions for practical application. In [5], a model predictive control-based method is presented to coordinate the OLTC and CBs considering RES power forecasting errors. The problem is formulated as a stochastic mixed-integer nonlinear programming (MINLP) model. However, the MINLP model may not be efficiently solved especially for large-scale problems. Ref. [6] introduces a hybrid VVC framework: the central control loop dispatches the VVC devices in a hourly interval under normal operating condition; when large variation of PV output occurs, the decentralised control loop is activated and CBs and SVC become self-controlled based on local voltage measurements. While this method is effective to counteract the transient cloud movements, the central loop does not consider the uncertainties from PV output and load demand, and the two control loops are not optimally coordinated. Reference [7] proposes a model predictive control-based multi-objective VVC scheme to coordinate STATCOM and OLTC against small and large wind power fluctuations. However, STATCOMs or SVCs are still very expensive, and difficult to be widely deployed in distribution grids.

In addition to the above VVC devices, the power-electronic inverters of the RES units are also able to provide reactive power support [8]. In practice, an RES unit usually generates less active power than the rated capacity of its inverter, which leaves more space for reactive power support. Especially when the voltage drops are caused by RES power output decrease, there will be more capacity for reactive power generation/absorption. To promote its practical implementation, the IEEE 1547.8 working groups have revised the standards to allow the inverters to contribute in voltage regulation [9]. In [10], a local droop controller is used for the inverters to absorb/generate reactive power to mitigate voltage rise. In [11], an adaptive proportional-integral controller is proposed for multiple inverters to achieve plug-and-play VVC. However, given the local control nature, a system-wide coordination is missing in the above works. In [12], an adaptive control scheme is proposed for PV inverters to balance the power loss and voltage deviation minimisation objectives. However, the uncertainties from RES generation and load demand are not fully modelled. In [13], under a limited communication infrastructure, a distributed method is designed to dispatch the OLTCs and CBs for slow voltage variation, and dispatch inverters and demand response for fast voltage variation via neighbour-to-neighbour communication.

This chapter presents an optimal multi-timescale coordinated VVC approach considering uncertain RES power outputs and load demands. Given different driven forces for voltage variation and the response speeds of these devices, the voltage is regulated at two different timescales: the slow-acting devices, CBs and OLTCs, are used to manage the slow voltage variations over a longer time interval (e.g. every hour); within each interval, the RES inverters act in a shorter timescale (e.g. every 15 min), responding to fast voltage deviation caused by stochastic RES output and load fluctuations. The two timescales are coordinated through a two-stage stochastic programming model, where the first stage makes an initial decision irrespective of the uncertainties and then a recourse action is made in the second stage to compensate the first-stage decision after the uncertainty is realised or better observed. The optimisation objective is to minimise the total expected power loss while satisfying voltage limits as well as other operational constraints. A sampling average approximation combined with scenario reduction is employed to solve the model. The proposed method is tested on the IEEE 33-bus distribution grid.

12.2 Multi-timescale coordinated voltage/var regulation

12.2.1 Timescale decomposition

For a distribution network with high-level RES, the voltage variation can be decomposed into a slow component caused by regular variation in load demand and a fast component caused by stochastic fluctuation in RES power output.

Figure 12.1 illustrates the net load (load demand minus PV power generation) of an Australian residential feeder. It can be observed that the hourly timescale follows a regular load change pattern, while in the 10-min time interval, the net load fluctuates rapidly due to the combined uncertainties of PV generation and load demand. Correspondingly, the voltage will fluctuate in a similar pattern.

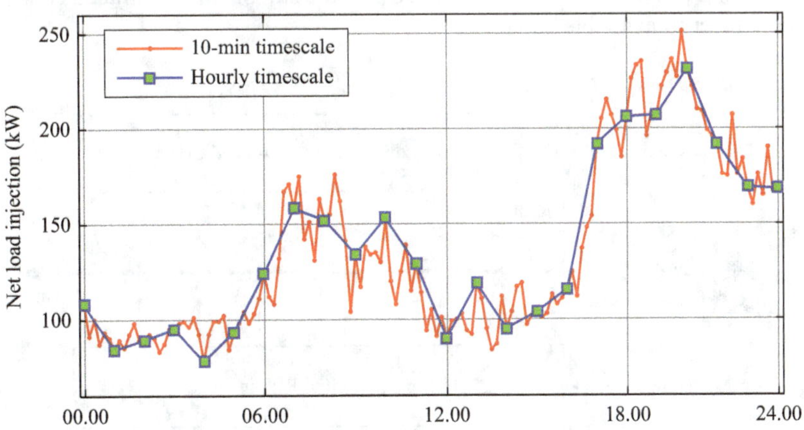

Figure 12.1 Illustration of the net load of a real-world residential feeder

Based on this fact, this chapter decomposes the voltage regulation into two timescales: hourly and 15-min (note that other timescale lengths can also be considered depending on practical need). For each timescale, different VVC devices with different response speeds are used for voltage regulation. While the conventional CBs and OLTCs can well manage the slow voltage variation, they are less competent to handle faster and uncertain voltage deviations caused by stochastic power output from RES. To counteract the fast voltage fluctuations, RES inverters are used to provide faster reactive power support against rapid voltage deviations.

12.2.2 Proposed coordination framework

The proposed VVC coordination framework is shown in Figure 12.2: the first stage aims to schedule the status of OLTCs and CBs every hour (slow timescale) based on the overall power profile of load and RES power output; within each hour, the second stage dispatches the RES inverters every 15 min (fast timescale) in response to faster power fluctuations in load and RES output. Note that even shorter time, e.g. 5 min, can also be considered wherever necessary since RES inverters as power-electronic devices can act in millisecond level.

For practical implementation, at each first-stage decision cycle, the next hour's load and RES generation are forecasted and used to determine the VVC reference (i.e. base set points). Once optimised, the first-stage decision is fixed until next hour. Then, the second-stage compensates the first-stage decision for the adverse effects of fast and stochastic load and RES generation outputs. At each second-stage decision cycle within each hour, the next 15-min load and RES generation are forecasted and their random variations are used to optimise inverters' reactive output for each 15-min interval. Note that the 15-min ahead forecasting tends to be more accurate than the hourly ahead one; therefore, the uncertainty can be better estimated.

This timescale separation is also to address the fact that frequent movements of conventional VVC devices could dramatically reduce their lifetime, while the inverters, as a power-electronic device, can provide smooth and continuous reactive power subject to available capacity.

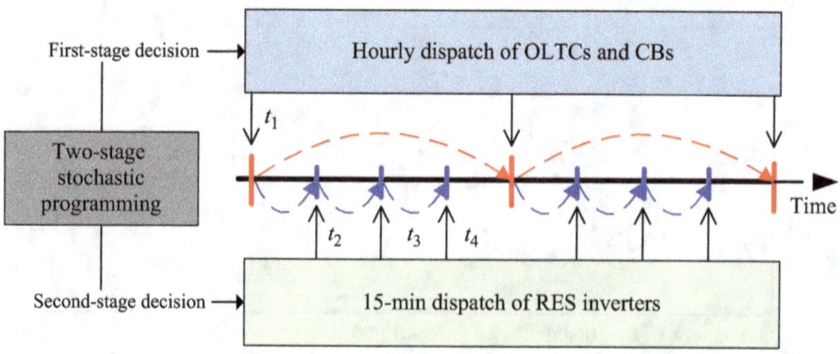

Figure 12.2 Proposed VVC coordination framework

12.2.3 Load and RES generation forecasting

For load forecasting, conventional methods only work for active power and the reactive load is usually determined by a typical power factor. However, in practice, due to the increasingly complicated mix of load components and large share of power-electronic loads, the power factor is also time-varying and the reactive load does not always follow the active load profile [14]. For VVC purposes, more accurate reactive power load forecasting is necessary. In [14], a practical tool for reactive power forecasting at distribution level has been reported. The tool combines several high-performance machine learning techniques, and its accuracy has been verified on a load database of a distribution network in US considering different forecasting lead times.

For RES generation forecasting, a number of methods have been developed in the literature with satisfactory performance [15,16]. All of the methods show that shorter the lead time, the more accurate forecasting results can be obtained. These tools can be integrated within the proposed VVC framework for hourly and 15-min ahead load and RES generation forecasting.

12.3 Mathematical formulation

12.3.1 Distribution network power flow model

A typical n-node radial distribution network is shown in Figure 12.3, and its power flow relationship can be described by the Dist-Flow model [17]:

$$P_{i+1} = P_i - r_i \frac{P_i^2 + Q_i^2}{V_i^2} - p_{i+1} \tag{12.1}$$

$$Q_{i+1} = Q_i - x_i \frac{P_i^2 + Q_i^2}{V_i^2} - q_{i+1} \tag{12.2}$$

$$V_{i+1}^2 = V_i^2 - 2(r_i P_i + x_i Q_i) + \left(r_i^2 + x_i^2\right) \frac{P_i^2 + Q_i^2}{V_i^2} \tag{12.3}$$

$$p_{i+1} = p_{i+1}^l - p_{i+1}^g, q_{i+1} = q_{i+1}^l - q_{i+1}^g \tag{12.4}$$

where r_i and x_i denote the resistance and reactance of between the nodes i and $i+1$; p_i^l and p_i^g are active power load and RES generation at node i, respectively; q_i^l

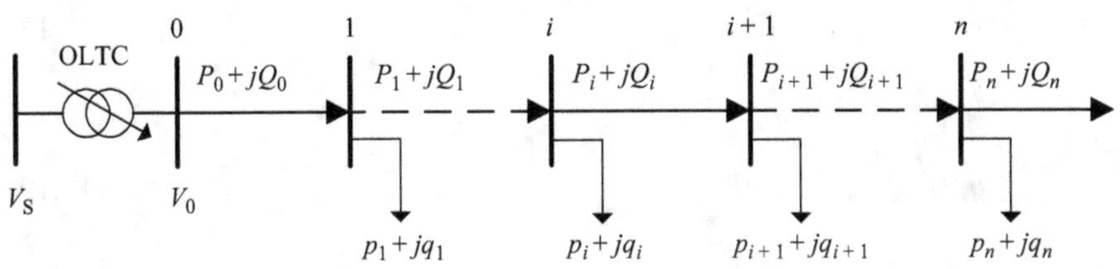

Figure 12.3 Schematic diagram of a radial distribution network

and q_i^g are reactive power load and reactive power injection by reactive power compensation devices at node i, respectively; P_i and Q_i are active and reactive power from node i to $i+1$, respectively; and V_i is the voltage magnitude of node i.

A meshed network can be converted to a radial network by breaking the loops through adding dummy buses [18]; therefore the Dist-Flow can still be applied. However, for the sake of developing efficient numeric programming algorithms, the Dist-Flow equations are usually linearised and the corresponding equations will be introduced in the sequel.

12.3.2 Optimisation model for volt/var control

The objective of the VVC is to minimise the total expected power loss while satisfying the voltage limits over different timescales in presence of stochastic RESs and loads. The mathematical model is formulated as follows:

$$\min \sum_t P_{\text{loss}}(t) \tag{12.5}$$

$$\text{s.t.} P_{i+1}(t) = P_i(t) - p_{i+1}(t) \tag{12.6}$$

$$Q_{i+1}(t) = Q_i(t) - q_{i+1}(t) \tag{12.7}$$

$$V_{i+1}(t) = V_i(t) - \frac{r_i P_i(t) + x_i Q_i(t)}{V_S} \tag{12.8}$$

$$p_{i+1}(t) = p_{i+1}^l(t) - p_{i+1}^g(t) \tag{12.9}$$

$$q_{i+1}(t) = q_{i+1}^l(t) - q_{i+1}^g(t) \tag{12.10}$$

$$V_0(t) = V_S + Tap(t) \times V^{\text{Tap}} \tag{12.11}$$

$$P_{\text{loss}}(t) = \sum_{i=0}^{n} r_i \frac{P_i^2(t) + Q_i^2(t)}{V_S^2} \tag{12.12}$$

$$q_i^g(t) = q_i^C(t) + q_i^{\text{Inv}}(t) \tag{12.13}$$

$$q_i^C(t) = I_{i,C}(t) \times Q_{i,C} \tag{12.14}$$

$$p_i^g(t) = \widehat{p}_i^g(t) + \Delta_i^g(t) \tag{12.15}$$

$$p_i^l(t) = \widehat{p}_i^l(t) + \Delta_i^l(t) \tag{12.16}$$

$$\Delta_i^g(t) \in D^g, \Delta_i^l(t) \in D^l \tag{12.17}$$

$$\left|q_i^{\text{Inv}}(t)\right| \le \sqrt{S_i^2 - \left[p_i^g(t)\right]^2}, \forall i \tag{12.18}$$

$$0 \le q_i^C(t) \le q_i^{C\max}, \forall i \tag{12.19}$$

$$Tap^{\min} \le Tap(t) \le Tap^{\max} \tag{12.20}$$

$$V_i^{\min} \le V_i(t) \le V_i^{\max}, \forall i \tag{12.21}$$

where the variables are defined as follows:

V_s primary voltage of the substation
$Tap(t)$ tap position variable of the OLTCs
V^{Tap} voltage adjustment size of one tap step
$q_i^C(t)$, $q_i^{\mathrm{Inv}}(t)$ reactive power output of the CB and RES inverter, respectively, of node i at time t
$l_{i,C}$ position level of CB at bus i
$Q_{i,C}$ unit size of a CB at bus i
$\widehat{p}_i^g(t)$, $\widehat{p}_i^l(t)$ forecasted values of RES generation and load demand at interval t, respectively
$\Delta_i^g(t)$, $\Delta_i^l(t)$ stochastic variations of RES generation and load demand, respectively
D^g, D^l probabilistic distributions for RES generation and load demand, respectively
$P_{\mathrm{loss}}(t)$ power loss during the tth short-time interval (15-min) in a hour
S_i apparent power size of the RES inverter at bus i.

Equations (12.6)–(12.12) are the linearised version of the Dist-Flow model, which are based on the fact that the non-linear terms in (12.1)–(12.4) are much smaller than the linear terms. Note that in (12.6)–(12.10), the power loss is temporarily neglected for solving the branch flow in order to retain the model linearity. After the branch flow is determined, the power loss is re-calculated in (12.12). The accuracy and effectiveness of the linearised Dist-Flow model have been demonstrated in the literature [6,12,17]; (12.13) means that the total reactive power injection to a bus is the sum from the CBs and the inverters; (12.14) implies that the reactive output from CBs is discrete depending on its tap position level; (12.15)–(12.17) denote the active power generation from the RES units and the load demands randomly vary from their prediction values; (12.18)–(12.20) are the operational limits on the VVC devices; (12.21) imposes the voltage limits for each bus.

Note that (12.18) means that the available reactive power support capacity of an RES inverter depends on its apparent power capacity and the current active power output. Due to the stochastic variable $p_i^g(t)$, this constraint is non-linear and makes the whole model more difficult to handle. In this chapter, (12.18) is approximated to $|q_i^{\mathrm{Inv}}(t)| \le \sqrt{S_i^2 - [\widehat{p}_i^g(t) \times \kappa]^2}$, where κ is a factor that reflects the average deviation of RES active power output from its forecasted value. For example, to be conservative, it can be set to 1.05 assuming that the average deviation is 5% higher than the forecasted value. This value can be set according to statistical studies of the load and RES generation forecasting results [14,16].

12.4 Two-stage stochastic programming model

12.4.1 Stochastic programming model

The VVC is realised at two different but coordinated timescales. The mathematical model of (12.5)–(12.21) is re-formulated as a two-stage stochastic programming model:

$$\min_{x \in F} \{f(x) + \mathrm{E}[Q(x, \xi)]\} \tag{12.22}$$

where $f(x)$ is the first-stage problem, i.e. the long-term (hourly timescale) VVC, and x is the first-stage decision vector; $Q(x, \xi)$is the optimal value of the second-stage problem, i.e. the short-term (15-min timescale) VVC, $\min_{y \in \Omega(x,\xi)} g(y)$, where y is the second-stage decision vector, ξ is the random vector, and $\mathrm{E}[Q(x, \xi)]$ is the expected value of the second-stage problem.

12.4.2 Deterministic equivalent

Assuming ξ has a finite number of possible realisations, which is called scenarios, denoted as $\xi_1, \ldots, \xi_K$ with respective possibilities of $\rho_1, \ldots, \rho_k$, then the expectation term in (12.22) can be written as

$$\mathrm{E}[Q(x, \xi)] = \sum_{k=1}^{K} \rho_k Q(x, \xi_k) \tag{12.23}$$

Then, the original two-stage stochastic programming model can be converted to be the following *deterministic equivalence*:

$$\min_{x, y_1, \ldots, y_K} f(x) + \sum_{k=1}^{K} \rho_k g(y_k) \tag{12.24}$$

$$\text{s.t.} x \in F \tag{12.25}$$

$$y_k \in \Omega(x, \xi_k), \forall k \tag{12.26}$$

It can be seen from the formulation that, at the first stage a '*here-and-now*' decision x is made before the realisation of the uncertain parameter ξ is known. Then, at the second stage, after a realisation of uncertain scenario ξ becomes available, a '*wait-and-see*' decision y is made to compensate for a possible inconsistency between the forecasting and the reality in the first stage, y is also called as *recourse action*.

Corresponding to the VVC problem, for the first stage, the objective function $f(x)$ corresponds to the power loss at the first 15-min interval of a hour, i.e. $t = 1$ for (12.5); the 'here-and-now' decision x corresponds to the OLTC position variables $Tap(t)$ and CB position variables $q_i^C(t)$ are optimised for the first interval and fixed in the whole hour, i.e. $t = 1, \ldots, 4$ for $Tap(t)$ and $q_i^C(t)$, as well as the inverter output $q_i^{\mathrm{Inv}}(t)$ for the first interval; constraint set F corresponds to (12.6)–(12.14) and (12.19)–(12.21) at the first 15-min interval, i.e. $t = 1$ for (12.6)–(12.14) and (12.19)–(12.21).

For the second-stage, the objective function $g(y_k)$ corresponds to the total power loss of the last three 15-min intervals, i.e. $t = 2, 3, 4$ for (12.5); the 'wait-and-see' decision y corresponds to the inverter outputs $q_i^{\mathrm{Inv}}(t)$ at the second-stage 15-min interval, i.e. $t = 2, 3, 4$ for $q_i^{\mathrm{Inv}}(t)$; the random variable vector ξ corresponds to the stochastic variation in RES generation and load demand; constraint set $\Omega(x,\xi)$ corresponds to (12.6)–(12.21) for the last three 15-min intervals, i.e. $t = 2$, 3, 4 for (12.6)–(12.21).

The deterministic equivalence (12.24)–(12.26) is an MIQP model which can be efficiently solved by off-the-shelf solvers such as GUROBI and CPLEX. Note that the quadratic terms can be further linearised leading to an MILP model and be solved even faster. Nowadays, such high-performance solvers are already fast enough for MIQP problems [19].

12.4.3 Scenario construction and reduction

The stochastic variations of RES power output and load from their forecasted values are assumed to follow the Beta distribution and the normal distribution, respectively [4,5,20].

The Beta distribution is defined by two shape parameters: α and β which represent the forecasting error (stochastic variation) for a forecasted power $\widehat{P}$ [20]:

$$f_{\widehat{P}}(y) = y^{\alpha-1} \cdot (1-y)^{\beta-1} \cdot N \tag{12.27}$$

where f is the Beta distribution function, y is the occurrence of the active power value, and N is the normalisation factor.

The relationship between the shape parameters with the mean value (which is the forecasted power) and the variance value of the Beta distribution function is as follows:

$$\text{mean} : P_{\text{norm}} = \frac{\widehat{P}}{P_{\text{inst}}} = \frac{\alpha}{\alpha+\beta} \tag{12.28}$$

$$\text{variance} : \sigma^2 = \frac{\alpha \cdot \beta}{(\alpha+\beta)^2 \cdot (\alpha+\beta+1)} \tag{12.29}$$

where P_{norm} denote the normalised power, P_{inst} is the installed capacity of the RES unit, and σ is the forecasting error variance.

The relationship between P_{norm} and σ is as follows [20]:

$$\sigma = 0.2 \times P_{\text{norm}} + 0.21 \tag{12.30}$$

Given the forecasted power, the shape parameters of the Beta distribution can be calculated using (12.28)–(12.30). Typical Beta distribution curves are illustrated in Figure 12.4 [20].

The computational burden of the model (12.24)–(12.26) depends on the number of scenarios. For practical use, it is usually needed to reduce the scenario numbers to save the computational time while maintaining necessary accuracy. In this chapter, an efficient algorithm based on backward method [21] is used to select a smaller yet representative number of scenarios to approximate the original whole scenarios.

Let $\xi_s (s = 1, \ldots, K)$ denote K initial scenarios, each with a probability ρ_s, and $d_{s,s'}$ denote the distance of a scenario pair (s, s'):

$$d_{s,s'} = \max\{1, \|s-\bar{s}\|, \|s'-\bar{s}\|\}\|s-s'\| \tag{12.31}$$

where $\bar{s}$ is the average value of scenarios.

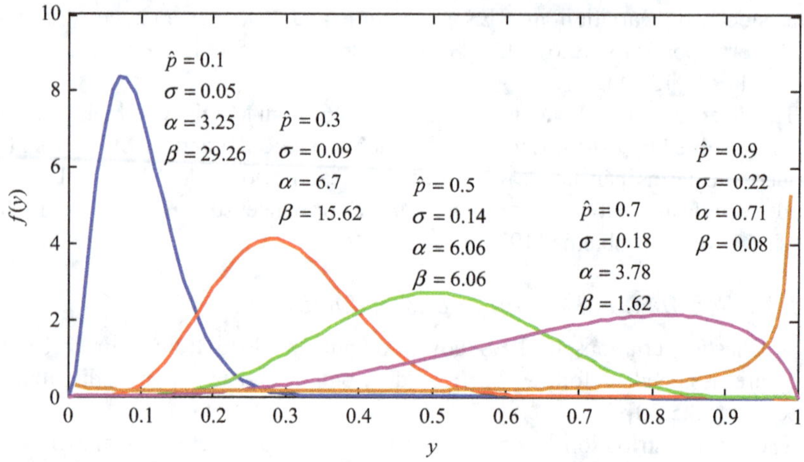

Figure 12.4 Typical Beta distribution functions

Let S be the initial set of scenarios, and J be the set of scenarios to be deleted (initially null). The simultaneous backward reduction consists of the following steps [21]:

Step 1. Compute the distances of all scenario pairs $d_{s,s'} = d(\xi_s, \xi_{s'}), (s, s' = 1, \ldots, K)$.

Step 2. For each scenario k, $dt_k(r) = \min d_{k,s'}$, $s', k \in S$ and $s' \neq k$, r is the scenario index that has the minimum distance with scenario k.

Step 3. Compute $\rho d_k(r) = \rho_k \times dt_k(r), k \in S$. Choose d so that $\rho d_d = \min \rho d_k, k \in S$.

Step 4. $S = S - \{d\}, J = J + \{d\}; \rho_r = \rho_r + \rho_d$.

Step 5. Repeat Steps 2–4 until the required number (R) to be deleted is met.

After scenario reduction, the number of scenarios is R and their probabilities are $\rho'_r, r = 1, \ldots, R$. The original model becomes

$$\min_{x,y_1,\ldots y_R} f(x) + \sum_{r=1}^{R} \rho'_r g(y_r) \tag{12.32}$$

$$\text{s.t.} x \in F; y_r \in \Omega(x, \xi_r), \forall r \tag{12.33}$$

as $R << K$, the above model has a much smaller dimension and therefore can be solved much faster.

12.4.4 Mapping relationship between VVC and programming model

The proposed VVC method is a general approach, which is not limited to a specific distribution network or utility code; rather, the problem is formulated as a

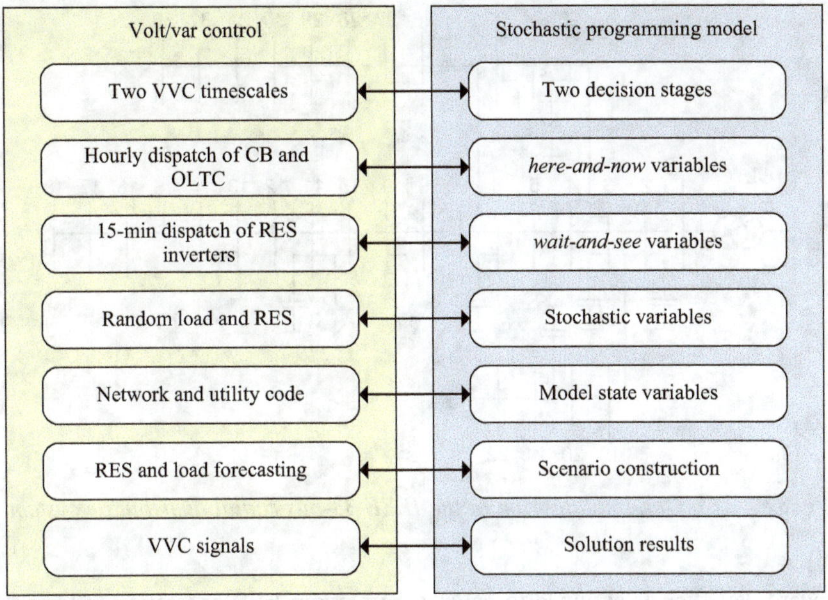

Figure 12.5 Mapping relationship between the VVC and programming models

two-stage stochastic programming model, where the network parameters, utility codes, and control timescales are all modelled as the state and decision variables in the programming model. The mapping relationship between the two is shown in Figure 12.5.

12.5 Simulation test results

12.5.1 Test system and parameter settings

The proposed VVC method is tested on the IEEE 33-bus radial distribution grid [17]. Its one-line diagram is shown in Figure 12.6.

In this test, the load demand of the test system is increased by 1.2 times, the substation voltage V_S is set to 1.0 p.u., and the voltage limit is considered to be $\pm 5\%$ of the nominal level, so $V^{\min}$ and $V^{\max}$ are 0.95 and 1.05 p.u., respectively; the substation transformer has 5% tap range and 10 tap positions, so $V^{\mathrm{Tap}} = 0.005$, $Tap = [-10,10]$; the CBs are installed at buses 2, 3, 6, 11, 21, and 23, each with the same capacity of 300 kVar; the RES units are assumed to be installed at buses 9, 11, 13, 21, 22, 25, 27, and 29, each with the capacity of 550 kVA. It should be indicated that the location and size of the reactive compensation devices have an impact on the performance of the VVC; however, optimal placement of them is beyond the scope of this chapter – some references can be found in [19,22,23].

For uncertainty modelling, it is assumed that the random load variation follows the normal distribution where the mean value is the forecasted load and the

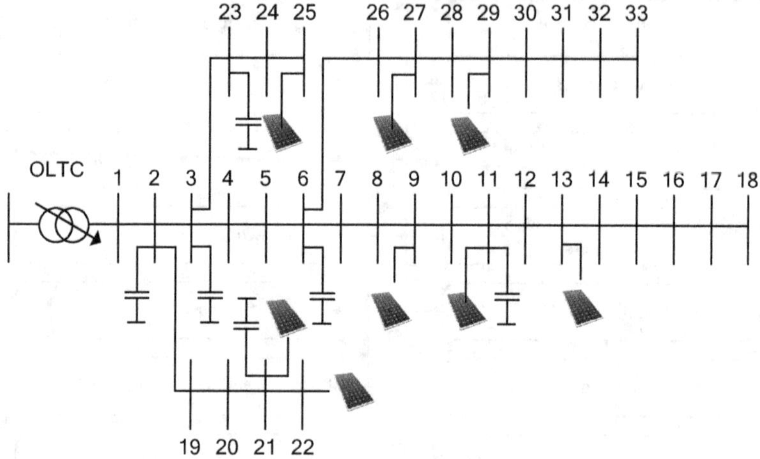

Figure 12.6 One-line diagram of the IEEE 33-bus radial distribution system

standard deviation is assumed to be 4%; the random RES generation follows the Beta distribution where the shaping parameters α and β are set to 6.06, and 6.06 with the forecasted power output 0.5 p.u.as shown in Figure 12.4 (the green profile). Note that other values can also be considered depending on the statistical performance of the forecasting tool. As discussed in Section 12.3.2, the reactive power support capacity of the RES inverters can be conservatively determined as $\pm\sqrt{S_i^2 - \left[\widehat{p}_i^g(t) \times \kappa\right]^2}$, and here κ is set to 1.03 considering an upper deviation of 3%.

The commercial mixed integer programming (MIP) solver GUROBI is employed to solve the mathematical model and all the simulations are conducted on an ordinary PC with 3.10 GHz CPU and 4.0 GB RAM.

12.5.2 Initial case

For the initial case (i.e. the initial single operating point), the voltage profile without VVC is shown in Figure 12.7, where it can be seen that the bus voltage magnitudes near the end of the feeder are close to the lower limit. The power loss of the network is 110.3 kW.

12.5.3 Deterministic VVC

For comparison, the conventional deterministic VVC is also tested by dispatching all the CBs and inverters simultaneously without considering the uncertainties of the RES and load. The resulting voltage profile is shown in Figure 12.7. It can be seen that without considering stochastic load and RES variation (see the profile of deterministic VVC, i.e. blue line), all the buses can satisfy the voltage limits. The power loss of the deterministic VVC is 43.8 kW. However, in practice, both the RES power output and the load demand will randomly vary from time to time, which may make the

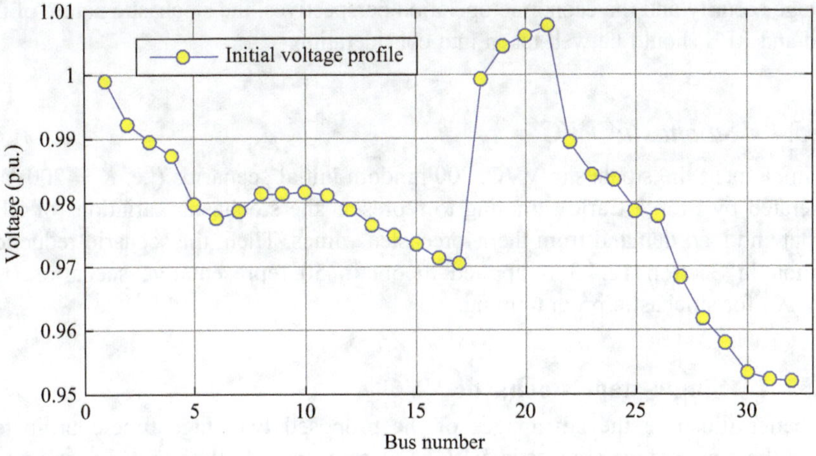

Figure 12.7 Initial voltage profile without VVC

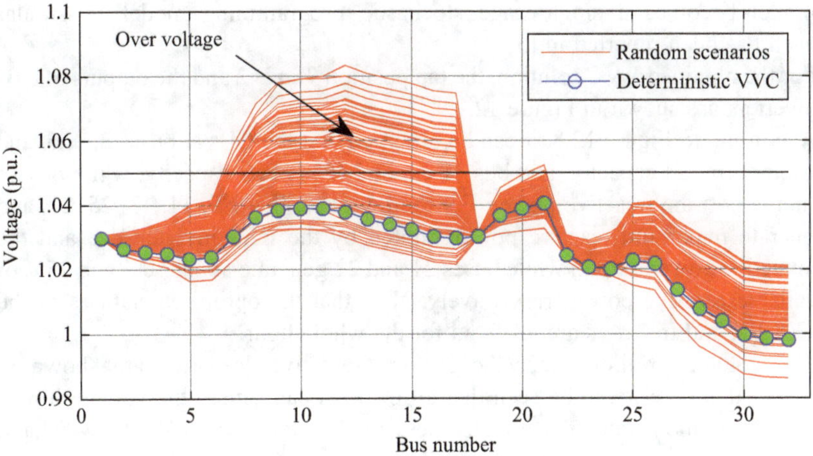

Figure 12.8 Voltage profiles under deterministic VVC solution and random scenarios

deterministic VVC solution fail to sustain a secure voltage profile. To test this, 100 random load and RES generation scenarios are generated and checked against the deterministic VVC solution. The 100 corresponding random voltage profiles are also shown in Figure 12.8, where it can be seen that about 50% scenarios violate the voltage upper limits due to the stochastic variation in RES power generation. Besides, for the 100 random scenarios, the average power loss becomes 62.7 kW, which is much larger than that obtained under the deterministic scenario. Therefore, from both voltage security and the economic operation perspectives, the stochastic nature of the load and RES should be well taken into consideration.

12.5.4 Stochastic VVC

To implement the stochastic VVC, 200 random initial scenarios (i.e. $K = 200$) are generated by Monte Carlo sampling to represent the stochastic variations of RES output and load demand from their forecasted values. Then, the scenario reduction method in Section 12.4.3 is applied to obtain 50 representative scenarios (i.e. $R = 30$) for stochastic programming.

12.5.4.1 Single-stage stochastic VVC

To better illustrate the advantages of the proposed two-stage time-coordinated VVC, the single-stage stochastic VVC is also compared. Note that this approach considers the stochastic variations of load demand and RES output during scheduling of VVC devices (CBs, OLTC, and inverters), but it coordinates them in a single timescale. To this end, the second-stage modelling parts in (12.5)–(12.26) are removed and the VVC devices are only optimised for the first time step. Then, the model becomes a single-stage stochastic programming model. A similar approach has been reported in [5].

For the obtained VVC solution, the tap position *Tap* is 2 and the outputs of CBs and inverters are shown in Figure 12.9.

According to Figure 12.8, it can be observed that the CBs at buses 2, 3, 6, and 23 export their full capacity, and bus 11 exports 40% of its capacity, while bus 21 does not export any reactive power. For the inverters, buses 9, 11, 13, 25, 27, and 29 generate maximum reactive power limited by the RES inverter size and the current active power output, while buses 21 and 22 generate about 50% and 25% of its available reactive power, respectively. Note that the optimised statuses of the OLTC, CBs, and inverters are all fixed for the whole hour.

The voltage profiles under the 30 representative scenarios are shown in Figure 12.10, where it can be seen that all of them can satisfy the voltage limits. This clearly demonstrates the effectiveness of the stochastic VVC to withstand random variations of load demand and RES generation. The average power loss under the random scenarios is 48.4 kW, which is higher than the deterministic VVC due to the consideration of a large number of stochastic scenarios.

However, although the single-stage stochastic VVC can sustain a secure voltage level under the considered random scenarios, it cannot track the fast intra-hour load and RES generation ramping since all the VVC devices are fixed for the whole hour. To examine this, a 20% RES generation increment is considered for the decision hour, and 1,000 random scenarios are sampled and checked for the solution of the single-stage stochastic approach. The voltage profiles are shown in Figure 12.11, where it can be seen that about 20% scenarios experience voltage violations. Under these 1,000 random RES incremental scenarios, the average power loss becomes 57.7 kW.

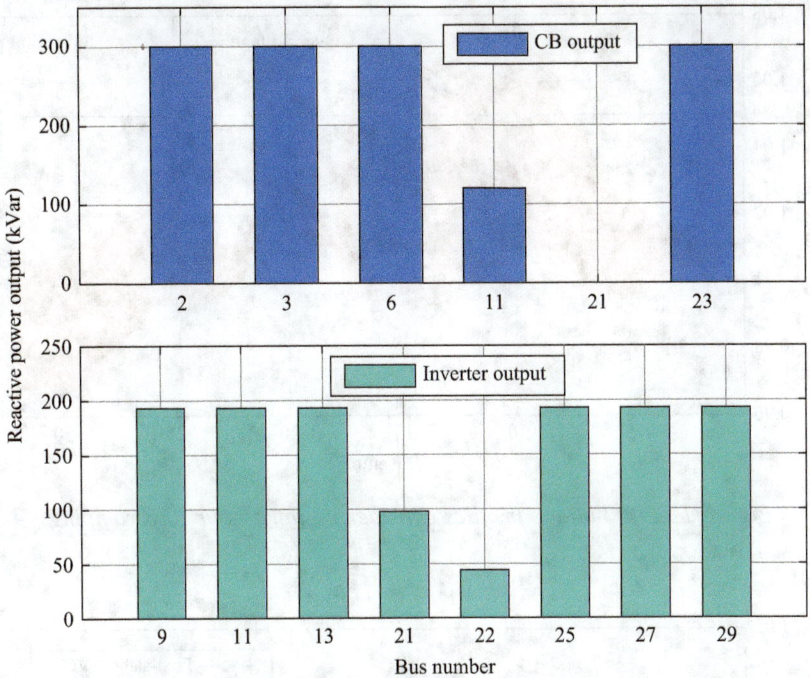

Figure 12.9 Reactive power outputs of CBs (upper window) and STACOMs (lower window) of the single-stage stochastic VVC solution

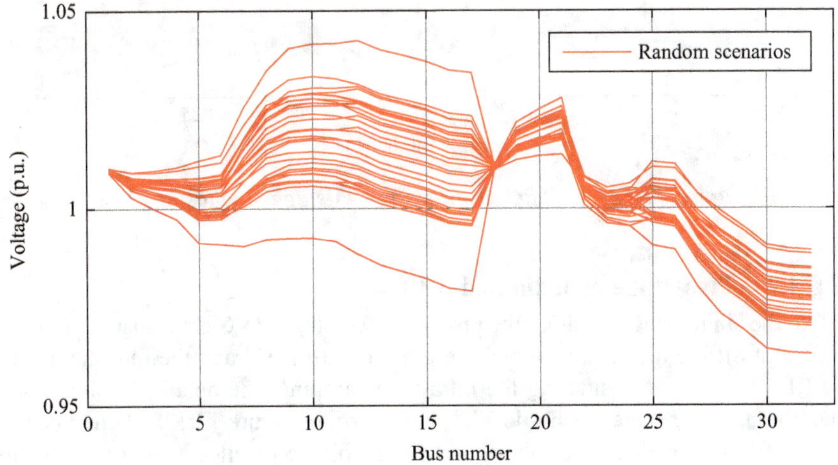

Figure 12.10 Voltage profiles from single-stage stochastic VVC

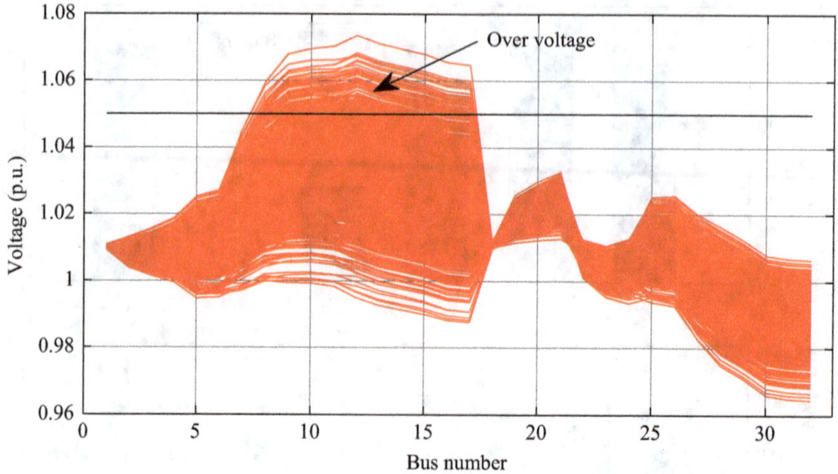

Figure 12.11 Voltage profiles considering intra-hour RES ramping

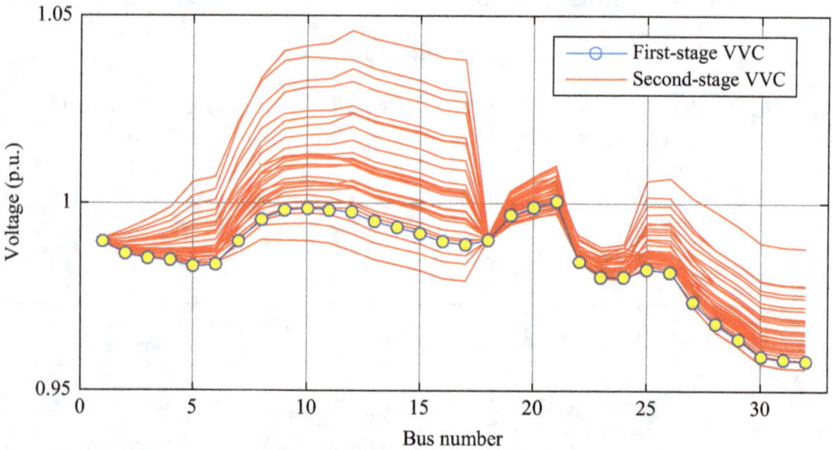

Figure 12.12 Voltage profiles for two-stage stochastic VVC

12.5.4.2 Two-stage coordinated VVC

Unlike the single-stage method, the proposed two-stage VVC coordinates the VVC devices at different timescales, and therefore can timely track the intra-hour load and RES ramping. Considering the intra-hour ramping, the obtained voltage profiles from the two-stage stochastic VVC are shown in Figure 12.12, where it can be seen that both the first- and the second-stage voltage profiles satisfy the voltage limits. The outputs of the CBs are shown in Figure 12.13 and the OLTC tap position is −2. Note that they are fixed in the whole hour.

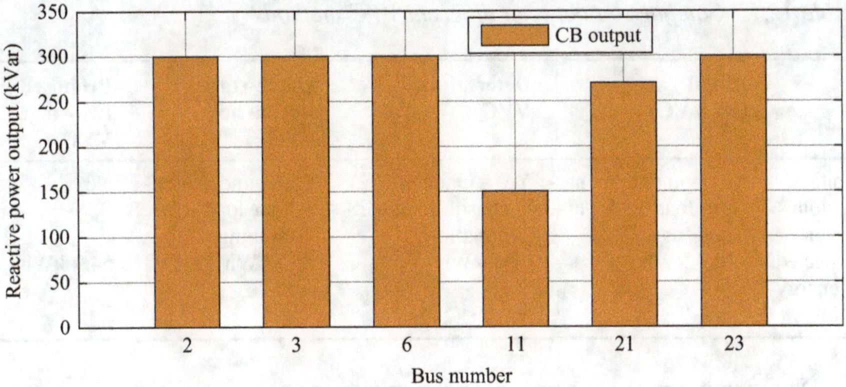

Figure 12.13 First-stage CB outputs (wait-and-see decisions)

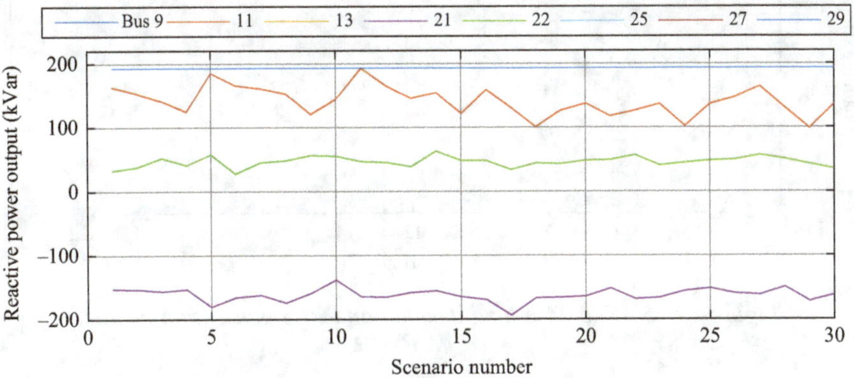

Figure 12.14 Second-stage inverter output (here-and-now decisions) for different scenarios

The reactive power output of the inverters for each scenario is shown in Figure 12.14, where it can be seen that the inverters' generation is dynamically adjusted to compensate for the first-stage decisions. In particular, it can be observed that the inverters follow the RES generation to inject and absorb reactive power during fast RES power variations. Therefore, the multi-timescale coordinated VVC approach can better manage the stochastic voltage fluctuations.

The power loss is 43.8 kW at the first stage and is 58.6 kW at the second stages, and the total expected power loss is 54.9 kW. The computation speed of the proposed VVC method is very fast, considering 30 scenarios, and the total solution time is only 1.6 s which is fully compatible with online implementation.

Table 12.1 compares different VVC methods in terms of voltage violation and expected energy loss in the hour. It can be seen that the proposed multi-timescale coordinated VVC outperforms all other methods in both terms.

Table 12.1 Comparison between different VVC methods

	Initial case (no VVC)	**Deterministic VVC**	**Single-stage stochastic VVC**	**Proposed two-stage VVC**
Voltage violation	Yes – under stochastic load/RES variations	Yes – under stochastic load/RES variations	Yes – under intra-hour load/RES ramping	No
Expected energy loss	110.3 kWh	62.7 kWh	57.7 kWh	54.9 kWh

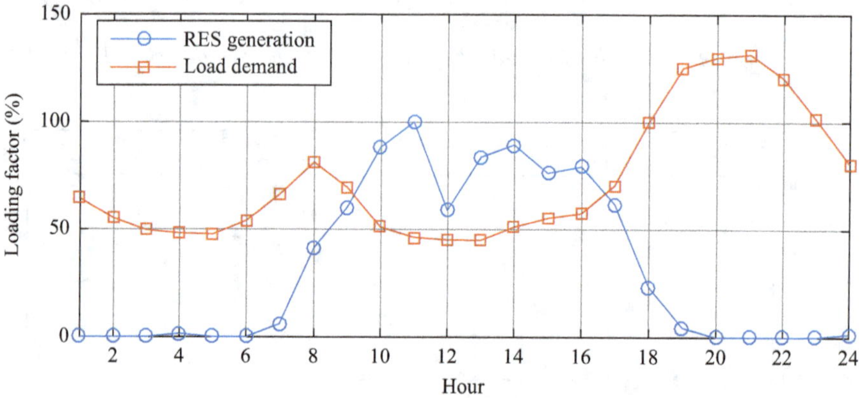

Figure 12.15 The 24-h load demand and RES generation profile

12.5.5 Time-series test

The above test is for a single operating hour of the distribution network, and the purpose is to straightforwardly show the effectiveness of the proposed method and its advantage over other methods. This section conducts a time-series test to better illustrate the time-varying coordination between the two timescales and the different VVC devices in the proposed method. To this end, a 24-h load demand and RES generation profile is used (see Figure 12.15). Note that the loading factor means the percentage value of the initial operating point.

Given the 24-h profile, the VVC solutions are obtained by the two-stage model. The obtained hourly results for power loss, OLTC tap position, and CB output are shown in Figure 12.16. Note that the OLTC tap and CB output are first-stage decision variables, and act on an hourly basis.

The inverter outputs are shown in Figure 12.17. Note that the inverter outputs are the second-stage decision variables, and are in the 15-min timescale. According to Figures 12.16 and 12.17, it can be easily seen that the two types of VVC devices coordinate at different timescales. The CB and OLTC act for the slow (hourly) load/RES generation change, and the inverters respond to the fast (15 min) load/RES generation

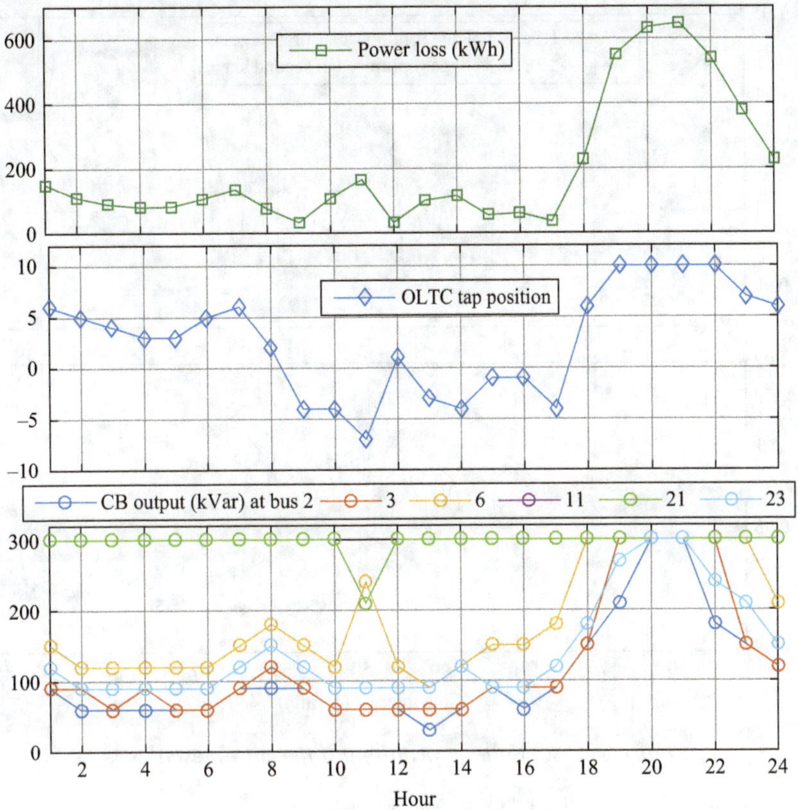

Figure 12.16 Hourly profile results

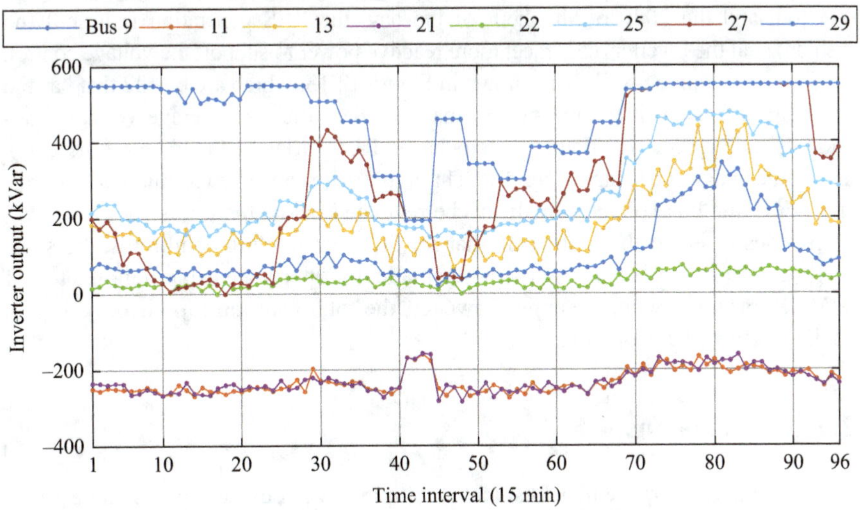

Figure 12.17 The 15-min profile for Inverter output

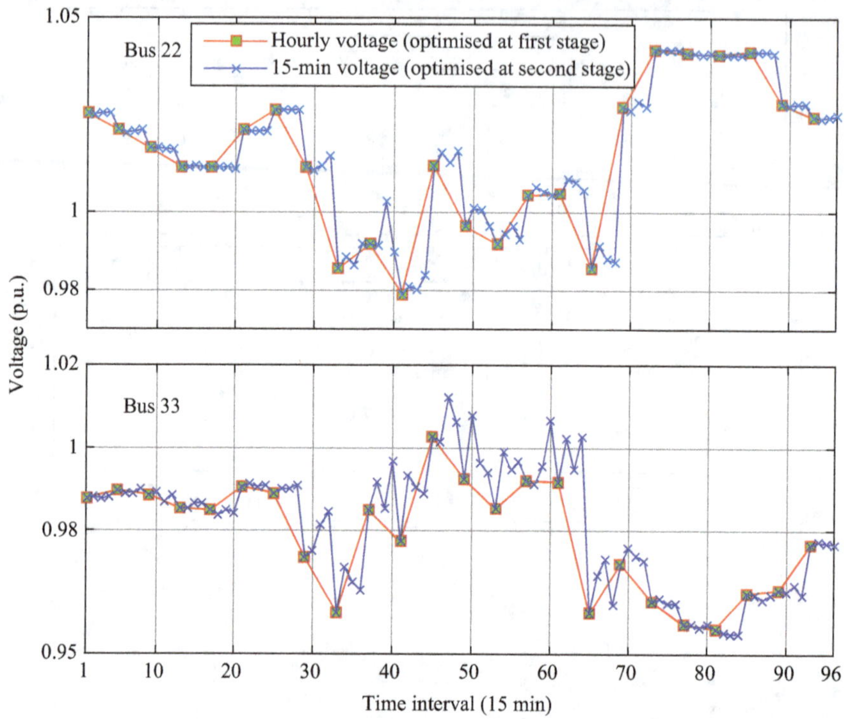

Figure 12.18 Voltage profile at different timescales

change. It is also worth noting that, with the RES generation change, the available reactive power capacity of the inverters will change accordingly; for example, during the 1st to 6th and the 20th to 24th hour of the day, the RES generation is near 0 (no sunshine), and the inverters can inject more reactive power to support the voltage control.

Two bus voltage profiles are shown in Figure 12.18, which are bus 22 that has the maximum average voltage and bus 33 that has the minimum average voltage magnitude. It can be seen that due to the stochastic RES variation (at 15 min timescale), the voltage magnitude varies rapidly. Through the proposed two-timescale coordinated VVC, the bus voltage can always be regulated below the limits.

Besides, it is worth mentioning that for different loading conditions, the solution time varies, that is because the mathematical property (feasible region, etc.) of the MIQP problem would change; however, the total solution time is only 43.6 s, which is very fast for practical use.

12.6 Conclusions

Unlike previous works which dispatch multiple VVC devices in the same timescale, this chapter coordinates them at different timescales for distribution network voltage regulation in the presence of stochastic RES outputs: in the first stage, CBs and OLTCs are scheduled before stochastic RES output and load variations are observed. In the second stage, RES inverters act to supplement the first-stage decision after the uncertainty is realised. The proposed VVC is modelled as a two-stage stochastic programming problem and converted to a deterministic MIQP model that can be efficiently solved by commercial solvers. Compared with conventional deterministic VVC methods, it can sustain a secure voltage level under stochastic load and RES variations; compared with the single-stage stochastic VVC approach, it fully utilises different response speeds of multiple VVC devices and dynamically track the voltage fluctuations and reduce energy loss.

References

[1] K. Turitsyn, P. Sulc, S. Backhaus, and M. Chertkov, 'Distributed control of reactive power flow in a radial distribution circuit with high photovoltaic penetration', *Proc. 2010 IEEE Power Energy Soc. Gen. Meet.*, pp. 1–6, 2010.

[2] M. Farivar, R. Neal, C. Clarke, and S. Low, 'Optimal inverter VAR control in distribution systems with high PV penetration', *Proc. IEEE PES General Meeting*, pp. 1–7, Jul. 2012.

[3] J.-Y. Park, S.-R. Nam, and J.-K. Park, 'Control of a ULTC considering the dispatch schedule of capacitors in a distribution system', *IEEE Trans. Power Syst.*, vol. 22, pp. 755–761, 2007.

[4] T. Niknam, M. Zare, and J. Aghaei, 'Scenario-based multiobjective volt/var control in distribution networks including renewable energy sources', *IEEE Trans. Power Syst.*, vol. 27, no. 4, pp. 2004–2019, 2012.

[5] Z. Wang, J. Wang, B. Chen, M. Begovic, and Y. He, 'MPC-based voltage/var optimization for distribution circuits with distributed generators and exponential load models', *IEEE Trans. Smart Grid*, vol. 5, no. 5, pp. 2412–2420, 2014.

[6] Z. Wang, H. Chen, J. Wang, and M. Begovic, 'Inverter-less hybrid voltage/var control for distribution circuits with photovoltaic generators', *IEEE Trans. Smart Grid*, vol. 5, no. 6, pp. 2718–2728, 2014.

[7] Z. Tang, D.J. Hill, and T. Liu, 'Coordinated voltage control of weak sub-transmission networks considering wind power variability', *Proc. 9th IFAC Sym., Control of Power and Energy Systems, India*, Dec. 2015.

[8] J. Smith, W. Sunderman, R. Dugan, and B. Seal, 'Smart inverter volt/var control functions for high penetration of PV on distribution systems', *Proc. IEEE Power Systems Conf. Exposition*, Phoenix, AZ, USA, pp. 1–6, Mar. 2011.

[9] 1547 Series of Interconnection Standards, IEEE SCC21 Standards Coordinating Committee on Fuel Cells, Photovoltaics, Dispersed Generation, and Energy Storage, IEEE Standards Association. [Online]. Available: http://grouper.ieee.org/groups/scc21/dr_shared/.

[10] P. Jahangiri and D.C. Aliprantis, 'Distributed volt/var control by PV inverters', *IEEE Trans. Power Syst.*, vol. 28, no. 3, pp. 1–11, 2013.

[11] H. Li, F. Li, Y. Xu, D. T. Rizy, and S. Adhikari, 'Autonomous and adaptive voltage control using multiple distributed energy resources', *IEEE Trans Power Syst.*, vol. 28, no.2, pp. 718–730, 2013.

[12] H.-G. Yeh, D. F. Gayme, and S. H. Low, 'Adaptive VAR control for distribution circuits with photovoltaic generators', *IEEE Trans Power Syst.*, vol. 27, pp. 1656–1663, 2012.

[13] B. Zhang, A. Lam, A. Domínguez-Garcíz, and D. Tse, 'An optimal and distributed method for voltage regulation in power distribution systems', *IEEE Trans. Power Syst.*, vol. 30, no. 4, pp. 1714–1726, 2015.

[14] Y. Xu, S. Liu, Z.Y. Dong, R. Zhang, M. Koenig, and A. Wigington, 'A hybrid model for forecasting time-varying reactive power load of power systems', *17th Int. Conf. on Intelligent System Applications to Power Systems (ISAP)*, Tokyo, Japan, Jul. 2013.

[15] H.-T. Yang, C.-M. Huang, Y.-C. Huang, and Y.-S. Pai, 'A weather-based hybrid method for 1-day ahead hourly forecasting of PV power output', *IEEE Trans. Sust. Energy*, vol. 5, no. 3, pp. 917–926, 2014.

[16] W. Liu and Y. Xu, 'Randomized learning-based hybrid ensemble model for probabilistic forecasting of PV power generation', *IET Gen. Trans. & Dist.*, vol. 14, no. 24, pp. 5909–5917, 2020.

[17] M.E. Baran and F.F. Wu, 'Network reconfiguration in distribution systems for loss reduction and load balancing', *IEEE Trans. Power Del.*, vol. 4, no. 2, pp. 1401–1407, 1989.

[18] M. Haque, 'Efficient load flow method for distribution systems with radial or mesh configuration', *IET Gen., Transm. Distrib.*, vol. 143, no. 1, pp. 33–38, 1996.

[19] Y. Xu, Z.Y. Dong, K.P. Wong, E. Liu, and B. Yue 'Optimal capacitor placement to distribution transformers for power loss reduction in radial distribution systems', *IEEE Trans. Power Syst.*, vol. 28, no. 4, pp. 4072–4079, 2013.

[20] A. Fabbri, T.G. San Román, J. Rivier Abbad, and V.H. Méndez Quezada, 'Assessment of the cost associated with wind generation prediction errors in a liberalized electricity market', *IEEE Trans. Power Syst.*, vol. 20, no. 3, pp. 1440–1446, 2005.

[21] L. Wu, M. Shahidehpour, and T. Li, 'Stochastic security-constrained unit commitment', *IEEE Trans. Power Syst.*, vol. 22, no. 2, pp. 800–811, 2007.

[22] H.N. Ng, M.M. A. Salama, and A.Y. Chikhani, 'Classification of capacitor allocation techniques', *IEEE Trans. Power Del.*, vol. 15, no. 1, pp. 387–392, 2000.

[23] Y. Xu, Z.Y. Dong, K. Meng, W.F. Yao, R. Zhang, and K.P. Wong, 'Multi-objective dynamic VAR planning against short-term voltage instability using a decomposition-based evolutionary algorithm', *IEEE Trans. Power Syst.*, vol. 29, no. 6, pp. 2813–2822, 2014.

Chapter 13

Three-stage robust inverter-based voltage/var control optimisation

This chapter introduces a novel three-stage robust inverter-based voltage/var control (TRI-VVC) optimisation approach for high photovoltaic (PV)-penetrated microgrids. The approach aims to coordinate three different control stages from central to local VVC to reduce network energy loss and mitigate voltage deviations. In the first stage, capacitor banks (CBs) and an on-load tap changer (OLTC) are scheduled hourly in a rolling horizon. In the second stage, PV inverters are dispatched in a short time window. In the third stage, the inverters respond to real-time voltage violations through local droop controllers. A new PV inverter model is developed to support both the central var dispatch and the local var droop control. To address uncertain PV output and load demand, a robust optimisation (RO) model is proposed to optimise the first two stages while taking into account the droop voltage control support from the third stage. A linearised distribution power flow model with branch power loss is developed and applied in the RO. The simulation results show high efficiency and robustness of the proposed TRI-VVC optimisation method.

Nomenclature

A Acronyms

CB	Capacitor bank
C&CG	Column and constraint generation
OA	Outer approximation
OLTC	On-load tap changer
PV	Photovoltaic
RO	Robust optimisation
SO	Stochastic optimisation
SSC-VVC	Single-stage centralised voltage/var control
TRI-VVC	Three-stage robust inverter-based voltage/var control
TSRO	Two-stage robust optimisation
VVC	Voltage/var control

B Sets

N	Set of all network buses
M	Set of all CB units at a bus
T	Set of all periods (1 h/15 min) in the rolling horizon/the current hour
U_{PV}, U_{PL}, U_{QL}	Uncertainty set for PV power generation, active and reactive power loads

C Parameters

$CB_{i,j}^{max}$	Allowed maximal switch changing times for the *j*th unit of the CB at bus *i*
n_{bus}	Number of the buses except the substation
$OLTC_{(t)}^{max}$	Allowed maximal OLTC position changing times during the whole horizon/period *t*
P_i^{cap}	Active power capacity of the branch from bus *i*
$P_{i,t}^{PV,pr}$, $P_{i,t}^{L,pr}$, $Q_{i,t}^{L,pr}$	Predicted PV power generation/active power load/reactive power load at bus *i* during period *t* (W/W/Var)
$\underline{P}_{i,t}^{PV}$, $\overline{P}_{i,t}^{PV}$, $\underline{P}_{i,t}^{L}$, $\overline{P}_{i,t}^{L}$, $\underline{Q}_{i,t}^{L}$, $\overline{Q}_{i,t}^{L}$	Predicted lower/upper bound of uncertain PV power generation/active power load/reactive power load at bus *i* during period *t* (W/W/Var)
Q_{unit}^{CB}	Reactive power capacity of one CB unit (Var)
$Q_{i,t}^{disp}$	Reactive power capacity which the inverter can dispatch in the loss reduction operation at bus *i* during period *t* (Var)
$Q_{i,t}^{max}$	Maximal reactive power which the inverter can generate/consume at bus *i* during period *t* (Var)
$Q_{i,t}^{res}$	Reactive power reserve for the third-stage droop control at bus *i* during period *t* (Var)
R_i, X_i	Resistance/reactance of the branch from bus *i*
S_i^{max}	Maximal apparent power which the inverter can generate/consume at bus *i* (VA)
V_0	Reference voltage at the substation (p.u.)
$\underline{V}_i$, $\overline{V}_i$	Lower/upper bound of the allowed voltage range at bus *i* (p.u.)
$\underline{V}^{cri}$, $\overline{V}^{cri}$	Lower/upper bound of the critical bus voltage range used for the reactive power reserve design (p.u.)
V^{Tap}	Voltage difference between two adjacent positions of the OLTC (p.u.)
β^{min}, β^{max}	Minimal and maximal positions of the OLTC
$\underline{\mu}^{PV/PL/QL}$, $\overline{\mu}^{PV/PL/QL}$	Lower/upper uncertainty budget of PV power generation/active power load/reactive power load
τ, π	Time length of the first two stages (h)

D Variables

$P_{i,t}$, $Q_{i,t}$	Active/reactive power through the main branch from bus i during period t (W/Var)
$P^{\mathrm{L}}_{i,t}$, $Q^{\mathrm{L}}_{i,t}$	Uncertain active/reactive power load at bus i during period t (W/Var)
$P^{\mathrm{lat}}_{i,t}$, $Q^{\mathrm{lat}}_{i,t}$	Active/reactive power through the lateral branch from bus i during period t (W/Var)
$P^{\mathrm{loss}}_{i,t}$, $Q^{\mathrm{loss}}_{i,t}$	Active/reactive power loss through the branch from bus i during period t (W/Var)
$P^{\mathrm{PV}}_{i,t}$	Uncertain PV power generation at bus i during period t (W)
$Q^{\mathrm{CB}}_{i,t}$	Reactive power output of the CB at bus i during period t (Var)
$Q^{\mathrm{lr}}_{i,t}$	Reactive power generated/consumed by the inverter in the loss reduction operation at bus i during period t (Var)
$Q^{\mathrm{inv}}_{i,rt}$	Real-time reactive power generated/consumed by the inverter at bus i in the third stage (Var)
$V_{i,t}$	Voltage of bus i during period t (p.u.)
$V_{i,rt}$	Real-time voltage of bus i in the third-stage droop control (p.u.)
$\alpha_{i,j,t}$	Binary on/off decision of the jth unit of the CB at bus i during period t
β_t	Integer position of the OLTC during period t

13.1 Introduction

PV power generation has been increasingly deployed in today's microgrids. While high-level PV penetration can alleviate fast depletion of fossil fuels and reduce carbon emissions from conventional power plants, it has created significant technical challenges to operation and control of the microgrids. In particular, PV power generation is highly variable and stochastic depending on solar radiance; for example, output power of a PV system can ramp up and down on the order of 15% of its capacity per minute with intermittent cloud coverage [1], which causes fast and uncertain voltage deviation and fluctuation issues.

For voltage/var control (VVC), traditional devices such as CBs and OLTCs are mechanical equipment that can only act slowly (10 s to minutes); their lifetime will reduce rapidly with excessive operation. Thus, they are becoming less effective to cope with the fast and uncertain voltage fluctuations caused by high PV penetration in the microgrids. On the other hand, PV inverters are power-electronics equipment. They have much higher response speeds and flexible reactive power injection ranges, especially when PV active power output is low to allow more space for reactive power compensation. In Ref. [2], inverter-based VVC has been studied and suggested due to its high efficiency, flexibility, scalability, and reliability. Therefore, PV inverters have been considered as a promising VVC supplementary device to address the voltage deviation and fluctuation issues.

In general, VVC methods can be divided into local and centralised methods. For local methods, the fast-responding VVC devices, e.g. PV inverters, respond to local measurements to control local bus voltage at an acceptable level. Reference [2] proposes three local VVC schemes, which are, respectively, based on bus voltage, power flow, and both. Reference [3] introduces several straightforward local rules for inverters to generate or consume reactive power. In [4], a piecewise droop controller is designed for the inverters to counteract bus voltage rises. In [5], an adaptive proportional-integral controller is designed for multiple inverters to achieve a plug-and-play VVC function. Considering limitation on regional communication links, Ref. [6] proposes a two-stage distributed control method where each inverter acts locally and then adjusts its output based on local information exchange among neighbouring inverters. These local control schemes have high response speeds and do not require full communication links. However, given their local control nature, they prohibit system-wide coordination and optimisation, thus not guaranteeing optimal control.

By contrast, the centralised schemes aim to coordinate multiple VVC devices based on system-wide information (e.g. network parameters as well as PV power generation and load predictions) to achieve optimal VVC objectives [7]. The centralised methods generally optimise tap positions of the OLTCs, as well as reactive power set points of the CBs and the PV inverters. In [8], a multi-objective programming model is proposed for voltage regulation and network power loss minimisation. Depending on voltage deviations, an adaptation law is used to switch between the two optimisation objectives. In [9], considering limited communication infrastructure, a distributed control method is proposed to coordinate CBs/OLTCs and inverters/demand response in two stages, where the first group of devices act for slow voltage variations and the second group for faster voltage variations. However, in the above works, the stochastic nature of PV power generation and loads is not addressed in the optimisation, such that the VVC decisions may not be robust under uncertainties. To deal with the uncertainties, Ref. [10] proposes a scenario-based modelling method to optimise day-ahead VVC decisions through coordination of CBs and an OLTC for four objectives. However, the day-ahead VVC decisions may not be able to robustly accommodate intraday stochastic variations of the PV power generation. In [11], a model predictive control-based method is proposed to coordinate CBs and an OLTC for VVC considering PV power generation prediction errors by a stochastic mixed-integer non-linear programming model. However, the stochastic programming model needs a large number of scenarios and a large-scale mixed-integer non-linear programming problem cannot be efficiently solved for online use. Reference [12] proposes a hybrid VVC architecture against transient cloud movements. The centralised loop hourly dispatches CBs and an OLTC. When a large variation of PV output power occurs, the local control loop is triggered and the CBs and static VAR compensators become self-controlled with local measurements. However, the centralised loop does not consider the uncertainties of PV power generation and loads, and there is a lack of optimal coordination between these two loops. Importantly, the above works [11–13] do not consider optimisation of inverter reactive power for VVC support, and rely on the conventional VVC devices, i.e. CBs and OLTCs.

Based on the above literature survey, the following unsolved issues are noted: (1) There is a lack of systematic coordination between centralised and local VVC strategies, and coordination among different control timescales, especially for real-time control where intermittent cloud coverage may significantly affect bus voltages. (2) For centralised VVC optimisation methods, there is a lack of coordination between different VVC devices to address the uncertainties of PV power generation and loads. (3) Most of the existing optimisation methods address the PV output power uncertainty based on stochastic programming which relies on accurate knowledge of probability distribution of the PV output power, and the stochastic programming usually has heavy computational burdens.

This chapter proposes a TRI-VVC optimisation method to address fast and uncertain voltage deviation and fluctuation issues caused by high PV penetration. Compared to the existing works, the proposed method aims to systematically coordinate different timescales, devices, and VVC priorities for more effective voltage regulation and network energy loss reduction: the first stage schedules slow-acting devices, CBs and an OLTC, in a relatively long period (e.g. 1 h), and the objective is to minimise network energy losses and track regular bus voltage deviation patterns; the second stage dispatches fast-acting devices, PV inverters in a short period (e.g. 15 min), and the objective is to compensate the first-stage decisions in case of significantly uncertain PV power generation and load variations; in the third stage, the PV inverters locally respond to sudden voltage fluctuations in real time. The first two stages correspond to the centralised VVC, and the priority is to minimise the network energy losses while satisfying bus voltage constraints. To address the uncertain PV power generation and loads, they are coordinated through a rolling horizon-based robust optimisation (RO) method, where the worst case of uncertainty realisation is considered to ensure solution robustness against any uncertainty realisation. With a reserved reactive capacity from the first two stages, the third stage corresponds to a local droop VVC, and the priority is to alleviate the bus voltage fluctuations in case of sudden and unexpected PV output power fluctuations, such as those during transient cloud movements. The TRI-VVC optimisation method is tested and compared with the existing methods. Simulation results have verified the high optimisation efficiency and solution robustness of the proposed TRI-VVC method.

13.2 Three-stage robust inverter-based voltage/var control

This chapter proposes a multi-timescale multi-device coordinated VVC framework covering centralised and local control methods. The TRI-VVC method aims to robustly minimise network energy losses while maintaining secure bus voltages with fast and uncertain deviations and fluctuations of PV power generation and loads. Its principle is to address uncertain bus voltage deviations and fluctuations at different timescales by different VVC devices. The TRI-VVC framework is illustrated in Figure 13.1.

13.2.1 First stage – capacitor bank and on-load tap changer (OLTC) scheduling in a rolling horizon

To accurately reflect the stochastic nature of PV power generation and loads, an interval prediction technique is applied to provide a mean value (i.e. expected

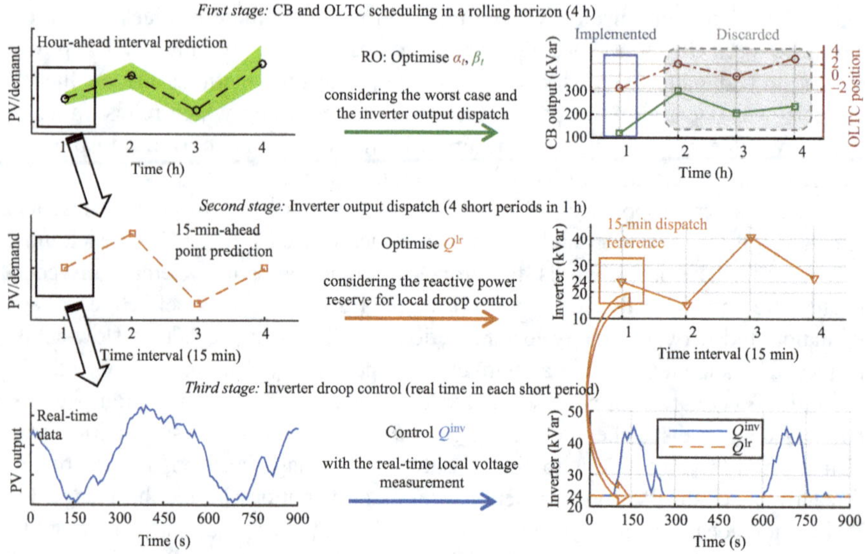

Figure 13.1 Three-stage robust inverter-based voltage/var control strategy

value) and a variation range (i.e. upper and lower bounds) of the PV output power and loads, where they realise randomly between the bounds. The related interval prediction technique can be found in [14].

The first stage aims to optimally schedule CBs and an OLTC in a microgrid. In this stage, the PV power generation and load intervals are predicted for a finite prediction horizon T, e.g. 4 h in this chapter. The hourly CB outputs and OLTC position are optimised for the whole horizon to minimise the total network energy loss while satisfying the bus voltage constraints. But only the decision of the first hour is implemented and the remaining is discarded. This aims to address coupling constraints (e.g. the maximal switching operation times of the CBs and the OLTC over a certain period) of the whole optimisation horizon and reserve sufficient capacity for the following hours. Thus, in the following hours, the CBs and the OLTC can operate effectively with the reserved capacity to handle the future uncertainties. The optimisation procedure is rolled forward to benefit from more accurate forecasting of PV output power and load in coming future with shorter leading time of prediction. In other words, in the following horizon, updated and more accurate interval predictions for the following horizon are available and applied. With the more accurate predictions, the scheduling of the CBs and the OLTC can be optimised more efficiently, since impacts of prediction errors are reduced.

The first stage is expected to be optimised considering uncertainty realisation of PV power generation and loads as well as compensating capability of PV inverters in the second stage. As would be detailed in Section 13.4, the first two stages are coordinated through a two-stage RO (TSRO) method which optimises operation decisions based on the worst case of uncertainty realisation. In the first-stage RO model, the CB and OLTC states as well as the PV inverter output dispatch are optimised as control variables. Note that the inverter dispatch is optimised as a recourse operation under the worst case. Hence, only the CB and OLTC scheduling decisions for the current hour are implemented in the first stage, and the inverter dispatch is re-optimised again in the second stage according to uncertainty realisation.

13.2.2 Second stage – inverter output dispatch

The first-stage decisions are optimised considering the worst case of PV output power and load realisation. The purpose is to achieve full solution robustness against any uncertainty realisation in the second stage. In the second stage, the PV inverters are dispatched to further reduce the total network energy loss in a shorter period p, e.g. 15 min, as a recourse operation for the first-stage decisions after the uncertainties are realised.

In this stage, more accurate, 15-min-ahead, point predictions of PV power generation and loads are used, and with them, the inverter reactive power outputs are optimised and implemented for each 15-min period within the current hour. The optimised inverter outputs are also set as reference set points for the third stage.

13.2.3 Third stage – inverter droop control

The first two stages minimise the total network energy loss while satisfying the bus voltage constraints under the uncertainties in the 1-h to 15-min periods. However, within each 15-min period, the PV power generation can still dramatically vary under special conditions, e.g. transient cloud movements, where the bus voltage limits may be violated. Thus, the third stage provides real-time reactive power compensation to alleviate the bus voltage violations if occurs.

In this stage, a droop controller is designed: when a real-time local bus voltage is out of the allowed operating range due to the significant PV output power fluctuation, the PV inverters generate or consume reactive power linearly with the bus voltage fluctuation. On the other hand, if the bus voltage is still in the allowed operating range, the inverter reactive power outputs are kept to the values optimised in the second stage, i.e. the reference set points. Note that in such a real-time period, e.g. 1 s in this chapter, the primary VVC objective is to mitigate the bus voltage fluctuations; thus, the associated network power losses are not considered.

13.3 Mathematical formulation

13.3.1 Photovoltaic (PV) inverter operation model

In the proposed TRI-VVC framework, the PV inverters have two operation stages: in the second stage, they are dispatched for the network energy loss reduction; in the third stage, they provide local reactive power compensation against the real-time bus voltage fluctuations.

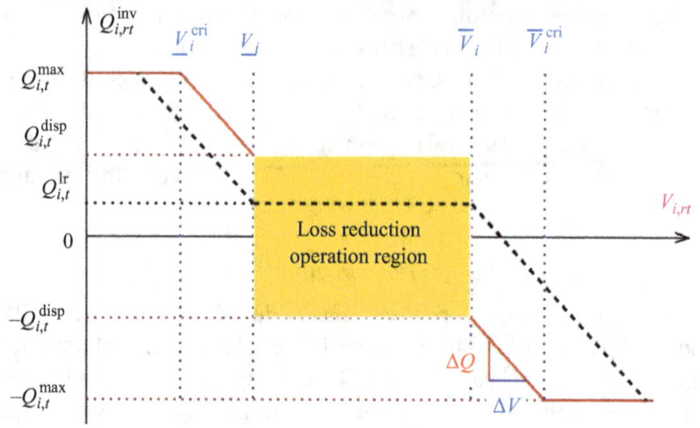

Figure 13.2 Piecewise inverter operation and reserved reactive power design

A piecewise droop control function for the PV inverters is employed in the third stage. As shown in Figure 13.2, the reactive power capacity of an inverter is $[-Q_{i,t}^{\max}, Q_{i,t}^{\max}]$, depending on the inverter apparent power capacity and the active power generation. In the first two stages, the upper bound of the predicted PV active power output interval is considered to avoid overcurrent inside the inverter, so the inverter maximum reactive power capacity can be calculated as

$$Q_{i,t}^{\max} = \sqrt{(S_i^{\max})^2 - \left(\overline{P}_{i,t}^{\text{PV}}\right)^2}, \forall i, t \tag{13.1}$$

It is worth noting that in the third stage, the real-time PV active power output is used instead of the upper bound. Thus, the inverter reactive power capacity varies with the real-time PV active power output in the third stage.

A dispatchable range $[-Q_{i,t}^{\text{disp}}, Q_{i,t}^{\text{disp}}]$ is set for the first two stages. Within this range, the PV inverter reactive power output $Q_{i,t}^{\text{lr}}$ is optimised to minimise network energy losses while keeping the local bus voltage between limits $\underline{V}_i$ and $\overline{V}_i$. During the real-time stage, if the bus voltage falls in $[\underline{V}_i, \overline{V}_i]$, the inverter output is kept as $Q_{i,t}^{\text{lr}}$. If the bus voltage is out of this range, the inverter droop control is enabled: the inverter controls its reactive power output proportional to the real-time bus voltage fluctuation from the limit, until the output reaches the reactive power capacity. Thus, the inverter reactive power output is controlled as follows:

$$Q_{i,rt}^{\text{inv}} = \begin{cases} Q_{i,t}^{\text{lr}} + \omega L_{i,t}\left(\underline{V}_i - V_{i,rt}\right), & V_{i,rt} < \underline{V}_i \\ Q_{i,t}^{\text{lr}}, \quad \underline{V}_i \le V_{i,rt} \le \overline{V}_i \\ Q_{i,t}^{\text{lr}} - \omega L_{i,t}\left(V_{i,rt} - \overline{V}_i\right), & V_{i,rt} > \overline{V}_i \end{cases}, \forall i, t$$

$$-Q_{i,t}^{\max} \le Q_{i,rt}^{\text{inv}} \le Q_{i,t}^{\max} \tag{13.2}$$

Here, ω is the control gain and it can be set as $1/n_{\text{bus}}$ by default, and $L_{i,t}$ which stands for the sensitivity of the bus voltage to the bus reactive power injection is calculated as $\partial Q_{i,t}/\partial V_{i,t}$.

It should be noted that, for one inverter, its droop control capacity in the third stage is the total capacity subtracting the reactive power dispatched in the second stage. Thus, to reserve sufficient reactive power capacity for the third-stage droop control, the inverter reactive power output should be only dispatched within a certain range in the second stage. Given the droop slope $\omega L_{i,t}$, for a bus voltage increment (ΔV), the required local increment of the inverter reactive power output (ΔQ) in the droop control can be calculated as $\Delta Q = -\omega L_{i,t}\Delta V$. For determining the reserved capacity for the droop control, two critical bus voltage values $\overline{V}^{\text{cri}}$, $\underline{V}^{\text{cri}}$ are designed first. These two values can form two linear droop control ranges $[\underline{V}^{\text{cri}}, \underline{V}_i]$ and $[\overline{V}^{\text{cri}}, \overline{V}_i]$. Note that these two ranges are designed as minimal droop control ranges (the red lines with the droop slope in Figure 13.2) before the inverter reactive power dispatch optimisation. After the inverter reactive power output ($Q^{\text{lr}}_{i,t}$) is optimised in the second stage, two actual real-time droop control ranges (the black dashed lines with the droop slope in Figure 13.2) are set in the third stage. Based on the minimal droop control ranges, the inverter reactive power reserve for the third-stage droop control can be calculated as

$$Q^{\text{res}}_{0,i,t} = \omega L_{i,t}\left(\max\{\underline{V}_i - \underline{V}^{\text{cri}}, \overline{V}^{\text{cri}} - \overline{V}_i\}\right), \forall i, t \tag{13.3}$$

Finally, considering the maximal reactive power capacity $Q^{\max}_{i,t}$, the inverter reactive power reserve $Q^{res}_{i,t}$ is modified as

$$Q^{res}_{i,t} = \begin{cases} Q^{\text{res}}_{0,i,t}, & Q^{\text{res}}_{0,i,t} \le Q^{\max}_{i,t} \\ Q^{\max}_{i,t}, & Q^{\text{res}}_{0,i,t} > Q^{\max}_{i,t} \end{cases}, \forall i, t \tag{13.4}$$

These minimal droop control ranges (the red lines with the droop slope in Figure 13.2) mean the worst droop control capacity. As the critical bus voltage range $[\underline{V}^{\text{cri}}, \overline{V}^{\text{cri}}]$ enlarges, i.e. $\underline{V}^{\text{cri}}$ decreases and $\overline{V}^{\text{cri}}$ increases, more reactive power is reserved for the third stage and the droop control capacity is increased. However, by doing so, the dispatchable reactive power for the second stage is less and the network energy loss reduction objective may be impacted negatively. After the reactive power reserve is set, the inverter dispatch capacity $Q^{\text{disp}}_{i,t}$ can be obtained as limits for the inverter reactive power output $Q^{\text{lr}}_{i,t}$ in the second stage, i.e. $-Q^{\text{disp}}_{i,t} \le Q^{\text{lr}}_{i,t} \le Q^{\text{disp}}_{i,t}$.

13.3.2 Coordination optimisation model

With the developed inverter operation model, the proposed TRI-VVC method is formulated as the following optimisation model. This model aims to optimise the first two centralised control stages with full consideration of the third local control stage.

$$\min \sum_{t\in T}\sum_{i\in N} P^{\text{loss}}_{i,t}\tau \tag{13.5}$$

$$\text{s.t.} \quad \alpha_{i,j,t} \in \{0, 1\}, \forall i, j, t \tag{13.6}$$

$$\beta_t \in \{\beta^{\min}, \beta^{\min} + 1, \ldots, -1, 0, 1, \ldots, \beta^{\max} - 1, \beta^{\max}\}, \forall t \tag{13.7}$$

$$Q_{i,t}^{\text{CB}} = Q_{\text{unit}}^{\text{CB}} \sum_{j \in M} \alpha_{i,j,t}, \forall i, t \tag{13.8}$$

$$V_{1,t} = V_0 + \beta_t V^{\text{Tap}}, \forall t \tag{13.9}$$

$$\sum_{t \in T} \left| \alpha_{i,j,t} - \alpha_{i,j,t-1} \right| \leq CB_{i,j}^{\max}, \forall i,j \tag{13.10}$$

$$\sum_{t \in T} |\beta_t - \beta_{t-1}| \leq OLTC^{\max} \tag{13.11}$$

$$|\beta_t - \beta_{t-1}| \leq OLTC_t^{\max}, \forall t \tag{13.12}$$

$$Q_{i,t}^{\text{disp}} = Q_{i,t}^{\max} - Q_{i,t}^{\text{res}}, \forall i, t \tag{13.13}$$

$$-Q_{i,t}^{\text{disp}} \leq Q_{i,t}^{\text{lr}} \leq Q_{i,t}^{\text{disp}}, \forall i, t \tag{13.14}$$

$$P_{i+1} = P_i - P_i^{\text{loss}} + P_{i+1}^{\text{G(PV)}} - P_{i+1}^{\text{L}} - P_{i+1}^{\text{lat}}, \forall i \tag{13.15}$$

$$Q_{i+1} = Q_i - Q_i^{\text{loss}} + Q_{i+1}^{\text{G(CB/lr)}} - Q_{i+1}^{\text{L}} - Q_{i+1}^{\text{lat}}, \forall i \tag{13.16}$$

$$V_{i+1} = V_i - \frac{R_i P_i + X_i Q_i}{V_0}, \forall i \tag{13.17}$$

$$P_i^{\text{loss}} = \sum_{k \in Ki} a_{i,k} \left(P_{i,k} - P_{i,k}^* \right) + \sum_{l \in Li} b_{i,l} \left(Q_{i,l} - Q_{i,l}^* \right) \tag{13.18}$$

$$Q_i^{\text{loss}} = \sum_{k \in Ki} c_{i,k} \left(P_{i,k} - P_{i,k}^* \right) + \sum_{l \in Li} d_{i,l} \left(Q_{i,l} - Q_{i,l}^* \right) \tag{13.19}$$

$$P_i = \sum_{k \in Ki} \left(P_{i,k} + P_{i,k}^* \right), Q_i = \sum_{l \in Li} \left(Q_{i,l} + Q_{i,l}^* \right), \forall i \tag{13.20}$$

$$0 \leq P_{i,k} \leq P_i^{(k)} - P_i^{(k-1)}, \forall i, k \tag{13.21.a}$$

$$P_i^{(k-1)} - P_i^{(k)} \leq P_{i,k}^* \leq 0, \forall i, k \tag{13.21.b}$$

$$0 \leq Q_{i,l} \leq Q_i^{(l)} - Q_i^{(l-1)}, \forall i, l \tag{13.21.c}$$

$$Q_i^{(l-1)} - Q_i^{(l)} \leq Q_{i,l}^* \leq 0, \forall i, l \tag{13.21.d}$$

$$f_i(x) = \frac{R_i}{V_0^2} x^2, g_i(x) = \frac{X_i}{V_0^2} x^2, \forall i \tag{13.22}$$

$$a_{ik} = \frac{f_i\left(P_i^{(k)}\right) - f_i\left(P_i^{(k-1)}\right)}{P_i^{(k)} - P_i^{(k-1)}}, \forall i, k \tag{13.23.a}$$

$$b_{il} = \frac{f_i\left(Q_i^{(l)}\right) - f_i\left(Q_i^{(l-1)}\right)}{Q_i^{(l)} - Q_i^{(l-1)}}, \forall i, l \tag{13.23.b}$$

$$c_{ik} = \frac{g_i\left(P_i^{(k)}\right) - g_i\left(P_i^{(k-1)}\right)}{P_i^{(k)} - P_i^{(k-1)}}, \forall i, k \tag{13.23.c}$$

$$d_{il} = \frac{g_i\left(Q_i^{(l)}\right) - g_i\left(Q_i^{(l-1)}\right)}{Q_i^{(l)} - Q_i^{(l-1)}}, \forall i, l \tag{13.23.d}$$

$$\underline{V}_i \le V_{i,t} \le \overline{V}_i, \forall i, t \tag{13.24}$$

$$-P_i^{\mathrm{cap}} \le P_{i,t} \le P_i^{\mathrm{cap}}, \forall i, t \tag{13.25}$$

$$\underline{P}_{i,t}^{\mathrm{PV}} \le P_{i,t}^{\mathrm{PV}} \le \overline{P}_{i,t}^{\mathrm{PV}}, \forall i, t \tag{13.26}$$

$$\underline{P}_{i,t}^{\mathrm{L}} \le P_{i,t}^{\mathrm{L}} \le \overline{P}_{i,t}^{\mathrm{L}}, \forall i, t \tag{13.27}$$

$$\underline{Q}_{i,t}^{\mathrm{L}} \le Q_{i,t}^{\mathrm{L}} \le \overline{Q}_{i,t}^{\mathrm{L}}, \forall i, t \tag{13.28}$$

Objective function (13.5) is to minimise the total energy loss in the current prediction horizon T.

Constraints (13.6) and (13.7) indicate that $\alpha_{i,j,t}$ is a binary variable and β_t is an integer variable within an allowed integer range. Equation (13.8) calculates the CB outputs and (13.9) computes the substation voltage with the OLTC position. Constraint (13.10) denotes the allowed maximal changing times for the CB module switch in the current horizon T, while (13.11) and (13.12) present the allowed maximal times for the OLTC position changes during the whole horizon T and each period t. Constraints (13.10)–(13.12) are non-linear due to the absolute value form. In order to reduce the non-linearity of such constraints, $Z = |X - Y|$ can be transformed into $Z \ge X - Y$ and $Z \ge Y - X$ as linear constraints. Thus, (13.10)–(13.12) can be expressed in linear forms.

Equation (13.13) calculates the reactive power dispatch capacity of the inverter in the second stage; that is, the maximal reactive power the inverter can inject/absorb $Q_{i,t}^{\mathrm{max}}$ subtracts the reserved reactive power for the third stage $Q_{i,t}^{\mathrm{res}}$. Constraint (13.14) defines the dispatchable range of the inverter reactive power output $Q_{i,t}^{\mathrm{Ir}}$ where the limit $Q_{i,t}^{\mathrm{disp}}$ is calculated by (13.13).

The fully linearised Dist-Flow formulae (13.15)–(13.23) are applied for each period t. Figure 13.3 illustrates the formulation for a certain quadratic calculation. x is divided into several intervals with a series of breaking points ($X^{(b)}$). Then, a slope ($s^{(b)}$) for each linear equation between the adjacent breaking points is calculated. Finally, y can be calculated by summing all the separated linear equation results, i.e. $y = \sum y^{(b)}, y^{(b)} = s^{(b)} x^{(b)}$.

Equations (13.15)–(13.17) provide the linear branch flow equations with branch power losses. Equations (13.18) and (13.19) indicate the linear calculation for the complex power losses, and (13.20) denotes the complex power flow variables are divided into pieces. Constraint (13.21) expresses that each piecewise power flow variable can vary only within its corresponding interval. Here, $P_i^{(0)} = 0$, $Q_i^{(0)} = 0, P_i^{(k)} = P_i^{\mathrm{max}}, Q_i^{(l)} = Q_i^{\mathrm{max}}$. $P_{i,k}^{*}$ and $Q_{i,l}^{*}$ are the negative piecewise power flow variables and they are utilised to calculate the power losses when the power

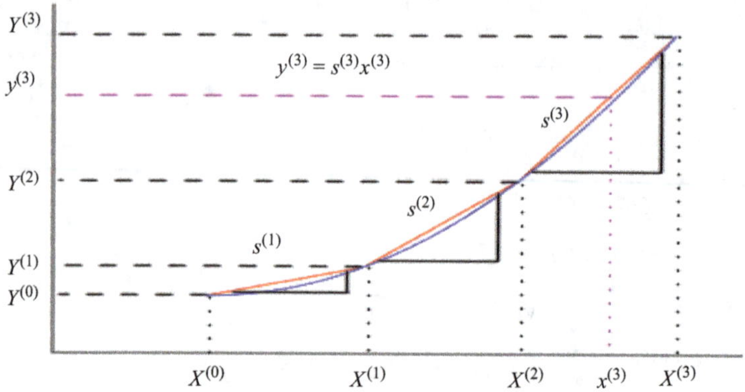

Figure 13.3 Piecewise linear formulation

flows are in the reverse direction. Formulae (13.22) and (13.23) are the calculations of all the linear equation slopes.

It is presented in (13.24) that the voltage of each bus must be kept within the allowed deviation range. The active power flow is limited within the transmission capacity of each branch in (13.25).

It is expressed in (13.26)–(13.28) that the PV power generation as well as active and reactive power loads can vary within the predicted lower and upper bounds.

The coordination optimisation model (13.5)–(13.28) is a mixed-integer linear programming with uncertainty variables. The first-stage decisions include the CB switch states $\alpha_{i,j,t}$ and the OLTC tap positions β_t, and the second-stage decisions include the inverter reactive power output for network energy loss reduction $Q^{\mathrm{lr}}_{i,t}$. Besides, $P^{\mathrm{PV}}_{i,t}$, $P^{\mathrm{L}}_{i,t}$, and $Q^{\mathrm{L}}_{i,t}$ are the uncertainty variables.

13.4 Two-stage robust optimisation method

The first two control stages are expected to be coordinately optimised with consideration of reserve for the third control stage. To deal with the uncertainties, the TSRO method is applied to robustly solve the coordination optimisation problem of the first two stages.

The principle of the TSRO is introduced in Section 7.4.1.

13.4.1 Uncertainty set modelling

The uncertainty variables $P^{\mathrm{PV}}_{i,t}$, $P^{\mathrm{L}}_{i,t}$, and $Q^{\mathrm{L}}_{i,t}$ are modelled into polyhedral uncertainty sets made up by predicted ranges (13.26)–(13.28) and designed uncertainty budgets. The uncertainty sets are formulated as follows:

$$U_{\mathrm{PV}} = \left\{ P^{\mathrm{PV}}_{i,t} \in \mathbb{R}^N : \underline{P}^{\mathrm{PV}}_{i,t} \le P^{\mathrm{PV}}_{i,t} \le \overline{P}^{\mathrm{PV}}_{i,t}, \forall i,t, \underline{\mu}^{\mathrm{PV}} \le \frac{\sum_{i\in N}\sum_{t\in T} P^{\mathrm{PV}}_{i,t}}{\sum_{i\in N}\sum_{t\in T} P^{\mathrm{PV,pr}}_{i,t}} \le \overline{\mu}^{\mathrm{PV}} \right\} \tag{13.29}$$

$$U_{\rm PL} = \left\{ P^{\rm L}_{i,t} \in \mathbb{R}^N: \underline{P}^{\rm L}_{i,t} \le P^{\rm L}_{i,t} \le \overline{P}^{\rm L}_{i,t}, \forall i,t, \underline{\mu}^{\rm PL} \le \frac{\sum_{i\in N}\sum_{t\in T} P^{\rm L}_{i,t}}{\sum_{i\in N}\sum_{t\in T} P^{\rm L,pr}_{i,t}} \le \overline{\mu}^{\rm PL} \right\} \tag{13.30}$$

$$U_{\rm QL} = \left\{ Q^{\rm L}_{i,t} \in \mathbb{R}^N: \underline{Q}^{\rm L}_{i,t} \le Q^{\rm L}_{i,t} \le \overline{Q}^{\rm L}_{i,t}, \forall i,t, \underline{\mu}^{\rm QL} \le \frac{\sum_{i\in N}\sum_{t\in T} Q^{\rm L}_{i,t}}{\sum_{i\in N}\sum_{t\in T} Q^{\rm L,pr}_{i,t}} \le \overline{\mu}^{\rm QL} \right\} \tag{13.31}$$

In the proposed TRI-VVC, the upper and lower bounds of the uncertainty variables are obtained from the interval predictions of the PV power generation and the loads over the following rolling horizon.

Furthermore, for each group of the uncertainty variables, two uncertainty budgets in pair of $\underline{\mu}$ and $\overline{\mu}$ are designed. They are the lower and upper bounds of the ratio between the summation of the possible uncertainty realisation and the summation of the predicted values. They can indicate prediction errors and determine solution robustness and conservativeness. Detailed discussion of the uncertainty budget design can be found in Section 7.4.2.

13.4.2 Two-stage robust optimisation model

The TSRO model for the proposed TRI-VVC method can be formulated in the following compact matrix form:

$$\min_{\boldsymbol{x}} \max_{\boldsymbol{u}} \min_{\boldsymbol{y}} \boldsymbol{a}^T\boldsymbol{y} \tag{13.32}$$

$$\text{s.t.} \quad \boldsymbol{Bx} \ge \boldsymbol{c} \tag{13.33}$$

$$\boldsymbol{Dx} + \boldsymbol{Ey} \le \boldsymbol{f} \tag{13.34}$$

$$\boldsymbol{Gx} + \boldsymbol{Hy} + \boldsymbol{Iu} = \boldsymbol{j} \tag{13.35}$$

$$\boldsymbol{u} \in \boldsymbol{U} \tag{13.36}$$

According to (13.5), the '*min-max-min*' objective of the TRI-VVC problem can be expressed as follows:

$$\min_{\alpha,\beta} \quad \max_{P^{\rm PV},P^{\rm L},Q^{\rm L}} \quad \min_{Q^{\rm inv},P,Q,V} \sum_{t\in T}\sum_{i\in N} P^{\rm loss}_{i,t}\tau \tag{13.37}$$

Here, the second-stage objective is to minimise the total network power loss. It is worth noting that there is no objective for the first operation stage, but the first-stage variables have impacts on the second-stage decision and the corresponding objective.

There are three types of variable vectors: (1) $\boldsymbol{x}$ is the vector of the first-stage decision variables, including the CB switch status α and the OLTC position β; (2) $\boldsymbol{y}$ stands for the vector of the second-stage decision variables, i.e. the inverter reactive power output for the loss reduction Q^{lr}, as well as the dependent power flow variables P, Q, V; (3) $\boldsymbol{u}$ is the vector of the uncertainty variables which include the PV power generation $P_{i,t}^{\mathrm{PV}}$ as well as the active and reactive loads $P_{i,t}^{\mathrm{L}}$ and $Q_{i,t}^{\mathrm{L}}$.

The TRI-VVC constraints (13.6)–(13.12) form the matrix constraint (13.33); constraints (13.13), (13.14), (13.21), (13.24), and (13.25) make up (13.34); constraints (13.15)–(13.20), (13.22), and (13.23) form (13.35); and (13.36) is constrained by (13.26)–(13.28).

The maximisation in this '*min-max-min*' form is to search for the worst case of the uncertainty realisation. Then, the first-stage decision variables are optimised to minimise the total power loss under the worst case and the corresponding optimal second-stage decision.

Detailed descriptions of the TSRO model can be found in Section 7.4.3.

13.4.3 Solution algorithm

To solve the '*min-max-min*' TSRO problem, the column-and-constraint generation (C&CG) algorithm which has a convergence speed [15] is applied in this chapter.

In the C&CG algorithm, the '*min-max-min*' TSRO problem is divided into a '*min*' master problem and a '*max-min*' sub-problem which can formulate the two operation stages, respectively. The C&CG algorithm solves the master problem and the sub-problem iteratively until the gap between the lower bound given by the master problem and the upper bound given by the sub-problem is reduced to an allowed termination threshold. Finally, the first-stage decision variable vector, $\boldsymbol{x}$, is robustly optimised. The complete procedure of the C&CG algorithm is provided in Section 7.4.4.

It is worth noting that exactly solving the '*max-min*' sub-problem with polyhedral uncertainty sets is significantly challenging [15]. This problem can be converted into a bilinear problem by strong duality, which is described in Section 7.4.4. To improve the computing efficiency, the chapter applies an outer approximation (OA) algorithm [16] to solve the bilinear problem. In this algorithm, the bilinear problem is decomposed into two linear programming problems. The OA algorithm can linearise the bilinear items around intermediate solution points and pass the linear approximation to the OA formulation. Then, the linear programming problems are solved iteratively at a fast computing speed even though the variables are numerous.

The complete procedure of the OA algorithm to solve the bilinear problem can be found in [16].

13.5 Case study

13.5.1 Test system description

In this chapter, a 33-bus radial distribution network as a microgrid with CBs, an OLTC, and PVs installed is used, and the system data can be found in [17]. Each CB has ten capacitor units of 30 kVar.

In this test system, $V_0 = 1$ p.u., $V^{\mathrm{Tap}} = 0.005$ p.u., the allowed position range for the OLTC is form -10 to 10, the allowed bus voltage range $[\underline{V}_i, \overline{V}_i]$ is set as [0.95, 1.05] p.u., and the critical bus voltage range $[\underline{V}^{\mathrm{cri}}, \overline{V}^{\mathrm{cri}}]$ is set as [0.94, 1.06] p.u.

The simulations are conducted on a 64-bit PC with 3.30-GHz CPU and 8 GB RAM using MATLAB® platform. The proposed problems are implemented by YALMIP toolbox [18] and solved by GUROBI solver [19].

13.5.2 One-hour simulation

13.5.2.1 First-stage results

In the first-stage optimisation ($\tau = 1$ h), interval predictions of PV output power and loads including the mean values and the intervals for an optimisation horizon, i.e. 4 h, are obtained first. Then, they are applied to construct the uncertainty sets (13.29)–(13.31), which are utilised in the TSRO model. The prediction data are shown in Figure 13.4 and it is applied in this first-stage simulation. Only the active power load is shown, and the reactive power load has the same variation tendency with the active one. The lower and upper uncertainty budgets for the PV output power are, respectively, 80% and 120% and those for the loads are 90% and 110%.

With the uncertainty sets constructed by this prediction data, the proposed TSRO problem (13.32)–(13.36) is solved, and the first-stage control variables including the CB outputs and the OLTC status for the whole horizon (4 h) are optimised. The CB and OLTC decisions in the first stage are shown in Figure 13.5. In this figure, CB 2 means the CB at bus 2 and so forth. Note that only the first-hour CB outputs and OLTC position are implemented and the remaining 3-h decisions are discarded. By doing this, the first-stage optimisation considers the inter-hour limits of the CB switches and the OLTC position, as well as the minimisation of the overall network energy loss during the whole horizon of 4 h.

The total network energy loss in the first hour under the worst case of uncertainty realisation is 114.4 kWh. Note that this energy loss is not the actual loss but the expected loss if the worst case occurs. With the complement of the inverter output dispatch in the second stage, the actual energy loss may be lower, since the worst case may not occur.

For bus voltages, under the worst case, the maximum bus voltage absolute deviation in the first hour is 0.0413 p.u., verifying effectiveness of the first-stage operation to keep the bus voltages within the allowed range of [0.95, 1.05] p.u.

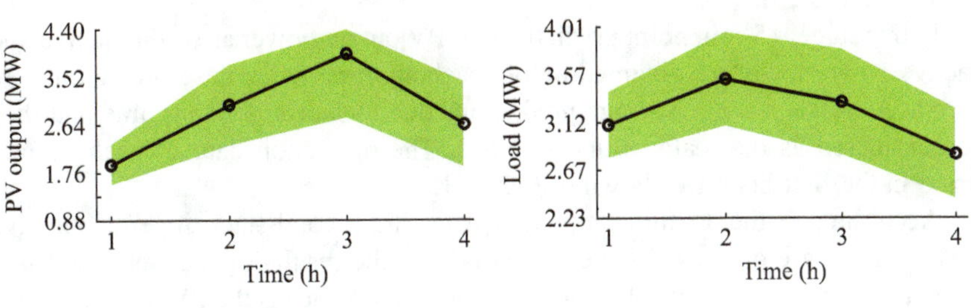

Figure 13.4 First-stage PV output power and load interval predictions

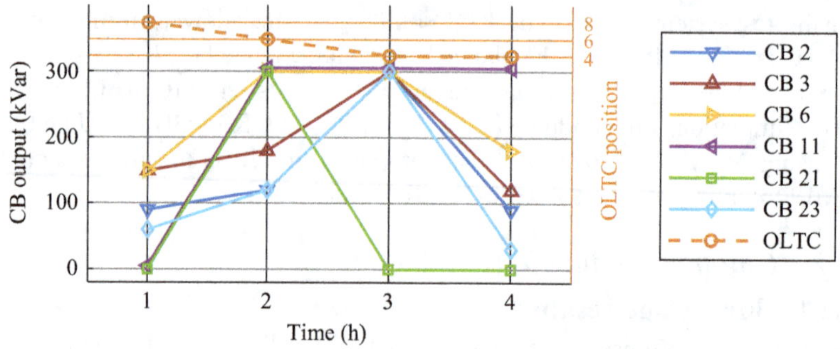

Figure 13.5 CB and OLTC decisions in the first stage

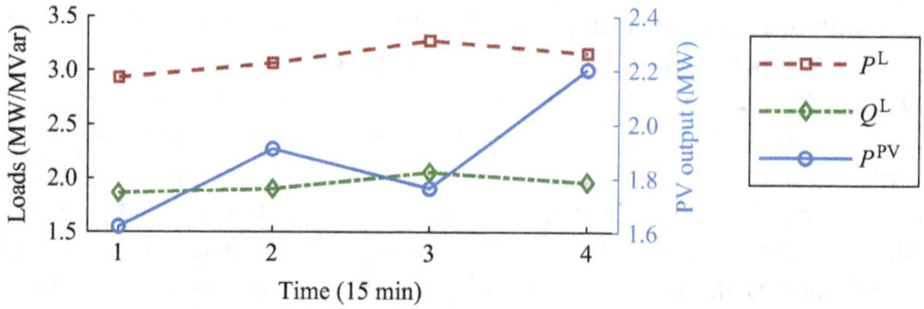

Figure 13.6 Second-stage PV output power and load data

In terms of computational efficiency, the solver time is 4.8 s, which is fully compatible for online use. This shows high computational efficiency of the TSRO method.

13.5.2.2 Second-stage results

In the second stage, given short-period ($\pi = 15$ min) predictions of PV output power and loads, the inverter reactive power outputs are dispatched through the following model:

$$\min_{Q^{\text{inv}},P,Q,V} \sum_{t\in T} \sum_{i\in N} P_{i,t}^{\text{loss}} \pi \tag{13.38}$$

s.t. (13.13)–(13.25).

In this stage, 15-min point predictions of PV output power as well as active and reactive power loads are obtained before each short period. These predictions are more accurate than the first-stage predictions due to shorter leading time and they are considered as the realised uncertainties. The prediction data of each 15-min period in the first hour are shown in Figure 13.6.

According to the 15-min prediction data, the second-stage inverter reactive power outputs are optimised for each period and the results of four representative PV units are shown in Figure 13.7. In this figure, PV 9 means the PV inverter at bus 9 and so forth. The positive values mean reactive power injection to the network.

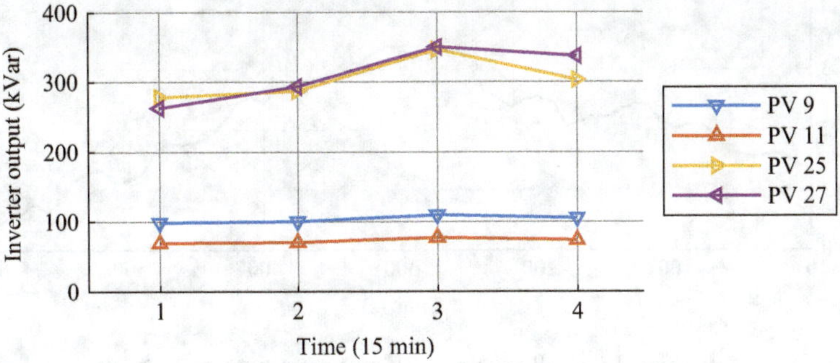

Figure 13.7 Inverter output dispatch decisions in the second stage

After the second-stage operation, the actual total network energy loss in this hour is 80.6 kWh, which is less than that under the worst case. It indicates that the second-stage operation can further reduce the network energy loss efficiently based on the short-period predictions. For each 15-min period, the maximum bus voltage absolute deviation is 0.0446 p.u., and all the short-period bus voltages are kept in the allowed range. It shows that the second-stage inverter output dispatch can keep the bus voltages in the allowed range efficiently.

In terms of computing efficiency, the average solver time is 0.2 s, which is fully compatible for online use.

13.5.2.3 Third-stage results

In the third stage, when the local bus voltages fluctuate significantly from the allowed range by sudden PV output power changes, the inverter local droop control is activated. When the local bus voltage is measured and updated at a certain time length (1 s in this chapter), the corresponding inverter reactive power output is controlled as (13.2) to alleviate the bus voltage fluctuations. Thus, all the inverters are in the local control for each 1-s period. In this simulation, a 600-s real-time PV output power profile is applied for the first 10 min of the second stage as shown in Figure 13.8.

Based on this real-time PV output power profile, real-time bus voltages are obtained through power flow computation. Then the third-stage local droop control is implemented according to the bus voltages. The voltages of buses 9, 11, and 13 are out of the allowed range, and the inverters at these three buses are triggered to provide reactive power compensation. Taking bus 11 as an example, the bus voltage profile and the inverter reactive power output profile are shown in Figure 13.9.

From Figure 13.9, it can be seen that when the bus voltage rises over 1.05 p.u., the inverter decreases its reactive power injection to lower the bus voltage. However, it is worth noting that as a primary voltage control method via the reactive power compensation, the third-stage inverter local droop control is a non-zero control method. It means that the droop control cannot bring the bus voltage fully back to 1.05 p.u., but it can reduce the bus voltage fluctuations effectively, the maximum value from near 1.06 p.u. to below 1.055 p.u.

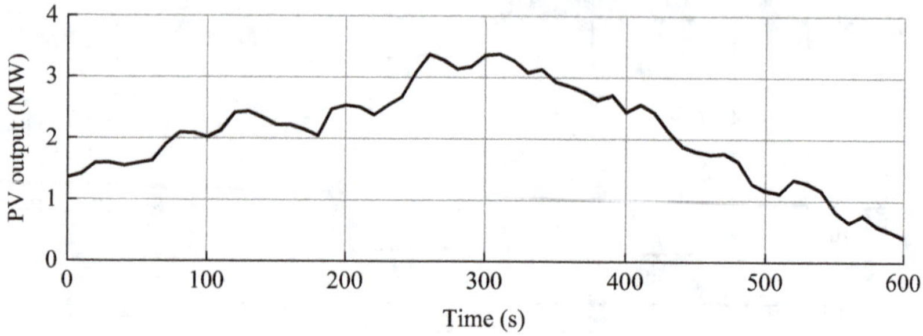

Figure 13.8 Real-time PV output power profile

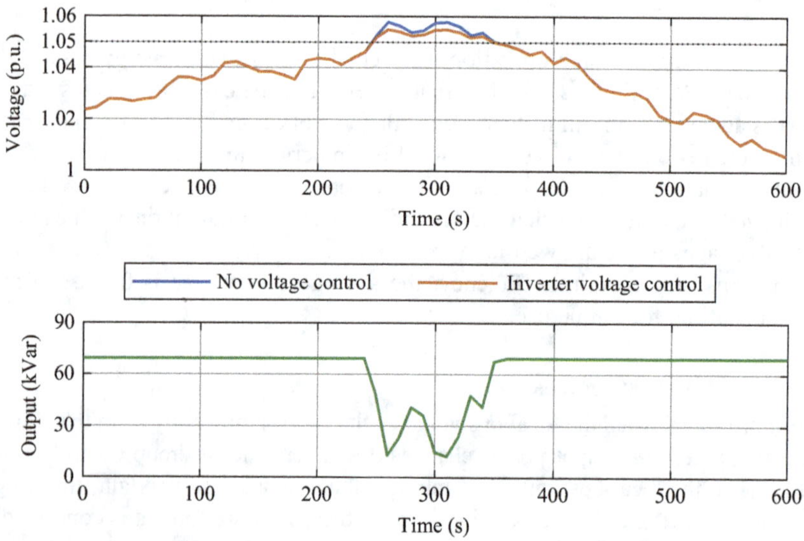

Figure 13.9 Bus voltage and inverter output results at bus 11

13.5.3 24-hour time-series simulation

The proposed TRI-VVC method is applied for 24-hour. The 24-hour rolling horizon predictions of PV output power and loads are generated based on the real data in Newington area in Australia [20]. These predictions include hourly mean values and intervals like Figure 13.4 for each hour, and they are used in the TSRO method for the first-stage optimisation. Based on the rolling horizon predictions, the first-stage CB and OLTC decisions are optimised for each hour. Then, the real 24-hour short-period data, i.e. Newington data on 9 December 2011, is applied for the second-stage optimisation as the short-period predictions. Figure 13.10 shows the profiles of PV output power and active power loads in the short periods. Based on the short-period predictions, the second-stage inverter output dispatch decisions are optimised for each short period, and the network energy loss and the bus voltage

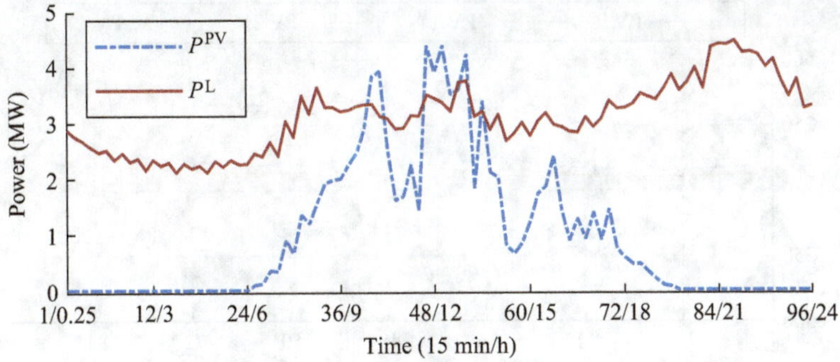

Figure 13.10 The 24-hour PV output power and load profiles

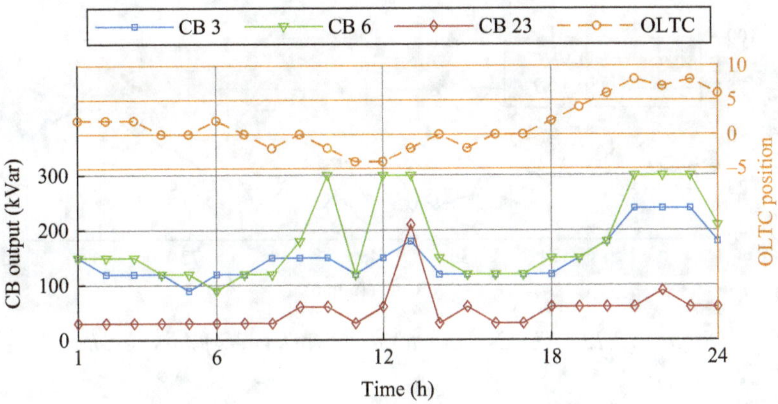

Figure 13.11 The 24-hour first-stage decisions

results of each short period are obtained. In addition, to show the efficiency of the proposed TRI-VVC method, the system without VVC is also tested by applying the short-period real profiles, i.e. Figure 13.10. The network energy loss and the bus voltage results without VVC are obtained and compared with the TRI-VVC results.

Figure 13.11 shows the first-stage decisions of three specified CBs and the OLTC for the 24-hour. From Figure 13.11, during low-load hours, i.e. 0.00–8.00, the CBs generate less reactive power, while during high-load hours, i.e. 20.00–24.00, they generate more reactive power. When the PVs generate high power during the period from 9.00 to 14.00, the CBs also increase their reactive power outputs. In addition, when the PV output power has a drop from 10.00 to 11.00, the CBs decrease the reactive power outputs.

Figure 13.12 shows the inverter output dispatch decisions of four specified PV units for the 96 15-min periods. The positive and negative values mean reactive power generation and consumption, respectively. It can be seen that the inverters dispatch the reactive power outputs dynamically to reduce the network energy

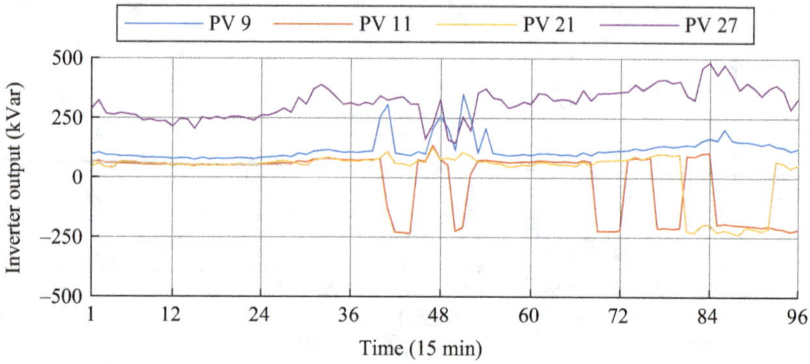

Figure 13.12 The 24-hour second-stage decisions

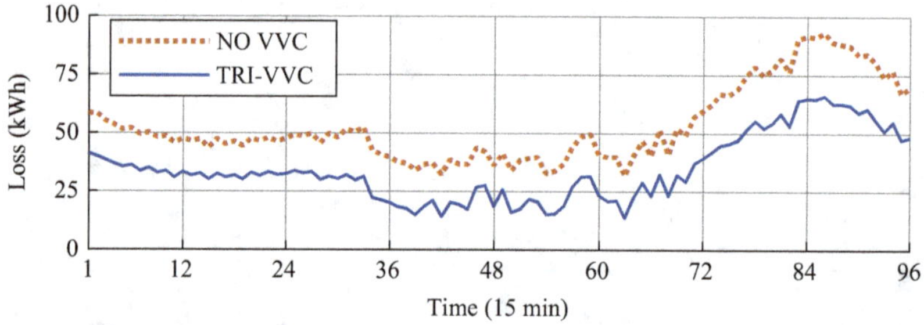

Figure 13.13 The 24-hour total network energy loss results

losses and regulate the bus voltages within the allowed range. When the PV output power varies heavily between 10.00 and 14.00, the inverter reactive power outputs at buses 9, 11, and 27 are dispatched in relatively larger ranges.

The network energy losses for all the short periods are shown in Figure 13.13. During PV high-generation hours, i.e. 8.00–16.00, the network energy losses are relatively low, since the loads are locally fulfilled by the PV power generation. After 18.00, as the PVs decrease the output power to 0 and the loads increase to the peak, the network energy losses increase sharply. It shows that for each period, the network energy losses achieved by the TRI-VVC method are much less than those without VVC. The 24-hour total network energy loss with the TRI-VVC method is 3.300 MWh which is significantly less than that without VVC, i.e. 5.116 MWh. It indicates that the proposed TRI-VVC method can highly efficiently reduce the network energy losses.

The voltage profile of bus 11 for all the short periods are shown in Figure 13.14. Compared to the voltage without VVC, the TRI-VVC method can keep the bus voltage in each short period within the allowed range. Without the VVC, the bus voltage violations occur during the periods 0.00–2.00 and 17.30–24.00. In addition, the average bus voltage deviation of all the buses with the TRI-VVC method is 0.0198 p.u., while that without VVC is 0.0362 p.u.

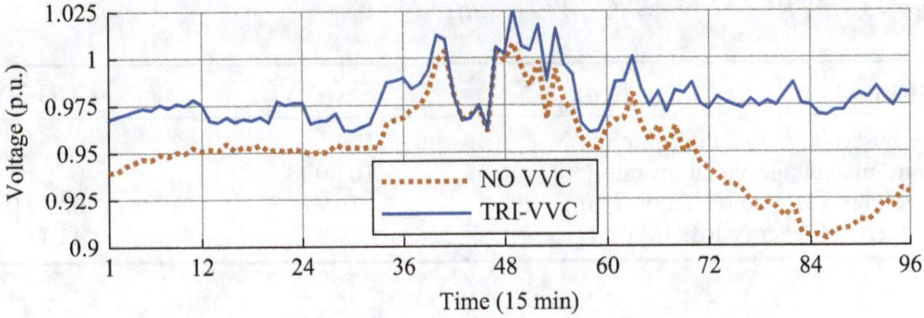

Figure 13.14 The 24-hour bus voltage results at bus 11

13.5.4 TRI-VVC vs. Single-Stage Centralised VVC

Last, the proposed TRI-VVC optimisation method is compared with a conventional single-stage centralised VVC (SSC-VVC) method. In this conventional method, the operation decisions of the CBs, the OLTC, and the PV inverters are optimised together with rolling point predictions where only mean values are predicted. This method is a centralised VVC method which can achieve a global optimum for each rolling horizon. However, due to the prediction errors of the uncertainties, the efficiency of this conventional method is impacted negatively. On the other hand, since the local inverter droop control is not enabled in this SSC-VVC method, the reactive power capacity of the inverters is fully used in the network energy loss reduction. However, such fast and uncertain PV output power fluctuations caused by transient cloud movements impact the operation negatively and bus voltage violations may occur.

To comprehensively check the efficiency of the proposed TRI-VVC optimisation method and the conventional SSC-VVC one, Monte Carlo sampling is applied to generate random scenarios as uncertainty realisations to test the methods; 500 random daily scenarios are generated with *Gaussian* probability distribution, and each daily scenario has 96 15-min sub-scenarios, such that there are 48,000 sub-scenarios in total. In this *Gaussian* probability distribution, the standard deviation values are 5% and 2.5% of the predicted mean values for the PV output power and the loads, respectively.

For the SSC-VVC method, the operation decisions are optimised with the rolling point predictions of the uncertainties. Then, these sub-scenarios are applied as realisations of the uncertainties to check the performance of this conventional method for each daily scenario. For the TRI-VVC method, the first-stage operation decisions are optimised with the rolling interval predictions which use the point prediction values as the mean values. Then, the second-stage operation decisions are optimised with these sub-scenarios to obtain final operation results of the proposed TRI-VVC optimisation method for each daily scenario. The comparison results are shown in Table 13.1. The bus voltage violation rates are calculated on the daily scenario basis and the 15-min sub-scenario basis, respectively.

Table 13.1 TRI-VVC vs. SSC-VVC results

Strategy	SSC-VVC	TRI-VVC
Daily bus voltage violation rate (%)	100	0
15-min bus voltage violation rate (%)	10.80	0
Bus voltage absolute deviation (p.u.)	0.0194	0.0171
Daily network energy loss (MWh)	4.048	4.013

In Table 13.1, for the SSC-VVC method, the bus voltage violations occur in all the daily scenarios and in the 10.80% of the 15-min sub-scenarios. By contrast, no bus voltage violation occurs in any daily scenario when the TRI-VVC method is applied. It denotes that the TRI-VVC method is fully robust against the uncertainty realisations to avoid the bus voltage violations. Besides, the proposed TRI-VVC method achieves the smaller average bus voltage absolute deviation than the SSC-VVC one, which indicates the TRI-VVC method has higher efficiency in alleviation of the bus voltage deviations.

In terms of the network energy loss reduction, even though the SSC-VVC method uses the full capacity of the inverters for the centralised optimisation of the energy loss reduction, the TRI-VVC method attains the less total network energy loss than the SSC-VVC one.

In conclusion, the TRI-VVC optimisation method can achieve effectively robust solutions under the uncertainties to avoid bus voltage violations while carrying out relatively low network energy losses.

13.6 Conclusion

This chapter proposes a novel TRI-VVC optimisation method in microgrids to reduce network energy losses and to mitigate bus voltage deviations and fluctuations considering uncertainties of PV power generation and loads. The TRI-VVC framework consists of three stages of different timescales, i.e. hourly CB and OLTC scheduling in a rolling horizon, 15-min inverter output dispatch, and real-time inverter droop control. A new operation model of PV inverters is developed. In this model, PV inverters can dispatch reactive power to reduce network energy losses in the centralised VVC and act droop control to alleviate the bus voltage fluctuations in the local VVC. The TSRO method is utilised to optimise the first-stage operation considering the worst case of uncertainty realisation and the second-stage operation corresponding to the worst case. Thus, the first-stage operation is coordinated with the second-stage operation, and solutions are robust against the uncertainties.

In the case study, first, the proposed TRI-VVC optimisation method is simulated on the 33-bus system for the three stages, respectively, to demonstrate the procedure with detailed uncertainty data. Second, the TRI-VVC optimisation method is simulated for 24 h to illustrate its efficiency for the whole day. Third, the proposed TRI-VVC optimisation method is compared with a conventional SSC-VVC method via Monte Carlo simulations. According to the simulation results, the three stages of the TRI-VVC optimisation method can coordinate efficiently and robustly against any uncertainty realisation, to reduce the network energy losses as well as mitigate the bus voltage deviations and fluctuations.

References

[1] K. Turitsyn, P. Šulc, S. Backhaus, and M. Chertkov, 'Distributed control of reactive power flow in a radial distribution circuit with high photovoltaic penetration', *Proc. IEEE PES General Meeting*, pp. 1–6, Minneapolis, MN, USA, 2010.

[2] K. Turitsyn, P. Sulc, S. Backhaus, and M. Chertkov, 'Options for control of reactive power by distributed photovoltaic generators', *Proc. of the IEEE*, vol. 99, no. 6, pp. 1063–1073, 2011.

[3] J.W. Smith, W. Sunderman, R. Dugan, and B. Seal, 'Smart inverter volt/var control functions for high penetration of PV on distribution systems', *Proc. IEEE PES Power Syst. Conf. and Expo.*, pp. 1–6, Phoenix, AZ, USA, 2011.

[4] P. Jahangiri and D.C. Aliprantis, 'Distributed volt/var control by PV inverters', *IEEE Trans. Power Syst.*, vol. 28, no. 3, pp. 3429–3439, 2013.

[5] H. Li, F. Li, Y. Xu, D.T. Rizy, and S. Adhikari, 'Autonomous and adaptive voltage control using multiple distributed energy resources', *IEEE Trans. Power Syst.*, vol. 28, no. 2, pp. 718–730, 2013.

[6] B.A. Robbins, C.N. Hadjicostis, and A.D. Domínguez-García, 'A two-stage distributed architecture for voltage control in power distribution systems', *IEEE Trans. Power Syst.*, vol. 28, no. 2, pp. 1470–1482, 2013.

[7] J.Y. Park, S.R. Nam, and J.K. Park, 'Control of a ULTC considering the dispatch schedule of capacitors in a distribution system', *IEEE Trans. Power Syst.*, vol. 22, no. 2, pp. 755–761, 2007.

[8] H.G. Yeh, D.F. Gayme, and S.H. Low, 'Adaptive VAR control for distribution circuits with photovoltaic generators', *IEEE Trans. Power Syst.*, vol. 27, no. 3, pp. 1656–1663, 2012.

[9] B. Zhang, A.Y.S. Lam, A.D. Domínguez-García, and D. Tse, 'An optimal and distributed method for voltage regulation in power distribution systems', *IEEE Trans. Power Syst.*, vol. 30, no. 4, pp. 1714–1726, 2015.

[10] T. Niknam, M. Zare, and J. Aghaei, 'Scenario-based multiobjective volt/var control in distribution networks including renewable energy sources', *IEEE Trans. Power Del.*, vol. 27, no. 4, pp. 2004–2019, 2012.

[11] Z. Wang, J. Wang, B. Chen, M.M. Begovic, and Y. He, 'MPC-based voltage/var optimisation for distribution circuits with distributed generators and exponential load models', *IEEE Trans. Smart Grid*, vol. 5, no. 5, pp. 2412–2420, 2014.

[12] Z. Wang, H. Chen, J. Wang, and M. Begovic, 'Inverter-less hybrid voltage/var control for distribution circuits with photovoltaic generators', *IEEE Trans. Smart Grid*, vol. 5, no. 6, pp. 2718–2728, 2014.

[13] G.M. Gilbert, D.E. Bouchard, and A.Y. Chikhani, 'A comparison of load flow analysis using DistFlow, Gauss-Seidel, and optimal load flow algorithms', *Proc. IEEE Canadian Conf. on Electr. and Comput. Eng.*, vol. 2, pp. 850–853, Waterloo, Ont., Canada, 1998.

[14] C. Wan, Z. Xu, P. Pinson, Z.Y. Dong, and K.P. Wong, 'Probabilistic forecasting of wind power generation using extreme learning machine', *IEEE Trans. Power Syst.*, vol. 29, no. 3, pp. 1033–1044, 2014.

[15] B. Zeng and L. Zhao, 'Solving two-stage robust optimisation problems using a column-and-constraint generation method', *Operations Research Letters*, vol. 41, no. 5, pp. 457–461, 2013.

[16] D. Bertsimas, E. Litvinov, X.A. Sun, J. Zhao, and T. Zheng, 'Adaptive robust optimisation for the security constrained unit commitment problem', *IEEE Trans. Power Syst.*, vol. 28, no. 1, pp. 52–63, 2013.

[17] Y. Xu, Z.Y. Dong, R. Zhang, and D.J. Hill, 'Multi-timescale coordinated voltage/var control of high renewable-penetrated distribution systems', *IEEE Trans. Power Syst.*, vol. 32, no. 6, pp. 4398–4408, 2017.

[18] J. Lofberg, 'YALMIP: a toolbox for modelling and optimisation in MATLAB', *Proc. IEEE Inter. Conf. on Robotics and Automation*, pp. 284–289, Taipei, Taiwan, 2004.

[19] GUROBI, Available: www.gurobi.com, accessed March 2018.

[20] Ausgrid, 'Distributed generation and distributed storage – SGSC technical compendium', Available: www.ausgrid.com.au, accessed March 2018.

Part IV

Coordinated real-time control of DERs: distributed controller design and hardware-in-the-loop tests

Chapter 14
Power system frequency control by aggregated energy storage systems

The widespread distributed energy storage systems (ESSs) can be aggregated to provide grid ancillary services. In this chapter, an improved load frequency control scheme with the energy storage aggregators (ESAs) and a power disturbance observer is proposed. The disturbance observer is designed to assist the original automatic generation control (AGC) scheme for generators and ESAs, which improves the system frequency response and recovery. In each ESA, a finite-time leader–follower consensus control is proposed to aggregate the small-scale ESSs via sparse communication networks. By using such an algorithm, the ESA can track the frequency control signal and the state-of-charge (SoC) among each ESS can be balanced in finite time. The external characteristics of the ESA will resemble that of one large-scale ESS. The whole frequency control scheme is validated under a series of scenarios including contingency and normal operation.

14.1 Introduction

In modern power systems, ESSs have been widely integrated into different system levels, which can be utilised for frequency ancillary services. As the number of distributed ESSs (DESSs) such as community ESSs, plug-in electric vehicles, and home ESSs keep growing, they are expected to have large portions in the future frequency ancillary services market. However, the DESSs located at the demand side usually have large numbers but small capacity, which makes their contributions to the power system hard to account. The concept of aggregator offers the feasibility for a large number of DESSs to participate in ancillary services (such as frequency regulation and restoration) to gain additional income [1,2]. In an aggregator, the geographically dispersed units can cooperate to achieve a common objective through information exchange among communication networks. The ESA will have comparable power and capacity rating as a large centralised ESS and it can be viewed as one entity for the system operator. In this chapter, the distributed control algorithms to enable the aggregation of ESSs will be introduced, and the dynamic response including the external and internal behaviour of ESAs will be evaluated.

How to incorporate ESSs into existing frequency control architecture and effectively utilise the timely but limited capacity are important problems to be investigated. In the context of primary frequency control, researchers have investigated the utilisation of both grid ESSs and DESSs for local/decentralised frequency support [3]. As for secondary frequency control by small-scale DESSs, the control architecture can be designed in either a centralised or distributed structure [4]. The centralised secondary control structures of DESSs have been reported by existing works. For example, in [5], a master–slave centralised control of DESS has been proposed for mitigating voltage and frequency deviations. The major concern of the centralised control structure is the requirement of a central communication infrastructure, and it is inherently vulnerable to any communication failure. The bottleneck limits the support for plug-and-play (PnP) functions, which is highly desired for widespread distributed devices.

As an alternative approach, distributed control has raised considerable attention in power and energy research areas. Unlike centralised control, this control scheme does not need full communication access as each controllable device only communicates with its neighbours, which is more suitable for widespread distributed devices in power systems. The core idea of distributed control based on consensus algorithms is to achieve fair utilisation of available devices by exchanging information through sparse communication networks. A major application of distributed control approaches is to coordinate photovoltaic (PV) units and ESSs for voltage regulation in power distribution networks [6,7]. Besides, the distributed control methods for distributed devices such as responsible loads, ESSs, and wind turbines to achieve aggregation are reported by recent research [8–10]. However, the SoC, as the key performance indicator of ESSs, is not considered in [10]. Besides, the convergence of the above methods in finite time is not guaranteed [8–10]. As a result, the convergence time may be very long, and the frequency performance cannot be satisfactory.

This chapter introduces a distributed aggregation control approach with finite-time consensus for multiple ESSs to provide frequency control services. The method has the following features:

- A novel frequency control scheme is designed to integrate the ESA into the multi-area microgrids. To fully exploit the capability of ESA for frequency control, a system disturbance observer is designed to supplement the original AGC signal. Then the high-frequency system disturbance in one control area can be extracted and compensated by the ESA. Compared with the conventional proportional-integral (PI)-based secondary frequency control methods, the proposed control scheme takes full advantage of fast response speed from ESAs, thereby improving system frequency recovery speed.
- A leader–follower finite-time consensus control is designed to aggregate ESSs into the ESA. This control ensures power tracking and SoC balancing of ESSs in finite time, allowing for desired dynamic performance to be obtained. In the proposed ESA, a virtual leader updates the reference power and energy states based on the frequency control signal. The reference states are received by the pinning ESSs and states of each ESS are shared to neighbours through the predefined communication network. Both power and energy states of each ESS will reach consensus in steady state.

14.2 Proposed frequency control scheme

14.2.1 System overview

An overview of the proposed frequency control scheme is shown in Figure 14.1. As highlighted in Figure 14.1, there are two main contributions of this chapter: (i) a novel frequency control scheme that incorporates a system disturbance observer to supplement the secondary frequency control and (ii) a leader–follower finite-time consensus algorithm to control ESSs within an ESA. With the proposed frequency control scheme, the frequency control signal of the ESA is comprised of two parts: (i) The AGC signal from the secondary frequency control with a predefined participation factor α_{ESA}; it should be noted that the AGC signal is shared by both generators and ESAs in each control area. (ii) The system disturbance signal is estimated by the system disturbance observer. The objective of this design is to accelerate the frequency recovery speed as compared to the conventional PI-based secondary frequency control methods. The frequency control signal of the ESA is further shared among each ESS through the proposed finite-time consensus algorithm. The proposed disturbance observer with the band-pass filter is introduced in Section 14.3. The detailed design of the ESA based on the finite-time consensus algorithm is presented in Section 14.4.

14.2.2 Multi-area microgrids

Consider an interconnected microgrid with M control areas indexed by $i = 1, 2, \ldots, M$. The load frequency control (LFC) model, including RESs and ESAs, of the ith control area is presented in Figure 14.1. The objective of primary control is to contain the frequency deviation caused by power imbalance in any control area, while the secondary control, i.e. AGC, aims to recover the frequency back to the

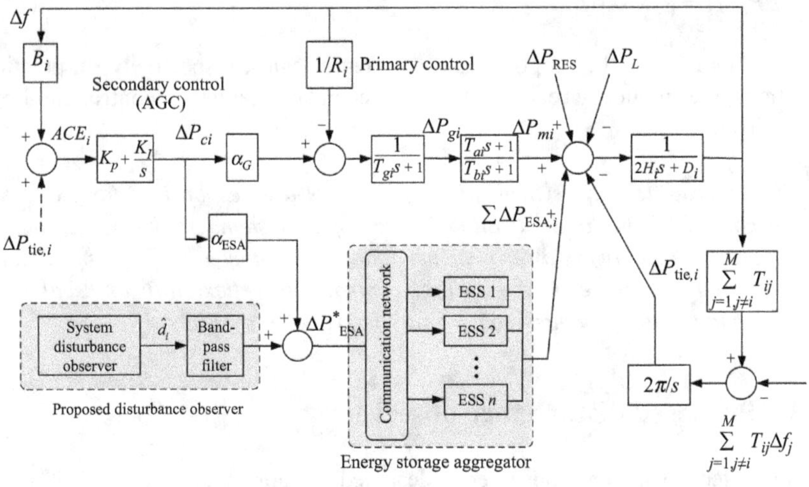

Figure 14.1 An overview of the proposed frequency control scheme

nominal value. The system dynamics of the ith area can be represented by the following differential equations:

$$\Delta \dot{f}_i(t) = -\frac{D_i}{2H_i}\Delta f_i(t) + \frac{1}{2H_i}(\Delta P_{mi}(t) - \Delta P_{L,i}(t) + \Delta P_{\mathrm{RES},i}(t) - \Delta P_{\mathrm{tie},i}(t) + \Delta P_{\mathrm{ESA},i}(t)) \tag{14.1}$$

$$\Delta \dot{P}_{mi}(t) = -\frac{1}{T_{bi}}\Delta P_{mi}(t) + \frac{1}{T_{bi}}\Delta P_{gi}(t) + \frac{T_{ai}}{T_{bi}}\Delta \dot{P}_{gi}(t) \tag{14.2}$$

$$\Delta \dot{P}_{gi}(t) = -\frac{1}{T_{gi}}\Delta P_{gi}(t) + \frac{1}{T_{gi}}\Delta P_{ci}(t) - \frac{1}{R_i T_{gi}}\Delta f_i(t) \tag{14.3}$$

$$\Delta \dot{P}_{\mathrm{tie},i}(t) = 2\pi \cdot \left[\sum_{j=1, j\neq i}^{M} T_{ij}(\Delta f_i(t) - \Delta f_j(t))\right] \tag{14.4}$$

where Δf_i is the change in system frequency, ΔP_{ci} denotes control effort of secondary control, ΔP_{mi} and ΔP_{gi} are the deviations of generator mechanical output and valve position, respectively. H_i, D_i, R_i, T_{gi}, T_{ai}, and T_{bi} are the system inertia, load damping coefficient, speed droop, governor, and turbine time constants, respectively. $\Delta P_{\mathrm{L},i}$, $\Delta P_{\mathrm{RES},i}$, $\Delta P_{\mathrm{tie},i}$, and $\Delta P_{\mathrm{ESA},i}$ are power variations of loads, RESs, tie line, and ESAs, respectively, which can be viewed as external disturbances to the system.

The control input for the secondary control is called area control error (ACE) and can be defined as

$$ACE_i(t) = B_i\Delta f_i(t) + \Delta P_{\mathrm{tie},i}(t) \tag{14.5}$$

where B_i the frequency bias factor. In the LFC model, PI control is commonly utilised for the secondary control, which is to eliminate the ACE for each control area. The AGC controller based on PI control is represented as

$$\Delta P_{ci}(t) = -K_P ACE_i(t) - K_I \int ACE_i(t) \tag{14.6}$$

where K_P and K_I are the proportional and integral gains, respectively. In practical conditions, the frequency recovery time by secondary frequency control can be as long as 10 min.

Remark 1: Traditionally, synchronous generators are responsible for frequency control caused by the load variation in the power system. As the penetration of RESs increases rapidly in power systems, additional disturbances have been added to the original LFC model. Given the high controllability, flexibility, and scalability of ESAs, additional frequency support can be provided by ESAs.

14.3 Proposed disturbance observer

The proposed disturbance observer is designed to supplement the secondary frequency control for the ESA; therefore, the system frequency response and recovery can be improved. The proposed disturbance observer is comprised of a system disturbance observer in series with a band-pass filter, as shown in Figure 14.1. The following subsections explain the detailed design of the proposed disturbance observer.

14.3.1 System disturbance observer

The design objective of the disturbance observer is to estimate the real power variations of the control area within which it is deployed in real time. Based on system dynamical equations (14.1)–(14.6) in *i*th control area, a state–space model *i*th control area can be obtained as follows:

$$\dot{x}_i(t) = A_i x_i(t) + B_i \mu_i(t) + F_i \Delta P_{di}(t) \tag{14.7}$$

where the state variable $x_i = [\Delta f_i \quad \Delta P_{mi} \quad \Delta P_{gi}]^T$, the system input $\mu_i = \Delta P_{ci}$, the system disturbance $\Delta P_{di} = \Delta P_{\mathrm{RES},i} - \Delta P_{L,i} - \Delta P_{\mathrm{tie},i} + \Delta P_{\mathrm{ESA},i}$. It is noted that the power output of ESA is also included in the system disturbance. Thus, with proper control of the ESA, the observed system disturbance can be reduced.

$$A_i = \begin{bmatrix} -D_i/2H_i & -1/2H_i & 0 \\ -T_a/R_i T_g T_b & -1/T_{bi} & \left(1/T_{bi} - T_{ai}/T_{gi} T_{bi}\right) \\ -1/R_i T_{gi} & 0 & -1/T_{gi} \end{bmatrix},$$

$$B_i = \begin{bmatrix} 0 \\ 0 \\ 1/T_{gi} \end{bmatrix}, F_i = \begin{bmatrix} -1/2H_i \\ 0 \\ 0 \end{bmatrix}$$

With the state–space model (14.7) of the *i*th area, the system disturbance observer can be designed as follows [11]:

$$\begin{cases} \dot{\hat{x}}_i(t) = A_i \hat{x}_i(t) + B_i \mu_i(t) + F_i \hat{d}_i(t) + R(y_i(t) - \hat{y}_i(t)) \\ \hat{y}_i(t) = C \hat{x}_i(t) \end{cases} \tag{14.8}$$

where $\boldsymbol{C}$ is an identity matrix, $\boldsymbol{C} = \mathrm{diag}\{1, 1, 1\}$, $\boldsymbol{R}$ is a constant gain matrix, $\hat{d}_i$ is the estimated system disturbance.

With the estimated states $\hat{\boldsymbol{x}}_i$, the system disturbance can be estimated as follows:

$$\begin{cases} \hat{d}_i(t) = \xi_i(t) + M\hat{x}_i(t), \\ \dot{\xi}_i(t) = -M\left(A_i \hat{x}_i(t) + B_i \mu_i(t) + F_i \hat{d}_i(t)\right) + \dot{\hat{d}}_i(t) \end{cases} \tag{14.9}$$

where ξ_i is an auxiliary variable, and $\boldsymbol{M}$ is a constant gain matrix.

Defining the state estimation error and disturbance estimation error as $\overline{x}_i = x_i - \hat{x}_i$ and $\overline{d}_i = d_i - \hat{d}_i$, the output estimation error can be obtained with $C\overline{x}_i$.

With (14.8) and (14.9), the estimated state error dynamics become

$$\dot{\overline{x}}_i(t) = (A_i - RC)\overline{x}_i(t) + F_i \overline{d}_i(t) \tag{14.10}$$

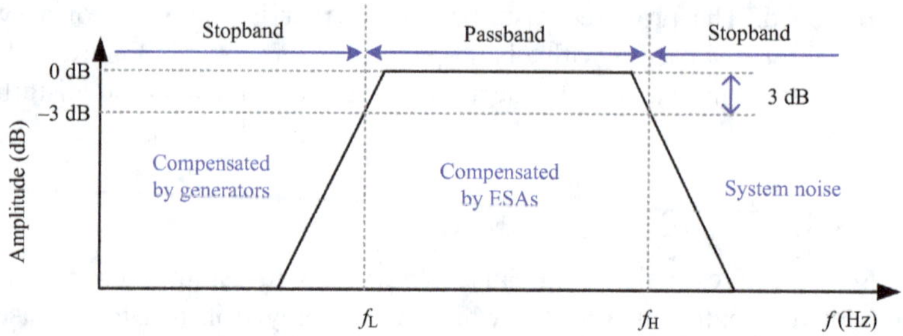

Figure 14.2 The function of the band-pass filter

Combining the disturbance estimation error and state estimation error,

$$\dot{\omega}_i(t) = \overline{A}_i \omega_i(t) \tag{14.11}$$

where $\omega_i = \begin{bmatrix} \overline{x}_i^T & \overline{d}_i \end{bmatrix}$ and $\overline{A}_i = \begin{bmatrix} A_i - RC & F_i \\ -MRC & 0 \end{bmatrix}$.

The eigenvalues of the matrix $\overline{A}_i$ can be placed arbitrarily by selecting the gain matrixes $\boldsymbol{R}$ and $\boldsymbol{M}$ as

$$\det\left[sI - \overline{A}_i\right] = \prod_{m=1}^{4} (s + \lambda_i) \tag{14.12}$$

where λ_i are the desired eigenvalues of the system. Through the eigenvalue placement in (14.12), the desired dynamics of system disturbance observer can be designed.

14.3.2 Band-pass filter

A band-pass filter is designed to extract the magnitude of system disturbance to be compensated by the ESA within the control area. The function of the band-pass filter is illustrated in Figure 14.2. The key idea is to decompose the estimated system disturbance, where the high-frequency components of system disturbance within the passband [f_L, f_H] will be sent to the ESA. Therefore, the ESAs can respond to sudden system disturbance and improve the system frequency response. The very high-frequency components of system disturbance [f_H, ∞] can be viewed as system noise, while the low-frequency components within [0, f_L] can be compensated by conventional generators. Generally, f_H is selected to be larger than the reciprocal of the response time of the ESA, and f_L is selected to cover the most sensitive frequency band.

14.4 Distributed finite-time control of ESA

In a realistic power grid, the small-scale ESSs are usually owned by customers that lack direct communication with the AGC controller. In this condition, the ESSs are aggregated into the ESA with only peer-to-peer communications, which can participate in system frequency services as an entity. In this section, a detailed design of the ESA based on the finite-time consensus approach and its characteristics will be introduced.

14.4.1 Communication graph

The communication network of an aggregated ESSs (denoted by $i = 0, 1, \ldots, N$) can be described by a graph, which is defined as $\mathcal{G} = (\mathcal{V}, \mathcal{E})$, with a set of nodes $\mathcal{V} = v_1, v_2, \ldots, v_N\}$ and a set of edges $\mathcal{E} \subseteq \mathcal{V} \times \mathcal{V}$. Each node is assigned to an ESS in the ESA, and edges represent communication links for data exchange. If communication links are bidirectional, $(v_i, v_j) \in \mathcal{E} \Rightarrow (v_j, v_i) \in \mathcal{E}\ \forall i, j$ the graph is said to be undirected. Otherwise, it is directed. A directed graph is said to have a spanning tree, if there is a root node, such that there is a directed path from the root to any other node in the graph. A matrix called adjacency matrix $A_{\mathcal{G}} = [a_{ij}]$, where a_{ij} can be defined as

$$a_{ij} = \begin{cases} 1, & \text{if}\,(v_i, v_j) \in \mathcal{E} \\ 0, & \text{otherwise} \end{cases} \tag{14.13}$$

The in-degree matrix of $\mathcal{G}$ is defined as $\Delta = \text{diag}\{\Delta i\}$ with $\Delta_i = \sum_{j=1}^{N} a_{ij}$. The Laplacian matrix can be represented as

$$\boldsymbol{L} = \boldsymbol{\Delta} - \boldsymbol{A}_{\mathcal{G}} \Rightarrow l_{ij} = \begin{cases} -a_{ij} & ,i \neq j \\ \sum_{j \in N_i}^{n} a_{ij} & ,i = j \end{cases} \tag{14.14}$$

where the Laplacian matrix $\boldsymbol{L}$ has properties that all row sums are zero.

For the system with a leader (labelled as node 0), the interaction topology can be expressed by a directed graph $\overline{\mathcal{G}}$, which contains original graph $\mathcal{G}$, node v_0, and edges (v_i, v_0) from node v_0 to other nodes. The leader can send information to followers, but not vice versa. The matrix $\boldsymbol{G} = \text{diag}\{g_1, g_2, \ldots, g_N\}$ is used to describe whether each follower directly receive information from the leader, where

$$g_i = \begin{cases} 1, & if\,\exists (v_i, v_0) \\ 0, & \text{otherwise} \end{cases} \tag{14.15}$$

One important property to guarantee that the leader–follower consensus algorithm with graph $\overline{\mathcal{G}}$ will convergence is that the graph $\overline{\mathcal{G}}$ contains a spanning tree rooted at node 0 and all the eigenvalues of matrix $\boldsymbol{L}+\boldsymbol{G}$ have positive real parts.

14.4.2 Finite-time consensus control

In this part, a finite-time leader–follower consensus algorithm is proposed to control each ESS in the ESA. Compared to other consensus algorithms, the finite-time consensus approach guarantees the power tracking and SoC balancing of each ESS in finite time. The modelling and finite-time consensus control of ESA are illustrated in the following.

The real power output of ith ESS is determined by

$$P_{\text{ESS},i}(t) = P_i(t) P_{ESS,i}^{\max} \tag{14.16}$$

where $P_{\mathrm{ESS},i}^{\max}$ is the power rating (MW) of ith ESS. p_i is the power state of the communication agent associated with ith ESS. $p_i \in [-1, 1]$ considering the power limits of ESSs.

Based on the basic Coulomb counting method [12], the SoC of ith ESS can be estimated as

$$SoC_i(t) = SoC_i(0) - \int_{t=0}^{T} \frac{\eta_i P_{\mathrm{ESS},i}(t)}{3,600 \times C_{\mathrm{ESS},i}} dt \tag{14.17}$$

where $C_{\mathrm{ESS},i}$ and $SoC_i(0)$are the capacity (MWh) and initial SoC of the ith ESS. η_i is the charging or discharging efficiency of the ith ESS, $\eta_i = \eta_i^{\mathrm{ch}}$if $P_{\mathrm{ESS},i}(t) < 0$, $\eta_i = \eta_i^{\mathrm{dis}}$if $P_{\mathrm{ESS},i}(t) > 0$. It is considered that $\eta_i^{\mathrm{ch}} = \eta_i^{\mathrm{dis}} = 0.95$ for all ESSs. $C_{\mathrm{ESS},i}$ is multiplied by 3,600 to convert the units from hours to seconds.

Here, we define an energy state e_i which indicates the SoC_i of ith ESS. $e_i \in [0, 1]$ considering the capacity limits of ESSs. Combining (14.17) with (14.16) and differentiating it with respect to time, the relationship between p_i and e_i is expressed as

$$\dot{e}_i(t) = -\frac{\eta_i P_{\mathrm{ESS},i}^{\max} p_i(t)}{3,600 \times C_{\mathrm{ESS},i}} = K_{\mathrm{ESS},i} p_i(t) \tag{14.18}$$

where $K_{\mathrm{ESS},i}$ is a coefficient between p_i and e_i. It is assumed that the power output of each ESS is rated at 1 C, which means $P_{\mathrm{ESS},i}^{\max}$ and $C_{\mathrm{ESS},i}$ have the same value but different units.

Considering the control input dynamics of power state, the ESA can be modelled as a group of homogeneous double integral systems as follows:

$$\begin{cases} \dot{e}_i(t) = K_{\mathrm{ESS}} p_i(t) \\ \dot{p}_i(t) = u_i(t) \end{cases}, i = 1, 2..., N \tag{14.19}$$

where e_i and p_i are the energy and power states and u_i is the control input of ith ESS.

A leader is defined as a controller which can receive the upper-level control signal (frequency control signal). The leader will update the reference power state and energy state of the ESA as follows:

$$\begin{cases} \dot{e}_0(t) = K_{\mathrm{ESS}} p_0(t) \\ p_0(t) = \dfrac{P_{\mathrm{ESA}}^{*}(t)}{P_{\mathrm{ESA}}^{\max}} \end{cases} \tag{14.20}$$

where e_0 and p_0 are reference energy and power states of the leader, respectively. P_{ESA}^{*} is the system frequency control signal and $P_{\mathrm{ESA}}^{\max}$ is the power rating of the ESA, which is equal to $\sum P_{\mathrm{ESS},i}^{\max}$.

Therefore, the research problem can be formulated as designing a leader–follower finite-time consensus control for a group of ESSs with the models (14.19) and (14.20) in an ESA.

For an ESA formulated by (14.19) and (14.20), the leader–follower finite-time consensus is achieved, if for any initial states, there exists a $T_0 \in [0, \infty]$, such that

$$\lim_{t \to T_0} \|e_i(t) - e_0(t)\| = 0, \lim_{t \to T_0} \|p_i(t) - p_0(t)\| = 0$$
$$e_i(t) = e_0(t), p_i(t) = p_0(t), \forall t \geq T_0, i = 1, 2, ..., N \tag{14.21}$$

In order to achieve (14.21), the leader–follower finite-time consensus control law is designed as follows:

$$\begin{aligned} u_i(t) = & \sum_{j=1}^{N} a_{ij}\left(\mathrm{sig}\left(e_i(t) - e_j(t)\right)^{\alpha}\right) - g_i(\mathrm{sig}(e_i(t) - e_0(t))^{\alpha}) \\ & -\gamma \sum_{j=1}^{N} a_{ij}\left(\mathrm{sig}(p_i(t) - p_j(t))^{\beta}\right) - g_i\left(\mathrm{sig}(p_i(t) - p_0(t))^{\beta}\right) \end{aligned} \tag{14.22}$$

where the function $\mathrm{sig}(x)^{\alpha} = |x|^{\alpha}\mathrm{sign}(x)$, $|x|$ denotes the absolute value of variable x, and sign(.) denotes the sign function. α, β, and γ are control gains to be selected, $\alpha, \beta, \gamma \in (0, 1), \mathrm{and} \beta = 2 \cdot \alpha/(1 + \alpha)$.

The settling time T_0 satisfies the following inequality [13]:

$$T \leq \frac{V^{1-\alpha}(0)}{c(1 - \alpha)} \tag{14.23}$$

where $V(x)$ is a positive definite function defined in a neighbourhood of the origin, and real number $c > 0$. The stability of the systems (14.19) and (14.20) with control law (14.22) can be proved using Lyapunov's second method by properly selecting the Lyapunov function.

Remark 2: The proposed ESA with finite-time power tracking and SoC balancing has the following advantages: (i) it avoids unintentional switching off of certain ESSs when they reach their SoC limits; (ii) the ESA can maintain a larger power rating for a longer time; (iii) it is convenient for the upper-level controller to manage the SoCs of the entire aggregator; and (iv) it offers PnP capability – i.e. when a new ESS joins the aggregator for frequency control, its SoC can autonomously be balanced with other ESSs.

14.4.3 Numerical illustrations

This section aims to show the dynamics of power and energy states and validate control effectiveness under different communication graphs. The control algorithm is tested under two directed communication graphs ($\overline{\mathcal{G}}_1$ and $\overline{\mathcal{G}}_2$), as shown in Figure 14.3. In $\overline{\mathcal{G}}_1$, Node 1 (ESS 1) is considered as a pinning node that can receive the information from the leader labelled as 0. Other ESSs are connected in a directed line as shown in Figure 14.3(a). In $\overline{\mathcal{G}}_2$, Nodes 1 and 8 are both pining nodes while other ESSs are connected in two directed lines as shown in Figure 14.3 (b). The convergence of each ESS with communication graphs $\overline{\mathcal{G}}_1$ and $\overline{\mathcal{G}}_2$ are

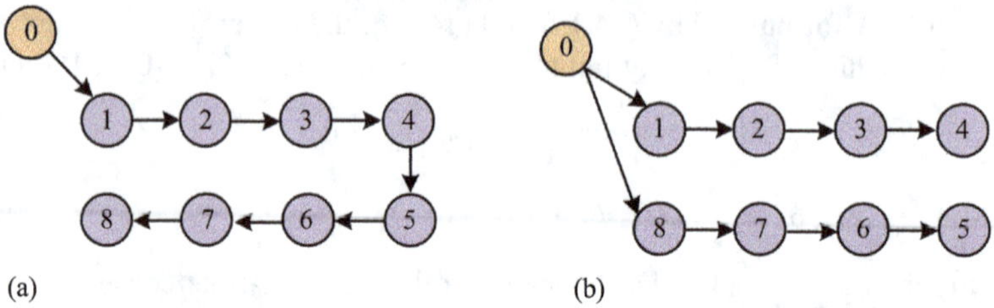

Figure 14.3 Communication network represented by two directed graphs (a) $\overline{\mathscr{G}}_1$ *and (b)* $\overline{\mathscr{G}}_2$

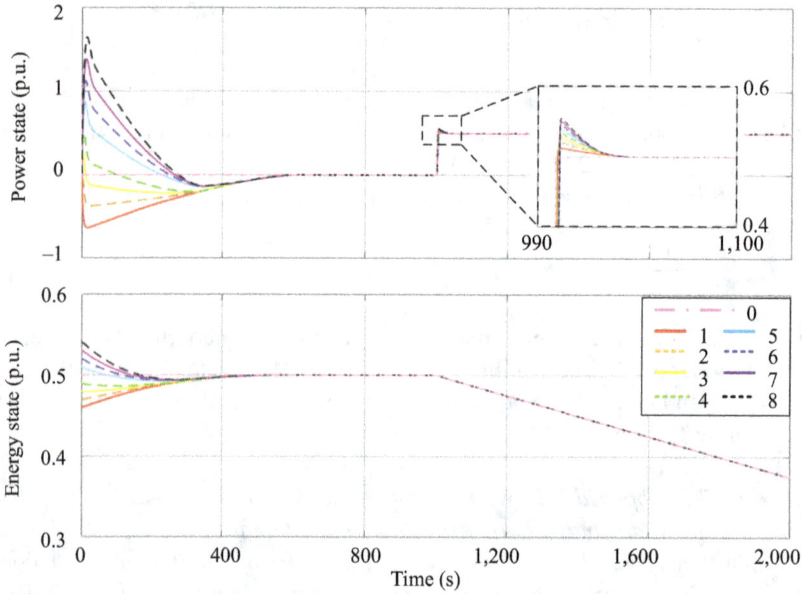

Figure 14.4 Convergence speed of ESA with communication graph $\overline{\mathscr{G}}_1$

shown in Figures 14.4 and 14.5, respectively. The figures are drawn with $T_{ca} = 0.01$ s, $e_0 = 0.5$, and $\alpha = 0.2$, $\beta = 0.33$, $\gamma = 0.6$. There is a step change in p_0 from 0 to 0.5 at 1,000 s. The initial energy states of each ESS are [0.46, 0.47, 0.48, 0.49, 0.51, 0.52, 0.53, 0.54] and initial power state for all is 0.

The results in Figures 14.4 and 14.5 are further illustrated in time sequence:

1. $0 < t < 1{,}000$ s. At the beginning, the energy state of each ESS is not equal. Although the reference power state is 0 during this period, each ESS still needs to charge or discharge in order to balance the energy state. As shown in Figures 14.4 and 14.5, both $\overline{\mathscr{G}}_1$ and $\overline{\mathscr{G}}_2$ will converge to consensus and satisfy (14.21) in finite time. In Figure 14.4, the convergence of ESSs is not

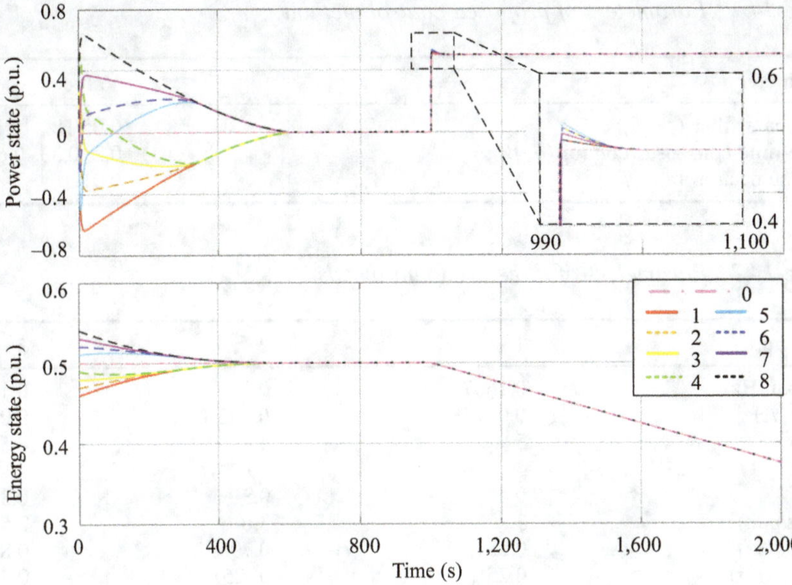

Figure 14.5 Converge speed of ESA with communication graph $\overline{\mathscr{G}}_2$

symmetrical as the graph $\overline{\mathscr{G}}_1$ is unbalanced. While in Figure 14.5, the convergence of ESSs 1–4 and ESSs 5–8 becomes symmetrical as these two subgroups have the same topologies in the graph $\overline{\mathscr{G}}_2$. The sum of power state for all eight ESSs in the graph $\overline{\mathscr{G}}_2$ remains 0, which is preferred as it is not desired that the SoC balancing influences the external characteristics of the ESA.

2. 1,000 s $< t <$ 2,000 s. As the consensus is reached in the first time period, the power and energy states of each ESS are equalised. When there is a step change of p_0 from 0 to 0.5 at 1,000 s, all ESSs start to track the new reference power state. Because the energy states are balanced during this period, the power states of ESSs will converge to p_0 much faster. The overshoot of $\overline{\mathscr{G}}_2$ is smaller than $\overline{\mathscr{G}}_1$, as the connection of the last node in $\overline{\mathscr{G}}_2$ is actually shorter than $\overline{\mathscr{G}}_1$.

Remark 3: From the numerical illustration, it is validated that both $\overline{\mathscr{G}}_1$ and $\overline{\mathscr{G}}_2$ can converge in the finite time. However, $\overline{\mathscr{G}}_2$ is preferred as it can reach consensus with smaller overshoot and does not influence the external power response of the ESA. $\overline{\mathscr{G}}_2$ will be adopted in the simulation tests in Section 14.5. Generally, the dynamics of the ESA are influenced by factors such as communication rate, graph, and control law.

14.5 Results and discussions

In this section, a variety of scenarios are tested to validate the proposed frequency control scheme. The LFC model and the proposed ESA are implemented in

Table 14.1 Parameters of proposed control method

Parameter	Value
Band-pass filter (f_L, f_H)	0.05 Hz, 10 Hz
Finite-time consensus control (α, β, γ)	0.2, 0.33, 0.6
Communication rate (T_{ca})	0.01 s

Table 14.2 Parameters of three-area microgrids

Area no.	1	2	3
$2H$ (p.u./Hz)	0.1667	0.2	0.15
D (p.u./Hz)	0.0015	0.002	0.001
T_a (s)	5	4	4.5
T_b (s)	40	30	40
T_g (s)	0.4	0.3	0.35
R (Hz/p.u.)	3	3.6	2.5
B (p.u./Hz)	0.8675	0.795	0.87
T_{ij} (p.u./Hz)	0.25	0. 25	0.25
C_{ESA} (MWh)	3.2	0	4.8

MATLAB®/Simulink®. To evaluate the performance of the proposed method, both system contingency and normal operation scenarios have been considered. The simulation has been first conducted in a single LFC area, and then extended to multi-LFC areas. The LFC system under consideration is a per-unit system with $S_{base} = 80$ MVA. The parameters of the proposed method are shown in Table 14.1. The parameters of the three-area LFC are shown in Table 14.2. It is assumed that in one control area, all the ESSs are aggregated into one ESA. The capacity of the ESA (C_{ESA}) in Area 1 is 3.2 MWh, with eight ESSs of capacity [0.68, 0.6, 0.52, 0.44, 0.36, 0.28, 0.2, 0.12] MWh, while the capacity of the ESA in Area 3 is scaled up to 4.8 MWh.

14.5.1 Case 1: system contingency

In Case 1, the dynamic performance of the proposed method is investigated under system contingency conditions in a single area microgrid. Area 1 with 3.2 MWh ESA, as in Table 14.2, is considered with initial SoCs of ESSs as 50%. A load disturbance of 6.4 MW at 5 s emulates the contingency.

The system frequency responses under the conventional LFC scheme, LFC with ESA but without the observer, and the proposed control scheme are compared in Figure 14.6. Upon the occurrence of the disturbance at 5, the frequency of the system begins to drop. The dynamics of the ESA is mainly governed by the proposed disturbance observer other than the AGC signal in this condition. As a result, it can be observed that both the frequency nadir and the frequency deviation recovery within the allowable range ($\pm$0.2 Hz) are improved with the proposed control scheme.

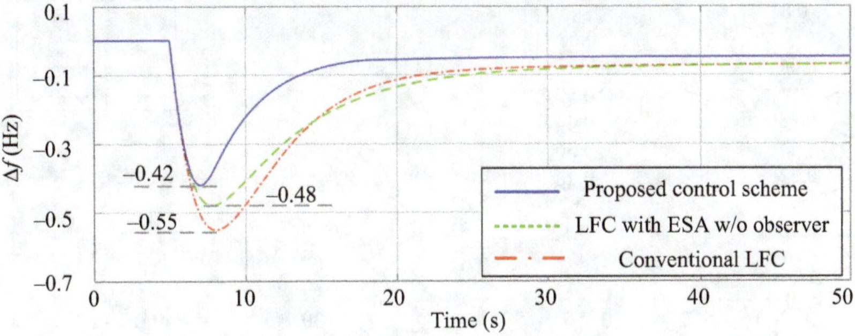

Figure 14.6 The frequency response under system contingency

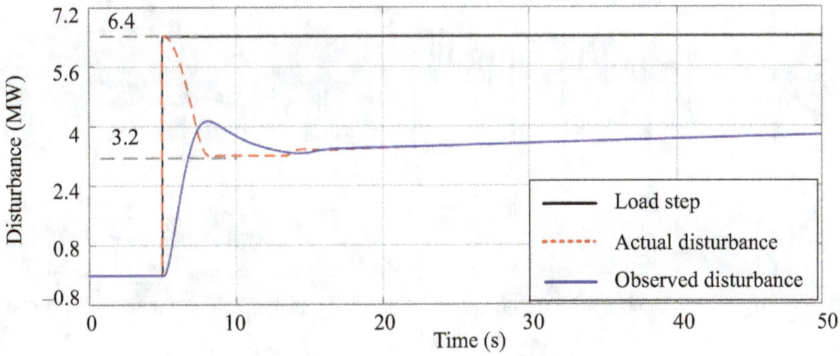

Figure 14.7 Performance of the proposed disturbance observer

The performance of the system disturbance observer is shown in Figure 14.7. The parameters of matrixes $\boldsymbol{M}$ and $\boldsymbol{R}$ are chosen as $\boldsymbol{M} = [-1, -1, -1]$ and $\boldsymbol{R} = \text{diag}\{10, 10, 10\}$ according to (14.11) and (14.12). In this condition, the eigenvalues $\lambda_1 = -22.4$, $\lambda_2 = -10$, and $\lambda_{3,4} = -6.4\pm4.3j$ are assigned to the disturbance observer and the system (14.11) is stable. It can be noticed that the estimated system disturbance can track the actual system disturbance after a short time. The participation of the ESA in conjunction with the proposed disturbance observer actually reduces the total system disturbance as compared to the load step of 6.4 MW, thereby improving the frequency response.

The power output of each ESS in the ESA during this process is shown in Figure 14.8. As shown in Figure 14.8, each ESS can respond fast to the load change and reach its maximum power output when there is a large system disturbance. The response speed of each ESS is determined by control law (14.22) as well as communication graphs. The maximum power rating of each ESS is also reflected in Figure 14.8, which is the same as the capacity of each ESS mentioned before. As the ESA is designed to compensate for the power variation rather than the steady-state disturbances, the band-pass filer will reduce the power output of each ESS

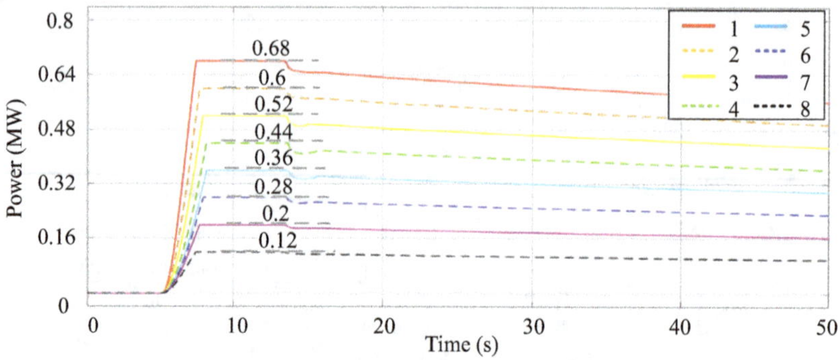

Figure 14.8 The power output of each ESS in the ESA

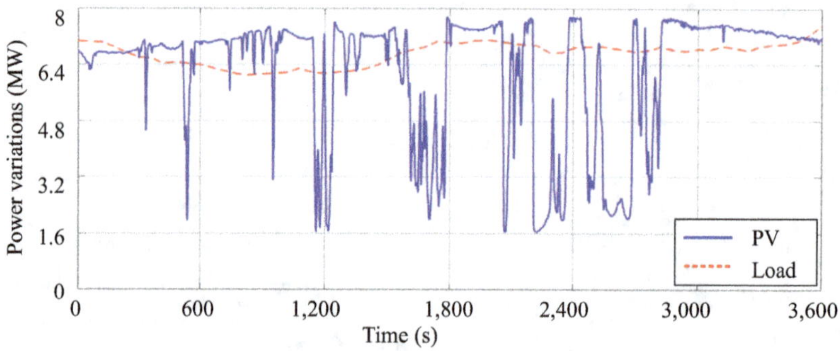

Figure 14.9 The 1-h PV and load variation profiles

from 14 s. Correspondingly, the disturbance in Figure 14.7 will gradually increase after 14 s due to the reduction of power output from the ESA.

14.5.2 Case 2: normal operation

In Case 2, the normal operation study is conducted to further investigate the dynamic performance of the proposed method. The same single area system is considered as in Case 1. The 1 h PV and load variation profiles used in this case are shown in Figure 14.9. The PV data with 1-s resolution were measured in June 2012 by EPRI [14]. The PV profile has very high fluctuations while the load profile is much smoother. The initial SoC of each ESS in the ESA is [46%, 47%, 48%, 49%, 51%, 52%, 53%, 54%]. In this case, the AGC participation factors of ESAs and generators are $\alpha_{\mathrm{ESA}} = 0.1$ and $\alpha_{\mathrm{G}} = 0.9$, respectively.

Figure 14.10(a)–(c) shows system frequency deviations for conventional LFC scheme, LFC with ESA but without the observer, and with the proposed control scheme, respectively. Comparing Figure 14.10(a)–(c), it can be observed that frequency deviations are mitigated much more effectively with the proposed method. The frequency deviations can be regulated in an allowable range (±0.2 Hz) with

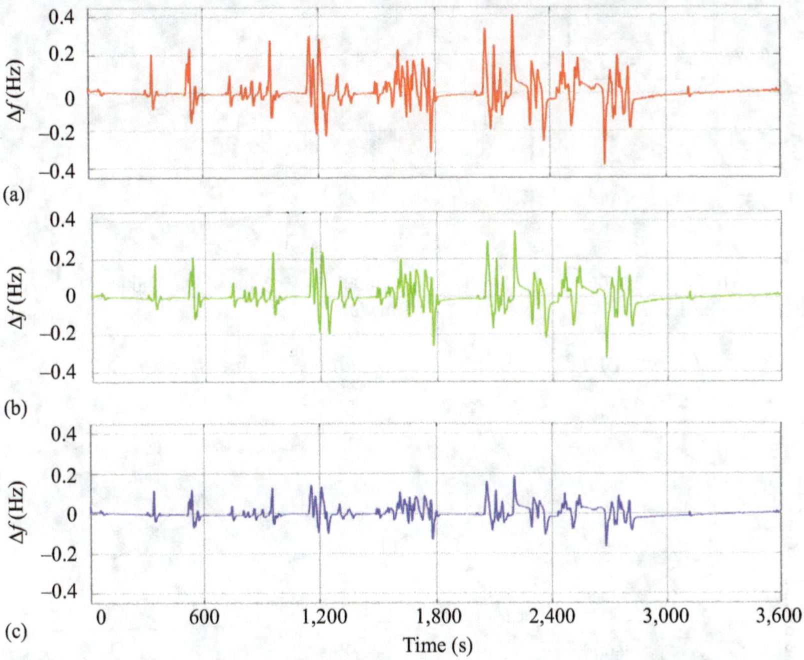

Figure 14.10 Comparison of system frequency deviations under (a) conventional LFC scheme (without ESA), (b) LFC with ESA without observer, and (c) proposed control scheme

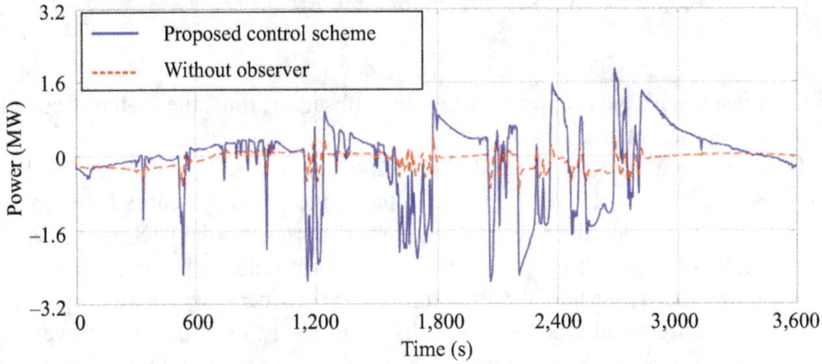

Figure 14.11 Total power output of the ESA with and without the proposed disturbance observer

the 3.2 MWh ESA using the proposed method. Furthermore, the total power output of the ESA with and without the proposed disturbance observer is compared in Figure 14.11. The results are corresponding to (b) and (c) in Figure 14.10. The ESA

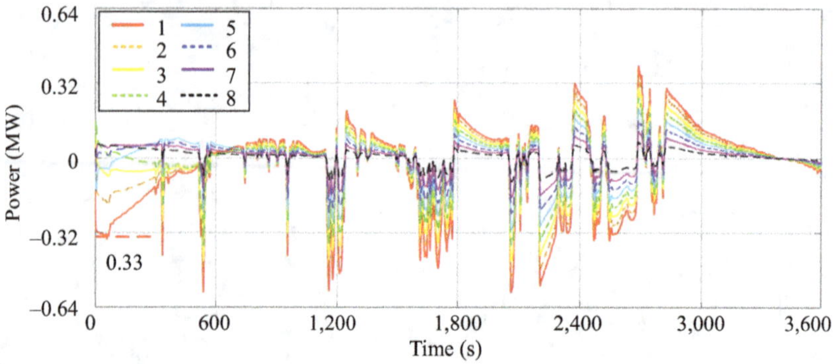

Figure 14.12 Power output of each ESS in test Case 2

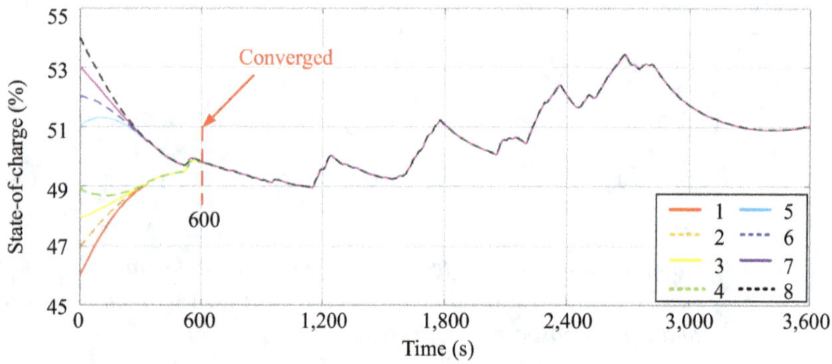

Figure 14.13 SoC profile of each ESS in test Case 2

will react faster with the proposed disturbance observer; thus, the system frequency deviation becomes smaller.

The power output and SoC profiles of each ESS in the ESA, in this case, is shown in Figures 14.12 and 14.13, respectively. Initially, the SoC differences between each ESS are large. Therefore, the SoC balancing effect dominates the ESS dynamics. The power sharing among each ESS is much influenced in order to balance SoCs. While the total power output of the ESA still follows the frequency control signal. As the SoC of each ESS gradually becomes equalised during the operation, the power sharing among each ESS becomes proportional to their power rating, same as Figure 14.8 in Case 1. It can be also observed from Figure 14.11 that each ESS in the ESA responds to sudden PV fluctuation to mitigate the frequency deviations.

14.5.3 Case 3: multi-area microgrids

In Case 3, the system contingency study is extended to multi-area LFC to further investigate the performance of the proposed method. The structure of the studied

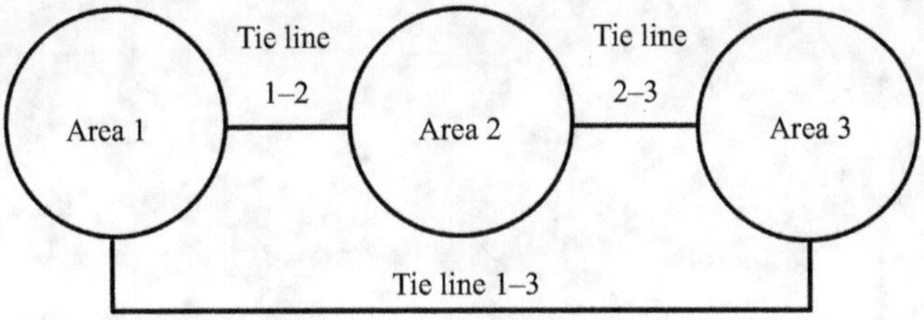

Figure 14.14 The structure of the studied three-area power system

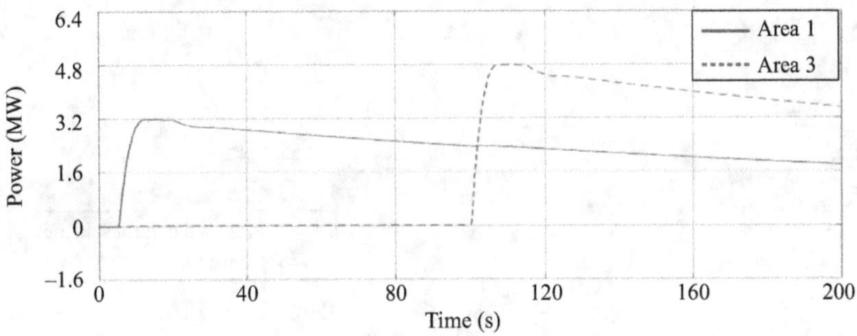

Figure 14.15 Total power output of ESAs in Areas 1 and 3

three-area interconnected microgrid is depicted in Figure 14.14. In Case 3, it is assumed that there is 3.2 MWh ESA in Area 1 and 4.8 MWh ESA in Area 3. There are step load disturbances of 6.4 MW at 5 s in Area 1 and 9.6 MW at 100 s in Area 3. The proposed disturbance observer will estimate the system distance in each control area for the corresponding ESA. Other simulation conditions are the same as in Case 1.

Figure 14.15 shows the total power output of ESAs under step load changes in Areas 1 and 3, respectively. The ESAs in Areas 1 and 3 will respond to the system disturbance in the respective control area, as shown in Figure 14.15. Figure 14.16 shows the frequency response in each control area of Case 3. As shown in Figure 14.16, the sudden load changes will influence the frequency in not only the same area but also the interconnected microgrids. The frequency response of each control area can be improved with the proposed observer and finite-time consensus-based ESAs.

14.5.4 Case 4: comparison with linear control algorithm

In this subsection, the proposed finite-time consensus control law is compared to a widely utilised linear consensus control law for the second-order multi-agent system reported in [15]. Compared to the proposed finite-time approach, the linear control law in [15] cannot ensure consensus in finite time. For systems (14.19) and

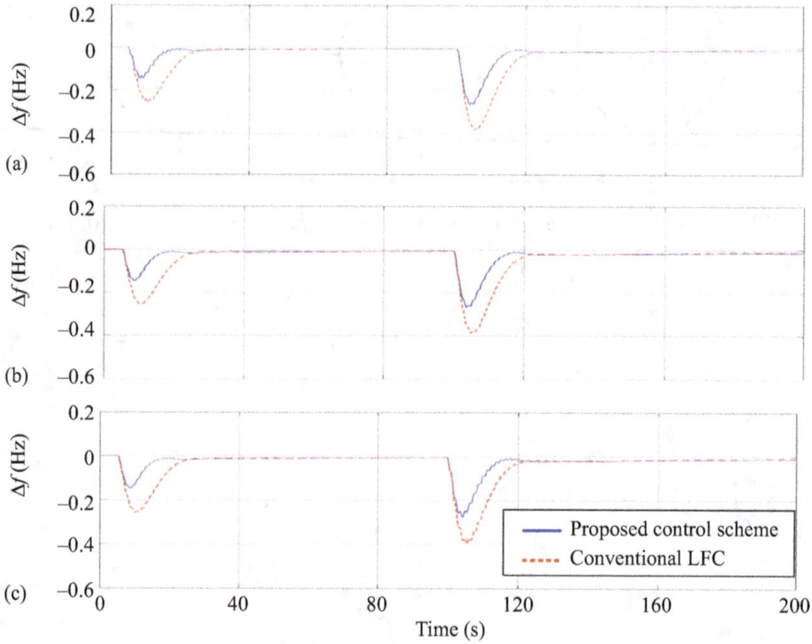

Figure 14.16 Frequency response of each control area: (a) Area 1; (b) Area 2; (c) Area 3

(14.20), the linear control law in [15] can be expressed as

$$\begin{aligned} u_i(t) = {} & k_1 \sum_{j=1}^{N} a_{ij}\big(e_i(t) - e_j(t)\big) - g_i(e_i(t) - e_0(t)) \\ & -k_2 \sum_{j=1}^{N} a_{ij}\big(p_i(t) - p_j(t)\big) - g_i(p_i(t) - p_0(t)) \end{aligned} \tag{14.24}$$

where k_1 and k_2 are control gains to be selected.

First, the performance of the linear control law (14.24) and the proposed control law (14.22) are compared under system contingency condition in Case 1. The comparison results of frequency response and ESA power output are shown in Figure 14.17. The control gains of (14.24) are selected as $k_1 = 35$, $k_2 = 1$. As shown in Figure 14.17(a), it can be found that the proposed method improves the frequency nadir as compared to the linear method. Correspondingly, the power output of the ESA under the proposed method also reacts faster than the linear method, as shown in Figure 14.17(b).

Second, the linear control in (14.24) is compared against the proposed control law under the normal operation scenario in Case 2. The results of Case 2 serve as a base reference; that is, the convergence time for (14.22) is 600 s (as seen in Figure 14.13) and the maximum power overshoot observed is 0.33 MW (for ESS 1

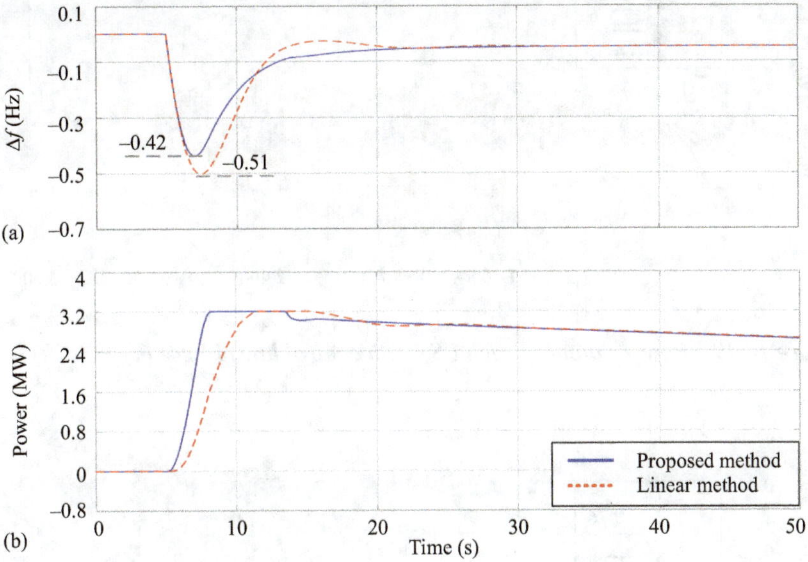

Figure 14.17 Comparison with the method in [15]. (a) Frequency response. (b) ESA power output

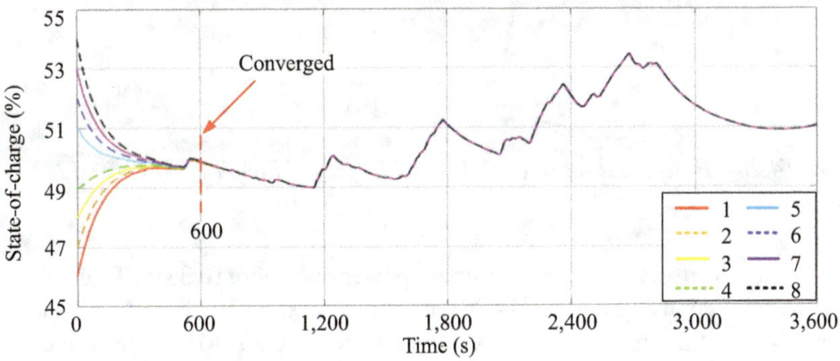

Figure 14.18 SoC profile of each ESS with control law (14.24) ($k_1 = 35$, $k_2 = 1$)

as seen in Figure 14.12). The study is undertaken in two steps and the performance in terms of convergence time and system overshoot are compared.

In order to ensure the same convergence time (600 s) as in Case 2, the control gains of (14.24) are tuned as $k_1 = 35$, $k_2 = 1$. As shown in Figure 14.18, the SoC of each ESS can converge at 600 s, the same as in Figure 14.13. However, the power overshoots in Figure 14.19 during the convergence are much higher than in Figure 14.12, which even reach the power limits of the ESSs. It can therefore be said that the proposed method presents a smaller system overshoot under the same convergence time.

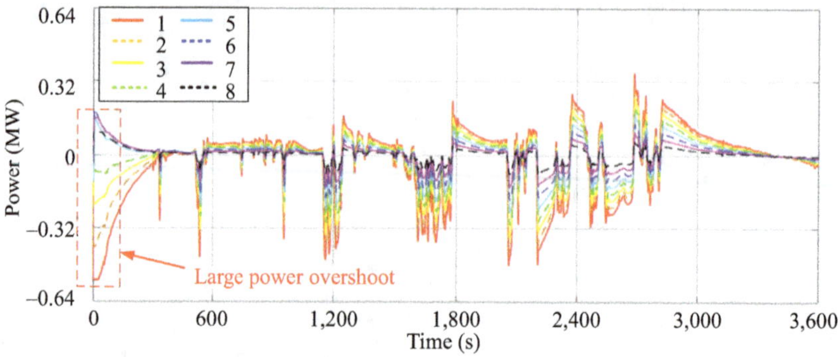

Figure 14.19 Power output of each ESS with control law (14.24) ($k_1 = 35$, $k_2 = 1$)

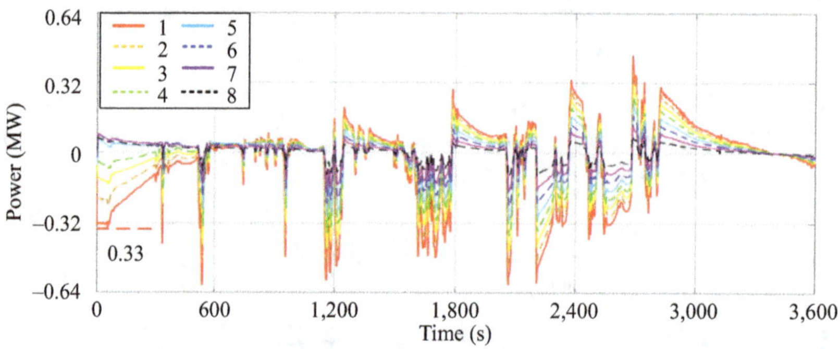

Figure 14.20 Power output of each ESS with control law (14.24) ($k_1 = 12$, $k_2 = 1$)

In order to ensure the same maximum power overshoot (0.33 MW of ESS-1) as in Case 2, the control gains of (14.24) are tuned as $k_1 = 12$, $k_2 = 1$. As shown in Figure 14.20, the maximum power overshoot is 0.33 MW, the same as in Figure 14.12. However, the SoC (or system) in Figure 14.21 converges much slower as compared to Figure 14.13. Namely, the proposed method can provide faster convergence speed under the same system overshoot.

The advantages of the finite-time approach compared with the linear method in [15] are summarised as follows. In case of a system contingency, the ESA with the proposed finite-time approach has a faster response of power output, thus far improves the system frequency response. Under the normal operation, the ESS with the proposed finite-time approach presents smaller power overshoot and faster convergence. Smaller power overshoot and faster convergence are desired for operation of ESA as (i) ESSs within the ESA have a lower chance to reach the power limits during the operation and (ii) any newly joined ESS can attain faster SoC balancing, ensuring PnP capability.

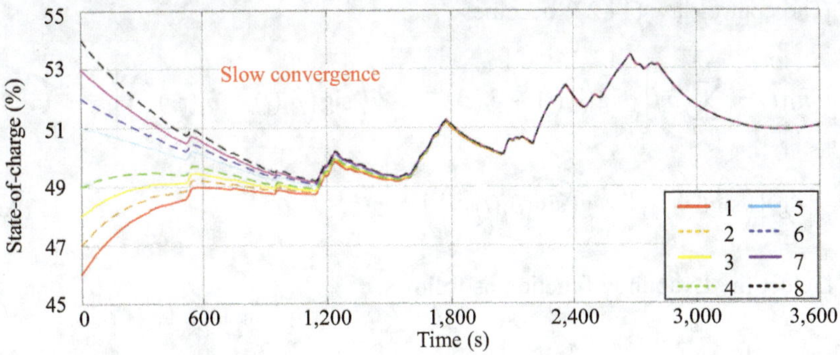

Figure 14.21 *SoC profile of each ESS with control law (14.24) ($k_1 = 12$, $k_2 = 1$)*

14.6 Conclusion

In this chapter, a new frequency control scheme with finite-time consensus-controlled ESA is proposed to improve the microgrid frequency response in the presence of stochastic renewable power generation. The total frequency control signal for ESA is determined by both AGC commands and the proposed disturbance observer. The control law using a leader–follower finite-time consensus algorithm is developed to aggregate ESSs in one control area into an ESA. The stability proof of the proposed method for ESSs is given. The internal and external characteristics of the ESA can be adjusted by its communication graph and control law settings.

The simulation results have demonstrated that the system frequency response can be improved by using the proposed disturbance observer and finite-time consensus-based ESA. The proposed method is effective for both system contingency and normal operating conditions. The ESA can track the frequency control signal and the SoC of each ESS can be equalised in finite-time during the operation. The proposed method is further validated in a three-area microgrid. The comparative studies validate the advantages and necessity of the proposed method. In summary, the proposed method provides an effective way to aggregate widespread DESSs for system frequency control.

Appendix A

In this subsection, the stability of the ESS represented by (14.19) and (14.20) with the finite-time control law in (14.22) is proved. First, the globally asymptotically stability is validated as follows:

Let $\widehat{p}_i(t) = p_i(t) - p_0(t), \widehat{e}_i(t) = e_i(t) - e_0(t)$, then the error system of ESS depicted by (14.19) and (14.20) becomes

$$\begin{cases} \dot{\widehat{e}}_i(t) = K_{\text{ESS}}\widehat{p}_i(t) \\ \dot{\widehat{p}}_i(t) = \widehat{u}_i(t) \end{cases}, i = 1, 2, \ldots, N \tag{A.1}$$

The control law (14.22) becomes

$$\widehat{u}_i(t) = \sum_{j=1}^{N} a_{ij}\left[\left(\mathrm{sig}\left(\widehat{e}_i(t) - \widehat{e}_j(t)\right)^{\alpha}\right) - \gamma\left(\mathrm{sig}\left(\widehat{p}_i(t) - \widehat{p}_j(t)\right)^{\beta}\right)\right]$$

$$-g_i\left[\left(\mathrm{sig}(\widehat{e}_i(t))^{\alpha} - \gamma\left(\mathrm{sig}(\widehat{p}_i(t))^{\beta}\right)\right)\right] \tag{A.2}$$

Select the Lyapunov function as follows:

$$V = \frac{1}{2}\sum_{i=1}^{n}\widehat{p}_i^2 + \sum_{i=1}^{n}\sum_{j=1}^{n}\int_0^{\widehat{e_i}-\widehat{e_j}} a_{ij}(\mathrm{sig}(s)^{\alpha})\mathrm{d}s + \sum_{i=1}^{n}\int_0^{\widehat{e_i}} \widehat{p}_i(\mathrm{sig}(s)^{\alpha})\mathrm{d}s \tag{A.3}$$

Then take the derivative of V

$$\begin{aligned}
\dot{V} &= \sum_{j=1}^{N}\widehat{p}_i\left[\sum_{j=1}^{N}a_{ij}(sig(\widehat{e}_i - \widehat{e}_j)^{\alpha}) - \gamma(sig(\widehat{p}_i - \widehat{p}_j)^{\beta}) - g_i(sig(\widehat{e}_1)^{\alpha} - \gamma sig(\widehat{p}_1)^{\beta})\right] \\
&+ \frac{1}{2}\sum_{i=1}^{N}\sum_{j=1}^{N}a_{ij}(\widehat{p}_i - \widehat{p}_j)sig(\widehat{e}_1 - \widehat{e}_j)^{\alpha} + \sum_{i=1}^{N}g_i\widehat{p}_i sig(\widehat{e}_1)^{\alpha} \\
&= \sum_{i=1}^{N}\widehat{p}_i\gamma\left[\sum_{j=1}^{N}a_{ij}sig(\widehat{p}_i - \widehat{p}_i)^{\beta} - sig(\widehat{p}_i)^{\beta}\right] \\
&= \frac{1}{2}\sum_{i=1}^{N}\sum_{j=1}^{N}a_{ij}(\widehat{p}_i - \widehat{p}_i)sig(\widehat{p}_i - \widehat{p}_i)^{\beta} - \sum_{i=1}^{n}g_i\widehat{p}_i sig(\widehat{p}_1)^{\beta} \\
&\le \frac{1}{2}\sum_{i=1}^{N}\sum_{j=1}^{N}a_{ij}(\widehat{p}_i - \widehat{p}_i)sig(\widehat{p}_i - \widehat{p}_i)^{\beta} \\
&\le 0
\end{aligned} \tag{A.4}$$

According to Lyapunov's second method for stability, $V \ge 0, \dot{V} \le 0$. Therefore, the system (14.24) with the control law (A.1) is stable.

Furthermore, note that $\dot{V} = 0$, if and only if $\widehat{p}_i = \widehat{p}_j = 0$, which implies $\dot{\widehat{p}}_i = 0, \forall i,j = 1,2,\ldots,N$.

$$\begin{aligned}
\dot{\widehat{p}}_i(t) &= \sum_{j=1}^{N}a_{ij}\left(\mathrm{sig}(\widehat{e}_i(t) - \widehat{e}_j(t))^{\alpha} - \gamma\mathrm{sig}\left(\widehat{p}_i(t) - \widehat{p}_j(t)\right)^{\beta}\right) \\
&\quad -g_i\left(\mathrm{sig}(\widehat{e}_i(t))^{\alpha} - \gamma\mathrm{sig}(\widehat{p}_i(t))^{\beta}\right)] \\
&= \sum_{j=1}^{N}a_{ij}\mathrm{sig}\left(\widehat{e}_i(t) - \widehat{e}_j(t)\right)^{\alpha} - g_i\mathrm{sig}(\widehat{e}_i(t))^{\alpha} = 0
\end{aligned} \tag{A.5}$$

It follows

$$\sum_{i=1}^{N}\widehat{p}_i\left[\sum_{j=1}^{N}a_{ij}\mathrm{sig}(\widehat{e}_i(t)-\widehat{e}_j(t))^{\alpha}-g_i\mathrm{sig}(\widehat{e}_i(t))^{\alpha}\right]$$

$$=-\frac{1}{2}\sum_{i=1}^{N}\sum_{j=1}^{N}a_{ij}(\widehat{e}_i-\widehat{e}_j)\mathrm{sig}(\widehat{e}_i-\widehat{e}_j)^{\alpha}-\sum_{i=1}^{N}g_i\widehat{e}_i\mathrm{sig}(\widehat{e}_i)^{\alpha}=0 \quad \text{(A.6)}$$

Equality above indicates that $\widehat{e}_i=\widehat{e}_j=0$, which means the system will converge to $e_i\to e_0, p_i\to p_0$in steady state. Thus, the system (14.24) with control law (A.1) is globally asymptotically stable.

Next, continuing with the previous proof of globally asymptotic stability, the locally finite-time convergence of the system (14.24) with control law (A.1) is proved as follows. The following lemmas and definition from [16] are used in the proof.

Lemma 1 Lasalle's Invariance Principle. *Let* x(t) *be a solution of* $\dot{x}=f(x)$, $x(0)=x_0\in R^k$, *where* $\mathrm{f}:\mathrm{U}\to\mathrm{R}^{\mathrm{k}}$ *is continuous with* U *an open subset of* R^{k}, *and let* $\mathrm{V}:\mathrm{U}\to\mathrm{R}^{\mathrm{k}}$ *be a locally Lipschitz function such that* $\mathrm{D}^+\mathrm{V}$ *(x(t))* $\leq$ *0, where* D^+ *denotes the upper Dini derivative. Then, denoting the positive limit set as* $\Lambda^+(\mathrm{x}_0)$, $\Lambda^+(\mathrm{x}_0)\cap\mathrm{U}$ *is contained in the union of all solutions that remain in* $\mathrm{S}=\{\mathrm{x}\in\mathrm{U}:\mathrm{D}^+\mathrm{V}(\mathrm{x})=0\}$.

Definition 1 *Homogeneity with dilation*. A function $V(x)$ of $x\in R^k$ is homogeneous of degree $\sigma\geq 0$ with dilation coefficients $(r_1, \ldots, r_k)$, if $V(\varepsilon^{r_1}x_1,\ldots,\varepsilon^{r_k}x_k)=\varepsilon^{\sigma}V(x), \varepsilon>0$. If $r_1=\ldots=r_k=1$, then the dilation is called trivial.

Consider k-dimensional system

$$\dot{x}=f(x), x=(x_1,\ldots,x_k)^T\in R^k \quad \text{(A.7)}$$

A continuous vector field $f(x)=(f_1(x), \ldots, f_k(x))^{\mathrm{T}}$ is homogeneous of degree $\sigma\in R$ with dilation $(r_1, \ldots, r_k)$, if $f_i(\varepsilon^{r_1}x_1,\ldots,\varepsilon^{r_k}x_k)=\varepsilon^{\sigma}f_i(x), \varepsilon>0$.

System (31) is called homogeneous if its vector field is homogeneous. Moreover,

$$\dot{x}=f(x)+\widehat{f}(x), \widehat{f}(0)=0, x\in R^k \quad \text{(A.8)}$$

is called locally homogeneous if f is homogeneous of degree $\sigma\in R$ with dilation $(r_1, \ldots, r_k)$ and $\widehat{f}$ is a continuous vector field satisfying

$$\lim_{\varepsilon\to 0}\frac{\widehat{f}_i(\varepsilon^{r_1}x_1,\ldots,\varepsilon^{r_k}x_k)}{\varepsilon^{\sigma+r_i}}=0, \forall x\neq 0 \quad \text{(A.9)}$$

Lemma 2 *Suppose system (A.7) is homogeneous of degree* σ *with dilation* $(\mathrm{r}_1, \ldots, \mathrm{r}_k)$, f *is continuous and* x = *0 is its asymptotically stable equilibrium. If homogeneity* *degree $\sigma < 0$, the equilibrium of system (A.7) is finite-time stable. Moreover, if (A.9) holds, then the equilibrium of system (A.9) is locally finite-time stable.*

Then we can get the system with variables $(\widehat{e}_1, \widehat{e}_2, ..., \widehat{e}_n, \widehat{p}_1, \widehat{p}_2, ..., \widehat{p}_n)$ is homogeneous of degree $\kappa = \alpha - 1 < 0$ with dilation $(2, 2, ..., 2, 1 + \alpha, 1 + \alpha, ..., 1 + \alpha)$. Therefore, according to Lemma 2, the system with control law is locally finite-time stable.

If the equilibrium of a control is globally asymptotically stable and locally finite-time convergent, then the control is globally finite-time stable. This follows the principle that globally asymptotical stability implies finite-time convergence to any given bounded neighbourhood of the equilibrium. Therefore, the system (14.24) with control law (A.1) is globally finite-time stable. In other words, we have $\widehat{e}_i - \widehat{e}_j \to 0, \widehat{p}_i - \widehat{p}_j \to 0$, $\forall i, j = 1, \ldots, N$ in finite time. This completes the proof.

References

[1] Y. Xu, L. Xie, and C. Singh, 'Optimal scheduling and operation of load aggregator with electric energy storage in power markets', *North American Power Symposium (NAPS), 2010*, IEEE, 2010.

[2] L. Gkatzikis, K. Iordanis, and S. Theodoros, 'The role of aggregators in smart grid demand response markets', *IEEE Journal on Selected Areas in Communications*, vol. 7, no. 31, pp. 1247–1257, 2013.

[3] H. Zhao, Q. Wu, S. Hu, H. Xu, and C.N. Rasmussen, 'Review of energy storage system for wind power integration support', *Applied Energy*, vol. 137, pp. 545–553, 2015.

[4] T. Morstyn, B. Hredzak, and V.G. Agelidis, 'Control strategies for microgrids with distributed energy storage systems: an overview', *IEEE Transactions on Smart Grid*, vol. 9, no. 4, pp. 3652–3666, 2018.

[5] S.J. Lee, J.H. Kim, C.H. Kim, *et al.*, 'Coordinated control algorithm for distributed battery energy storage systems for mitigating voltage and frequency deviations', *IEEE Transactions on Smart Grid*, vol. 7, no. 3, pp. 1713–1722, 2017.

[6] B.A. Robbins, C.N. Hadjicostis, and A.D. Domínguez-García, 'A two-stage distributed architecture for voltage control in power distribution systems', *IEEE Transactions on Power Systems*, vol. 28, no. 2, pp.1470–1482, 2013.

[7] Y. Wang, M.H. Syed, E. Guillo-Sansano, Y. Xu, and G. Burt, 'Inverter-based voltage control of distribution networks: a three-level coordinated method and power hardware-in-the-loop validation', *IEEE Transactions on Sustainable Energy*, 2019.

[8] S. Baros and M.D. Ilic, 'Distributed torque control of deloaded wind DFIGs for wind farm power output regulation', *IEEE Transactions on Power Systems*, vol. 32, no. 6, pp. 4590–4599, 2017.

[9] J. Hu, J. Cao, J.M. Guerrero, T. Yong, and J. Yu, 'Improving frequency stability based on distributed control of multiple load aggregators', *IEEE Transactions on Smart Grid*, vol. 8, no. 4, pp. 1553–1567, 2017.

[10] Y. Wang, Y. Tang, Y. Xu, and Y. Xu, 'A distributed control scheme of thermostatically controlled loads for the building-microgrid community', *IEEE Transactions on Sustainable Energy*, vol. 11, no. 1, pp. 350–360, 2020.

[11] D. Ginoya, P.D. Shendge, and S.B. Phadke, 'State and extended disturbance observer for sliding mode control of mismatched uncertain system', *Journal of Dynamic Systems, Measurement, and Control*, vol. 137, no. 7, 2015.

[12] K.S. Ng, C.S. Moo, Y.P. Chen, and Y.C. Hsieh, 'Enhanced Coulomb counting method for estimating state-of-charge and state-of-health of lithium-ion batteries', *Applied Energy*, vol. 86, no. 9, pp. 1506–1511, 2009.

[13] S. Yu and X. Long, 'Finite-time consensus for second-order multi-agent systems with disturbances by integral sliding mode', *Automatica*, vol. 54, pp. 158–165, 2015.

[14] '1-second resolution data', EPRI, Available: http://dpv.epri.com/measurement_data.html.

[15] Z. Li, Z. Duan, and L. Huang. 'Leader-follower consensus of multi-agent systems', American Control Conference, 2009, ACC'09, pp. 3256–3261. IEEE, 2009.

[16] X. Wang and Y. Hong, 'Finite-time consensus for multi-agent networks with second-order agent dynamics', *IFAC Proceedings Volumes*, vol. 41, no. 2, pp. 15185–15190, 2008.

Chapter 15

Power system frequency support by grid-interactive smart buildings

Grid-interactive smart buildings (GISBs) with thermostatically controlled loads (TCLs) can be modelled as virtual energy storage systems (ESSs) with dissipation. They have large potentials for providing grid ancillary services such as frequency support. In this chapter, a distributed aggregation control method based on leader–follower sliding mode consensus control is designed for multiple GISBs in a residential community to provide primary frequency support. A leader control is designed to provide power and comfort/energy level references for the smart building aggregator based on the area frequency deviation. The references are tracked by each smart building by using the proposed aggregation control. The Lyapunov method is used to prove the stability of the proposed control method for GISBs. By adopting the proposed method, the aggregated smart buildings can provide good power tracking and energy recovery capability, which effectively improves the system frequency response. Fair and efficient power and comfort/energy level sharing can be achieved among all participating GISBs. The proposed control method is validated on a three-area power system frequency control model considering smart buildings.

15.1 Introduction

In recent years, TCLs such as air conditioning systems, water heaters, and refrigeration systems have become a promising candidate for demand response [1]. The share of TCLs in total electricity energy consumption keeps increasing, especially for tropical areas such as Singapore. A major difference between TCLs and conventional load shedding is that the room temperature or customers comfort level has to be taken into account. As the power of a single TCL unit is quite small, a group of TCLs needs to be aggregated for grid ancillary services. The GISB, as an aggregator of TCLs, can be treated as one entity from the grid perspective. The GISB can provide a power response in a short time scale, by managing the power consumption of each TCL while maintaining the temperature within the user's comfort zone. A GISB with controllable TCLs can be viewed as virtual ESSs, which is able to contribute actively for system frequency control. In this chapter, how to effectively utilise GISBs for timely frequency support is investigated. In state of the art, the TCL aggregators/smart buildings have been proposed for various grid ancillary services such as primary/secondary frequency control [2], voltage regulation [3], load following [4], fulfilling dispatch order [5], system reserves [6], etc. It is found that there are several issues to be investigated.

First, the capacity from one aggregator or GISB may still be too small for supporting grid operations. Therefore, it is sensible to further aggregate the GISB in a community or coordinate them at the higher control level to make a bigger impact, which leads to 'aggregator of aggregators'. The existing works [7,8] mainly focus on distributed power sharing among multiple load/TCL aggregators. The distributed aggregation control for multiple GISBs while considering temperature or comfort levels needs to be explored.

Second, some practical issues regarding infrastructures to implement the GISBs or demand-side aggregators for frequency regulation services are to be further considered. In a residential community, most buildings/houses may not have a frequency measurement device. As discussed in [9], the frequency measurement units are still too expensive for large-scale applications in power distribution level networks, such as a community. Compared to install a new frequency measurement, the control implementation based on existing communication infrastructures is a more flexible, scalable, and economical choice. In this condition, a community only needs to have one leader communication node, which can measure local frequency and receive secondary control signals from grid operators. This leader node can be located at the community substation, which generates the control reference signals for all GISBs in the community. Therefore, the distributed aggregation control of GISBs has a significant necessity in residential communities.

Third, different from [8], which discusses the distributed controlled load/TCL aggregators for secondary frequency control, the capability of GISBs for timely primary frequency support is investigated. The decentralised/local control is a straightforward way to deploy primary frequency control by thermal loads [10]. As the limitation of frequency measurement infrastructure discussed above, a community of GISBs may need to participate in the frequency support in a distributed way based on sparse communication networks. When the system frequency deviations exceed certain limits, the distributed controlled GISBs can also provide timely frequency support. In addition, the power and comfort levels can also be fairly shared among all participating GISBs, which is hard to be achieved by fully decentralised control.

This chapter introduces a distributed aggregation control approach for multiple GISBs to provide fast frequency support to the microgrid. The proposed method has the following features:

- An aggregation control for a group of GISBs is proposed based on the distributed sliding mode control (DSMC) method. This control only relies on peer-to-peer communications and sparse networks. The stability of the proposed DSMC method is proved via the Lyapunov method.
- The GISB aggregator can operate in the *frequency support mode* (FSM) or *comfort recovery mode* (CRM) (i.e. energy recovery mode) depending on area frequency deviations. The power and comfort/energy level reference for each GISB in the aggregator is calculated by a leader control.

- The GISB aggregator with the proposed distributed aggregation method has both superior internal and external characteristics. Internally, fair power and comfort/energy level sharing among all participated GISBs are realised. Externally, the entire aggregator can track leader control references to achieve finite-time power tracking and comfort/energy level recovery.

15.2 System modelling

In a community area microgrid, the GISBs can be aggregated for fast primary frequency support. The block diagram of the multi-area load frequency control (LFC) with GISBs is shown in Figure 15.1. In this chapter, the GISBs refer to buildings with controllable air conditioning systems. The GISBs in each area are connected by both electrical and communication networks. The detailed modelling of LFC, GISB, and communication network is further introduced.

15.2.1 Multi-area microgrids with GISBs

It is considered that the multi-area microgrid has M interconnected control areas (indexed by $k = 1, 2, \ldots, M$). The multi-area microgrid can be represented by an undirected graph $\mathcal{G}_P = (\mathcal{V}_P, \mathcal{E}_P)$, where the nodes $V_P = \{v_{P1}, v_{P2}, ..., v_{PM}\}$ represent the control areas and the edges $\mathcal{E}_P \subseteq \mathcal{V}_P \times \mathcal{V}_P$ represent the tie lines connecting the control areas. The topology and coupling factors of the multi-area microgrids are depicted by an adjacency matrix $\boldsymbol{T} = [T_{kj}] \subseteq \mathbb{R}^{M \times M}$, where $T_{kj} > 0$ if two areas have a tie-line connection, otherwise $T_{kj} = 0$.

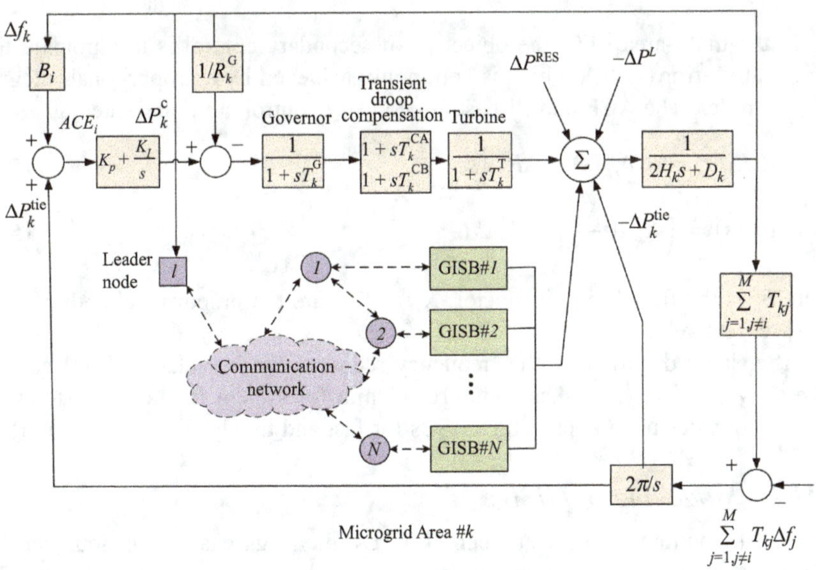

Figure 15.1 An overview of the multi-area LFC with GISBs

The generators are controlled by primary and secondary control, while RESs power output and load demands are uncontrollable disturbances for the LFC. Based on the famous swing equation of generators [11], the power mismatch and frequency deviation in kth control area can be represented as

$$\Delta f_k(s) = \frac{1}{2H_k s + D_k} \times \left\{ -\Delta P_k^{\mathrm{L}}(s) + \Delta P_k^{\mathrm{RES}}(s) + M_k(s)\left[\Delta P_k^{\mathrm{c}}(s) - \frac{1}{R_k^G}\Delta f_k(s)\right] - \Delta P_k^{\mathrm{tie}}(s) + \Delta P_k^{\mathrm{agg}}(s) \right\} \tag{15.1}$$

where Δf_k is the frequency deviation in ith control area, and H_k, D_k, and R_k^{G} are the system inertia, load damping, and speed droop coefficient, respectively. ΔP_k^{L}, $\Delta P_k^{\mathrm{RES}}$, $\Delta P_k^{\mathrm{tie}}$, and $\Delta P_k^{\mathrm{agg}}$ are power variations of loads, RESs, tie line, and aggregator of GISBs, respectively. ΔP_k^{c} denotes the control effort from secondary control. M_k denotes the dynamics of the generators, which is described by

$$M_k(s) = \frac{1}{1 + sT_k^{\mathrm{G}}} \cdot \frac{1}{1 + sT_k^{\mathrm{T}}} \cdot \frac{1 + sT_k^{\mathrm{CA}}}{1 + sT_k^{\mathrm{CB}}} \tag{15.2}$$

where T_k^{G} and T_k^{T} are the time constants of the generator and turbine. T_k^{CA} and T_k^{CA} are the time constants of a transient droop compensator [12].

The power exchange of tie line $\Delta P_k^{\mathrm{tie}}$ is calculated as

$$\Delta P_k^{\mathrm{tie}}(s) = \frac{2\pi}{s} \cdot \left[\sum_{j=1}^{M} T_{kj}\left(f_k(s) - \Delta f_j(s)\right) \right] \tag{15.3}$$

In the multi-area LFC, the objective of secondary control is to eliminate the area control error (ACE), which is commonly achieved by a proportional-integral (PI) controller. The ACE and PI-based secondary control are represented as

$$ACE_k(s) = B_k \Delta f_k(s) + \Delta P_k^{\mathrm{tie}}(s) \tag{15.4}$$

$$\Delta P_k^{\mathrm{c}}(s) = \left(K_P + \frac{K_I}{s}\right) ACE_k(s) \tag{15.5}$$

where B_k is the frequency bias factor. K_{P} and K_{I} are the proportional and integral gains, respectively.

After large disturbance, the frequency recovery by secondary control can be quite slow. Typically, it takes about 10–15 min for system frequency restoration [13]. It motivates us to exploit the GISBs for fast and timely frequency support.

15.2.2 Modelling of GISBs

For a smart building with a number of TCLs, their aggregate behaviour can be approximated by using a continuous model [1], which functions as a virtual ESS with dissipation. It is considered there are N GISBs (indexed by $i = 1, 2, \ldots, N$) in each control area. The average value of internal temperature in the ith GISB can be estimated by the following differential equation:

$$C_i^{\text{th}}\dot{\theta}_i(t) = \frac{\theta^{\text{amb}}(t) - \theta_i(t)}{R_i^{\text{th}}} - \eta_i \frac{p_i(t)}{\lambda_i} + \omega_i(t) \tag{15.6}$$

where θ_i (°C) and θ^{amb} (°C) are the internal and ambient temperatures, respectively; C_i^{th} and R_i^{th} are the thermal capacitance (kWh/°C) and thermal resistance (°C/kW), respectively; λ_i is the number of TCLs in ith GISB; $p_i \in [0, \lambda_i \overline{P}_i]$ is the power consumption of all air conditioners in ith GISB; η_i is the thermal coefficient, which is positive for cooling TCLs and negative for heating TCLs; and ω_i is treated as a Gaussian disturbance with zero means. This model provides an estimation of the average value of internal temperature in a GISB regarding power consumption of all its air conditioners.

In practice, the internal temperature of each building should be kept in a customer comfort zone. Namely, each GISB should be maintained within a temperature range $[\theta^{\text{set}} - \Delta\theta, \theta^{\text{set}} + \Delta\theta]$, where θ^{set} is the nominal temperature set point and $\Delta\theta$ is the temperature tolerance. As different buildings may have different comfort temperature ranges, they are unified into a common index between [0, 1] for convenience. This index is called the *comfort level* (i.e. energy level) of a GISB, which is calculated as follows:

$$\varepsilon_i(t) = \frac{\theta_i(t) - \theta_i^{\text{set}} + \Delta\theta_i}{2\Delta\theta_i} \tag{15.7}$$

where the comfort level of GISB should be maintained in the range of $\varepsilon_i \in [0, 1]$.

Substituting (15.7) into (15.6), then the dynamics of comfort level can be derived as

$$\dot{\varepsilon}_i(t) = -\frac{1}{C_i^{\text{th}} R_i^{\text{th}}}\varepsilon_i(t) - \frac{\eta_i}{2\Delta\theta_i C_i^{\text{th}}\lambda_i} p_i(t) + \frac{\theta^{\text{amb}}(t) - \theta^{\text{set}} + \Delta\theta_i + R_i^{\text{th}}\omega_i(t)}{2\Delta\theta_i C_i^{\text{th}} R_i^{\text{th}}} \tag{15.8}$$

In order to control the system (15.8), an auxiliary state variable $\xi_i(t)$ for feedback linearisation is defined here, which is equal to the right-hand side of (15.8):

$$\xi_i(t) = a_i\varepsilon_i(t) + b_i p_i(t) + d_i(t) \tag{15.9}$$

where a_i, b_i, and d_i represent the corresponding parameters in (15.9).

Differentiating $\xi_i(t)$ at both sides, a set of heterogeneous second-order systems representing the dynamics of GISBs can be formulated as follows:

$$\begin{cases} \dot{\varepsilon}_i(t) = \xi_i(t) \\ \dot{\xi}_i(t) = a_i\xi_i(t) + b_i u_i(t) + \dot{d}_i(t) \end{cases} \tag{15.10}$$

where $u_i = \dot{p}_i(t)$ is the control input to be designed in Section 15.3.

In addition, to maintain the comfort level of a GISB at a reference value ε^*, the GISB should consume a certain amount of power. This power consumption is defined as the *baseline power* of a GISB $p_i^{\text{base}}(t)$, which is obtained by substituting ε^* into (15.9):

$$p_i^{\text{base}}(t) = \lim_{t\to\infty} \frac{\xi_i(t) - a_i\varepsilon_i(t) - d_i(t)}{b_i} = \frac{-a_i\varepsilon^* - d_i(t)}{b_i} \tag{15.11}$$

Thus, $\sum_{i=1}^{N} p_i^{\text{base}}(t)$ is the baseline power for the aggregator of GISBs in one control area.

It is noted that the comfort level of a GISB indicates the thermal 'energy' stored in it. In this chapter, we only consider the cooling effect of the air conditioning system in tropical areas. $\varepsilon_i = 1$ means the GISB is at the highest allowable temperature, and the cooling power of the GISB cannot be further decreased. $\varepsilon_i = 0$ means the GISB is at the lowest allowable temperature, and the cooling power of the GISB cannot be further increased. The behaviour of a GISB is similar to an ESS with dissipation. If the GISB is working at the baseline power defined in (15.9), the GISB will be maintained at the desired temperature or comfort level. If the GISB decreases its power consumption below the baseline, it functions to inject power or discharge to the grid. On the contrary, if the GISB increases its power consumption above the baseline, it functions to absorb power or charge from the grid.

15.2.3 Communication network of GISBs

The communication networks can be depicted by a graph $\mathcal{G}_B = (\mathcal{V}_B, \mathcal{E}_B)$, where $\mathcal{V}_B = \{v_{B1}, v_{B2}, \ldots, v_{BN}\}$ is a set of nodes and $\mathcal{E}_B \subseteq \mathcal{V}_B \times \mathcal{V}_B$ is a set of edges [14]. Each node is associated with a communication agent, and edges represent communication links for data exchange. In this research, bidirectional communication links are considered, which means the graph is undirected, i.e. $(v_{Bi}, v_{Bj}) \in \mathcal{E}_B \Rightarrow (v_{Bj}, v_{Bi}) \in \mathcal{E}_B \forall i, j$. A matrix $\boldsymbol{A} = [a_{ij}]$ is called the adjacency matrix and a_{ij} can be defined as

$$a_{ij} = \begin{cases} 1, & \text{if}\,(v_{Bi}, v_{Bj}) \in \mathcal{E}_B \\ 0, & \text{otherwise} \end{cases} \tag{15.12}$$

The Laplacian matrix is represented as $L = [l_{ij}]$, and the element l_{ij} is calculated from

$$l_{ij} = \begin{cases} -a_{ij}, & i \neq j \\ \sum_{i=1}^{N} a_{ij}, & i = j \end{cases} \tag{15.13}$$

In this chapter, the communication network of GISBs is considered to have a *leader* (labelled as node 0) and the interaction topology is expressed by the graph $\overline{\mathcal{G}}_B$. This leader–follower communication network contains the original graph $\mathcal{G}_B$, node v_{B0}, and edges (v_{Bi}, v_{B0}) from node v_{B0} to other nodes. The leader can send information to followers, but not vice versa. The pinning matrix $\boldsymbol{G} = \text{diag}\{g_i\}$ is used to describe whether each follower directly receives information from the leader, where

$$g_i = \begin{cases} 1, & if \exists (v_{Bi}, v_{B0}) \\ 0, & \text{otherwise} \end{cases} \tag{15.14}$$

One important property which guarantees the convergence of a leader–follower consensus algorithm with the graph $\overline{\mathcal{G}}_B$ is that all the eigenvalues of matrix $\boldsymbol{L}+\boldsymbol{G}$ have negative real parts.

15.3 Proposed control framework for GISBs

The block diagram illustrates the proposed control framework, as shown in Figure 15.2. In Figure 15.2, the multi-area microgrid model is described by (15.1)–(15.5). The frequency signal of *k*th area is given to the leader control of the GISB aggregator to generate the tracking references by (15.16)–(15.19). Then the reference signals are shared to each GISB through the sparse communication network $\overline{\mathcal{G}}_B$ given in (15.12)–(15.14). In the meantime, the information exchange of system states among adjacent GISBs is also accomplished in the communication network $\overline{\mathcal{G}}_B$. With the received information, the DSMC control given in (15.20)–(15.26) will calculate the control inputs $(u_1, \ldots, u_N)$ for each GISB model. The GISB models will execute the control commands and then update the new system states to the DSMC control laws. Finally, the aggregated power of GISBs is imported to the microgrid model for frequency support. In the following subsections, the control objectives of the proposed method are introduced. The leader control, as the key linkage of the system frequency and the power consumption of GISBs, is proposed. After that, the proposed distributed sliding mode control of GISBs and its stability analysis are presented in detail.

15.3.1 Control objectives

The proposed control aims to utilise a group of GISBs for primary frequency support while considering users' comfort. Three control objectives are to be met: (i)

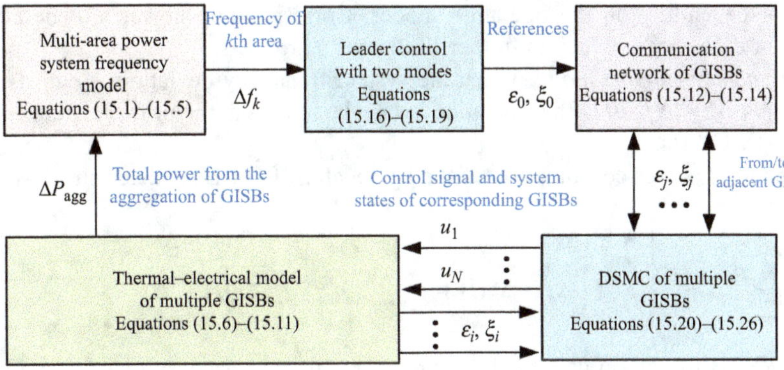

Figure 15.2 Block diagram of proposed control framework

The aggregation of GISBs can track the power of primary frequency control in the FSM. (ii) The comfort levels of GISBs are to be recovered to a reference value ε^* (e.g. 50%) in CRM. (iii) The comfort levels of GISBs are to be gradually balanced during the operation.

The control problems will be solved by two steps: (i) Design the leader control to update the reference states ε_0 and ξ_0 based on system frequency conditions. (ii) Design control input u_i in (15.10), so that system state variables ε_i and ξ_i can track ε_0 and ξ_0 via the sparse network $\overline{\mathcal{G}}_B$. The following condition will be reached by the controlled system in the steady state.

Consensus condition: For an aggregator of GISBs formulated by (15.10) with $\overline{\mathcal{G}}_B$, the second-order leader–follower finite-time consensus is achieved, if for any initial states, there exists a $T_0 \in [0, \infty]$, such that

$$\lim_{t \to T_0} \|\varepsilon_i(t) - \varepsilon_0(t)\| = 0, \lim_{t \to T_0} \|\xi_i(t) - \xi_0(t)\| = 0$$

$$\varepsilon_i(t) = \varepsilon_0(t), \xi_i(t) = \xi_0(t), \forall t \geq T_0, i = 1, 2, ..., N \tag{15.15}$$

Remarks 2 In this chapter, fair sharing *means that the power consumption is proportionally shared among GISBs concerning their parameters, so that the comfort level is equalised during the operation. As defined in* consensus condition *in (15.15), the system state variables ε_i and ξ_i in (15.10) will be converged to ε_0 and ξ_0. This also indicates that all the comfort level $\varepsilon_i(t)$ and auxiliary state variable $\xi_i(t)$ will be equalised in steady state, i.e $\lim_{t\to T_0} \|\varepsilon_i(t) - \varepsilon_j(t)\| = 0, \lim_{t\to T_0} \|\xi_i(t) - \xi_j(t)\| = 0$. In addition, the power consumption of each GISB $p_i(t)$ can be derived from (15.9).*

15.3.2 Leader control

This chapter proposes two operating modes for an aggregator of GISBs: FSM which is active when the system frequency is out of the acceptable range and CRM when the system frequency is within the acceptable range. This is illustrated in Figure 15.3.

The leader control to update the reference states ε_0 and ξ_0 is proposed to incorporate the GISB aggregator into LFC. The updating rules of ε_0 and ξ_0 are different depending on the operation modes. Corresponding power will be consumed in each mode, which is further illustrated as follows.

Frequency support mode: When the system frequency deviation is out of the thresholds $\Delta f \notin [\underset{\sim}{\Delta f}, \tilde{\Delta f}]$, the operation mode is activated for primary frequency support by GISBs.

The total power consumption of the aggregator of GISBs is determined as

$$\sum_{i=1}^{N} p_i(t) = \begin{cases} R^{agg}\left(\tilde{\Delta f} - \Delta f(t)\right) + \sum_{i=1}^{N} p_i^{base}(t), \Delta f(t) \geq \tilde{\Delta f} \\ R^{agg}\left(\underset{\sim}{\Delta f} - \Delta f(t)\right) + \sum_{i=1}^{N} p_i^{base}(t), \Delta f(t) \leq \underset{\sim}{\Delta f} \end{cases} \tag{15.16}$$

where R^{agg} is the droop gain of the aggregator. The power consumption is within $\sum_{i=1}^{N} p_i(t) \in [0, \sum_{i=1}^{N} \overline{p}_i(t)]$. As illustrated in Figure 15.3, the power consumption

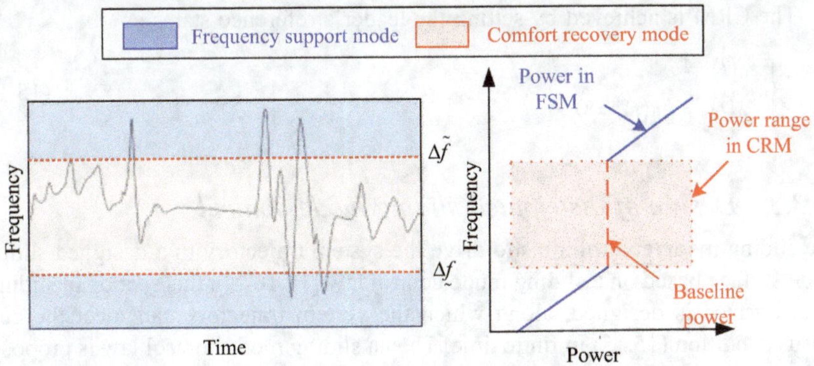

Figure 15.3 Operation modes and power consumption of an aggregator of GISBs

of the aggregator in this mode is the baseline power plus/minus the certain power defined in (15.16). The value of baseline power can be estimated by the leader control from (15.9).

Remark 3 The droop control gain R^{agg} and frequency thresholds $[\underset{\sim}{\Delta f}, \tilde{\Delta f}]$ are user-defined parameters. The system operator can adjust the contribution of the aggregator by alternating these values. Generally, the aggregator should devote its full capacity when the system frequency deviates to the safety limits. Suppose $[\underset{\sim}{\Delta f}, \bar{\Delta f}]$ is maximum allowable frequency range, therefore, R^{agg} can be selected as

$$R^{agg} \geq \min\left[\frac{-\sum_{i=1}^{N} p_i^{base}(t)}{\Delta f - \underset{\sim}{\Delta f}}, \frac{\sum_{i=1}^{N} \bar{p}_i(t) - \sum_{i=1}^{N} p_i^{base}(t)}{\bar{\Delta f} - \tilde{\Delta f}}\right] \tag{15.17}$$

Therefore, in the worst condition when $f = \underset{\sim}{\Delta f}$ or $f = \bar{\Delta f}$, the aggregator can provide the maximum allowable power support $\sum_{i=1}^{N} p_i(t) = 0$ or $\sum_{i=1}^{N} p_i(t) = \sum_{i=1}^{N} \bar{p}_i(t)$ for the system frequency support.

In steady state, $\xi_0(t) = \xi_i(t)$, and add the right- and left-hand terms of (15.9) from $i = 1$ to N, respectively. Thus, the reference states of the leader ε_0 and ξ_0 can be obtained as

$$\begin{cases} \xi_0(t) = \dfrac{\sum_{i=1}^{N} b_i p_i(t) + \sum_{i=1}^{N} [a_i \varepsilon_i(t) + d_i(t)]}{N} \\ \varepsilon_0(t) = \displaystyle\int_0^t \xi_0(t) dt \end{cases} \tag{15.18}$$

Comfort recovery mode: When the system frequency deviation is within the thresholds $f \in [\underset{\sim}{\Delta f}, \tilde{\Delta f}]$, the objective is to recover the comfort level of each GISB back to a reference value ε^*. As illustrated in Figure 15.2, the power consumption of the aggregator is within a range to recover the comfort level change in FSM, and the aggregator is operated at baseline power in steady state.

The CRM is achieved by setting the leader's reference states as

$$\begin{cases} \varepsilon_0(t) = \varepsilon^* \\ \xi_0(t) = \dot{\varepsilon}_0(t) = 0 \end{cases} \tag{15.19}$$

15.3.3 Design of distributed sliding mode control

The sliding mode control aims to drive the system trajectory to a designed sliding mode surface based on a sliding mode control law [15,16]. In this section, a sliding-mode surface is designed, along which the system trajectory can meet the consensus condition (15.15) in finite time. Then a sliding mode control law is proposed to force the system trajectory to reach the designed sliding-mode surface in finite time for the system (15.10) with $\overline{\mathscr{G}}_B$. As the states of agents are coupled on the sliding mode surface and the sliding mode states are decoupled out of this surface, the approach is called DSMC.

Sliding mode surface: The sliding mode surface should be established based on the desired control objectives. For the purpose of achieving finite-time consensus for system (15.10), a set of sliding mode states are represented by

$$\begin{aligned} s_i(t) = \ddot{\varepsilon}_i - \sum_{j=1}^{N} a_{ij}\left[\mathrm{sig}(\varepsilon_j - \varepsilon_i)^{\alpha} + \gamma \mathrm{sig}(\xi_j - \xi_i)^{\beta}\right] \\ - \sum_{j=1}^{N} g_i\left[\mathrm{sig}(\varepsilon_0 - \varepsilon_i)^{\alpha} + \gamma \mathrm{sig}(\xi_0 - \xi_i)^{\beta}\right] \end{aligned} \tag{15.20}$$

and accordingly, the sliding mode surface is

$$\{(\varepsilon_1^T, \varepsilon_2^T, \ldots, \varepsilon_N^T, \xi_1^T, \xi_2^T, \ldots, \xi_N^T) | s_i = 0\} \tag{15.21}$$

where the function $\mathrm{sig}(x)^{\alpha} = |x|^{\alpha}\mathrm{sign}(x)$, $|x|$ denotes the absolute value of variable x, and sign(.) denotes the sign function. α, β, and γ are the control gains to be selected, $\alpha \in (0,1)$, and $\beta = 2 \cdot \alpha/(1+\alpha)$.

Sliding mode control law: The sliding mode control law is designed to force the system trajectory to reach the sliding mode surface (15.21) in finite time. By using the sliding-mode states s_i, the control input of system (15.10) is controlled by the following control law:

$$u_i(t) = b_i^{-1}(u_i^{\mathrm{eq}}(t) + u_i^{\mathrm{sgn}}(t)) \tag{15.22}$$

$$\begin{aligned} u_i^{eq}(t) = \sum_{j=1}^{N} a_{ij}\left[\mathrm{sig}(\varepsilon_j - \varepsilon_i)^{\alpha} + \gamma \mathrm{sig}(\xi_j - \xi_i)^{\beta}\right] \\ + \sum_{j=1}^{N} g_i\left[\mathrm{sig}(\varepsilon_0 - \varepsilon_i)^{\alpha} + \gamma \mathrm{sig}(\xi_0 - \xi_i)^{\beta}\right] - a_i\xi_i - \dot{d}_i \end{aligned} \tag{15.23}$$

$$\dot{u}_i^{\text{sgn}}(t) = -\text{sgn}(s_i(t)) \tag{15.24}$$

In the designed control law only sgn($s_i(t)$) is used rather than the exact value $s_i(t)$. Thus the calculation can be further simplified by defining a new function $g_i(t)$ as in

$$g_i(t) = \int_0^t s_i(t) = \dot{\varepsilon}_i - \int_0^t \sum_{j=1}^N a_{ij}\left[\text{sig}(\varepsilon_j - \varepsilon_i)^{\alpha} + \gamma\text{sig}(\xi_j - \xi_i)^{\beta}\right]\text{d}t \tag{15.25}$$

Then, the value of sgn($s_i(t)$) can be calculated from the following equation:

$$\text{sgn}(s_i) = \text{sgn}(g_i(t) - g_i(t-\tau)) \tag{15.26}$$

where τ represents a time delay. Thus, instead of knowing the exact value of s_i, only whether g_i increases or decreases is to be known.

The finite-time consensus condition (15.15) of the system (15.10) is guaranteed based on Theorems 1 and 2 proved as follows.

Theorem 1 If the states of the system (15.10) can reach the designed sliding mode surface in (15.21), then second-order consensus can be achieved in finite time.

Theorem 2 The states of the second-order system (15.10) can reach the sliding-mode surface (15.21), if the sliding mode control law is designed as (15.22)–(15.24). The states of system (15.10) will achieve second-order consensus in finite time along the sliding-mode surface (15.21).

The stability proof of Theorems 1 and 2 can be found in the Appendix.

15.4 Results and discussions

In this section, simulation studies are conducted to validate the proposed control scheme in LFC. The multi-area LFC and GISBs with the proposed control scheme are implemented in MATLAB®/Simulink®. The topology of the test system for simulation studies is shown in Figure 15.4. It is assumed that there are two aggregators of GISBs with the total power rating of 6 MW in Area 1 and 8 MW in Area 2, respectively. The power rating (in kW) of each GISB and communication graphs are also given in Figure 15.4. The LFC system under consideration is a per-unit system with $S_{\text{base}} = 20$ MVA, which can also be scaled up/down in other applications. The parameters for a single TCL in GISB are given in Table 15.1 [1]. In simulation studies, these parameters are varied within [100%, 120%] to emulate the heterogeneous feature of each GISB. The parameters of the three-area LFC system are given in Table 15.2 [11]. The control parameters of the proposed method used in the following test cases are shown in Table 15.3. In Cases 1 and 2, the effectiveness and performance of the proposed method are demonstrated under system contingency (step response) and normal operation (time-varying photovoltaic (PV), load, and temperature) conditions, respectively. In Cases 3 and 4, the

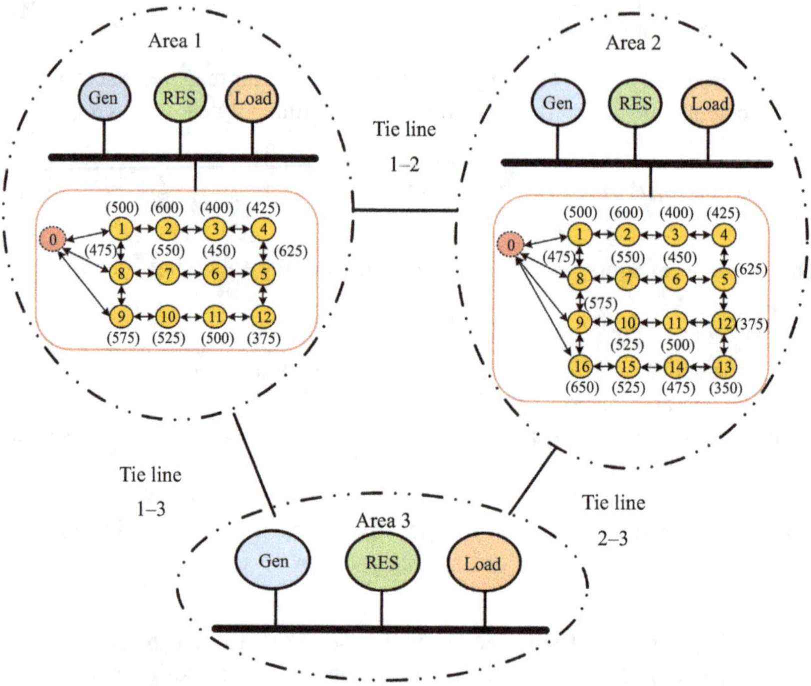

Figure 15.4 The topology of the test system in the simulation studies

Table 15.1 Typical parameter value for a residential TCL in GISB

Parameter	Symbol	Values
Thermal capacitance	C^{th}	2 kWh/°C
Thermal resistance	R^{th}	2 °C /kW
Rated power of an air conditioner	$\bar{P}$	5 kW
Thermal coefficient	η	2.5
Temperature tolerance	$\Delta\theta$	2 °C
Temperature set point	θ^{set}	23 °C

main purpose is to validate the advantages of the proposed method over linear methods in the aspects of converging speed and communication delay.

15.4.1 Case 1: system contingency

In Case 1, the performance of the proposed control scheme for GISBs is studied under a system contingency condition in Area 1. The initial comfort/energy levels of all GISBs are 50% and the ambient temperature is 30 °C. A load increment of 1 MW emulates the contingency at 5 s. The simulation results, in this case, are shown in Figures 15.5–15.8, which are illustrated as follows.

Table 15.2 Parameters of the three-area microgrid

Area no.	**1**	**2**	**3**
$2H$ (p.u./Hz)	0.1667	0.2	0.15
D (p.u./Hz)	0.0015	0.002	0.001
T^{CA} (s)	5	4	4.5
T^{tCB} (s)	40	30	40
T^{T}(s)	0.3	0.2	0.25
T^{G} (s)	0.4	0.3	0.35
R^{G} (Hz/p.u.)	3	3.6	2.5
B (p.u./Hz)	0.8675	0.795	0.87
T_{ij} (p.u./Hz)	0.25	0. 25	0.25
P^{agg} (MW)	6	8	0

Table 15.3 Parameters of the proposed control method

Parameters	**Symbols**	**Values**
Frequency threshold	$\pm\Delta\widetilde{f}$	±0.1 Hz
Droop gain of GISB in Area 1	R^{agg}	20 MW/Hz
Droop gain of GISB in Area 2	R^{agg}	30 MW/Hz
Reference comfort level	ε^*	50%
Control gains	α, β, γ	0.8, 0.889, 0.03

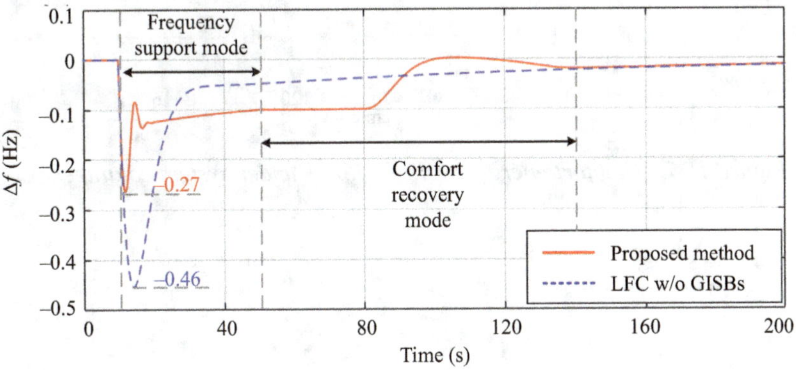

Figure 15.5 Frequency response under system contingency

In Figure 15.5, the system frequency responses under the original LFC and the proposed control scheme are compared. Upon the occurrence of the disturbance at 5 s, the system frequency sharply falls down. The aggregator of GISBs switches to frequency droop mode once the frequency drop exceeds 0.1 Hz. As a result, it can be observed that the maximum frequency deviation is evidently improved from −0.46 Hz to −0.27 Hz.

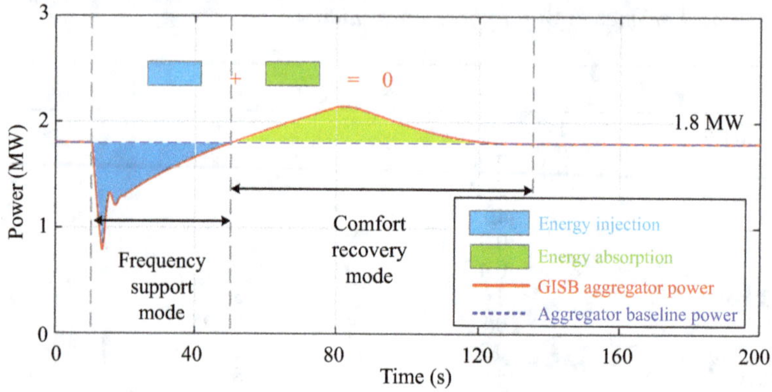

Figure 15.6 Power consumption, baseline power, energy absorption, and injection of the aggregator of GISBs in Area 1

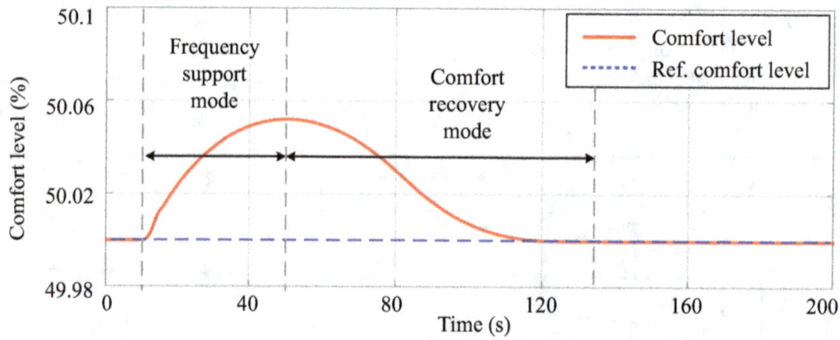

Figure 15.7 Comfort/energy level of GISB-1 under system contingency

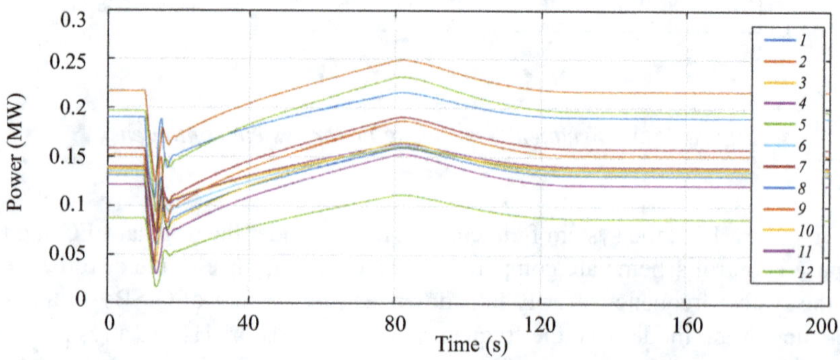

Figure 15.8 Power consumption of each GISB in Area 1

In Figure 15.6, the total power consumption, baseline power, energy absorption, and injection of the whole GISBs aggregator in Case 1 are shown. Here the calculation of the baseline power for GISB aggregator using (15.11) is explained. Based on typical TCL parameters given in Table 15.1, the baseline power of one GISB is calculated as $p_i^{\text{base}}(t) = 140$ kW, where $a_i = -0.25$, $b_i = -3.125 \times 10^{-3}$, $d_i = 0.5625$, and $\lambda_i = 100$. In Area 1, there are 12 GISBs with their parameters varying within [100%, 120%] of the typical values given in Table 15.1. Therefore, the baseline power of the whole GISB aggregator in Area 1 can be obtained by the sum of the baseline power of each GISB, i.e. $\sum_{i=1}^{N} p_i^{\text{base}}(t)$, which is equal to 1.8 MW as shown in Figure 15.6. The selection of R^{agg} in Area 1 is 20 MW/Hz, which is aligned with the criterion in (15.17), i.e. $R^{\text{agg}} \geq \min[18, 42]$ MW/Hz.

In Figure 15.7, the comfort level profile of GISB-1 is shown, as the comfort levels of all GISBs are equalised in this case. Besides, the power consumption of each GISB in Area 1 is demonstrated in Figure 15.8. From 5 to 44 s, the aggregator is operating below the baseline power and 'discharge' to the grid for primary frequency support, as shown in Figures 15.5–15.8. When the frequency is back to 0.1 Hz at 44 s, the aggregator starts to 'charge' from the grid to recover the comfort/energy level to the reference value of 50%.

15.4.2 Case 2: normal operation

In Case 2, the normal operation under half-hour PV, load demands, and ambient temperature variation (see Figure 15.9) is conducted to further investigate the performance of the proposed control scheme. The three-area system shown in Figure 15.4 is considered. The profiles of PV and load are scaled to 1, 1.1, and 0.9 for Areas 1, 2, and 3, respectively. The PV data of 1-s resolution were measured by The Electric Power Research Institute (EPRI) in June 2012 [17]. The initial comfort/energy level of each GISB in Area 2 is within [46%, 54%].

Figure 15.10(a)–(c) shows the system frequency deviations in each control area with and without the proposed control method. In Figure 15.10(a)–(c), it can be observed that frequency deviations are mitigated effectively with the proposed approach, especially for Areas 1 and 2 with DSMC controlled GISBs. The frequency deviations can be regulated in a safe range (±0.2 Hz) with the 6 MW GISB in Area 1 and 8 MW GISB in Area 2. The results prove that the proposed approach is effective for frequency support.

The power consumption and comfort/energy level profiles of each GISB in Area 2 are further shown in Figures 15.11 and 15.12, respectively. Initially, the difference in comfort level among each GISB is large. Therefore, the energy balancing effect dominates GISB dynamics. The power sharing among each GISB is substantially influenced to balance comfort levels. At 380 s, the comfort levels become equalised, which indicates the system reaches consensus. The power sharing becomes proportional to the parameters of GISB models. During the operation, the aggregator of GISB can 'charge' or 'discharge' depending on the system frequency condition. The total power consumption, baseline power, energy absorption, and injection of the aggregator of GISBs in Area 2 during this process

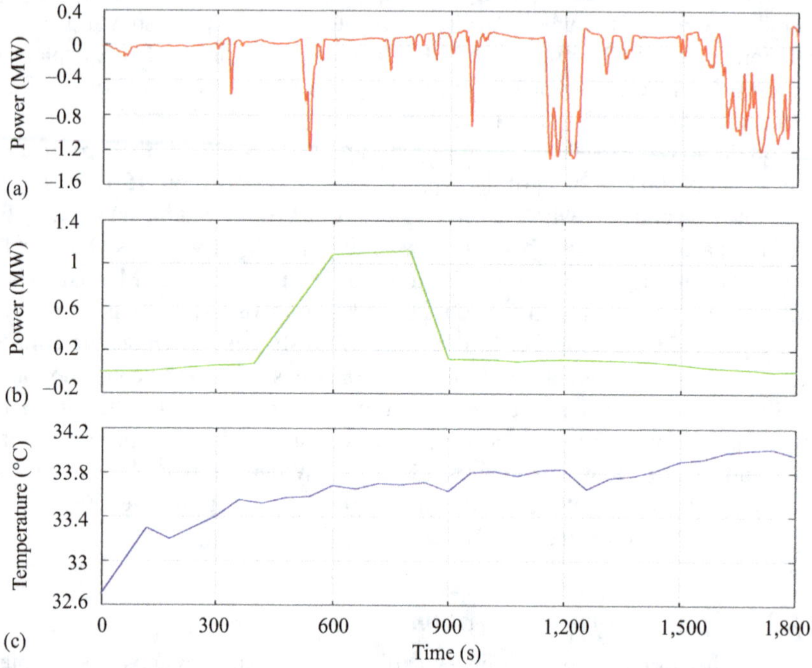

Figure 15.9 PV, load demand, and ambient temperature profiles of Case 2. (a) PV generation. (b) Non-dispatchable load. (c) Ambient temperature

are shown in Figure 15.13. The calculation of baseline power is similar to the process of Case 1. The selection of R^{agg} in Area 2 is 30 MW/Hz, which is also aligned with the criterion in (15.17). It should be noted that the baseline power is not a constant value due to the variations of the ambient temperature. The results validate that the aggregator of GISBs can operate autonomously for system frequency support under time-varying events.

15.4.3 Case 3: comparison with linear consensus control

In Case 3, the proposed DSMC is compared with a traditional linear consensus control law for the second-order multi-agent system proposed in [18]. Compared to the DSMC approach in this chapter, the control law in [18] is 'linear' and 'infinite-time'. For the GISB system considered in this chapter, the linear control law can be expressed as

$$u_i(t) = k_1 \sum_{j=1}^{N} a_{ij}\left[\varepsilon_j(t) - \varepsilon_i(t) + (\xi_j(t) - \xi_i(t)\right] + k_2 g_i[(\varepsilon_0(t) - \varepsilon_i(t) + (\xi_0(t) - \xi_i(t)] \tag{15.27}$$

where k_1 and k_2 are control gains to be selected.

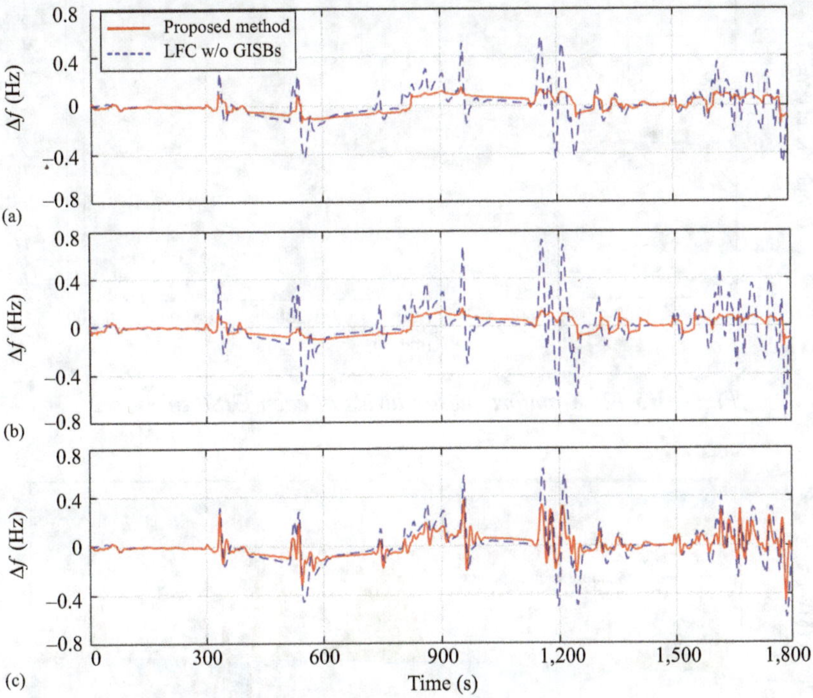

Figure 15.10 Frequency deviations of each control area with and without GISBs. (a) Area 1. (b) Area 2. (c) Area 3

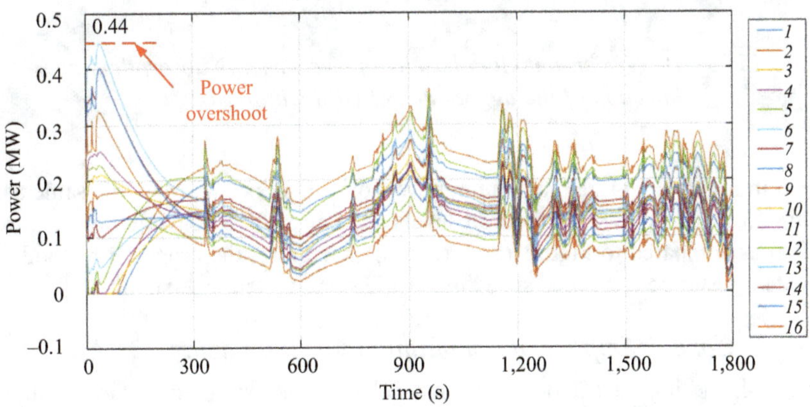

Figure 15.11 Power consumption of each GISB in Area 2

To ensure a fair comparison, the same test condition in Case 2 is used and only the control law is replaced with the linear consensus control law. The same maximum power overshoot (0.44 MW of GISB-13) is kept in these two cases. As

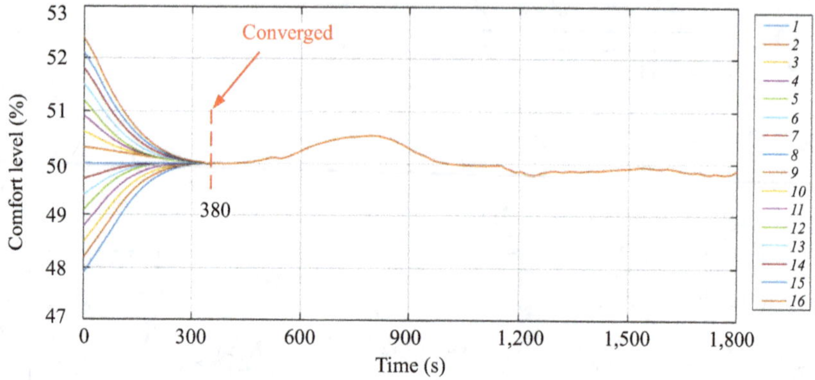

Figure 15.12 Comfort/energy levels of each GISB in Area 2

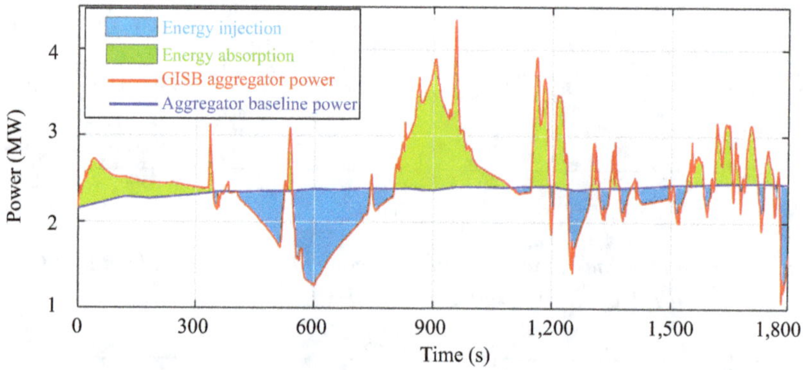

Figure 15.13 Power consumption, baseline power, energy absorption, and injection of the aggregator of GISBs in Area 2

shown in Figure 15.14, the maximum power overshoot is 0.44 MW, the same as in Figure 15.11. However, the comfort levels in Figure 15.15 converge much slower (900 s) as compared to Figure 15.12 (380 s). Therefore, the proposed method can provide a faster convergence speed under the same system overshoot.

15.4.4 Case 4: impact of communication time delay

In Case 4, the impact of communication delay on the proposed method is further investigated. The same condition as in Case 1 is considered in this case, where the influence of network latency for GISBs in Area 1 is evaluated. First, we consider the linear control law given in (15.27) and let $k_1 = k_2 = 1$. Based on Theorem 10 in [14], the upper boundary of tolerable communication delay can be estimated, which is inversely proportional to the largest eigenvalue of the matrix $(\boldsymbol{L}+\boldsymbol{G})$, i.e. $\tau^* = \pi/2\lambda_{\max}(\boldsymbol{L}+\boldsymbol{G})$. According to the topology of GISBs in Area 1, it can be calculated that

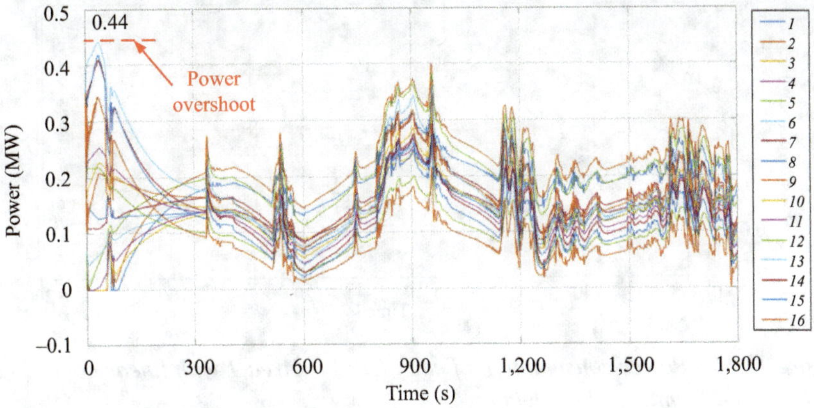

Figure 15.14 Power consumption of each GISB in Area 2 with linear control law

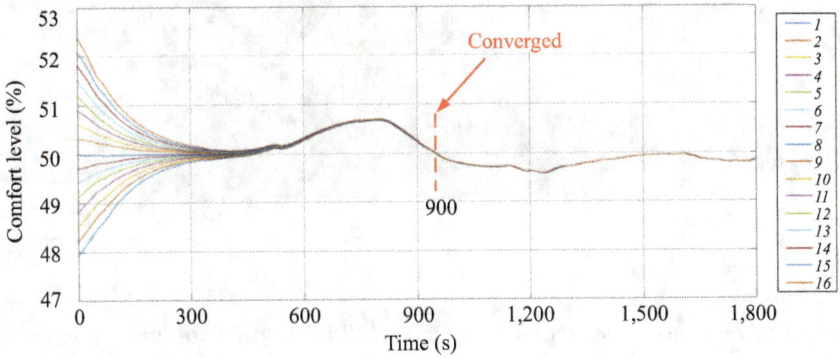

Figure 15.15 Comfort levels of each GISB in Area 2 with linear control law

$\lambda_{\max}(\boldsymbol{L}+\boldsymbol{G}) = 5.3387$ and $\tau^* = \pi/2\lambda_{\max}(\boldsymbol{L}+\boldsymbol{G}) = 0.2942$ s. Figures 15.16 and 15.17 show the power consumption of each GISB with linear control law in (15.27) under the communication delay of 0.29 and 0.3 s, respectively. It can be observed that the system is stable when the delay is smaller than τ^*, and the system starts to diverge when the delay is larger than τ^*. Due to the power limits of GISBs $p_i \in [0, \lambda_i \overline{P}_i]$, the power of each GISB oscillates within its limits. In Figure 15.18, the control law is changed back to the proposed DSMC. It can be found that the proposed DSMC can tolerate the communication delay of 0.3 s with slight oscillations. Since the focus of this chapter is to develop a new approach of distributed aggregation of GISBs, the upper boundary of tolerable delay is not discussed here but will be systematically considered in our future work. In Figure 15.19, the frequency response of Area 1 with the proposed method and linear method ($\tau = 0.3$ s) is compared. It shows that the proposed method outperforms the linear control law for system frequency response under large communication delay.

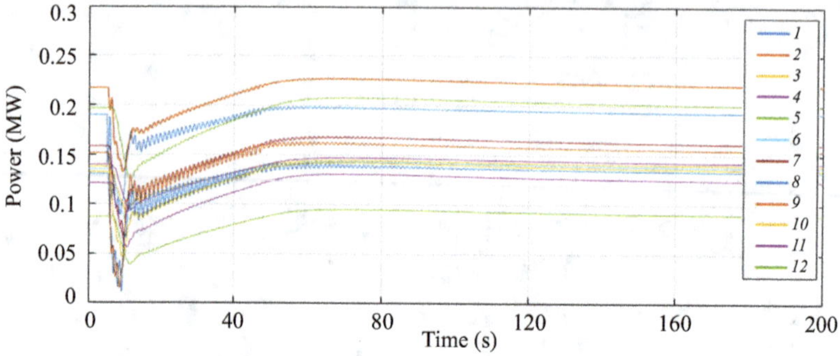

Figure 15.16 *Power consumption of each GISB in Area 1 with linear control law under 0.29-s delay*

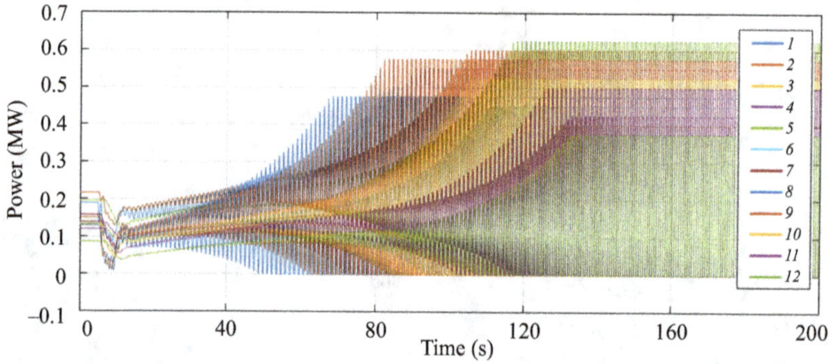

Figure 15.17 *Power consumption of each GISB in Area 1 with linear control law under 0.3-s delay*

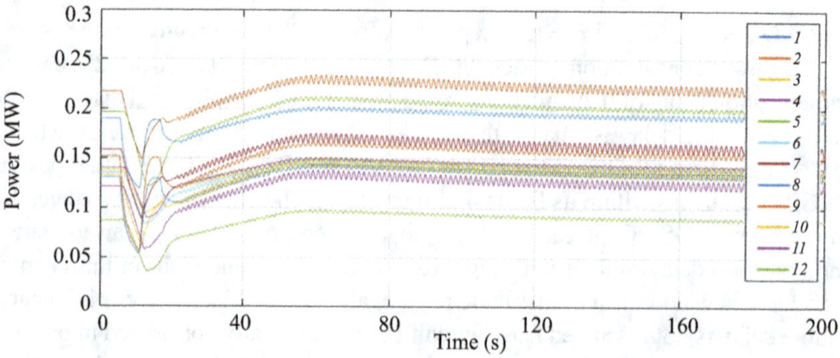

Figure 15.18 *Power consumption of each GISB in Area 1 with proposed control law under 0.3-s delay*

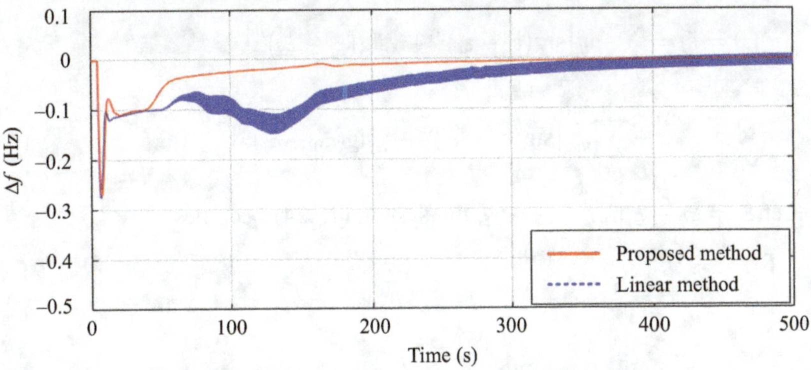

Figure 15.19 Comparison of system frequency response with the proposed method and linear method under 0.3-s delay

15.5 Conclusions

In this chapter, multiple GISBs at the demand side are utilised for timely frequency support to the microgrid. A new distributed sliding mode control, as well as its leader control design, has been proposed. The stability analysis proves the system variable can be driven to the designed sliding surface with the proposed sliding mode control law. With the proposed approach, the aggregation of GISBs can provide fast frequency support while considering the comfort/energy level of each building within a community aggregator.

The proposed control scheme has been tested in a multi-area microgrid with renewable penetration. Simulation results have shown that the system frequency response can be improved by using the proposed method under both system contingency and normal operating conditions. The mode of operations can be changed automatically based on system frequency. The aggregation of GISBs can be utilised to overcome the frequency variations. Comfort/energy level balancing and recovery can be achieved during the operation. A comparison with linear methods has been made to demonstrate the advantages of the proposed method. It is validated that the proposed method can provide fast coverage speed with stronger tolerance to the communication time delay. In summary, the results demonstrate that the proposed aggregation method for GISBs provides an effective way for fast microgrid frequency support.

Appendix

Proof of Theorem 1

Proof: Based on the sufficient condition of Theorem 1, the sliding mode control law is employed to enforce the system to achieve the sliding surface $s_i = 0$. The following system can be obtained:

$$\begin{cases} \dot{\varepsilon}_i = \xi_i, \\ \dot{\xi}_i = \tilde{u}_i, \end{cases} i = 1, 2, \ldots, N \tag{A.1}$$

$$\tilde{u}_i(t) = \ddot{\varepsilon}_i = \sum_{j=1}^{N} a_{ij}\left[\text{sig}\left(\varepsilon_j - \varepsilon_i\right)^{\alpha} + \gamma\text{sig}\left(\xi_j - \xi_i\right)^{\beta}\right]$$
$$+ \sum_{j=1}^{N} g_i\left[\text{sig}(\varepsilon_0 - \varepsilon_i)^{\alpha} + \gamma\text{sig}(\xi_0 - \xi_i)^{\beta}\right] \quad \text{(A.2)}$$

Let $\widehat{\xi}_i = \xi_0 - \xi_i$, $\widehat{\varepsilon}_i = \varepsilon_0 - \varepsilon_i$, the system (15.24) becomes

$$\begin{cases} \dot{\widehat{\varepsilon}}_i = \widehat{\xi}_i \\ \dot{\widehat{\xi}}_i = \widehat{u}_i \end{cases}, i = 1, 2 \ldots, N. \quad \text{(A.3)}$$

$$\widehat{u}_i(t) = \sum_{j=1}^{N} a_{ij}\left[\left(\text{sig}\left(\widehat{\varepsilon}_j - \widehat{\varepsilon}_i\right)^{\alpha}\right) + \gamma\left(\text{sig}\left(\widehat{\xi}_j - \widehat{\xi}_i\right)^{\beta}\right)\right]$$
$$+ g_i\left[\left(\text{sig}(\widehat{\varepsilon}_i)^{\alpha} + \gamma\left(\text{sig}\left(\widehat{\xi}_i\right)^{\beta}\right)\right)\right] \quad \text{(A.4)}$$

Choosing a Lyapunov function as

$$V_1 = \frac{1}{2}\sum_{i=1}^{n}\left(\widehat{\xi}_i\right)^2 + \sum_{i=1}^{n}\sum_{j=1}^{n}\int_0^{\widehat{\varepsilon}_j - \widehat{\varepsilon}_i} a_{ij}(\text{sig}(s)^{\alpha})ds + \sum_{i=1}^{n}\int_0^{\widehat{\varepsilon}_i} \widehat{\xi}_i(\text{sig}(s)^{\alpha})ds \quad \text{(A.5)}$$

Taking the derivative of V_1, it can be obtained that

$$\dot{V}_1 = \sum_{i=1}^{N}\widehat{\xi}_i\left[\sum_{j=1}^{N} a_{ij}\left(\text{sig}\left(\widehat{\varepsilon}_i - \widehat{\varepsilon}_j\right)^{\alpha} - \gamma\left(\text{sig}\left(\widehat{\xi}_i - \widehat{\xi}_j\right)^{\beta}\right) - g_i\left(\text{sig}(\widehat{\varepsilon}_i)^{\alpha} - \gamma\text{sig}\left(\widehat{\xi}_i\right)^{\beta}\right)\right]$$
$$+ \frac{1}{2}\sum_{i=1}^{N}\sum_{j=1}^{N} a_{ij}\left(\widehat{\xi}_i - \widehat{\xi}_j\right)\text{sig}\left(^{\varepsilon_i}\right) + \sum_{i=1}^{N} g_i\widehat{\xi}_i\text{sig}\left(^{\widehat{\varepsilon}_i}\right.$$
$$= \sum_{i=1}^{N}\widehat{\xi}_i\gamma\left[\sum_{j=1}^{N} a_{ij}\text{sig}\left(\widehat{\xi}_i - \widehat{\xi}_j\right)^{\beta} - g_i\text{sig}\left(\widehat{\xi}_i\right)^{\beta}\right]$$
$$= \frac{1}{2}\sum_{i=1}^{N}\sum_{j=1}^{N} a_{ij}\left(\widehat{\xi}_i - \widehat{\xi}_j\right)\text{sig}\left(^{\widehat{\xi}_i}\right) - \sum_{i=1}^{n} g_i\widehat{\xi}_i sig\left(^{\widehat{\xi}_i}\right.$$
$$\leq \frac{1}{2}\sum_{i=1}^{N}\sum_{j=1}^{N} a_{ij}\left(\widehat{\xi}_i - \widehat{\xi}_j\right)\text{sig}\left(^{\widehat{\xi}_i}\right) \leq 0 \quad \text{(A.6)}$$

According to Lyapunov's second method for stability, when V_1 is positive definite (except for original point) and $\dot{V}_1 \leq 0$ with $\dot{V}_1 = 0$, if and only if $\widehat{\xi}_i = \widehat{\xi}_j = 0, \widehat{\varepsilon}_i = \widehat{\varepsilon}_j = 0$, then the system will be stable. This means the system will converge to $\varepsilon_i \to \varepsilon_0, \xi_i \to \xi_0$ at steady state. Thus, the system (15.10) on surface (15.21) is globally asymptotically stable.

Besides, the local finite-time convergence of the system (15.10) on surface (15.21) is proved. Based on Lemma 1 (Lasalle's Invariance Principle), Lemma 2 (Finite-time stability), and definition of homogeneity with dilation in [19], it can be obtained the system with variables $(\widehat{\varepsilon}_1, \widehat{\varepsilon}_2, ..., \widehat{\varepsilon}_n, \widehat{\xi}_1, \widehat{\xi}_2, ..., \widehat{\xi}_n)$ is homogeneous of degree $\kappa = \alpha - 1 < 0$ with dilation $(2, 2, ..., 2, 1 + \alpha, 1 + \alpha, ..., 1 + \alpha)$. Therefore, the system (15.10) on surface (15.21) is locally finite-time stable.

If the equilibrium of a control is globally asymptotically stable and locally finite-time convergent, then the control is globally finite-time stable. This follows the principle that globally asymptotical stability implies finite-time convergence to any given bounded neighbourhood of the equilibrium. Therefore, the system (15.10) on surface (15.21) is globally finite-time stable. In other words, it can be got $\widehat{\varepsilon}_i - \widehat{\varepsilon}_j \to 0, \widehat{\xi}_i - \widehat{\xi}_j \to 0, \forall i, j = 1, \ldots, N$ in finite time. This completes the proof.

Proof of Theorem 2

Proof: Substituting control law (15.22)–(15.24) into the sliding mode states (15.20), it can be obtained that

$$\begin{aligned} s_i(t) &= a_i\xi_i + \dot{d}_i(t) + b_i u_i(t) \\ &= \sum_{j=1}^{N} a_{ij}\left[sig(\varepsilon_j - \varepsilon_i)^{\alpha} + \gamma sig(\xi_j - \xi_i)^{\beta}\right] \\ &\quad -\sum_{j=1}^{N} g_i\left[sig(\varepsilon_0 - \varepsilon_i)^{\alpha} + \gamma sig(\xi_0 - \xi_i)^{\beta}\right] \end{aligned} \tag{A.7}$$

Substituting the control law (15.22)–(15.24) into (A.6), it can be obtained that $s_i(t) = u_i^{\text{sgn}}(t)$. $\dot{s}_i(t) = \dot{u}_i^{\text{sgn}}(t)$.

A Lyapunov function can be established in the form of

$$V_2(t) = \frac{1}{2}\sum s_i^2(t) \tag{A.8}$$

Again taking the derivative of V_2, it can be obtained that

$$\dot{V}_2(t) = \sum s_i(t)\dot{s}_i(t) = -s_i(t)\text{sgn}(s_i(t)) = -|s_i(t)| \leq 0 \tag{A.9}$$

It guarantees that the states of the system can reach the sliding mode surface, if the control law is designed (15.22)–(15.24). Then following Theorem 1, the second-order consensus can be achieved in finite time along the surface. This completes the proof.

References

[1] B.M. Sanandaji, H. Hao, and K. Poolla, 'Fast regulation service provision via aggregation of thermostatically controlled loads', *System Sciences (HICSS), 2014 47th Hawaii International Conference*, pp. 2388–2397, 2015.

[2] M. Maasoumy, B.M. Sanandaji, A. Sangiovanni-Vincentelli, and K. Poolla, 'Model predictive control of regulation services from commercial buildings to the smart grid', *American Control Conference (ACC)*, 2015, IEEE, 2015.

[3] A. Zakariazadeh, O. Homaee, S. Jadid, and P. Siano, 'A new approach for real time voltage control using demand response in an automated distribution system', *Applied Energy*, vol. 117, pp.157–166, 2015.

[4] J.L. Mathieu, and D.S. Callaway, 'State estimation and control of heterogeneous thermostatically controlled loads for load following', *2012 45th Hawaii International Conference on System Sciences*, 2012.

[5] Y. Wang, Y. Xu, Y. Tang, and Y.L. Xu, 'A distributed control scheme of thermostatically controlled loads for the building-microgrid community', *IEEE Transactions on Sustainable Energy*, vol. 11, no. 1, pp. 350–360, 2020.

[6] D. Xie, H. Hui, Y. Ding, and Z. Lin, 'Operating reserve capacity evaluation of aggregated heterogeneous TCLs with price signals', *Applied Energy*, vol. 216, pp. 338–347, 2018.

[7] C. Chen, J. Wang, and S.A. Kishore, 'Distributed direct load control approach for large-scale residential demand response', *IEEE Trans. Power Systems*, vol. 29, no. 5, pp. 2219–2228, 2015.

[8] J. Hu, J. Cao, J.M. Guerrero, T. Yong, and J. Yu, 'Improving frequency stability based on distributed control of multiple load aggregators', *IEEE Trans. Smart Grid*, vol. 8, no .4, pp. 1553–1567, 2017.

[9] K. Dehghanpour and S. Afsharnia, 'Electrical demand side contribution to frequency control in microgrids: a review on technical aspects', *Renewable and Sustainable Energy Reviews*, vol. 41, pp.1267–1276, 2015.

[10] V. Trovato, I.M. Sanz, B. Chaudhuri, and G. Strbac, 'Advanced control of thermostatic loads for rapid frequency response in Great Britain', *IEEE Trans. Power Systems*, vol. 32, no. 3, pp. 2106–2117, 2017.

[11] P. Kundur, *Power System Stability and Control,* New York, NY, USA: McGraw-Hill, 1994.

[12] M. Cheng, J. Wu, S.J. Galsworthy, *et al.*, 'Microgrid frequency response from the control of bitumen tanks', *IEEE Trans. Power Systems*, vol. 31, no. 3, pp. 1769–1778, 2016.

[13] D. Gauthier, B. Francois, and G. Malarange, 'Dynamic frequency control support by energy storage to reduce the impact of wind and solar generation on isolated power system's inertia', *IEEE Trans. Sustainable Energy*, vol. 3, no. 4, pp. 931–939, 2012.

[14] R. Olfati-Saber and R.M. Murray, 'Consensus problems in networks of agents with switching topology and time-delays', *IEEE Trans. Automatic Control*, no. 49, vol. 9, pp. 1520–1533, 2004.

[15] Y. Wu, S.E. Li, Y. Zheng, and J.K. Hedrick, 'Distributed sliding mode control for multi-vehicle systems with positive definite topologies', *Decision and Control (CDC), 2016 IEEE 55th Conference on.*, 2016.

[16] W. Yu, H. Wang, F. Cheng, X. Yu, and G. Wen, 'Second-order consensus in multiagent systems via distributed sliding mode control', *IEEE Transactions on Cybernetics*, vol. 47, no. 8, pp. 1872–1881, 2017.

[17] '1- second resolution data', EPRI, Available: http://dpv.epri.com/measurement_data.html.

[18] S. Qiang, J. Cao, and W. Yu, 'Second-order leader-following consensus of nonlinear multi-agent systems via pinning control', *Systems & Control Letters*, vol. 59, no. 9, pp. 553–562, 2010.

[19] X. Wang and Y. Hong, 'Finite-time consensus for multi-agent networks with second-order agent dynamics', *IFAC Proceedings Volumes*, vol. 41, no .2, pp. 15185–15190, 2008.

Chapter 16

Decentralised-distributed hybrid voltage control by inverter-based DERs

Due to high-level penetration of renewable energy, voltage fluctuations and violations are becoming two main voltage quality issues in power distribution networks. In this chapter, a decentralised-distributed control strategy by using reactive power from inverter-based DERs is proposed for network voltage regulation. A decentralised voltage control is developed to mitigate voltage fluctuations by regulating the voltage ramp rate. The var capacity from the power inverters becomes smoothed as a by-product. A distributed voltage control is developed to fairly utilise the var capacity of power inverters to deal with the network voltage deviations. In addition, on-load tap changers (OLTCs) control can provide supplementary voltage regulation when there is a shortage of var capacity from power inverters. The effectiveness of the proposed voltage regulation method is validated by simulation on IEEE 33-bus distribution network with real-world data.

16.1 Introduction

The voltage control in power distribution systems is conventionally realised by mechanical devices such as step voltage regulators, OLTCs, and switched capacitors [1]. These mechanical devices with tap/switch actions are initially designed for distribution networks with low voltage fluctuations. In distribution networks with high renewable penetration, these components will subject to frequent tap/switch changing to regulate the network voltage, which will reduce their lifetime significantly. As a promising alternative, the power inverters of the distributed energy resources (DERs) are much flexible and have much faster responding speed for generating/absorbing reactive power [2,3]. The IEEE 1547.8 working group has been revising the standards to allow the power inverters to participate in the voltage control of distribution networks [4]. Herein, how to coordinate these power inverters for real-time voltage control in high DER penetrated distribution networks is becoming a hot topic.

The decentralised and centralised control schemes of DERs have been widely applied for voltage control in distribution systems [5–7]. The decentralised control only uses local measurement, which can be further divided into $Q(V)$, $P(V)$, and $Q(P)$ strategies. In $Q(V)$ or $P(V)$ strategy, the reactive or real power output of DER is dependent on the bus voltage, such as droop control [5]. While in $Q(P)$ strategy, the reactive power output of the DER is a function of its real power output, which mitigates the voltage fluctuation caused by PV generation intermittency [6]. The main limitation of decentralised control is that it cannot guarantee the network voltage variation range as the lack of system-wide coordination. By contrast, centralised control is widely proposed for voltage regulation based on optimisation techniques. For example, in [7], a centralised optimisation has been proposed for voltage regulation by using DERs and voltage regulation devices. The control signals are updated in two timescales (hourly and 15 min). The centralised control schemes have several limitations which make them not suitable for all conditions. (1) All the information of distribution networks, e.g. line impedance, topology generation, and load demand, should be collected by the central controller, which makes the security and data privacy of such system a concern. (2) When system scale becomes large, the central controller usually subjects to high computation and communication burdens. (3) The centralised control scheme is inherently vulnerable to a single-point failure. The distributed control is a trade-off between fully decentralised and centralised control. It can coordinate all control agents without the requirement of a central controller. The computation and communication burdens are shared to the distributed controllers, which also have more robustness against communication failures. The distributed control can achieve real-time coordination with only peer-to-peer communications. This motivates us to integrate the features of decentralised and distributed control and propose a voltage control method by using inverter-based DERs.

This chapter introduces a decentralised-distributed hybrid voltage control method to deal with both voltage fluctuations and limits violation issues. It has the following features:

- A decentralised voltage control is designed to regulate voltage ramp rate for mitigating voltage fluctuations. As a beneficial by-product, the var capacity from the power inverters becomes smooth.
- A distributed voltage control is developed to fairly utilise the var capacity of each power inverter to regulate the network voltage deviations.
- The OLTCs will provide supplementary voltage control when there is a shortage of var capacity from inverters. The simulation results have demonstrated the effectiveness and advantages of the proposed method.

16.2 Voltage control in distribution networks

16.2.1 Problem description

In this chapter, it is considered that the network voltage control is realised by ramp-rate voltage control and distributed voltage control from inverter-based DERs and OLTC as a backup. The ramp-rate voltage control aims to mitigate high-frequency voltage fluctuations. The distributed and OLTC voltage control aims to control the voltage deviations within the limits. The power inverters of DERs have fast response speed and flexible controllability, which makes them suitable for ramp-rate and distributed voltage control. When there is a lack of var capacity from inverters, the OLTCs are used to provide additional support. For a distribution network with $M+1$ buses denoted by $j=0, 1, ..., M$, bus $j=0$ is defined as the slack bus whose voltage can be adjusted by the OLTC. The remaining n buses are considered as PQ buses which inject or absorb real and reactive power. In vector form, the bus voltage magnitude deviation ($\Delta \boldsymbol{V} = [\Delta V_1, \ldots, \Delta V_n]^T$) with respect to real power deviation ($\Delta \boldsymbol{P} = [\Delta P_1, \ldots, \Delta P_n]^T$) and reactive power deviation ($\Delta \boldsymbol{Q} = [\Delta Q_1, \ldots, \Delta Q_n]^T$) can be represented as

$$\Delta \boldsymbol{V} = \boldsymbol{S}_{\mathbf{VP}} \Delta \boldsymbol{P} + \boldsymbol{S}_{\mathbf{VQ}} \Delta \boldsymbol{Q} \tag{16.1}$$

where S_{VP} and S_{VQ} are sensitivities of bus voltage magnitudes with respect to real and reactive power injection, obtained from the inverse of the Jacobian matrix [8]. In low-voltage distribution networks, the resistance to reactance ratio (R/X) is relatively high as compared to high-voltage transmission networks. Herein, the bus voltages are affected by both real $\Delta \boldsymbol{P}$ and reactive power $\Delta \boldsymbol{Q}$ changes in (16.1). When PV units experience shading effects, their power output fluctuation may become severe and create significant voltage fluctuations in high R/X networks. The power inverters can achieve fast voltage control by flexibly changing the $\Delta \boldsymbol{P}$ and $\Delta \boldsymbol{Q}$ at each bus.

The transformer of OLTC at the distribution substation will provide supplementary voltage control when the var capacity from power inverters is not enough. It functions to maintain the voltages along the distribution networks fed by the substation within the allowable range. The OLTCs consist of an autotransformer equipped with an on-load tap changing mechanism. The network voltages are regulated by changing taps on the series winding of the autotransformer. The control signals of these devices can be obtained by remote voltage measurements at voltage critical buses. The voltage deviation with OLTC in vector form is represented by

$$\Delta \boldsymbol{V} = \boldsymbol{S}_{\mathbf{VP}} \Delta \boldsymbol{P} + \boldsymbol{S}_{\mathbf{VQ}} \Delta \boldsymbol{Q} + T \Delta \boldsymbol{V}_{\mathrm{tap}} \tag{16.2}$$

where T is the tap position of the OLTC, and $\Delta \boldsymbol{V}_{\mathrm{tap}} = \left[\Delta \boldsymbol{V}_{\mathrm{tap}1}, \ldots, \Delta \boldsymbol{V}_{\mathrm{tap}N}\right]^T$ is a vector representing the voltage variation caused by each tap change to the downstream network, for the bus at the upstream network of the OLTC, $\Delta V_{\mathrm{tap},i} = 0$.

16.2.2 Var capacity from power inverters

We consider the distribution network has N power inverters denoted by $i = 0, 1, ..., N$. The time-varying var capacity of ith power inverter can be represented as follows:

$$\overline{Q}_{\mathrm{Inv},i}(k) = \sqrt{S_{\mathrm{Inv},i}2 - P_{\mathrm{Inv},i}2(k)} \tag{16.3}$$

where $\overline{Q}_{\mathrm{Inv},i}$ is the var capacity of ith power inverter when the real power output is $P_{\mathrm{Inv},i}$ at interval k. $S_{\mathrm{Inv},i}$ is the rated capacity of ith power inverter, which typically equals the maximum real power output of DC source.

The capacity of the power inverter can be oversized so that the inverter can generate or consume reactive power even when rated real power is being produced

[9]. The oversized inverters have the capability to provide more flexible var support. The oversized capacity of ith inverter $S'_{\mathrm{Inv},i}$ is given as follows:

$$S'_{\mathrm{Inv},i} = (1+\alpha_i)S_{\mathrm{Inv},i} \tag{16.4}$$

where α_i is the percentage of the inverter oversizing. According to (16.3) and (16.4), the variation range of var capacity $\overline{Q}_{\mathrm{Inv},i}$ after oversize becomes

$$\overline{Q}_{\mathrm{Inv},i} \in \left[\sqrt{\alpha_i 2 + 2\alpha_i} S_{\mathrm{Inv},i}, (1+\alpha_i)S_{\mathrm{Inv},i}\right] \tag{16.5}$$

In this chapter, the power inverters are divided into two types according to different DC sources: (1) photovoltaic (PV) type with unidirectional real power output and (2) electric vehicle (EV) type with bidirectional real power output. The relationship of var capacity, the real power output of PV and EV power inverters, and their operating points (OPs) are shown in Figure 16.1. For PV-type power inverter, the var capacity is minimum when the real power output reaches the peak point at OP 1. The var capacity is maximum when the real power reaches the bottom point at OP 2. Similarly, for EV-type power inverter, the var capacity is minimum when the real power output reaches the peak point at OP 1, and the var capacity is maximum when the real power becomes zero at OP 2, as from Figure 16.1(c) and (d). OP 3 provides the minimal var capacity point during the negative real power output (EV charging) period, as shown in Figure 16.1(c) and (d).

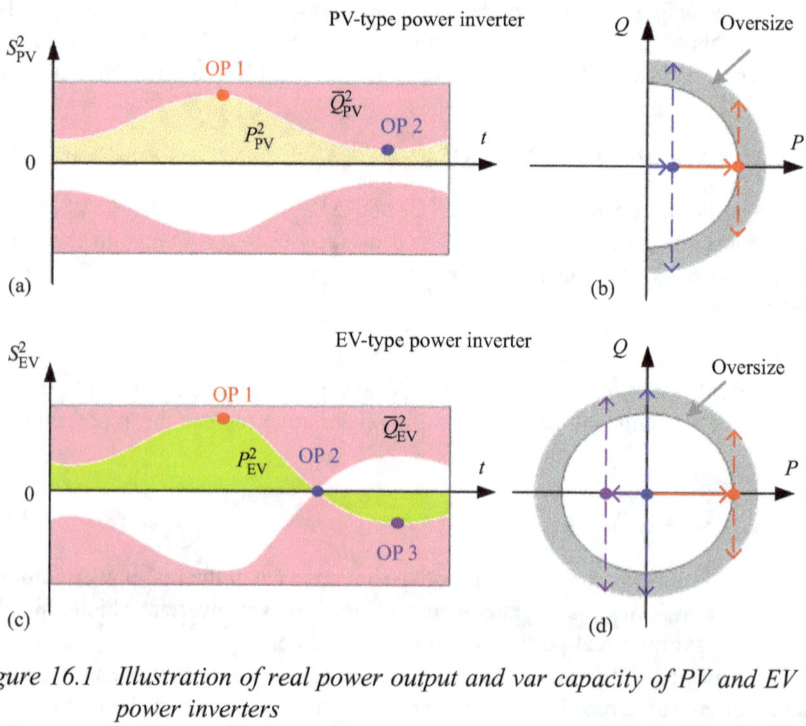

Figure 16.1 Illustration of real power output and var capacity of PV and EV power inverters

16.3 Proposed hybrid voltage control

A decentralised-distributed hybrid control strategy is proposed to address the network voltage control problem. The control scheme diagram in Figure 16.2 shows that a decentralised voltage control based on moving average filter is designed by measuring the real power output $P_{\text{Inv},i}$ of inverter i. Meanwhile, the decentralised control also determines the var capacity $\overline{Q}_{\text{Inv},i}$ of inverter i which can be utilised by distributed voltage control. The virtual leader control receives voltage information of critical buses to initiate the distributed control. The control signal u_0 of the virtual leader spreads to all available inverters through the communication network. The OLTC control takes effect when there is a lack of var capacity from the

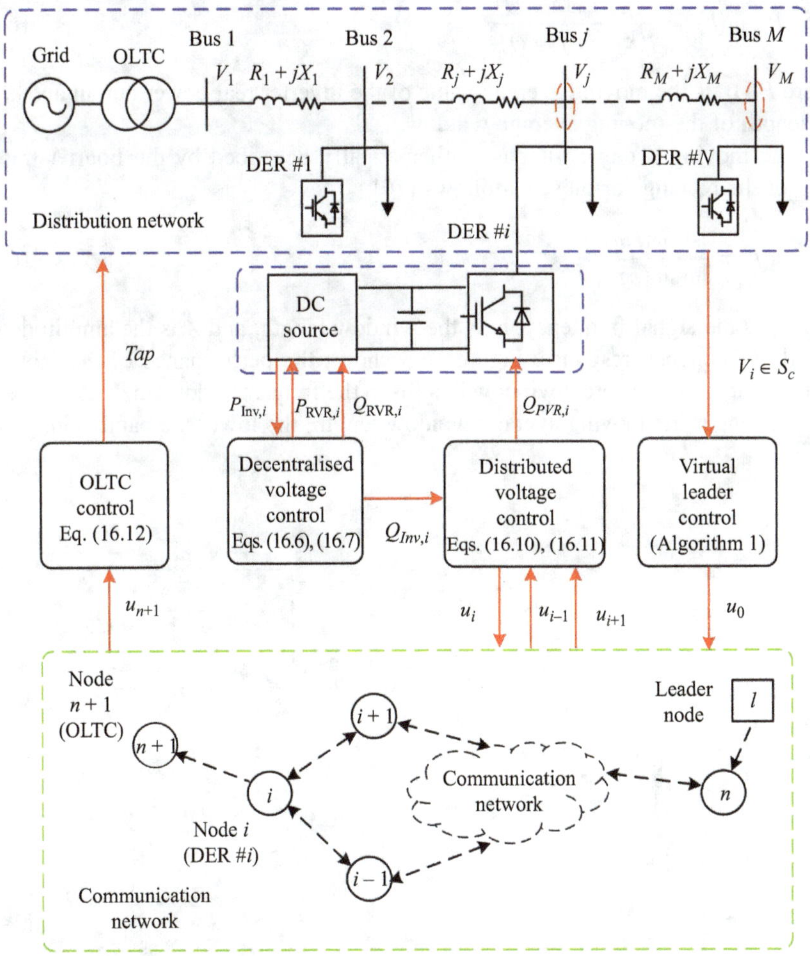

Figure 16.2 The overall control architecture of the proposed hybrid voltage control strategy

inverters. The details of the proposed control strategy are introduced in the following subsections.

16.3.1 Decentralised voltage control

The decentralised voltage control is designed for each power inverter to locally mitigate the network voltage fluctuations and provide smooth var capacity. A moving average filter is designed to extract high-frequency components from the power variations. Both real and reactive power of the inverter-based DERs are utilised to compensate for these variations. As a result, the voltage profile and remaining inverter capacity are smoothed. At each time interval k, the moving average of the real power output of ith power inverter for window length ω is calculated by

$$P_{\mathrm{MA},i}(k) = \frac{\sum_{j=k-\omega}^{k} P_{\mathrm{Inv},i}(j)}{t(k) - t(t-\omega)} \tag{16.6}$$

where $P_{\mathrm{MA},i}$ is the moving average value of the inverter real power output and ω is the length of the moving average window.

The moving average filter is mathematically described by the Fourier transform of the rectangular pulse as follows [10]:

$$A[f] = \frac{|\sin(\pi f \omega)|}{|\omega \sin(\pi f)|} \tag{16.7}$$

where f is the signal frequency, ω is the window length, and A is the amplitude of the filter frequency response. Figure 16.3 shows the performance of the moving average filter with different window lengths in the frequency domain. It can be seen that the longer the moving average window length, the lower the bandwidth. As a

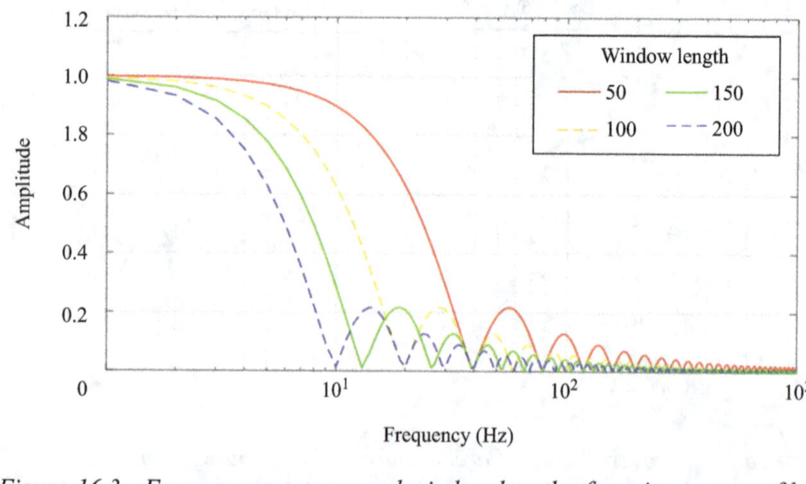

Figure 16.3 Frequency response and window length of moving average filter

consequence, fewer high-frequency components will pass through the filter. In other words, the longer the window length is, the smoother the $P_{MA,i}$ will be.

The high-frequency components are extracted by subtracting the real power variations $P_{Inv,i}$ from the moving average value $P_{MA,i}$. The voltage fluctuations caused by high-frequency components of real power variations can be mitigated by the corresponding compensating real and reactive powers. To summarise, the output of the decentralised voltage control P_{RVC} and Q_{RVC} are determined as

$$\begin{aligned} &\begin{cases} P_{RVC,i}(k) = P_{MA,i}(k) - P_{Inv,i}(k) \\ Q_{RVC,i}(k) = 0 \end{cases} , \text{ if } P_{MA,i}(k) - P_{Inv,i}(k) < 0 \\ &\begin{cases} P_{RVC,i}(k) = 0 \\ Q_{RVC,i}(k) = P_{MA,i}(k) - P_{Inv,i}(k) \end{cases} , \text{if } P_{MA,i}(k) - P_{Inv,i}(k) > 0 \end{aligned} \tag{16.8}$$

With (16.8), var capacity $\overline{Q}_{Inv,i}(k)$ can be smoothed by replacing $P_{Inv,i}(k)$ with $P_{MA,i}(k)$ in (16.3). The smoothed var capacity from each inverter will be utilised by the distributed voltage control.

16.3.2 Distributed voltage control

In this section, a distributed voltage control based on the leader–follower consensus algorithm is designed to regulate the bus voltage magnitudes within the required limits. The communication network to achieve leader–follower consensus is introduced.

16.3.2.1 Graph theory

The communication network among the power inverters can be described by an undirected graph, which is defined as $G = (V, E)$ with a set of N nodes $V = \{v_1, v_2, \ldots, v_N\}$ and a set of edges $E = V \times V$. Each node is assigned to a power inverter, and edges represent the communication links for data exchange. Undirected graph means communication links are bidirectional, $(v_i, v_j) \in E \Rightarrow (v_j, v_i) \in E\ \forall i, j$. A matrix called adjacency matrix $A = [a_{ij}]$ is associated with the edges. a_{ij} represents the weight for information exchanged between agents i and j, where $a_{ij} = 1$ if agents i and j are connected through an edge $(v_i, v_j) \in E$, otherwise, $a_{ij} = 0$. For the problem of this chapter, N power inverters have a virtual leader (index 0) to update their voltage control requirement. The interaction topology can be expressed by graph $\overline{G}$, which contains original graph G, node v_0 and edges (v_i, v_0) from node v_0 to other nodes. The leader can send information to power inverters, but not vice versa. A pining matrix $G = \text{diag}\{g_1, g_2, \ldots, g_N\}$ is used to describe whether each inverter directly receives information from the leader, where the pinning gain $g_i = 1$ if there is an edge (v_i, v_0), else $g_i = 0$.

16.3.2.2 Consensus condition

In this chapter, the distributed voltage control is designed using the leader–follower consensus algorithm, where N power inverters will track a reference utilisation

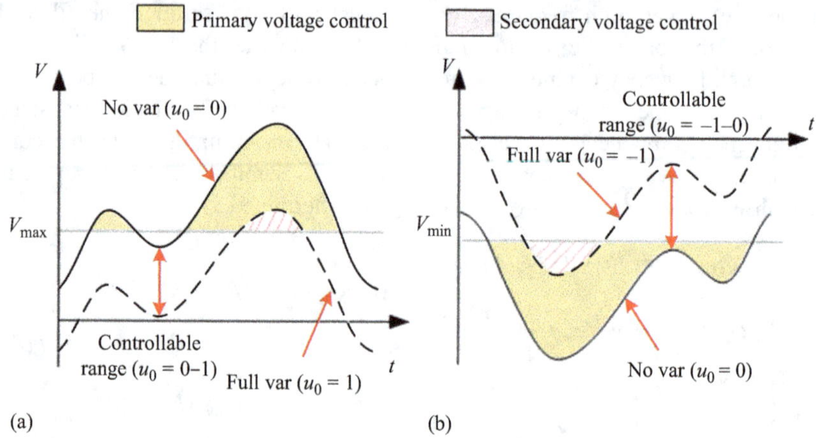

Figure 16.4 Illustration of the relationship between utilisation ratio and voltage profiles

state u_0 provided by a virtual leader. The utilisation state indicates the usage of var capacity by distributed control. The leader–follower consensus is achieved by information exchange via the communication graph defined above. The algorithm reaches a consensus so that each inverter operates at an equal utilisation state determined by the virtual leader. In steady state, the utilisation states between the reactive power output of distributed voltage control Q_{DVC} and var capacity $\overline{Q}_{Inv}$ of all power inverters are same; that is, the following condition will be held:

$$\frac{Q_{DVC,i}}{\overline{Q}_{Inv,i}} = \frac{Q_{DVC,j}}{\overline{Q}_{Inv,j}} = u_0 \text{ or } \lim_{t\to\infty} \|u_i - u_0\| = 0 \tag{16.9}$$

Figure 16.4(a) and (b) illustrates the relationship between the utilisation state of power inverters and network voltage profiles. As shown in Figure 16.4(a), $u_0 = 0$ means that there is no var support from the power inverters, while $u_0 = 1$ means that there is full var support from the power inverters. In order to regulate the voltage within the allowable ranges, u_0 is between 0 and 1 for over-voltage periods. Similarly, the under-voltage control is illustrated in Figure 16.4(b), where u_0 is between -1 and 0 for under-voltage periods. There are also areas where var capacity from inverters is not enough; therefore, additional voltage control from OLTC is needed.

16.3.2.3 Leader control

The reference utilisation state u_0 is determined by a virtual leader who measures the voltages of all critical buses in the distribution network. First, the leader control will choose the maximum voltage among critical buses (denoted as $V_l(k)$), where the subscript $l = \{i|\max_{i\in S_c}(|V_i(k)|)\}$, S_c is the set of the critical buses. In this chapter, the voltage critical bus refers to the bus which may have the highest/lowest voltage in the distribution network. The selection of critical buses can be based on the average highest/lowest voltage magnitudes over a certain period of time (e.g. one month or one season). When the information of critical buses is updated, the measurement set and communication links should also be changed correspondingly. This design ensures the voltages of critical buses are measured and controlled for network voltage control.

Then, based on the $V_l(k)$ defined above, the reference utilisation state u_0 is updated as follows:

$$u_0(k+1) = \begin{cases} u_0(k) + T_s\left(V_l(k) - V_{\text{up}}\right), \text{if } V_l(k) > V_{\text{up}} \text{ or } u_0(k) > 0 \\ u_0(k) + T_s\left(V_l(k) - V_{\text{low}}\right), \text{if } V_l(k) < V_{\text{low}} \text{ or } u_0(k) < 0 \\ 0, \text{otherwise} \end{cases} \tag{16.10}$$

where V_l is the maximum voltage of critical buses; V_{up} and V_{low} are voltage upper and lower references, respectively; and T_s is the updating time interval.

In (16.10), u_0 will become a non-zero value once the maximum voltage of critical buses is out of the safe operation range $[V_{\text{low}}, V_{\text{up}}]$ (i.e. [0.95, 1.05]). In the voltage rise period, u_0 will continue updating when the condition $V_l(k) > V_{\text{up}}$ or $u_0(k) > 0$ is satisfied. On the contrary, u_0 will become zero again if $V_l(k) \leq V_{\text{up}}$ and $u_0(k) \leq 0$, which means the voltage is back to safety range and no var support from inverters is needed. A similar process holds for the voltage drop period, u_0 will be updated when the condition $V_l(k) < V_{\text{low}}$ or $u_0(k) < 0$ is satisfied, and u_0 will become zero again if $V_l(k) \geq V_{\text{up}}$ and $u_0(k) \geq 0$.

16.3.2.4 Consensus control

After u_0 is calculated from (16.10), the utilisation state of each inverter u_i can be updated according to the leader–follower consensus protocol in discrete time as

$$u_i(k+1) = u_i(k) + \varepsilon\left[\sum_{j\in N_i} a_{ij}\left(u_j(k) - u_i(k)\right) + g_i\left(u_0(k) - u_i(k)\right)\right] \tag{16.11}$$

where $i, j \in G$; u_i and u_j are the utilisation states of inverter i and its neighbouring inverter j; ε is a constant weight; a_{ij} is the communication coefficient between inverters i and j; and g_i is the pinning gain of the inverter i.

Finally, the distributed voltage control output of inverter i $Q_{\text{DVC},i}$ is the product of its utilisation state and time-varying var capacity at time interval k:

$$Q_{\text{DVC},i}(k) = u_i(k)\overline{Q}_{\text{Inv},i}(k) \tag{16.12}$$

Note that the smoothed var capacity obtained from the decentralised control stage will benefit the performance of distributed control since it will reduce the overshoot of reactive power output in (16.12).

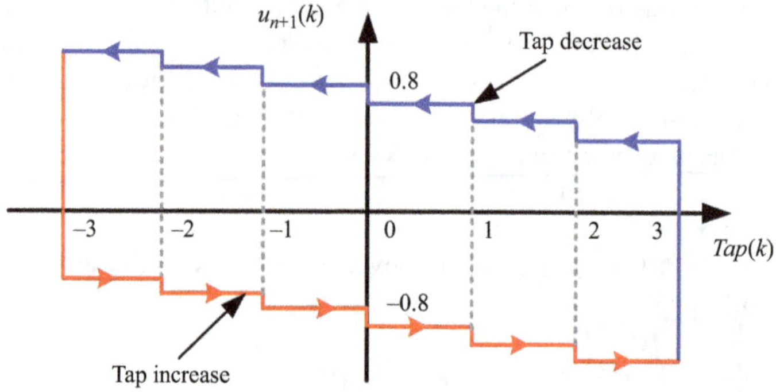

Figure 16.5 Demonstration of tap change and utilisation ratio when a = *0.8 and* b = *0.05*

16.3.3 Supplementary voltage control

The OLTC at the substation can change the voltage profiles of the entire downstream network in a discrete manner. The OLTC control will be activated, when var capacity from power inverters is to be exhausted. It is assumed that there is a directed communication link between OLTC and the adjacent inverter, which means that OLTC can receive the utilisation state from the adjacent inverter. The operation of the OLTC is defined using the following equation:

$$Tap(k+1) = \begin{cases} Tap(k) + 1, \text{if } u_{n+1}(k) > a + bTap(k) \\ Tap(k) - 1, \text{if } u_{n+1}(k) < -a + bTap(k) \\ Tap(k), \text{otherwise} \end{cases} \tag{16.13}$$

where u_{n+1} is the utilisation state received by the OLTC, and a and b are parameters in the tap changing thresholds of the OLTC.

According to (16.13), the tap of OLTC will increase when u_{n+1} is larger than the upper threshold $a + bTap(k)$, which indicates the remaining var capacity of power inverters is low and OLTC control will take effect. Similarly, the tap of OLTC deceases when u_{n+1} is smaller than the lower threshold $-a + bTap(k)$. In between the thresholds, the reactive power provided by power inverters is adequate for voltage control and therefore no tap change is required. Figure 16.5 shows an example of tap change and utilisation ratio when $a = 0.8$, $b = 0.05$, and $Tap(k) \in [-3, 3]$. The parameters in Figure 16.5 are also applied to the simulation study in the next section. In practice, these parameters can be empirically set by the operators.

16.4 Simulation studies

16.4.1 Test system

In this section, the proposed method is tested on the IEEE 33-bus distribution network. The test system is implemented in MATLAB®/Simulink®, and power flow is solved by the Load Flow Tool. The distribution network is modified with

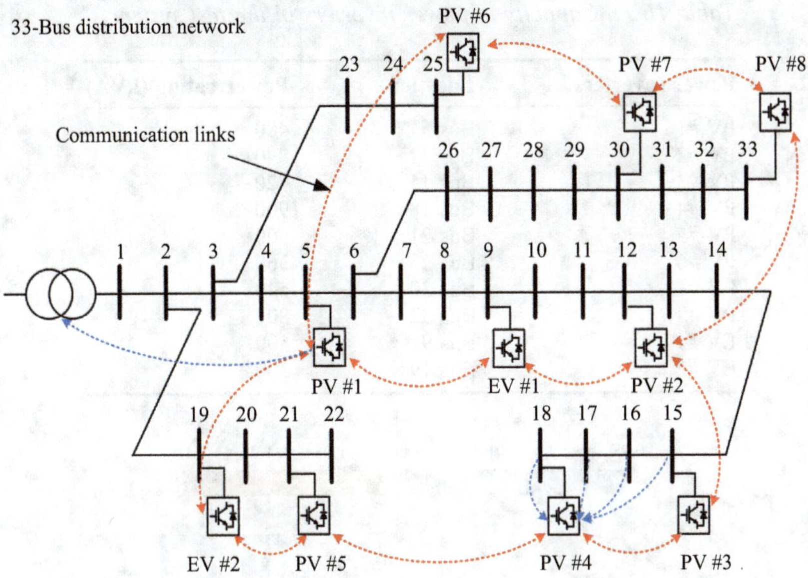

Figure 16.6 The configuration of the IEEE-33 bus system

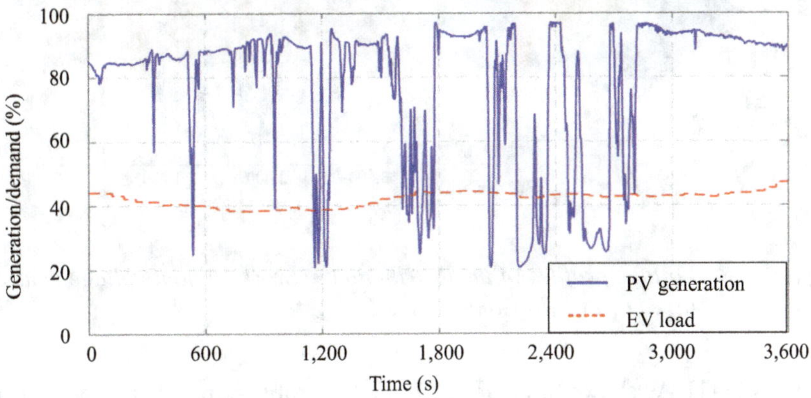

Figure 16.7 The 1-h PV and EV profiles

high PV and EV penetration at selected buses as shown in Figure 16.6. Besides, it is assumed that each power inverter only shares information with its adjacent inverters, which forms a sparse communication network. The additional communication links for voltage measurement of critical buses and OLTC control are also shown in Figure 16.6. The communication network is deployed considering the physical structure of the distribution network.

The 1-h PV generation and EV/load demand profiles used in the Scenarios 1–3 are shown in Figure 16.7. The PV data of 1-s resolution were measured by EPRI in

Table 16.1 Capacity of power inverters in the test system

Power inverters	Location	Power rating (kVA)
PV #1	Bus 5	480
PV #2	Bus 12	840
PV #3	Bus 15	720
PV #4	Bus 18	900
PV #5	Bus 21	600
PV #6	Bus 25	360
PV #7	Bus 30	540
PV #8	Bus 32	800
EV #1	Bus 9	500
EV #2	Bus 19	300

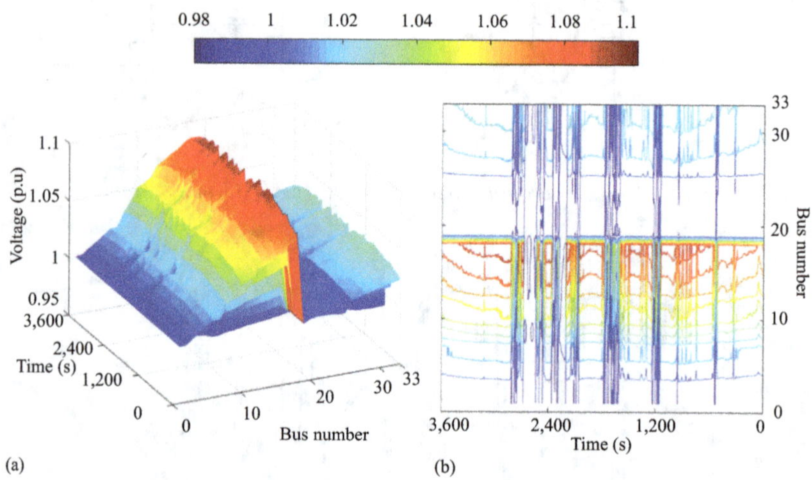

Figure 16.8 Voltage profiles of the distribution network without voltage control. (a) 3D view. (b) Contour line

June 2012 [11]. As shown in Figure 16.7, the PV profile ramps up and down from approximately 20% to 95% of the total PV system power rating several times within the span of an hour. The EV charging demand is assumed to be a typical residential load demand profile with a smaller variation. The rated power capacity of each PV and EV system is listed in Table 16.1. The real power output of each power inverter can be calculated from the PV/EV profile in percentage multiplying by its power rating in MVA.

Initially, it is assumed that there is no tap change of the OLTC, and the voltage at bus 1 is set to 1.0 p.u. In this chapter, the voltage limit range [V_{low}, V_{up}] is set to [0.95, 1.05] p.u. The initial voltage profiles of the 33-bus distribution network without any form of voltage control are shown in Figure 16.8(a) and (b). As shown in Figure 16.8(a) and (b), there are both voltage limit violations as well as severe voltage fluctuations in the distribution network. The maximum voltage deviation of bus 18 can reach 1.1 p.u. during the operation period, and the largest voltage fluctuation of bus 18 is as high as 1.1 p.u. for every 10 s. Buses 15–18 are selected as the critical buses to initiate the distributed control once there is a voltage limits violation. The performance of the proposed control method is demonstrated and discussed in the following.

16.4.2 Scenario 1

In Scenario 1, the performance of the decentralised voltage control is investigated. The simulation results of Scenario 1 are shown in Figures 16.9 and 16.10.

As a typical example, the voltages of bus 18 under different window lengths are shown in Figure 16.9. In Figure 16.9, the longer the ω is, the smoother the voltage profiles are. The profile from 2,000 to 2,300 s is zoomed-in in Figure 16.9, where the largest voltage fluctuation is reduced from 10% to 3% for every 10 s with $\omega > 100$. A conservative value of $\omega = 150$ is selected in the following part such that the voltage fluctuations are significantly smoothed.

Figure 16.10(a) and (b) shows the power output of decentralised voltage control and var capacity for distributed voltage control of PV #4 (bus 18) with $\omega = 150$, respectively. According to the power output calculated by (16.6) and (16.8), real power will be curtailed and reactive power will be injected by PV #4, as shown in Figure 16.10(a). Finally, the smoothed var capacity can be obtained from (16.3) by replacing $P_{\mathrm{Inv},i}(k)$ with moving average value in (16.6), as shown in Figure 16.10(b). The voltage profile becomes much smoother as compared to no decentralised control (i.e. $\omega = 1$) in Figure 16.10(b). This will reduce the reactive power overshoot from the distributed control and benefits the network voltage control.

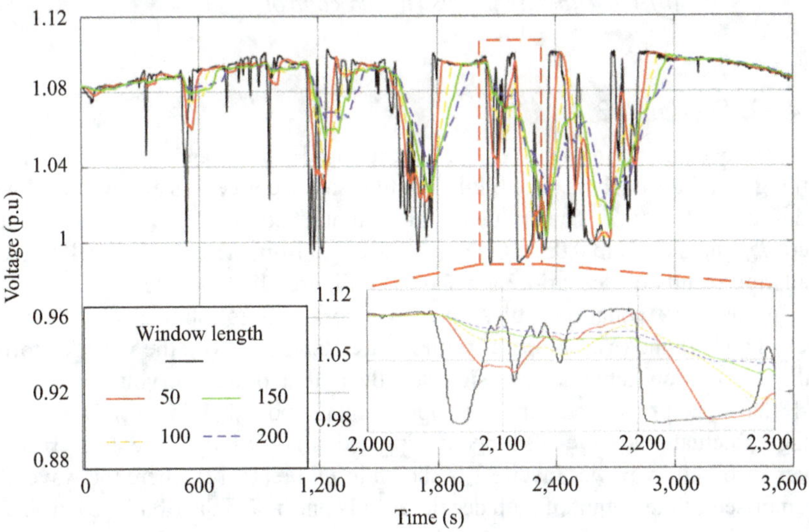

Figure 16.9 Voltage profile of bus 18 with different window lengths

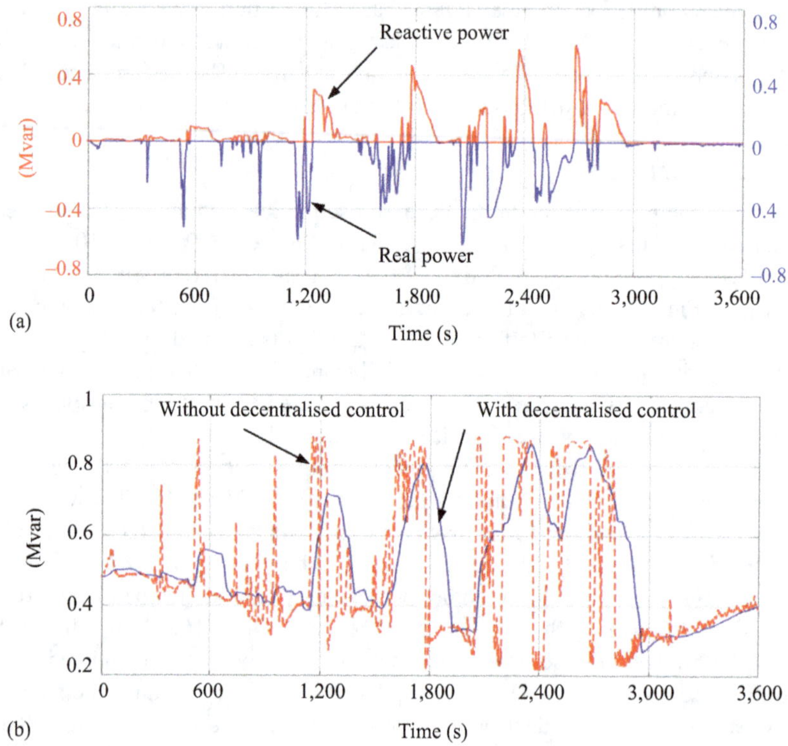

Figure 16.10 (a) Real and reactive power output of decentralised control. (b) Var capacity for distributed control

16.4.3 Scenario 2

In Scenario 2, the performance of the decentralised control and distributed control working simultaneously is tested. It is assumed that the inverters are oversized with a coefficient $\alpha = 0.2$. The influence of communication rate T_{ca} is also studied where T_{ca} changes from 0.05 to 0.1 s at 1,200 s and from 0.1 to 0.2 s at 2,400 s. The simulation results of Scenario 2 are shown in Figures 16.11–16.13.

The network voltage profiles with the hybrid control strategy are shown in Figure 16.11(a) and (b). Compared with Figure 16.8(a) and (b), the voltage profiles of the distribution network have significantly flattened and the voltages of buses 10–18 have been regulated within the upper voltage bound (1.05 p.u.). Besides, the voltage fluctuations have also been mitigated significantly. The total reactive power absorbed by power inverters is shown in Figure 16.12, where the waveform is comprised of the output of both decentralised control and distributed control. The corresponding utilisation state profiles of power inverters are shown in Figure 16.13. The utilisation states of all inverters track the utilisation state

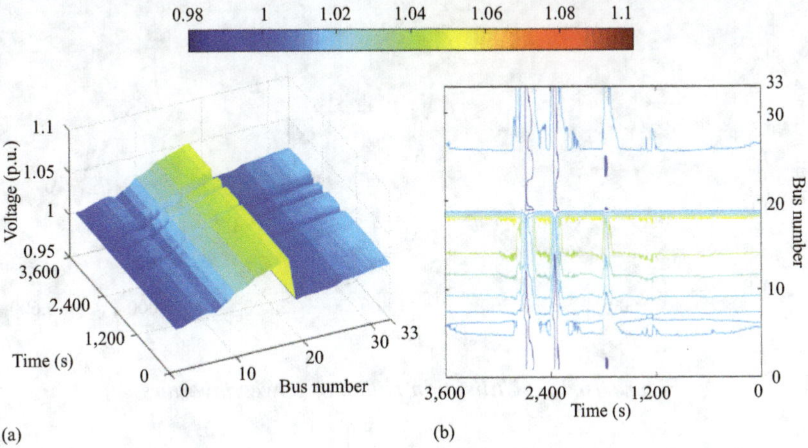

Figure 16.11 Voltage profiles of the distribution network with the proposed hybrid control strategy in Scenario 2. *(a) 3D view. (b) Contour line*

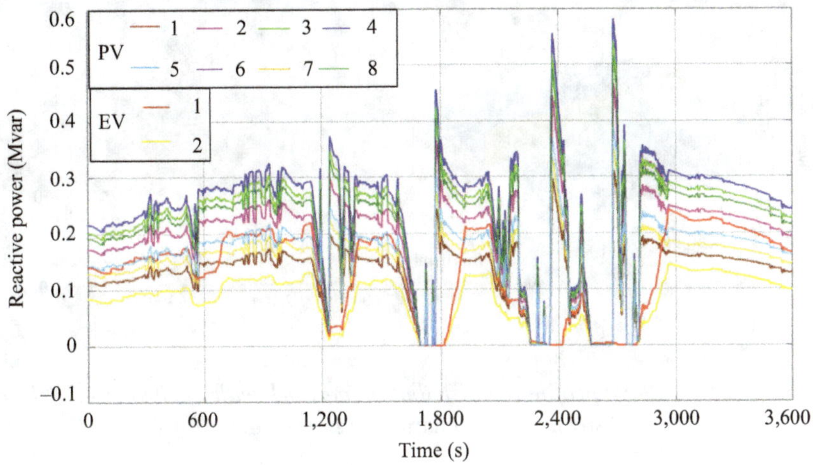

Figure 16.12 The total reactive power output of power inverters

determined by the leader. Besides, the utilisation state of each inverter can track the leader well even when there is communication rate change at 1,200 and 2,400 s, as shown in the enlarged view in Figure 16.13.

16.4.4 Scenario 3

In Scenario 3, the supplementary performance of OLTC voltage control is tested. The parameters of (16.13) are set as $a = 0.8$, $b = 0.05$, and $Tap(k) \in [-3, 3]$ in this scenario. It is assumed that inverters are not oversized and hence the var capacity of the inverters is not sufficient to regulate the network voltages within the limits.

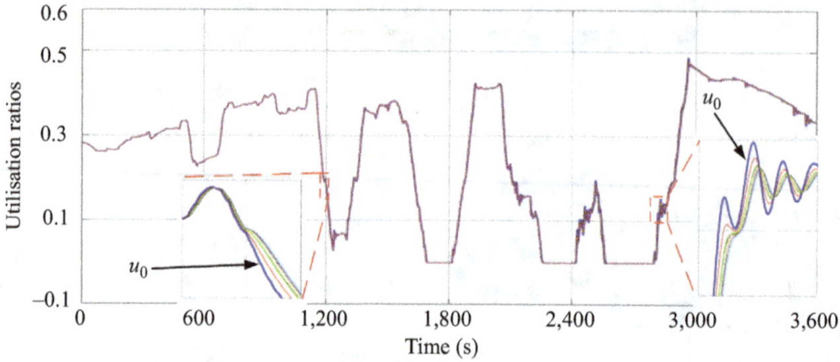

Figure 16.13 Utilisation ratios of power inverters

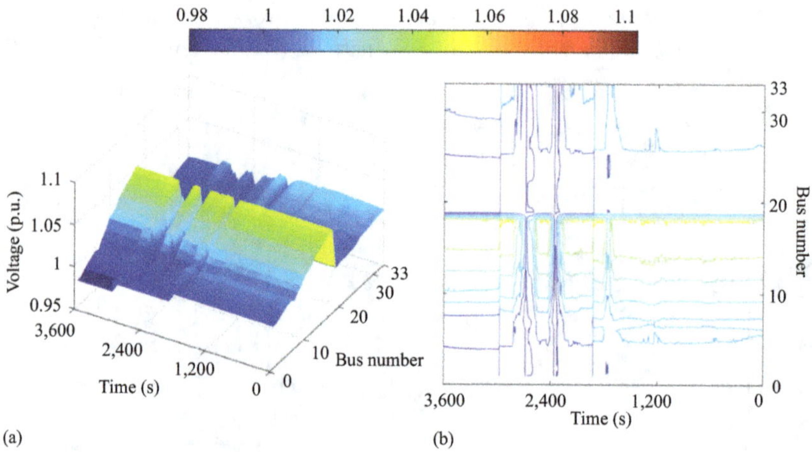

Figure 16.14 Voltage profile the distribution network with the proposed hybrid control strategy in Scenario 3. *(a) 3D view. (b) Contour line*

Therefore, when the utilisation state u_{n+1} received by OLTC reaches the upper threshold defined by (16.13), the tap of OLTC will be increased to provide additional voltage support. The simulation results of Scenario 3 are shown in Figures 16.14–16.16, where the bus voltages are regulated within the limits [0.95, 1.05]. As shown in Figures 16.15 and 16.16, once the utilisation states reach the threshold defined in (16.13), the tap of OLTC will be triggered. During 0 s $\leq t <$ 1,960 s, the var capacity of power inverters is adequate, and therefore no tap change is triggered (tap position remains at 0). During 1,960 s $\leq t <$ 3,000 s, the utilisation state reaches 0.8 at the 1,960 s and therefore tap changes to -1. It can be found in Figure 16.14(a) and (b) that there is a voltage step change of 0.1 p.u. when the control of OLTC is triggered. During 3,000 s $\leq t <$ 3,600 s, the utilisation state once again reaches 0.85 at 3,000 s and therefore the tap changes to -2, and another step change of 0.1 p.u. in the voltage can

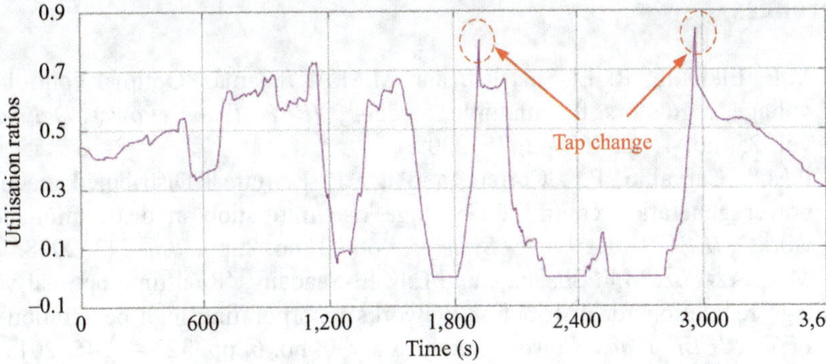

Figure 16.15 Utilisation ratios of power inverters

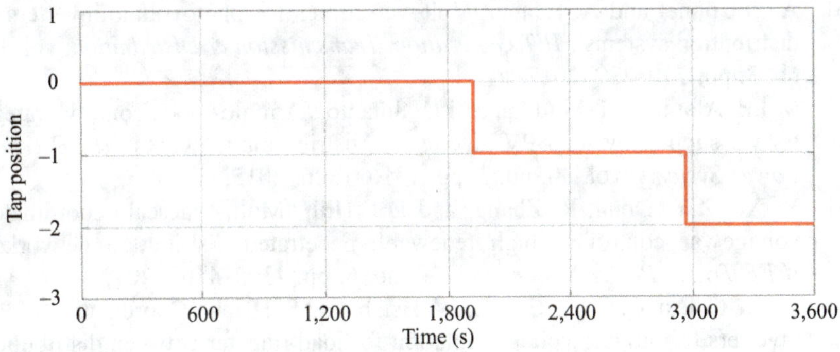

Figure 16.16 Tap positions of OLTC

be observed. Note that the proposed hybrid control strategy improves the voltage profiles in the distribution networks, while unnecessary tap changes caused by voltage fluctuation and limits violation are reduced.

16.5 Conclusion

This chapter introduces a decentralised-distributed hybrid control strategy for voltage control in distribution networks. The decentralised voltage control utilises the real and reactive power of power inverters to locally mitigate voltage fluctuations. The distributed voltage control coordinates all available var capacity from inverters to regulate network voltage within limits. The OLTC control as a backup has been incorporated into the voltage control architecture. The proposed control strategy has been tested in the IEEE 33-bus distribution network. The simulation results have shown that the proposed method can effectively and autonomously deal with voltage fluctuations and voltage rise/drop problems in distribution networks.

References

[1] M.E. Elkhatib, R. El Shatshat, and M.M.A. Salama, 'Optimal control of voltage regulators for multiple feeders', *IEEE Trans. Power Delivery*, vol. 25, no. 4, pp. 2670–2675, 2009.

[2] P.M.S. Carvalho, P.F. Correia, and L.A.F. Ferreira, 'Distributed reactive power generation control for voltage rise mitigation in distribution networks', *IEEE Trans. Power Systems*, vol. 23, no. 2, pp. 766–772, 2008.

[3] M.A. Azzouz, M.F. Shaaban, and E.F. El-Saadany, 'Real-time optimal voltage regulation for distribution networks incorporating high penetration of PEVs', *IEEE Trans. Power Systems*, vol. 30, no. 6, pp. 3234–3245, 2015.

[4] 1547 Series of Interconnection Standards, IEEE SCC21 Standards Coordinating Committee on Fuel Cells, Photovoltaics, Dispersed Generation, and Energy Storage, IEEE Standards Association, Available: http://grouper.ieee.org/groups/scc21/dr_shared/.

[5] A. O'Connel and A. Keane, 'Volt–var curves for photovoltaic inverters in distribution systems', *IET Generation, Transmission & Distribution*, vol. 11, no. 3, pp. 730–739, 2017.

[6] M.J.E. Alam, K.M. Muttaqi, and D. Sutanto, 'A multi-mode control strategy for var support by solar PV inverters in distribution networks', *IEEE Trans. Power Systems*, vol. 30, no. 3, pp. 1316–1326, 2015.

[7] Y. Xu, Z.Y. Dong, R. Zhang, and D.J. Hill, 'Multi-timescale coordinated voltage/var control of high renewable-penetrated distribution networks', *IEEE Trans. Power System*, vol. 32, no. 6, pp. 4398–4408, 2017.

[8] T. Ku, C. Lin, C. Chen, C. Hsu, W. Hsieh, and S. Hsieh, 'Coordination of PV inverters to mitigate voltage violation for load transfer between distribution feeders with high penetration of PV installation', *IEEE Trans. Industry Applications*, vol. 52, no. 2, pp. 1167–1174, 2016.

[9] S. Alyami, Y. Wang, C. Wang, J. Zhao, and B. Zhao, 'Adaptive real power capping method for fair overvoltage regulation of distribution networks with high penetration of PV systems', *IEEE Trans. Smart Grid*, vol. 5, no. 6, pp. 2729–2738, 2014.

[10] S.W. Smith, 'Chapter 15: Moving average filters', in *The Scientist and Engineer's Guide to Digital Signal Processing*. San Diego, CA, USA: California Technical Pub., 1997.

[11] '1-second resolution data', EPRI, Available: http://dpv.epri.com/measurement_data.html.

Chapter 17

Two-level distributed voltage/var control by aggregated PV inverters

The continuous increase of the photovoltaic (PV) penetration level in power distribution networks results in severe voltage limit violation problems. This chapter proposes a fully distributed two-level volt/var control (VVC) scheme which aggregates PV inverters for voltage regulation. In the lower-level VVC, the rooftop PV inverters are aggregated via a consensus algorithm and controlled by user-defined droop control in medium-voltage (MV) networks. The reactive power output of each PV aggregator can be adjusted in real time from its dispatched value depending on the bus voltage variations. In the upper-level VVC, the reactive power of PV aggregators is dispatched every 15 min to minimise the network power loss. The upper-level dispatch signals are set as base values in droop control for PV aggregators. This problem is formulated as second-order cone programming (SOCP) and solved by the alternating direction method of multipliers (ADMM). The effectiveness of the proposed method is demonstrated by simulation results in both short- and long-term scenarios.

17.1 Introduction

Considering all voltage control resources, the voltage control problem in distribution networks can be typically divided into two stages, where the slow and fast actuators are controlled separately [1,2]. In the first stage, the voltage control is realised by mechanical devices such as on-load tap changers, step voltage regulators, and switched capacitors [2]. These mechanical devices should be managed in a relatively long time interval, e.g. hourly or even daily, in order to reduce the frequency of taps/switches change and prolong their lifetime. In the second stage, the power inverters have much fast response speed and flexible operation ranges; therefore, their reactive power is suitable for faster voltage control. The mechanical devices function to shift the network voltage profile according to the PV/load predictions. The VVC of power inverters works to adjust the voltage profiles considering real-time PV/load fluctuations. This chapter focuses on the second-stage problem, where a distributed VVC scheme for aggregated PV inverters is proposed.

This chapter introduces a two-level distributed and timescale coordinated VVC scheme for PV aggregators. It aims to bridge the gap between the upper-level optimal reactive power dispatch and lower-level real-time control. The proposed method has the following features:

- The proposed VVC scheme is fully distributed and includes two timescales coordinated control: upper-level and lower-level VVC. The PV uncertainty is handled by two timescale coordination (15 min dispatch and real-time adjustment), while the control system resilience is enhanced by the fully distributed control structure.
- In the lower-level VVC, a distributed control approach is proposed to group each rooftop PV inverters at the low-voltage (LV) side into aggregators. Then the aggregators are governed by the droop control in MV networks. It is noted that most existing works only consider the inverters as one single inverter directly. The estimation of available reactive capacity from each aggregator is also realised in a distributed way.
- In the upper-level VVC, the aggregators are dispatched distributedly for network loss minimisation based on the network conditions and available reactive power capacity. It is formulated as a SOCP and solved by ADMM. These optimal dispatch signals are updated every 15 min and set as base values for lower-level droop control.

17.2 Proposed VVC architecture

17.2.1 System overview

This chapter considers a residential distribution network with high PV penetration, as shown in Figure 17.1. The rooftop PV inverters at the LV side are fed into a point of common coupling (PCC) and then connected to the MV network through an LV/MV transformer. In urban residual communities, it is reasonable to assume that the load demands are evenly distributed in each phase, and the building rooftop PVs are connected into the grid through three-phase AC/DC inverters. Each group of PV inverters in downstream of an MV/LV transformer is grouped into an aggregator via lower-level communication networks. The data exchange of each bus in the MV network is realised by an upper-level communication network. It motivates us to propose a VVC for this scenario with only peer-to-peer communications. The state-of-the-art communication infrastructure in residential distribution networks including both cable communications and wireless technologies can enable this application. Based on the above electrical and communication infrastructure, the research objective is to design a two-level fully distributed VVC scheme for aggregators of user-end PV inverters.

Next, notations of the studied system are defined. The radial MV distribution network is depicted by a tree graph $\mathcal{B}=(\mathcal{N}, \mathcal{E})$, which is assumed to be a three-phase balanced network. $\mathcal{N} = \{i \mid i = 1, 2, \ldots, N\}$ represents the set of buses and $\mathcal{E}$

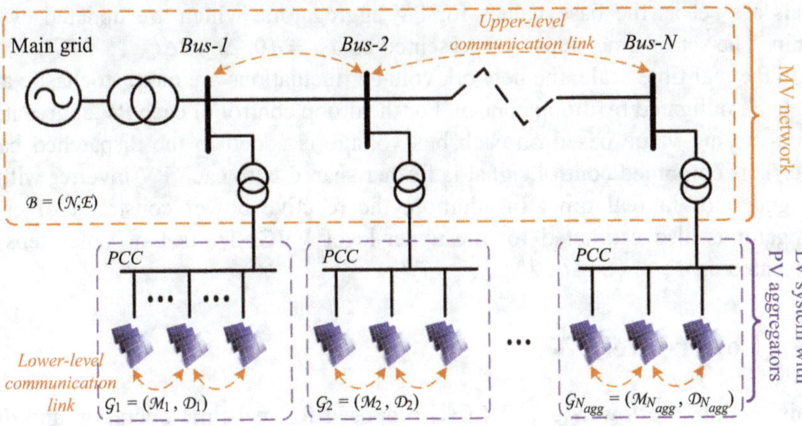

Figure 17.1 Overview of the studied system

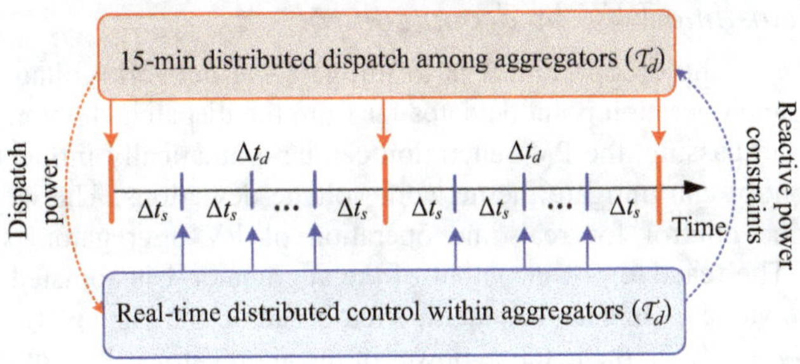

Figure 17.2 Timescale coordination of the proposed two-level VVC

represents the set of distribution lines. The feed bus of the distribution network is numbered as 0. It is noted that there is a set of $\mathcal{N}_{\text{agg}} = \{k|k = 1, 2, \ldots, N_{\text{agg}}\}$ ($\mathcal{N}_{\text{agg}} \subseteq \mathcal{N}$) PV aggregators in the MV network. The upper-level communication network has the same topology as the MV electrical network.

In aggregator k, the set of PV inverters are denoted as $\mathcal{M}_k$, which is denoted by $\mathcal{M}_k = \{j|j = 1, 2, \ldots, M_k\}$. The aggregation of PV inverters is realised by lower-level communication networks. The electrical network restraints are simplified as PV inverters are coupled by one PCC. The lower-level communication network of aggregator k is depicted by an undirected graph $\mathcal{G}_k = (\mathcal{M}_k, \mathcal{D}_k)$ with a set of nodes $\mathcal{M}_k$ and a set of edges $\mathcal{D}_k$. The nodes correspond to communication agents of PV inverters, and edges represent communication links for data exchange.

17.2.2 Architecture of proposed VVC

The proposed two-level VVC has two coordinated timescales 15-min and real-time, as illustrated in Figure 17.2. In 15-min timescale, the objective is to minimise the network power loss in the operational voltage limits. The optimised dispatch signals are set as the base values for PV aggregators, which are updated every 15 min. The set of time steps is represented as $\mathcal{T}_d = \{0, \Delta t_d, 2\Delta t_d, \ldots, T_d\}$.

In the real-time scale, the network voltage fluctuations are more stochastic and need to be mitigated by droop control. For the droop control of each PV aggregator, a time-varying value based on each bus voltage is added to the dispatched base value. This combined control signal is further shared with each PV inverter within the aggregator in real time. In addition, the reactive power constraint of each aggregator is also estimated for the upper-level VVC. The set of time steps is represented as $\mathcal{T}_s = \{0, \Delta t_s, 2\Delta t_s, \ldots, T_s\}$.

17.3 Lower-level VVC

In this section, the lower-level VCC is proposed for real-time operation and distributed reactive power sharing in each PV aggregator. The real-time control takes effect every 0.1 s in this chapter, and $t \in \mathcal{T}_s$.

17.3.1 Real-time VVC by droop control

The objective of the droop control is to mitigate the network voltage violations when the system operation point deviates far from the dispatched point. During the period of cloud passing, the PV generation can be dramatically influenced, where the droop control can mitigate the network voltage deviations. The working principle of droop control for real-time operation of PV aggregator is shown in Figure 17.3. The reactive power output of the aggregator k is adjusted around the dispatch base value q_k^d in each dispatch period based on the bus voltage in the MV network. If $V_k \in [\underline{V}, \overline{V}]$, the reactive power of the aggregator is kept at the dispatch base value q_k^d to minimise network loss. If $V_k \notin [\underline{V}, \overline{V}]$, a time-varying term is activated to mitigate network voltage deviations. The reactive power output q_k^{agg} of

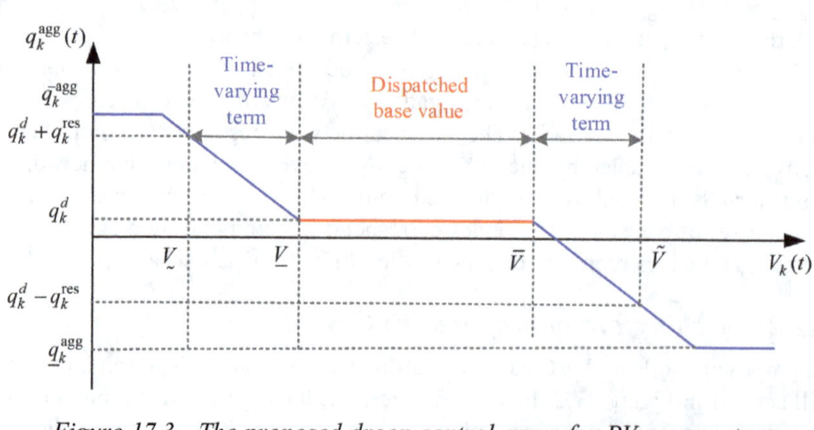

Figure 17.3 The proposed droop control curve for PV aggregator

the aggregator k is calculated by

$$q_k^{\mathrm{agg}}(t) = \begin{cases} q_k^d + G_k\left(V_k(t) - \overline{V}\right), V_k(t) > \overline{V} \\ q_k^d, \underline{V} \leq V_k(t) \leq \overline{V} \\ q_k^d + G_k\left(V_k(t) - \underline{V}\right), V_k(t) < \underline{V} \end{cases} \tag{17.1}$$

where G_k is the droop control gain. $\underline{V}$ and $\overline{V}$ are the lower and upper bounds to activate the time-varying term in droop control.

Remark 1: The droop control gains are selected as $G_k = \alpha_k L_{kk}$, where $L_{kk} = \partial Q_k / \partial V_k$ is the sensitivity of bus voltage magnitude to the reactive power injection. To ensure the stability of the droop-controlled distribution networks, $\alpha_k = 1/n_{\mathrm{bus}}$ are typically selected [1]. The voltage boundary $[\underline{V}, \overline{V}]$ is chosen as [0.95, 1.05].

It is noted that the available capacity for droop control is the overall reactive capacity of the aggregator subtracting the dispatched reactive power. Therefore, a certain amount of capacity can be reserved during the dispatch period. By defining a critical voltage range as $[\underset{\sim}{V}, \tilde{V}]$, the reactive power reserve of an aggregator q_k^{res} is calculated by

$$q_k^{\mathrm{res}} = \begin{cases} \overline{q}_k^{\mathrm{agg}}, q_k^{\mathrm{res}} > \overline{q}_k^{\mathrm{agg}} \\ G_k\left[\max\left(\underline{V} - \underset{\sim}{V}, \tilde{V} - \overline{V}\right)\right], \underline{q}_k^{\mathrm{agg}} \leq q_k^{\mathrm{res}} \leq \overline{q}_k^{\mathrm{agg}} \\ \underline{q}_k^{\mathrm{agg}}, q_k^{\mathrm{res}} < \underline{q}_k^{\mathrm{agg}} \end{cases} \tag{17.2}$$

where $[\underset{\sim}{V}, \tilde{V}]$ is selected as $[0.94, 1.06]$; $\overline{q}_k^{\mathrm{agg}}$ and $\underline{q}_k^{\mathrm{agg}}$ are the upper and lower reactive power limits of the aggregator k.

17.3.2 Distributed aggregation of PV inverters

The droop controller of an aggregator is associated with a pinning node (called the leader). The leader updates a reference state $\mu_{k,0}$ for the aggregator k (depicted by graph $\mathcal{G}_k$) based on the reactive power output from droop control in (17.1) as follows:

$$\mu_{k,0}(t) = \frac{q_k^{\mathrm{agg}}(t)}{M_k} \tag{17.3}$$

where $\mu_{k,0}$ is the reference state, and M_k is the number of PV inverters in the aggregator k.

This reference state is broadcasted to each PV inverter in the aggregator via communication graph $\mathcal{G}_k$ by the following leader–follower consensus control law:

$$\dot{\mu}_{k,j}(t) = \sum_{\hat{j}=1}^{M_k} a_{k,j\hat{j}}\left(\mu_{k,\hat{j}}(t) - \mu_{k,j}(t)\right) + g_{k,j}\left(\mu_{k,0}(t) - \mu_{k,j}(t)\right) \tag{17.4}$$

where $a_{k,j\widehat{j}}$ represents the weight for information exchanged between inverter j and its adjacent inverter $\widehat{j}$. $a_{k,j\widehat{j}} = 1$ if inverters j and $\widehat{j}$ are connected by an edge, otherwise, $a_{k,j\widehat{j}} = 0$. $g_{k,j}$ represents the pinning gain, $g_{k,j} = 1$ if inverter j directly exchanges information from the leader, otherwise $g_{k,j} = 0$.

The average values of upper and lower available var capacity for each inverter j ($\overline{\rho}_{k,j}$ and $\rho\, k,j$) of are estimated from

$$\begin{cases} \dot{\overline{\rho}}_{k,j}(t) = \overline{q}_{k,j}(t) + \sum_{\widehat{j}=1}^{M_k} a_{k,j\widehat{j}}\left(\overline{\rho}_{k,\widehat{j}}(t) - \overline{\rho}_{k,j}(t)\right) \\ \dot{\rho}_{k,j}(t) = q\, k,j(t) + \sum_{\widehat{j}=1}^{M_k} a_{k,j\widehat{j}}\left(\rho\, k,\widehat{j}(t) - \rho\, k,j(t)\right) \end{cases} \tag{17.5}$$

where $\overline{q}_{k,j}$ and $q\, k,j$ are, respectively, the upper and lower limits of available var capacity of inverter j, and $\overline{\rho}_{k,j}$ and $\rho\, k,j$ are, respectively, the estimated average values of upper and lower var limits.

Finally, based on (17.3)–(17.5), the reactive power output of each inverter j is calculated by

$$q_{k,j}(t) = \begin{cases} \dfrac{\mu_{k,j}(t)}{\overline{\rho}_{k,j}}(t)\overline{q}_{k,j}(t), \mu_{k,j}(t) > 0 \\ \dfrac{\mu_{k,j}(t)}{\rho\, k,j(t)} q\, k,j(t), \mu_{k,j}(t) < 0 \end{cases} \tag{17.6}$$

where $q_{k,j}$ is the reactive power output of inverter j in aggregator k, which is within its operational limits as $q_{k,j}(t) \in [q\, k,j(t), \overline{q}_{k,j}(t)]$.

Remark 2: For a set of initial values and connected communication graph $\mathcal{G}_k$, the steady-state value of $q_{k,j}$ converges to

$$\lim_{t\to\infty} q_{k,j}(t) = \begin{cases} \dfrac{M_k\mu_{k,0}(0)}{\sum_{j=1}^{M_k}\overline{q}_{k,j}}(0)\overline{q}_{k,j}(0), \mu_{k,j}(t) > 0 \\ \dfrac{M_k\mu_{k,0}(0)}{\sum_{j=1}^{M_k} q\, k,j(0)} q\, k,j(0), \mu_{k,j}(t) < 0 \end{cases} \tag{17.7}$$

For practical implementation in discrete time, there exists a $T_0 = k\Delta t_s$, for some accuracy level ε_1, such that

$$\begin{aligned} &\|\mu_{k,j}(t) - \mu_{k,0}(t)\| \le \varepsilon_1 \wedge \|\overline{\rho}_{k,j}(t) - \overline{\rho}_{k,\widehat{j}}(t)\| \le \varepsilon_1 \\ &\wedge \|\rho\, k,j(t) - \rho\, k,\widehat{j}(t)\| \le \varepsilon_1, \forall t \ge T_0, j, \widehat{j} \in M_k, t \in T_s \end{aligned} \tag{17.8}$$

and the equilibrium in (17.7) is also achieved. The convergence analysis is given in Appendix A.

17.3.3 Var capacity of PV aggregator

The available var capacity of each PV inverter j in aggregator k is calculated from

$$q_{k,j}^{\max}(t) = \sqrt{\left(s_{k,j}^{\mathrm{PV}}\right)^2 - p_{k,j}(t)^2} \tag{17.9}$$

where $s_{k,j}^{PV}$ and $p_{k,j}$ are the rated capacity and real power output of a PV inverter, respectively.

Equivalently, (17.9) can be written as

$$q_{k,j}^{\max}(t) = s_{k,j}^{\mathrm{PV}}\sqrt{1 - \xi_{k,j}(t)} \tag{17.10}$$

where $\xi_{k,j}(t) = p_{k,j}(t)/s_{k,j}^{\mathrm{PV}}$ is denoted as the PV generation factor. It is noted that the available var capacity is represented as $\left[q\,k,j(t), \overline{q}_{k,j}(t)\right]$, which is only a subset of $\left[-q_{k,j}^{\max}(t), q_{k,j}^{\max}(t)\right]$ considering the local var utilisation.

The upper and lower limits of var capacity $\overline{q}_k^{\mathrm{agg}}$ and $q\,k^{\mathrm{agg}}$ in aggregator k are time-varying, which is equal to the sum of all PV inverters' available var capacity. However, these values cannot be directly calculated in a distributed way, as each node only knows its neighbours' information. As mentioned above, the leader can exchange information with inverter j if $g_{k,j} = 1$. Therefore, the upper and lower limits of var capacity of aggregator k can be acquired as follows:

$$\begin{cases} \overline{q}_k^{\mathrm{agg}}(t) = \sum_{j=1}^{M_k} \overline{q}_{k,j}(t) \cong \dfrac{M_k \sum_{j=1}^{M_k} g_{k,j} \overline{\rho}_{k,j}(t)}{\sum_{j=1}^{M_k} g_{k,j}}, \\ q\,k^{\mathrm{agg}}(t) = \sum_{j=1}^{M_k} q\,k,j(t) \cong \dfrac{M_k \sum_{j=1}^{M_k} g_{k,j} \rho\,k,j(t)}{\sum_{j=1}^{M_k} g_{k,j}}. \end{cases} \tag{17.11}$$

Herein, the operational limits of an aggregator $\left[q\,k^{\mathrm{agg}}, \overline{q}_k^{\mathrm{agg}}\right]$ are obtained from (17.11), which is used to update the reactive power reserve in (17.2), and the distributed power dispatch in the next section.

17.4 Upper-level VVC

In the upper-level VVC, the reactive power output from PV aggregators is optimally dispatched in a longer timescale, i.e. 15 min, and $t \in \mathcal{T}_d$. The dispatch results will serve as the base value settings in droop control for each aggregator.

17.4.1 Power flow in MV distributed networks

The feed bus of the MV distribution network is numbered as 1. Each bus $i \in \mathcal{N}\backslash\{1\}$ has a unique ancestor bus A_i, and a set of children buses C_i. All the neighbour bus of bus i is represented by $\widehat{i} \in \mathcal{N}_i$. The line from bus A_i to bus i is labelled as line i, and $i \in \varepsilon = \{1, 2, \ldots., n\}$. For each bus $i \in \mathcal{N}$, $v_i = |V_i|^2$ represents the square of bus voltage. $s_i = p_i + jq_i$ is the complex power injection. For each line $i \in \mathcal{E}$, the line impedance is represented by $z_i = r_i + jx_i$. The complex power flow from bus A_i to bus i is $S_i = P_i + jQ_i$. The square of the line current is $l_i = |I_i|^2$.

For a radial MV distribution network $\mathcal{B}$, the power flow equations are represented as follows [2]:

$$v_{Ai} = v_i + 2(r_i P_i + x_i Q_i) + l_i(r_i^2 + x_i^2), i \neq 1 \tag{17.12a}$$

$$P_i = \sum_{\widehat{i} \in C_i} \left(P_{\widehat{i}} + r_{\widehat{i}} l_{\widehat{i}} \right) + p_i^L + p_i^d \tag{17.12b}$$

$$Q_i = \sum_{\widehat{i} \in C_i} \left(Q_{\widehat{i}} + x_{\widehat{i}} l_{\widehat{i}} \right) + q_i^L + q_i^d \tag{17.12c}$$

$$P_i^2 + Q_i^2 = v_i l_i \tag{17.12d}$$

The reactive power constraint of each PV aggregator is

$$\underline{q}_k^{agg} - q_k^{res} \leq q_k^d \leq \overline{q}_k^{agg} - q_k^{res} \tag{17.13}$$

The voltage magnitude should be within the constraints

$$\underline{v}_i \leq v_i \leq \overline{v}_i \tag{17.14}$$

The quadratic equality constraint (17.12d) can be relaxed to a second-order cone form, and the relaxation is also exact as discussed by [3]:

$$P_i^2 + Q_i^2 \leq v_i l_i \tag{17.15}$$

The upper-level VVC is to minimise the network loss subject to power flow equations and operational constraints. The objective function is defined as

$$\sum_{i \in \mathcal{N}} f_i(x_i) = \sum_{i \in N} r_i l_i \tag{17.16}$$

Therefore, the network loss minimisation problem can be formulated as

$$\min \sum_{i \in \mathcal{N}} r_i l_i \tag{17.17}$$

subject to (s.t.) (17.12a)–(17.12c), (17.13)–(17.15)

17.4.2 Distributed solution

The ADMM algorithm is applied to solve the relaxed optimisation problem (17.17) in a distributed manner [4]. The method iteratively minimises the augmented Lagrangian over three types of variables: (i) the primary variables; (ii) the auxiliary variables to enforce boundary conditions among neighbouring areas, and (iii) the multipliers for dualising the relaxed problem. The Lagrangian is designed to be separately relative to each type of variable so that we can cyclically minimise each variable while fixing the others. Thus, the problem can be solved distibutedly and asymptotically converge to the same solutions generated from centralised solvers.

For the specific problem in (17.17), the primary variable of node i is defined as $x_i = \{v_i, l_i, s_i, S_i\}$.

The auxiliary variable $y_{\widehat{i}i}$ denotes variables of node $\widehat{i}$ received by node i, which is defined as

$$y_{\widehat{i}i} = \begin{cases} \left(S_{ii}^{(y)}, l_{ii}^{(y)}, v_{ii}^{(y)}, s_{ii}^{(y)}\right) \widehat{i} = i \\ \left(S_{iC_i}^{(y)}, l_{iC_i}^{(y)}\right), \widehat{i} = C_i \\ \left(v_{iA_i}^{(y)}\right), \widehat{i} = A_i \end{cases} \tag{17.18}$$

By substituting the original variables with the auxiliary variables, the explicit distributed form of (17.17) can be represented as follows:

$$\min \sum_{i \in \mathcal{N}} r_i l_i$$

over:

$$\{x_i | i \in \mathcal{N}, y_{\widehat{i}i} | i \in \mathcal{N}, \widehat{i} \in \mathcal{N}_i\}$$

s.t.

$$v_{A_i}^{(y)} = v_i^{(y)} + 2\left(r_i P_i^{(y)} + x_i Q_i^{(y)}\right) + l_i^{(y)}\left(r_i^2 + x_i^2\right), i \neq 1 \tag{17.19a}$$

$$P_i^{(y)} = \sum_{\widehat{i} \in C_i} \left(P_{\widehat{i}}^{(y)} + r_{\widehat{i}} l_{\widehat{i}}^{(y)}\right) + p_i^{L(y)} + p_i^{d(y)} \tag{17.19b}$$

$$Q_i^{(y)} = \sum_{\widehat{i} \in C_i} \left(Q_{\widehat{i}}^{(y)} + x_{\widehat{i}} l_{\widehat{i}}^{(y)}\right) + q_i^{L(y)} + q_i^{d(y)} \tag{17.19c}$$

$$|P_i^{(x)}|^2 + |Q_i^{(x)}|^2 \leq v_i^{(x)} l_i^{(x)} \tag{17.19d}$$

$$q\,k^{\text{agg}} - q_k^{\text{res}} \leq \left(q_k^d\right)^{(x)} \leq \overline{q}_k^{\text{agg}} - q_k^{\text{res}} \tag{17.19e}$$

$$v_i \le v_i^{(x)} \le \bar{v}_i \tag{17.19f}$$

$$x_i^{(x)} = y_{\widehat{ii}}^{(y)} \tag{17.19g}$$

where the superscripts $(\cdot)^{(x)}$ and $(\cdot)^{(y)}$ on each variable denote whether the variable is updated in *X-update* or *Y-update*. The set of constraints in *X-update* is represented by $S_i^{(x)}$, and *Y-update* by $S_i^{(y)}$.

The augmented Lagrangian function of the above problem for each node is represented as

$$L_{\rho,i}\left(x_i, y_{\widehat{ii}}, \lambda_{\widehat{ii}}\right) = f_i(x_i) + \sum_{\widehat{i} \in \mathcal{N}_i} \left(\langle \lambda_{\widehat{ii}}, x_i - y_{\widehat{ii}} \rangle + \frac{\rho}{2} \| x_i - y_{\widehat{ii}} \|_2^2 \right) \tag{17.20}$$

where $\lambda_{\widehat{ii}}$ is the Lagrangian multiplier for (17.20).

Then the optimisation problem is solved by the following steps based on ADMM:

[S1: *X-update*] Update primary variables by

$$x_i^{m+1} = \arg\min_{x_i \in S_i^{(x)}} L_{\rho,i}\left(x_i, y_{\widehat{ii}}^m, \lambda_{\widehat{ii}}^m\right) \tag{17.21}$$

[S2: *Y-update*] Update auxiliary variables by

$$y_{\widehat{ii}}^{m+1} = \arg\min_{y \in S_i^{(y)}} L_{\rho,i}\left(x_i^{m+1}, y, \lambda_{\widehat{ii}}^m\right) \tag{17.22}$$

[S3: *λ-update*] Update the multiplier by

$$\lambda_{\widehat{ii}}^{m+1} = \lambda_{\widehat{ii}}^m + \rho\left(x_i^{m+1} - y_{\widehat{ii}}^{m+1}\right) \tag{17.23}$$

Two metrics are defined as follows to check the convergence:

$$r^{m+1} = \left[\sum_{i \in N} \left(x_i^{m+1} - y_i^{m+1}\right)^2 \right]^{1/2} \tag{17.24a}$$

$$s^{m+1} = \rho \left[\sum_{i \in N} \left(y_i^{m+1} - y_i^m\right)^2 \right]^{1/2} \tag{17.24b}$$

where r and s are primal and dual residuals. The iteration of [S1]–[S3] continues until r and s converges to the satisfied accuracy level ε_2.

Figure 17.4 demonstrates the information exchange between adjacent buses during the iteration process. In *S1: X-update*, $y_{\widehat{ii}}^m$ and $\lambda_{\widehat{ii}}^m$ of the previous step from both children buses and ancestor buses are received by bus i. The argument

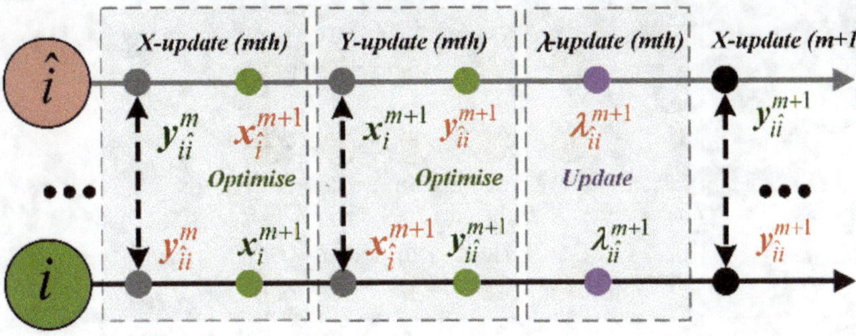

Figure 17.4 Data exchange among adjacent nodes in the ADMM

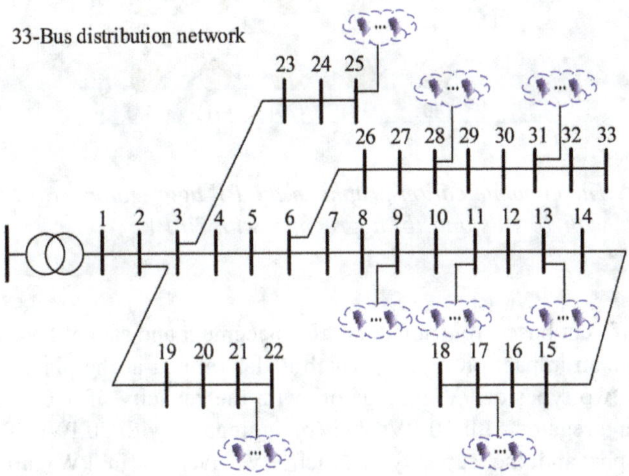

Figure 17.5 Diagram of the IEEE 33-bus radial distribution network

of the minimum regarding primary variables is calculated by (17.21). In *S2: Y-update*, the updated primary variable x_i^{m+1} is exchanged to its adjacent buses. The argument of the minimum regarding auxiliary variables $y_{\hat{i}i}^{m+1}$ is calculated by (17.22). Finally, in *S3: λ-update*, $\lambda_{\hat{i}i}^{m}$ is updated by (17.23) with only local information.

17.5 Simulation results

17.5.1 Test system and parameter settings

In this chapter, two IEEE test distribution networks (33-bus system and 69-bus system) are used to test the proposed VVC method [5]. The single-line diagram of the IEEE 33-bus system is shown in Figure 17.5. There are eight PV aggregators

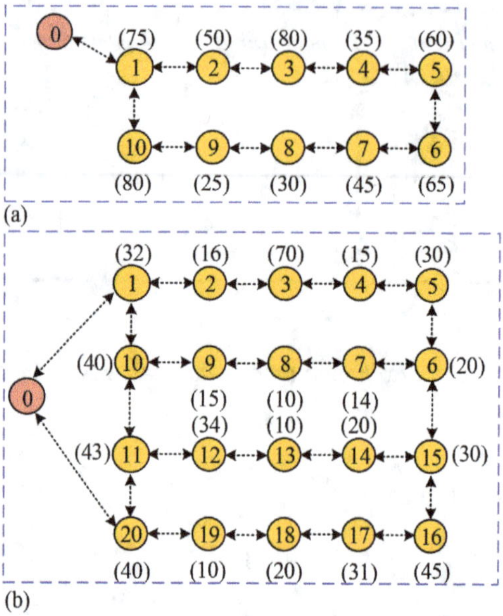

Figure 17.6 The communication graphs of the PV aggregators: (a) aggregator with 10 PVs and (b) aggregator with 20 PVs

located at different buses. It is noted that the placement and size of PV units and the inverters have an impact on results, which is determined at the planning stage. In this chapter, two types of PV aggregators with the capacity of 550 kVA are considered: (a) aggregator with 10 PVs and (b) aggregator with 20 PVs. The communication graphs and the capacity of each PV inverter (in kW) are shown in Figure 17.6. In this simulation, the substation voltage is 1.0 p.u., the operational voltage range is $[\underline{V}, \overline{V}] = [0.95, 1.05]$, and the critical voltage range is $[\underset{\sim}{V}, \tilde{V}] = [0.94, 1.06]$. In the meanwhile, the single-line diagram of the IEEE 69-bus system is shown in Figure 17.7, where there are 15 PV aggregators with the capacity of 550 kVA spread in the system. The simulation is conducted in MATLAB® with Gurobi solver.

In the following subsections, the convergence for each control level of the proposed VVC is discussed first. Then both long-timescale (24-h) and short-timescale (1-h) simulations are carried out to validate the proposed VVC. Two case studies are conducted with different test systems: (i) Case A: IEEE 33-bus system; (ii) Case B: IEEE 69-bus system.

17.5.2 Convergence

Convergence is an important concern for distributed algorithms, which is further discussed in this section. The convergence of the distributed consensus algorithms is highly related to the weights and topology of communication networks as well as

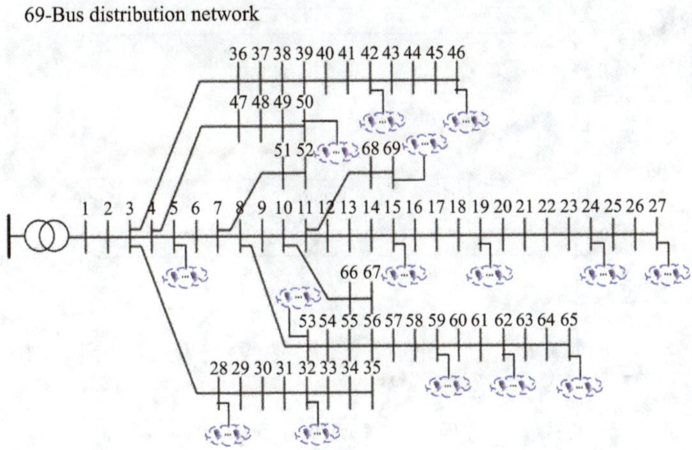

Figure 17.7 Diagram of the IEEE 69-bus radial distribution network

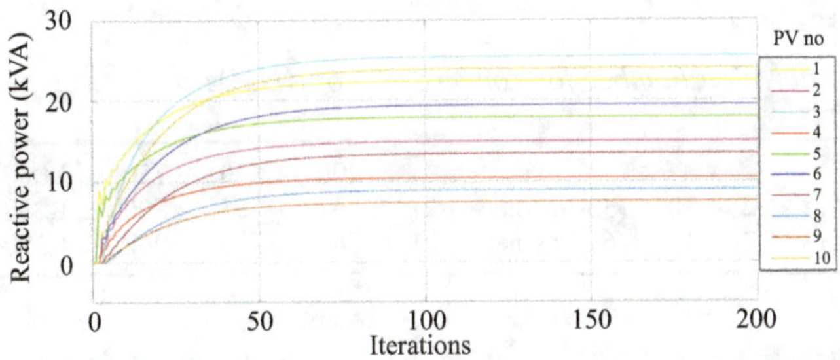

Figure 17.8. Convergence of distributed aggregation control for the aggregator with ten PVs

the location of the leader. The convergence of distributed aggregation control for PV aggregators with the communication graph in Figure 17.6(a) is shown in Figure 17.8. The PV generation factor is considered as 0.8; therefore, the reactive capacity of each aggregator is $550 \times \sqrt{1 - 0.8^2} = 330$ kVA according to (17.10). The overall reactive power increases from 0 to 165 kVA at the start of the simulation. The results demonstrate that the distributed control of PV aggregators reaches the steady state within 100 iterations, and the power sharing is proportional to their capacity given in Figure 17.6(a). As the lower-level VVC is a dynamic algorithm, the reference signals from droop control can be continuously updated without waiting for its convergence.

In the upper-level dispatch, the convergence of ADMM is related to the network scale and parameter of ρ. The primal and dual residuals (r and s) of the upper-level optimisation for the 33-bus system are shown in Figure 17.9. The results show that with a properly chosen value of ρ, the algorithm can reach a high accuracy in

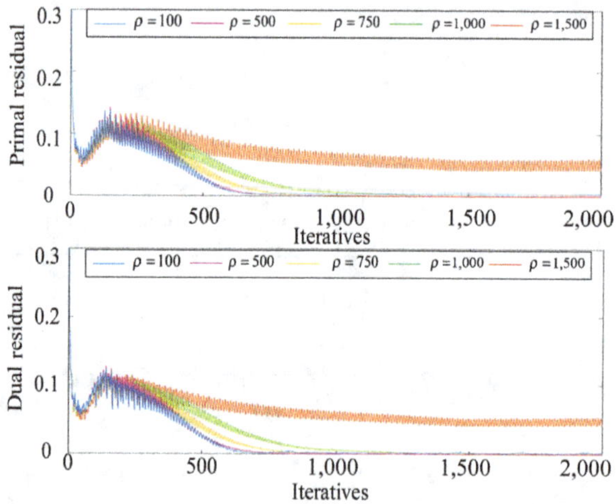

Figure 17.9 Primal and dual residuals of distributed optimisation based on ADMM

Table 17.1 Iteration number for convergence of ADMM

Value of ρ	200	500	750	1,000	1,500
Iterations for $\varepsilon_2 < 2 \times 10^{-3}$ (33-bus network)	691	705	890	1,238	>2,000
Iterations for $\varepsilon_2 < 2 \times 10^{-3}$ (69-bus network)	786	795	962	1,350	>2,000

hundreds of iterations. For the same signal exchange rate of 0.1 s as above, the algorithm can surely find the optimal solution within the 15-min dispatch interval of upper-level control. In Table 17.1, the iteration number with respect to different parameters ρ and two networks for primal and dual residuals to reach the accuracy of $\varepsilon_2 < 2 \times 10^{-3}$ are given. Generally, a smaller ρ will lead to fast convergence, but the solution accuracy will be influenced if ρ is too small.

17.5.3 Case A: 33-bus networks

The 24-h PV and load data with 15-min sampling rate are shown in Figure 17.10. The high-resolution PV and load data from 10:00 to 11:00 with 1-s sampling rate is shown in Figure 17.11. These two timescales of PV and load data are used for short- and long-term simulations.

17.5.3.1 Case A-1: long-term simulation

The results of 24-h simulation in *Case A-1* are shown in Figures 17.12–17.14. The voltage profiles of the 33-bus distribution network with and without the proposed control scheme are compared in Figure 17.12. All bus voltages are regulated [0.95, 1.05] by the upper-level dispatch. Accordingly, each aggregator will provide

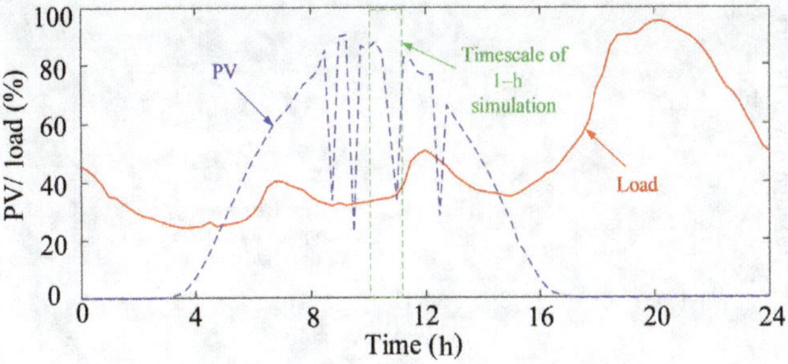

Figure 17.10 The 24-h PV and load profiles with 15-min sampling rate

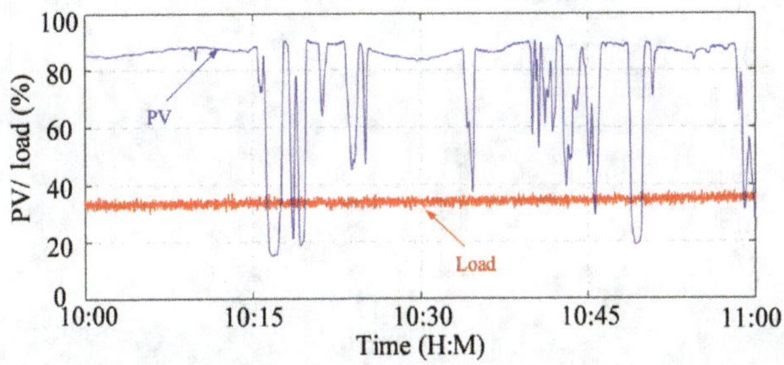

Figure 17.11 The 1-h PV and load profiles with 1-s sampling rate

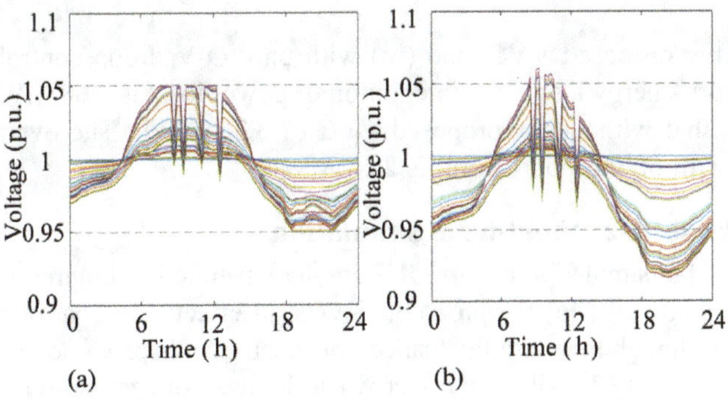

Figure 17.12 Voltage profiles of the 33-bus distribution network in Case A-1: (a) with the proposed VVC; (b) without the proposed method

the dispatched reactive power output, as shown in Figure 17.13. During high PV generation periods, a portion of aggregators in the distribution network needs to absorb some reactive power for voltage rise mitigation. The network power loss, as the optimisation objective of the upper level, is minimised during the operation period. Figure 17.14 compares the network power loss (i) with the proposed VVC,

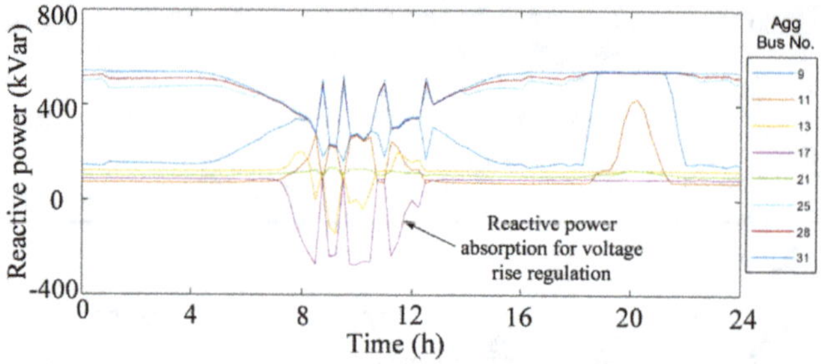

Figure 17.13 Reactive power dispatched to each aggregator in Case A-1

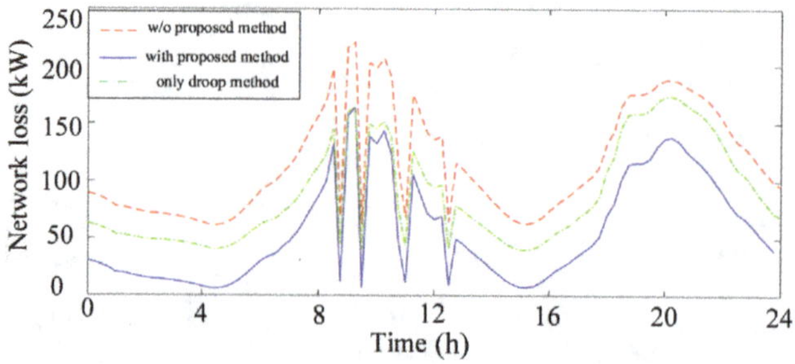

Figure 17.14 Network power loss (i) with the proposed VVC, (ii) without the proposed VVC, and (iii) with only droop control in Case A-1

(ii) without the proposed VVC, and (iii) with only Q-V droop control. The 24-h overall network energy loss (time integration of power loss) is 1.68 MWh, which is far less than that without the proposed VVC of 3.36 MWh. The overall network energy loss with only droop control is 2.65 MWh.

17.5.3.2 Case A-2: short-term simulation

The results of 1-h simulation in *Case A-2* are demonstrated in Figures 17.15–17.18. As shown in Figure 17.15, the proposed VVC can effectively control the network voltages, even though there are fluctuations between two dispatch points. The droop control defined in (17.1) will take effect when the bus voltage is out of the normal operational range $[\underline{V}, \overline{V}] = [0.95, 1.05]$. For each dispatch time instant (every 15 min), the optimal reactive power output of each aggregator will be dispatched by the upper control level. If the voltage cannot be maintained within [0.95, 1.05], the droop control is functioned to mitigate the PV fluctuations during one dispatch period. It can be observed that in Figure 17.16, the base value of each aggregator is changed every 15 min. The reactive power is further shared by each PV inverter within one aggregator. As an example, the reactive power sharing by each PV inverter of the PV aggregator at bus 17 (Aggregator-17) is shown in Figure 17.17. Each inverter in

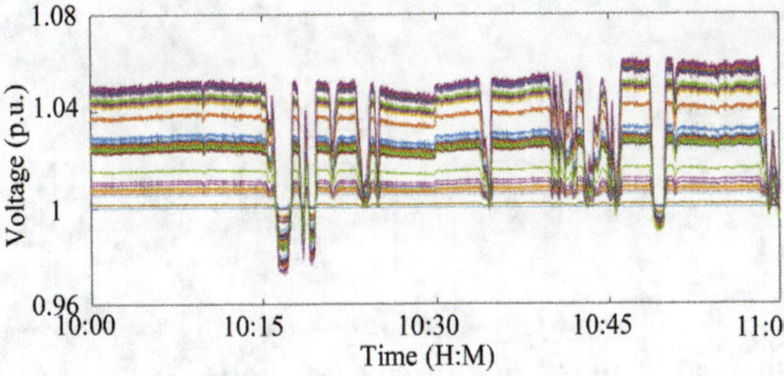

Figure 17.15 Voltage profiles of the 33-bus distribution network in Case A-2

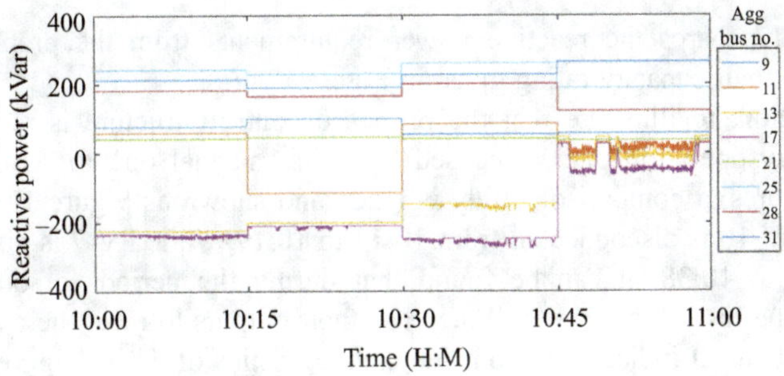

Figure 17.16 Reactive power injection of each aggregator in MV distribution networks in Case A-2

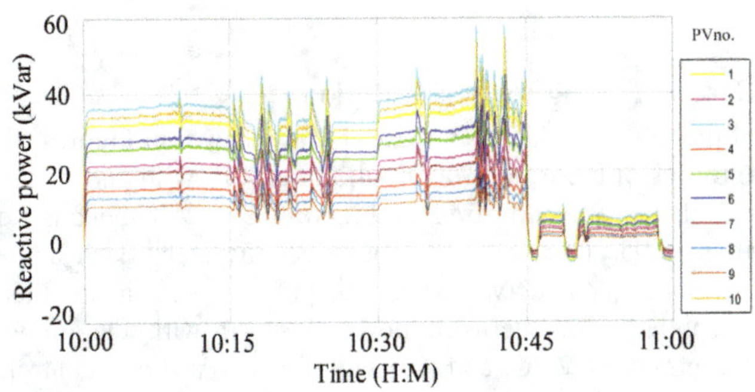

Figure 17.17 PV reactive power absorption in Aggregator-17 with ten PVs in Case A-2

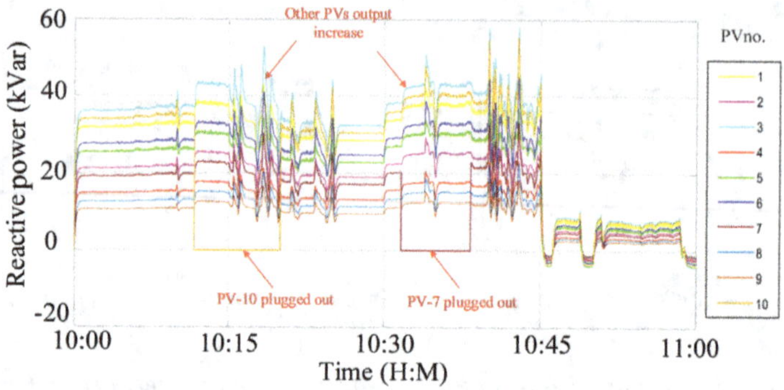

Figure 17.18 Plug-and-play operation of PVs in Case A-2

Aggregator-17 shares the reactive power requirements from the proposed VVC according to their capacity rating given in Figure 17.6(a).

In addition, to illustrate that the proposed control structure is resilient with plug-and-play operation of PVs caused by either manual operations or communication failures, a comparison study is made and shown as Figure 17.18. In this condition, PV-10 is disconnected from 10:13 to 10:19, while PV-7 is disconnected from 10:32 to 10:38. It can be found that during the periods of one PV disconnected, the rest of the PVs will increase their outputs to meet the overall reactive requirement. It indicates the outside characteristics of the PV aggregators can remain the same even though there are PV connected and disconnected operations.

17.5.4 Case B: 69-bus networks

In *Case B*, the proposed method is validated in a larger network IEEE 69-bus system with 15 PV aggregators, as shown in Figure 17.7. The 24-h and 1-h PV and load data in *Case A* given by Figures 17.9 and 17.10 are still utilised in this test case.

17.5.4.1 Case B-1: long-term simulation

The results of the 24-h simulation in *Case B-1* are shown in Figures 17.19–17.21. The voltage profiles of the 69-bus distribution network with and without the proposed control scheme are compared in Figure 17.19. All the bus voltages are regulated within [0.95, 1.05] by the upper-level dispatch. Each aggregator can provide the dispatched reactive power output, as shown in Figure 17.20. Figure 17.21 compares the network power loss (i) with the proposed VVC, (ii) without the proposed VVC, and (iii) with only Q-V droop control. The 24-h overall network energy loss (areas of power loss) is 2.16 MWh, which is far less than that without the proposed VVC of 3.1 MWh. The overall network energy loss of with droop control is 2.8 MWh.

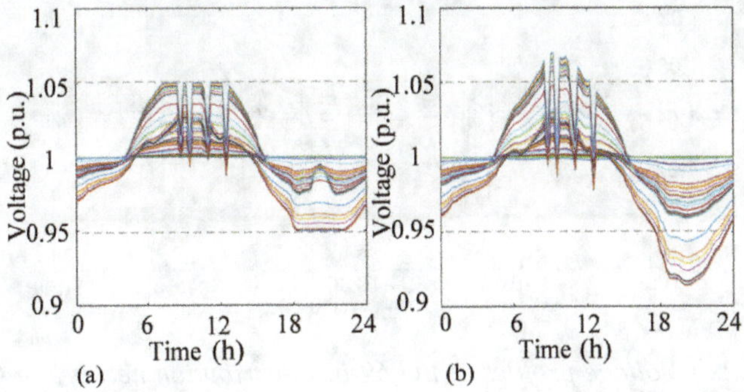

Figure 17.19 Voltage profiles of the 69-bus distribution network in Case B-1: (a) with the proposed VVC; (b) without the proposed method

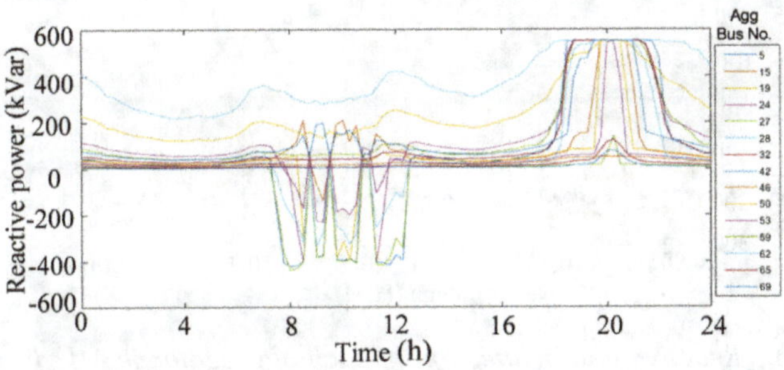

Figure 17.20 Reactive power dispatched to each aggregator in Case B-1

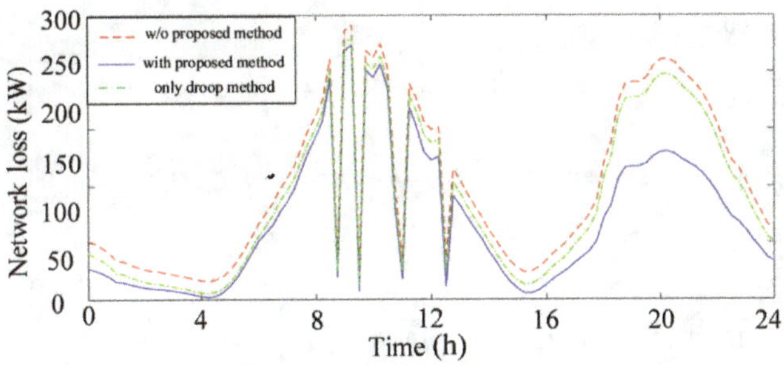

Figure 17.21 Network power loss (i) with the proposed VVC, (ii) without the proposed VVC, and (iii) with only droop control in Case B-1

17.5.4.2 Case B-2: short-term simulation

The results of 1-h simulation in *Case B-2* are shown in Figures 17.22–17.24. Similar to *Case A*, as shown in Figure 17.22, the proposed VVC can effectively control the network voltage variations. In Figure 17.23, the base value of each aggregator is changed every 15 min, and the droop control takes effect for the aggregators when there are bus voltage violations.

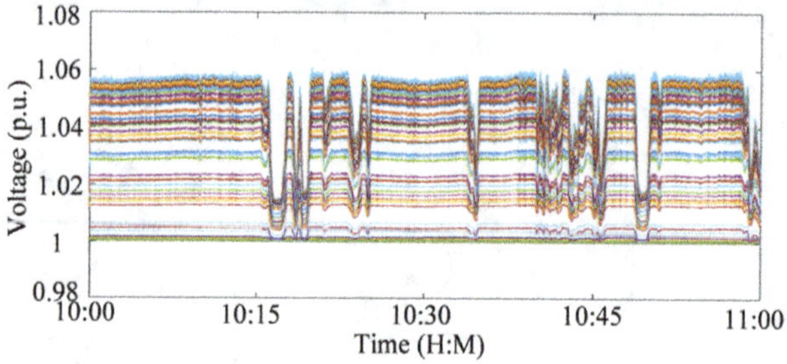

Figure 17.22 Voltage profiles of the 69-bus distribution network in Case B-2

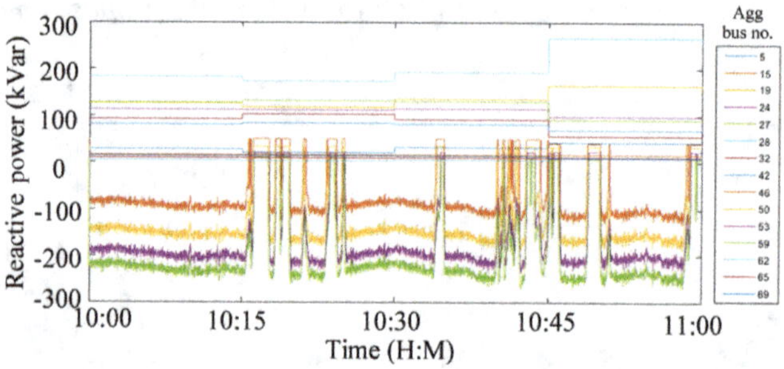

Figure 17.23 Reactive power injection of each aggregator in MV distribution networks in Case B-2

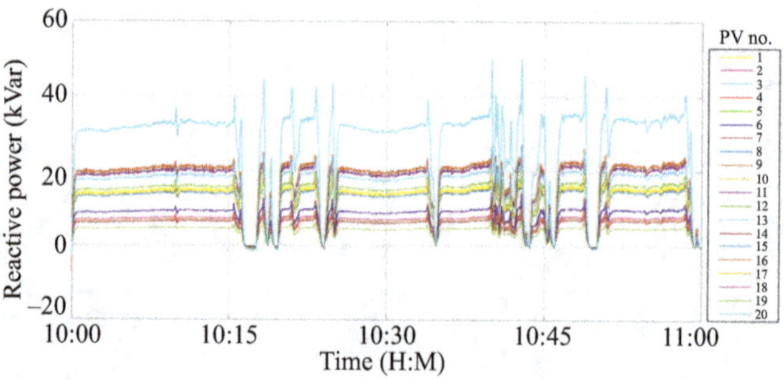

Figure 17.24 PV reactive power absorption in Aggregator-27 with 20 PVs in Case B-2

In addition, the larger PV aggregator with 20 PVs is further considered in this case. The capacity and communication graph of the studied PV aggregator are shown in Figure 17.6(b). As shown in Figure 17.24, each PV can provide reactive power proportional to its capacity to meet the overall reactive power requirement of Aggregator-27 in Figure 17.23.

In summary, the proposed VVC can handle both short- and long-term voltage control in a fully distributed manner. All the reactive power reserves from PV inverters are effectively utilised to handle the voltage problems bring along with high PV penetration. The overall network loss as an objective of the proposed VCC is minimised. The proposed method demonstrates a reliable and efficient way to manage and control PV inverters in future high-PV penetrated distribution networks.

17.6 Conclusion

In this chapter, a two-level distributed VVC scheme for high-PV penetrated distribution networks is proposed. Compared with centralised approaches, the proposed VVC enhances the system resilience as its distributed architecture. Various case studies are conducted to validate the effectiveness of the proposed control scheme. The results demonstrate that the network voltage control is achieved in two timescales by the proposed two-level VVC. The network loss is reduced with the upper-level dispatch of aggregators within the voltage constraints, while the PV fluctuations between the dispatch intervals are handled by the droop control in the lower level in real time. The reactive power of one aggregator is further shared by each PV inverter according to their capacity. Besides, the convergence, flexibility, and scalability issues are also discussed. The proposed method provides a feasible solution for fully distributed control and management of PV inverters in power distribution networks.

Appendix A

A.1 Converge analysis

The distributed control law in (17.4) is based on leader–follower consensus algorithms. If the communication graph $\mathcal{G}_k$ is connected, (17.4) can converge to a unique solution:

$$\lim_{t\to\infty} \mu_{k,j}(t) = \mu_{k,0}(0) \tag{A.1}$$

The distributed control law in (17.5) is based on dynamic average consensus algorithms. If the communication graph $\mathcal{G}_k$ is connected, (17.5) can converge to the unique solutions:

$$\lim_{t\to\infty} \overline{p}_{k,j}(t) = \frac{1}{M_k}\sum_{j=1}^{M_k} \overline{q}_{k,j}(0) \tag{A.2}$$

$$\lim_{t\to\infty} p\,k,j(t) = \frac{1}{M_k}\sum_{j=1}^{M_k} q\,k,j(0) \tag{A.3}$$

According to the proposed control in (17.6), if $\mu_{k,j}(t) > 0$, it can be derived that

$$\lim_{t\to\infty} q_{k,j}(t) = \frac{\lim_{t\to\infty} \mu_{k,j}(t)}{\lim_{t\to\infty} \overline{\rho}_{k,j}(t)} \overline{q}_{k,j}(0) = \frac{\mu_{k,0}(0)}{\frac{1}{M_k}\sum_{j=1}^{M_k} \overline{q}_{k,j}(0)} \overline{q}_{k,j}(0) \quad \text{(A.4)}$$

Similarly, if $\mu_{k,j}(t) < 0$, it can be derived that

$$\lim_{t\to\infty} q_{k,j}(t) = \frac{\lim_{t\to\infty} \mu_{k,j}(t)}{\lim_{t\to\infty} \rho\, k,j(t)} q\, k,j(0) = \frac{\mu_{k,0}(0)}{\frac{1}{M_k}\sum_{j=1}^{M_k} q\, k,j(0)} q\, k,j(0) \tag{A.5}$$

Therefore, combining (A.4) and (A.5), we can obtain the steady-state equilibrium in (17.7).

References

[1] B.A. Robbins, C.N. Hadjicostis, and A.D. Domínguez-García, ‘A two-stage distributed architecture for voltage control in power distribution systems’, *IEEE Transactions on Power Systems*, vol. 28, no. 2, pp. 1470–1482, 2013.

[2] Y. Xu, Z.Y. Dong, R. Zhang, and D.J. Hill, ‘Multi-timescale coordinated voltage/var control of high renewable-penetrated distribution networks’, *IEEE Transactions on Power Systems*, vol. 32, no. 6, pp. 4398–4408, 2017.

[3] Q. Peng and S.H. Low, ‘Distributed optimal power flow algorithm for radial networks, I: Balanced single phase case’, *IEEE Transactions on Smart Grid*, vol. 9, no. 1, pp. 111–121, 2016.

[4] S. Boyd, N. Parikh, E. Chu, B. Peleato, and J. Eckstein, ‘Distributed optimization and statistical learning via the alternating direction method of multipliers’, *Foundations and Trends® in Machine Learning*, vol. 3, no. 1, pp. 1–122, 2011.

[5] M.E. Baran and F.F. Wu, ‘Network reconfiguration in distribution systems for loss reduction and load balancing’, *IEEE Transactions on Power Delivery*, vol. 4, no. 2, pp. 1401–1407, 1989.

Chapter 18
Event-triggered control of DERs and controller hardware-in-the-loop validation

Microgrids (MGs) are cyber-physical systems with highly coupled electrical and communication networks. The centralised secondary control of MGs based on periodical communications restricts system efficiency and resilience. In this chapter, a distributed event-triggered secondary control scheme in islanded MGs is proposed and its cyber-physical implementation is introduced. The proposed control scheme reduces communications in secondary control, depending on the system state change 'events' (e.g. load variations and communication failures). The secondary control objectives, i.e. frequency/voltage regulation and accurate real/reactive power sharing, are decoupled into two timescales. Only event-triggering conditions (ETCs) for power sharing control in a slower timescale are designed. The overall communication times are significantly reduced as neighbour controllers only exchange information when the ETCs are satisfied. The proposed control is evaluated by means of the controller hardware-in-the-loop (CHIL) test. The physical system of MGs is operated in the OPAL-RT and the cyber system is implemented in Raspberry Pis (R-Pis). The experimental results validate the proposed method.

18.1 Introduction

The event-triggered mechanism has proved its capability to reduce the communication and computation burden of the distributed control framework [1]. The event-triggered controllers are not performed at predefined instants but only when needed. The event-triggered controllers are only updated when a local measurement error exceeds a tolerable bound, thereby significantly reducing communication burden while maintaining satisfactory control performance. For each secondary control function, one specific event-triggered controller and its triggering condition are designed [2–4]. In practical implementation, all the event-triggered secondary controllers should share only one common communication network. Therefore, in this chapter, the control of ω/P and V/Q are decoupled in different timescales and the event-triggered conditions for only slow dynamics are designed. Besides, previous works for the event-triggered secondary control of MGs only conducted simulation studies [2–4]. As distributed event-triggered control is inherently dependent on communications, the validation of the designed controllers on a realistic environment is meaningful. Thus, a multi-agent system (MAS) with multiple embedded systems is set up to test the performance of the proposed control method.

In this chapter, a distributed, event-triggered, and time-decoupled secondary control is proposed for frequency/voltage restoration and accurate power sharing in islanded cyber-physical MGs. Towards practical applications, a CHIL test is conducted to validate the proposed method in a realistic environment. It offers the following features:

- A distributed event-triggered secondary control scheme is proposed for islanded MGs. The secondary control problems are decoupled into two timescales, and ETCs are designed to significantly reduce the communication burdens in the cyber system.
- The design and implementation of MGs from the perspective of the cyber-physical system are presented. The physical system of MG is emulated in real time by OPAL-RT. The cyber system of MG is realised in a MAS, which consists of multiple R-Pis installed with Google remote process call (gRPC). The platform could emulate the MG governed by MAS more realistically.
- The proposed distributed event-trigged control is validated on the cyber-physical MG test bed. The CHIL results demonstrate the effectiveness of the proposed method under the events in both the physical system (load variations) and the cyber system (communication failures).

18.2 Cyber-physical Microgrids

The proposed distributed event-triggered control is designed and implemented from a cyber-physical system perspective. A cyber-physical MG test bed is built up with a real-time simulator OPAL-RT and R-Pi cluster-based CHIL platform, as shown in Figure 18.1. The electric network of MG and its local controllers, as the physical system, are simulated in OPAL-RT. The MAS and its communication network, as the cyber system, are realised by a cluster of R-Pis. Each R-Pi hosts an agent which is programmed to communicate with other agents and calculate the proposed event-triggered algorithm.

18.2.1 Physical system

18.2.1.1 Electric networks of MGs

Consider an MG with N distributed generators (DGs) (indexed by $i = 1, 2, \ldots, N$). The MGs can be represented by a complex-weighted graph $\mathcal{G}_{\rm MG} = (\mathcal{V}_{\rm MG}, \mathcal{E}_{\rm MG})$, where the nodes $\mathcal{V}_{\rm MG}$ represent the buses with DGs, and the edges $\mathcal{E}_{\rm MG} \subseteq \mathcal{V}_{\rm MG} \times \mathcal{V}_{\rm MG}$ represent the line connections, as demonstrated in Figure 18.1. The topology and line impedance of the MG network are depicted by an adjacency matrix $Z = [Z_{ik}] \subseteq \mathrm{C}$. If there is a line connection between ith and kth DGs, $Z_{ik} = R_{ik} + jX_{ik}$, where $R_{ik} + jX_{ik}$ are resistance and reactance, otherwise $Z_{ik} = 0$. Since

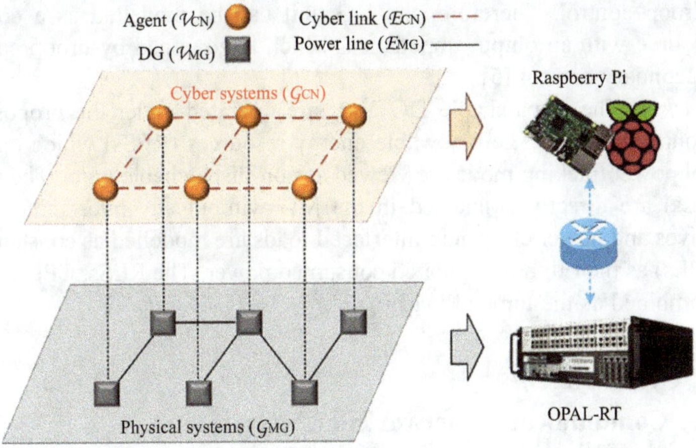

Figure 18.1 A cyber-physical MG test bed

the output impedance of DGs is inductive, the active and reactive power injections P_i and Q_i of DG i are represented by

$$P_i = \sum_{j=1}^{N} \frac{V_i V_j}{X_{ij}} \sin\left(\theta_i - \theta_j\right) + P_i^{\mathrm{L}} \tag{18.1}$$

$$Q_i = \frac{V_i^2}{X_{ij}} - \sum_{j=1}^{N} \frac{V_i V_j}{X_{ij}} \cos\left(\theta_i - \theta_j\right) + Q_i^{\mathrm{L}} \tag{18.2}$$

where V_i and θ_i are, respectively, bus voltage and angle of bus i, P_i^{L} and Q_i^{L} are, respectively, real and reactive power of the load at bus i.

18.2.1.2 Primary control

In a single MG, the difference of $\theta_{ik} = \theta_i - \theta_k$is small, which means that $\sin\left(\theta_i - \theta_k\right) \approx \theta_i - \theta_k$, $\cos\left(\theta_i - \theta_k\right) \approx 1$ in (18.1) and (18.2). Thus, the real power of DG i can be controlled by the difference of phase angles while the reactive power of DG i can be controlled by the difference of voltage magnitude. It comes out the basic mechanism of droop control (i.e. P versus ω and Q versus V) as follows:

$$\omega_i = \omega_i^{\mathrm{nom}} - K_i^P P_i \tag{18.3}$$

$$V_i = V_i^{\mathrm{nom}} - K_i^Q Q_i \tag{18.4}$$

where ω_i^{nom} and V_i^{nom} are the nominal set points of primary droop control. K_i^P and K_i^Q are droop coefficients of real and reactive power, which are typically chosen by the power rating of the corresponding DG. The inner voltage and current control loops also belong to the primary control, which have a much faster response than that of droop control. Therefore, the DG unit can be modelled as a controlled voltage source with an output impedance, which is governed by droop control as well as secondary control [5].

Remark 1: The dispatchable DG units are operated under the proposed secondary control scheme. The renewable energy resources (RESs) which operate in maximum power tracking mode are viewed as non-dispatchable units. The resistive loads (RLs) are directly connected in the MG without any interfacing devices. Motor drives and power electronic interfaced loads are modelled as constant power loads (CPLs) as their tightly regulated consumed power. The RESs, CPLs, and RLs can be combined as the lumped load $P_{\mathrm{L}} = P_{\mathrm{CPL}} + P_{\mathrm{CPL}} - P_{\mathrm{RES}}$.

18.2.2 Cyber system

18.2.2.1 Communication networks of MGs

The communication network of an MG with N DGs is depicted by a graph: $\mathcal{G}_{\mathrm{CN}} = (\mathcal{V}_{\mathrm{CN}}, \mathcal{E}_{\mathrm{CN}})$ with a set of nodes $\mathcal{V}_{\mathrm{CN}}$ and a set of edges $\mathcal{E}_{\mathrm{CN}} = \mathcal{V}_{\mathrm{CN}} \times \mathcal{V}_{\mathrm{CN}}$, as shown in Figure 18.1. The nodes in graph $\mathcal{G}_{\mathrm{CN}}$ (*communication agents*) are one to one corresponding to nodes in graph $\mathcal{G}_{\mathrm{MG}}$ (*DGs*). The edges in $\mathcal{G}_{\mathrm{CN}}$, which represent communication links for data exchange, can be different from the electrical network in $\mathcal{G}_{\mathrm{MG}}$. The adjacency matrix of $\mathcal{G}_{\mathrm{CN}}$ is defined as $\boldsymbol{A} = [a_{ij}]$. The element a_{ij} represents the information exchanged between agents i and j, where $a_{ij} = 1$ if agents i and j are connected, otherwise, $a_{ij} = 0$. The in-degree matrix is defined as $\boldsymbol{D} = \mathrm{diag}[d_i]$, where $d_i = \sum_{i=1}^{N} a_{ij}$. The Laplacian matrix is defined as $\boldsymbol{L} = [l_{ij}]$, and $\boldsymbol{L} = \boldsymbol{D} - \mathbf{A}$. The convergence rate is determined by the minimum non-zero eigenvalue of L, which is called algebraic connectivity [6].

18.2.2.2 Event-triggered mechanism

In the event-triggered control and communication scheme, the data of agent i are transmitted to neighbour agents only when the ETC is satisfied. The 'event' refers to any disturbance which alters the system states. The following two notations are introduced:

Triggered time sequence: $\{t_k | k = 1, 2, \ldots\}$. t_k denotes the time instant of the kth triggering event, where each agent broadcasts its data to neighbours. The transmitted data are assumed to be successfully stored by a zero-order holder.

State measurement error: $e(t) = x(t_k) - x(t), t \in [t_k, t_{k+1})$. $x(t)$ is the current system state variable, while $x(t_k)$ is the system state variable of the kth triggering event.

In centralised event-triggered mechanism, all agents are triggered by one ETC, the next time instant t_{k+1} can be calculated by

$$t_{k+1} = t_k + \min_{l}\{l | e^T(t)\boldsymbol{\Phi} e(t) \geq \sigma x^T(t_k)\boldsymbol{\Phi} x(t_k)\} \tag{18.5}$$

where $\boldsymbol{\Phi}$ and σ are matrix and constant to be designed.

It is noted that this triggering mechanism relies on a central controller to update the triggered time sequence for each agent. As only peer-to-peer communications are considered, a fully distributed secondary event-trigger control will be introduced in the cyber system in the next section.

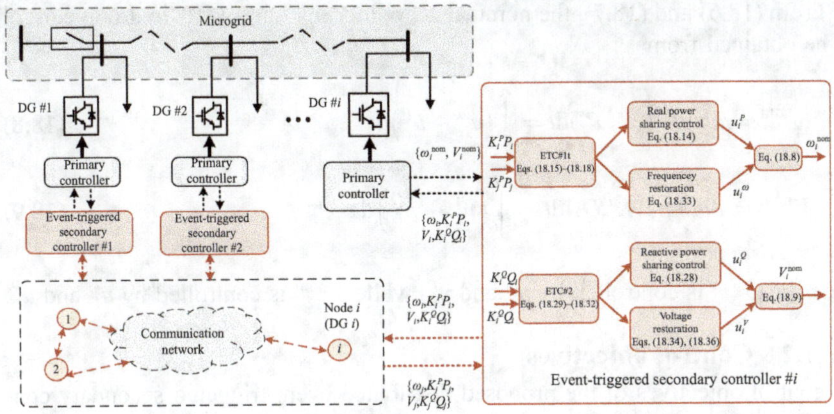

Figure 18.2 The framework of the proposed distributed event-triggered secondary control

18.3 Distributed event-triggered secondary control

The general framework of the proposed control scheme is shown in Figure 18.2. The secondary controller of a DG processes the distributed event-triggered and exchanges data with the local primary controller and neighbouring DGs. When either of the ETCs is satisfied, the secondary controller will update control signals to the primary control. In the meantime, the data packets of MG state variables at this event-triggered time will be sent to the neighbouring DG controllers. The detailed mechanism for each part will be elaborated in the following subsections.

18.3.1 Problem formation and control objectives

18.3.1.1 Problem formation

In order to formulate the problem of MG secondary control, differentiating (18.1) and (18.2) yields

$$\dot{\omega}_i = \dot{\omega}_i^{\text{nom}} - K_i^P \dot{P}_i \tag{18.6}$$

$$\dot{V}_i = \dot{V}_i^{\text{nom}} - K_i^Q \dot{Q}_i \tag{18.7}$$

Based on the feedback linearisation, the secondary controller for frequency and voltage restoration can be formulated as, $\dot{\omega}_i = u_i^{\omega}$, $\dot{V}_i = u_i^V$. In the meantime, frequency and voltage can be considered in steady state while performing the accurate power sharing control, and one can obtain that $K_i^P \dot{P}_i = u_i^P$, $K_i^Q \dot{Q}_i = u_i^Q$ [4].

From (18.6) and (18.7), the nominal set points ω_i^{nom} and V_i^{nom} in droop control can be obtained from

$$\omega^{\text{nom}} = om\dot{\omega}_i + K_i^P \dot{P}_i)\text{d}t = \int \left(u_i^{\omega} + u_i^P\right)\text{d}t \tag{18.8}$$

$$V^{\text{nom}} = om\dot{V}_i + K_i^P \dot{Q}_i)\text{d}t = \int \left(u_i^V + u_i^Q\right)\text{d}t \tag{18.9}$$

where the ω^{nom} is controlled by u_i^{ω} and u_i^P, while V^{nom} is controlled by u_i^{ω} and u_i^P.

18.3.1.2 Control objectives

The control objectives of the proposed distributed event-triggered secondary control are as follows:

1. Accurate real power sharing of DGs and frequency restoration of MGs, i.e.

$$\lim_{t\to\infty} |K_i^P P_i(t) - K_i^P P_j(t)| = 0, \forall i \neq j \tag{18.10}$$

$$\lim_{t\to\infty} |\omega_i(t) - \omega^{\text{ref}}| = 0, \forall i \in G \tag{18.11}$$

2. Accurate reactive power sharing of DGs and averaged voltage restoration of MGs, i.e.

$$\lim_{t\to\infty} |K_i^Q Q_i(t) - K_i^Q Q_j(t)| = 0, \forall i \neq j \tag{18.12}$$

$$\lim_{t\to\infty} |\frac{1}{N}\sum_{i=1}^{N} V_i(t) - V^{\text{ref}}| = 0 \tag{18.13}$$

The first objective will be achieved by control inputs $u_i^{\omega}\left(t_k^{Pi}\right)$ and $u_i^P\left(t_k^{Pi}\right)$ with the triggered time sequence $\{t_k^{Pi}|k = 1, 2, ...\}$ for each agent. The second objective will be achieved by control inputs $u_i^V(t_k^{Qi})$ and $u_i^Q(t_k^{Qi})$ with the triggered time sequence $\{t_k^{Qi}|k = 1, 2, ...\}$ for each agent.

Remark 2: In this chapter, the secondary control of ω/P and V/Q are decoupled into two timescales. Figure 18.3 is used to illustrate the frequency-domain response of the timescale decoupling. The frequency/voltage control can be realised with a faster response speed than power sharing control, i.e. the stabilisation time holds $T^P \approx 10T^{\omega}$ and $T^Q \approx 10T^V$.

There are mainly two advantages of the timescale decoupling in secondary control. (1) For each secondary control functionalities, the controllers can be separately designed for desired performance. (2) As all the secondary controllers share one common communication network, designing of four separate ETCs will increase the communication burden. Therefore, it is proposed to design only two ETCs based on the states of P and Q in a slower timescale. This design avoids the

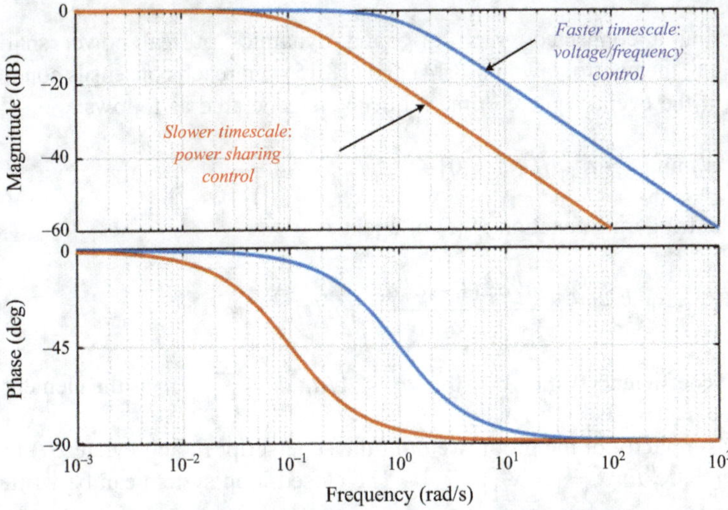

Figure 18.3 Timescale decoupled secondary control: frequency-domain response

overlap of triggering conditions in different timescales, which further reduces the communication burden in the cyber system.

18.3.2 Controller design

In this section, the distributed event-triggered secondary controllers for P, ω, Q, and V with two ETCs are presented.

18.3.2.1 Real power sharing control

The power sharing control among each DG owns the dynamics of a first-order system $K_i^P \dot{P}_i = u_i^P$. Accurate power sharing can be achieved by the following control laws:

$$u_i^P(t) = k_i^P \delta_i^P(t) \tag{18.14}$$

where the control gain $k_i^P > 0$, and $\delta_i^P(t)$ is defined as

$$\delta_i^P(t) = \sum_{j=1}^{N} a_{ij}\left[K_j^P P_j\left(t_k^{Pj}\right) - K_i^P P_i\left(t_k^{Pi}\right)\right] \tag{18.15}$$

Note that the control law in (18.14) and (18.15) is updated only at the event-triggered time rather than fixed periodical time interval.

The state measurement error is calculated from

$$e_i^P(t) = K_i^P P_i\left(t_k^{Pi}\right) - K_i^P P_i(t), t \in \left[t_k^{Pi}, t_{k+1}^{Pi}\right) \tag{18.16}$$

ETC#1

Considering a connected graph $\mathcal{G}_{\text{CN}}$ and dynamics of real power sharing as $K_i^P \dot{P}_i = u_i^P$, the control law in (18.14) and (18.15) can reach consensus condition in (18.10), if the event-triggered time sequence is generated as follows:

$$t_k^{Pi} = \inf\{t > t_{k-1}^{Pi} | f_i^P(t) = 0\} \tag{18.17}$$

where the triggering function $f_i^P(t)$ of agent i

$$f_i^P(t) = \|e_i^P(t)\|^2 - \frac{\sigma_i^P \alpha_i^P (1 - \alpha_i^P d_i)}{d_i} \|\delta_i^P(t)\|^2 \tag{18.18}$$

The coefficients $0 < \sigma_i^P < 1, 0 < \alpha_i^P < 1/d_i; d_i = \sum_{i=1}^{N} a_{ij}$ is the element in the in-degree matrix $\boldsymbol{D}$.

For simplicity of the proof, we omit the superscript P, abbreviate $x(t)$ to x and denote $p_i = K_i^P P_i$, $x = [x_1, x_2, \ldots, x_N]^T$. The closed-loop system can be written as

$$\dot{p} = -k\mathcal{L}(p + e) \tag{18.19}$$

Selecting the Lyapunov function candidate as

$$V = \frac{1}{2} p^T \mathcal{L} p \tag{18.20}$$

and then taking the derivative of (18.20), it can be derived that

$$\dot{V} = \frac{1}{2} p^T \mathcal{L} \dot{p} = -kp^T \mathcal{L}(\mathcal{L}p + \mathcal{L}e) = -k\delta^T \delta - k\delta^T \mathcal{L}e \tag{18.21}$$

Expanding the above equation,

$$\begin{aligned} \dot{V} &= -k_i \sum_{i=1}^{N} \delta_i^2 - k_i \sum_{i=1}^{N} \sum_{j \in N_i} \delta_i (e_i - e_j) \\ &= -k_i \sum_{i=1}^{N} d_i \delta_i e_i - k_i \sum_{i=1}^{N} \delta_i^2 + k_i \sum_{i=1}^{N} \sum_{j \in N_i} \delta_i e_j \end{aligned} \tag{18.22}$$

Using the inequality

$$|xy| \le \frac{\alpha}{2} x^2 + \frac{1}{2\alpha} y^2, \alpha > 0 \tag{18.23}$$

we can obtain the upper bound of $\dot{V}$ as follows:

$$\dot{V} \le -k_i \sum_{i=1}^{N} \delta_i^2 + k_i \sum_{i=1}^{N} \alpha d_i \delta_i^2 + k_i \sum_{i=1}^{N} \frac{1}{2\alpha} d_i e_i^2 + k_i \sum_{i=1}^{N} \sum_{j \in N_i} \frac{1}{2\alpha} e_j^2 \tag{18.24}$$

Since the graph is symmetric, by interchanging the indices of the last term, we get

$$\dot{V} \leq -k_i \sum_{i=1}^{N} (1 - \alpha d_i)\delta_i^2 + k_i \sum_{i=1}^{N} \frac{1}{\alpha} d_i e_i^2 \tag{18.25}$$

Assume that α satisfies $0 < \alpha < 1/d_i$. Then, if the following condition is met

$$e_i^2 \leq \frac{\sigma_i \alpha_i (1 - \alpha_i d_i)}{d_i} \delta_i^2 \tag{18.26}$$

then it can be further derived that

$$\dot{V} \leq k_i \sum_{i=1}^{N} (\sigma_i - 1)(1 - \alpha d_i)\delta_i^2 \leq 0 \tag{18.27}$$

which is negative definite for $0 < \sigma_i < 1$.

Therefore, the stability of the closed-loop system is proved. The minimal inter-event times are lower bounded away from zero as discussed in [1].

Remark 3: Figure 18.4 is used to illustrate the working principle of *ETC#1*. *ETC#1* actually compares the values of two terms $\|e_i^P(t)\|^2$ and $\frac{\sigma_i^P \alpha_i^P (1-\alpha_i^P d_i)}{d_i} \|\delta_i^P(t)\|^2$ in (18.18). The blue curve refers to the term of $\|e_i^P(t)\|^2$, which will keep increasing until the next event-triggered time. The upper boundary is given by the red curve, which will keep decreasing as the control eliminates the error among agents $\|\delta_i^P(t)\|$. Once the condition is satisfied, $\|e_i^P(t)\|^2$ will be updated to zero with the state $K_i^P P_i(t_k^{Pi})$ equalling to the current state $K_i^P P_i(t)$. During the intervals between two time instants, no communication is required. The ETC of each agent is only based on the information of its own and neighbour agents, so a fully distributed event-triggered control scheme is developed. In practical application, the triggering condition can be terminated when $\|e_i^P(t)\|$ is smaller than a certain value.

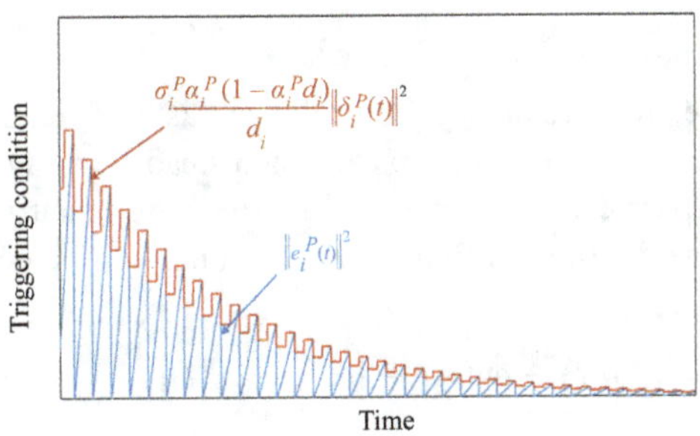

Figure 18.4 Illustration of the ETC

18.3.2.2 Reactive power sharing control

Aligning with the ideas of timescale decoupling, the event-triggered control of reactive power sharing and its ETC are designed. The reactive power sharing among each DG owns the dynamics of a first-order system $K_i^Q \dot{Q}_i = u_i^Q$. The accurate reactive power sharing can be achieved by the following control laws:

$$u_i^Q = k_i^Q \delta_i^Q(t) \tag{18.28}$$

where the control gain $k_i^Q > 0$ and $\delta_i^Q(t)$ is defined as

$$\delta_i^Q(t) = \sum_{j=1}^{N} a_{ij} \left[K_j^Q Q_j \left(t_k^{Qj} \right) - K_i^Q Q_i \left(t_k^{Qi} \right) \right] \tag{18.29}$$

The state measurement error is given as

$$e_i^Q(t) = K_i^Q Q_i \left(t_k^{Qi} \right) - K_i^Q Q_i(t), t \in \left[t_k^{Qi}, t_{k+1}^{Qi} \right) \tag{18.30}$$

ETC#2

Considering a connected graph $\mathcal{G}_{\mathrm{CN}}$ and dynamics of reactive power sharing as $K_i^Q \dot{Q}_i = u_i^Q$, the control law in (18.28) and (18.29) can reach consensus condition in (18.12), if the event-triggered time sequence is generated as follows:

$$t_k^{Qi} = \inf\{t > t_{k-1}^{Qi} | f_i^Q(t) = 0\} \tag{18.31}$$

where the triggering function $f_i^Q(t)$ of agent i

$$f_i^Q(t) = \|e_i^Q(t)\|^2 - \frac{\sigma_i^Q \alpha_i^Q (1 - \alpha_i^Q d_i)}{d_i} \|\delta_i^Q(t)\|^2 \tag{18.32}$$

The coefficients $0 < \sigma_i^Q < 1, 0 < \alpha_i^Q < 1/d_i; d_i = \sum_{i=1}^N a_{ij}$ is the element in in-degree matrix $\boldsymbol{D}$.

The stability proof has a similar process of *ETC#1*, which is not discussed here for simplicity.

18.3.2.3 Frequency control

As discussed above, the frequency control has faster dynamics of real power sharing control; thus, the system state ω can reach steady state faster than P. Instead of designing two separate ETCs for ω and P, frequency control can also be triggered by *ETC#1*. The distributed frequency controller can be designed as

$$u_i^\omega = k_i^\omega \left\{ \sum_{j=1}^{N} a_{ij} \left[\omega_j(t) - \omega_i(t) \right] + \left[\omega^{\mathrm{ref}} - \omega_i(t) \right] \right\} \tag{18.33}$$

where $k_i^\omega > 0$ and $t \in \{\mathrm{t}_k^{Pi} | \mathrm{k} = 1, 2, \ldots\}$.

18.3.2.4 Voltage control

The voltage control and the reactive power control share the *ETC#2*. In this chapter, the weighted average bus voltage is estimated and restored, as the trade-off between accurate reactive power sharing and voltage regulation. First, the average voltage value $\overline{V}_i$ of each DG is estimated by a distributed dynamic consensus algorithm. The estimator of DG i updates its own output $\overline{V}_i$ by processing the neighbours' estimation $\overline{V}_j$, and calculates the local voltage measurement V_i as follows:

$$\dot{\overline{V}}_i(t) = \dot{V}_i(t) + \sum_{j=1}^{N} a_{ij}\left[\overline{V}_j(t) - \overline{V}_i(t)\right] \tag{18.34}$$

where (18.34) can be represented in matrix form as $\dot{\overline{V}}(t) = \dot{V}(t) - L\overline{V}(t)$, and the transferred into the frequency domain as $\overline{V}(s) = s(sI_N + L)^{-1}V(s)$. For an undirected and connected $\mathcal{G}_{\mathrm{CN}}$, the system is stable. The estimator can steer each DG's voltage observations asymptotically converge to the weighted average value of all DGs' actual voltage magnitudes. Using the *final value theorem*, it gives

$$\lim_{t\to\infty} \overline{V}_i(t) = \lim_{t\to\infty} \frac{1}{N}\sum_{i=1}^{N} V_i(t) \tag{18.35}$$

Based on the observed average voltage of each DG, the voltage can be restored in a distributed way as

$$u_i^V(t) = k_i^{VP}\left[V^{\mathrm{ref}} - \overline{V}_i(t)\right] + k_i^{VI}\int \left[V^{\mathrm{ref}} - \overline{V}_i(t)\right]\mathrm{d}t \tag{18.36}$$

where $k_i^{VI}, k_i^{VP} > 0$ and $t \in \{t_k^{Q_i} | k = 1, 2, \ldots\}$.

18.4 Controller hardware-in-the-loop implementation

The proposed controllers in Section 18.3 are to be implemented on a cyber-physical MG platform. The details of this platform are introduced as follows:

1. *Physical system by OPAL-RT*: The electrical network of MGs and all primary controllers are viewed as the physical system and developed in RT-LAB version 11.2. In each control iteration, the data packet of local measurement $\{K_i^P P_i, K_i^Q Q_i, \omega_i, V_i\}$ from primary control will be sent to the corresponding R-Pi via user datagram protocol (UDP). The control inputs of $\{u_i^P, u_i^Q, u_i^\omega, u_i^V\}$ will be sent back the OPAL-RT to calculate set points ω_i^{nom} and V_i^{nom} by (18.8) and (18.9).
2. *Cyber system by Raspberry Pi*: A MAS and its communication network, as the cyber system, are realised by multiple R-Pis. The proposed control is implemented independently on each agent. The Raspberry Pi 3 Model B+ is applied in the CHIL experiment and the communications are realised by a network switch.

Each agent or R-Pi is installed with gRPC and has the structure illustrated in Figure 18.5. There are three asynchronous processes working concurrently on each

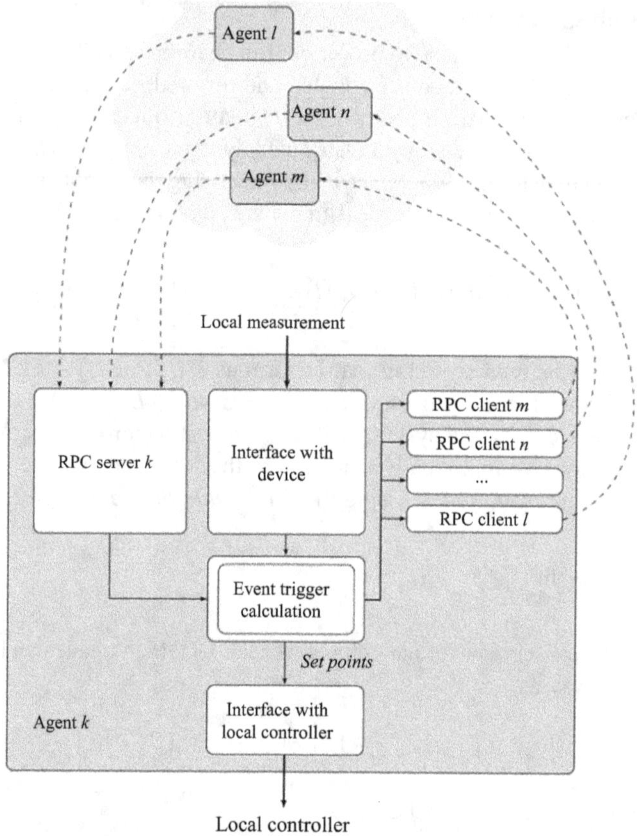

Figure 18.5 The structure of an agent implementing the proposed event-triggered algorithm

agent. In the first process, each agent receives local measurement data from OPAL-RT. In the second process, each agent processes the proposed control algorithm. When the ETC is satisfied, the states $\{K_i^P P_i, K_i^Q Q_i, \omega_i, V_i\}$ will be sent to neighbours using gRPC server/client structure over the TCP/IP protocol, and the control signals $\{u_i^P, u_i^Q, u_i^\omega, u_i^V\}$ will be updated. The third process sends the control signals from each agent to the corresponding local controller in OPAL-RT.

18.5 Experimental test results

The proposed control scheme is first tested on a 4-DG MG on the developed platform in *Cases 1* and *2*, and then tested on a larger MG system with 13 buses in *Case 3*.

The single-line diagram of the 4-DG MG and its associated communication topology is shown in Figure 18.6. The parameters of the test MG and the secondary

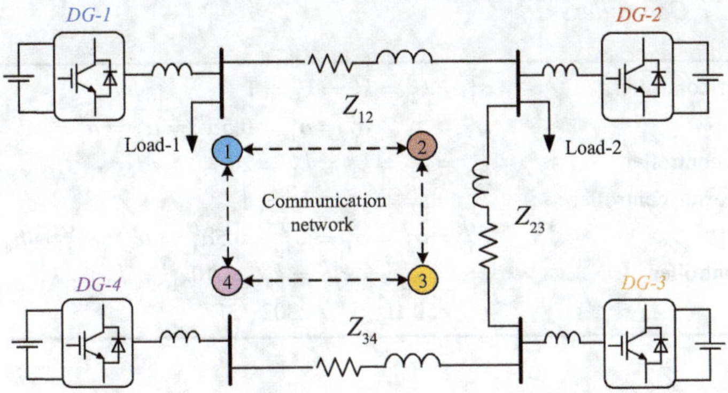

Figure 18.6 Single-line diagram of the 4-DG MG

Table 18.1 Parameter settings of the MG test bed

Line	Line 1		Line 2		Line 3			
	R_{12}	0.8 Ω	R_{23}	0.4 Ω	R_{34}	0.7 Ω		
	L_{12}	3.6 mH	L_{23}	1.8 mH	L_{34}	1.9 mH		
Load	Load-1				Load-2			
	P_1	18 kW	P_2	12 kW				
	Q_1	6 kVar	Q_2	4 kVar				
DG	DG-1		DG-2		DG-3		DG-4	
	K_1^P	2×10^{-4}	K_2^P	2×10^{-4}	K_3^P	10^{-4}	K_4^P	10^{-4}
	K_1^Q	4×10^{-3}	K_2^Q	4×10^{-3}	K_3^Q	2×10^{-3}	K_4^Q	2×10^{-3}
	R_1^o	0.1 Ω	R_2^o	0.1 Ω	R_3^o	0.1 Ω	R_4^o	0.1 Ω
	L_1^o	4.8 mH	L_2^o	4.8 mH	L_3^o	4.8 mH	L_4^o	4.8 mH

controllers are given in Tables 18.1 and 18.2, respectively. As an example, the adjacent matrix, in-degree matrix and Laplacian matrix of the communication topology in Figure 18.6 are shown as follows:

$$A = \begin{bmatrix} 0 & 1 & 0 & 1 \\ 1 & 0 & 1 & 0 \\ 0 & 1 & 0 & 1 \\ 1 & 0 & 1 & 0 \end{bmatrix}, \quad D = \begin{bmatrix} 2 & 0 & 0 & 0 \\ 0 & 2 & 0 & 0 \\ 0 & 0 & 2 & 0 \\ 0 & 0 & 0 & 2 \end{bmatrix} \quad L$$

$$= \begin{bmatrix} 2 & -1 & 0 & -1 \\ -1 & 2 & -1 & 0 \\ 0 & -1 & 2 & -1 \\ -1 & 0 & -1 & 2 \end{bmatrix}$$

Table 18.2 Controllers

Real power controller	$k_1^P = k_2^P = k_3^P = k_4^P = 1$
ETC#1	$\sigma_1^P = \sigma_2^P = \sigma_1^P = \sigma_1^P = 0.5\alpha_1^P = \alpha_2^P = \alpha_3^P = \alpha_4^P = 0.2$
Frequency controller	$k_1^\omega = k_2^\omega = k_3^\omega = k_4^\omega = 20$
Reactive power controller	$k_1^Q = k_2^Q = k_3^Q = k_4^Q = 1$
ETC#2	$\sigma_1^Q = \sigma_2^Q = \sigma_1^Q = \sigma_1^Q = 0.5\alpha_1^Q = \alpha_2^Q = \alpha_3^Q = \alpha_4^Q = 0.2$
Voltage controller	$k_1^{VP} = k_2^{VP} = k_3^{VP} = k_4^{VP} = 10k_1^{VI} = k_2^{VI} = k_3^{VI} = k_4^{VI} = 50$
Reference	ω^{ref}=50 Hz, V^{ref}=$2302\sqrt{2}V$

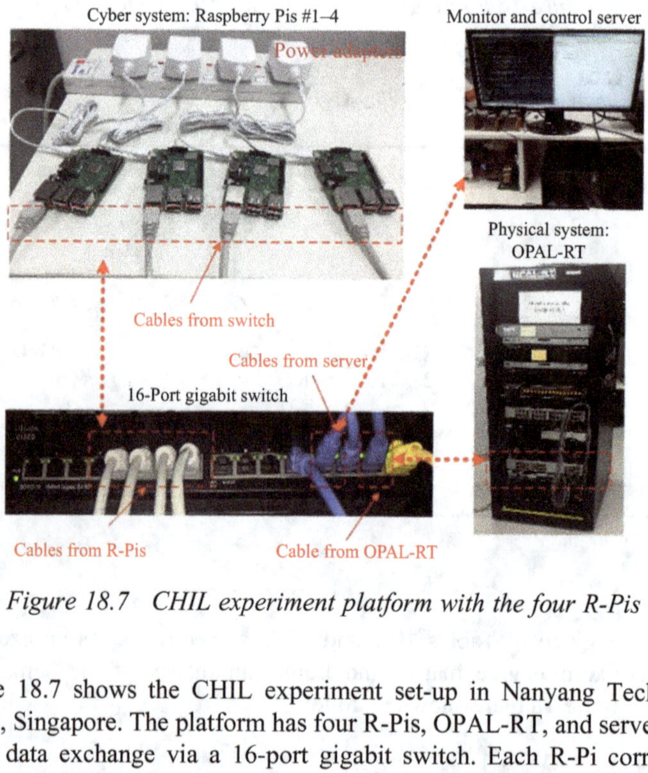

Figure 18.7 CHIL experiment platform with the four R-Pis

Figure 18.7 shows the CHIL experiment set-up in Nanyang Technological University, Singapore. The platform has four R-Pis, OPAL-RT, and server desktop, with their data exchange via a 16-port gigabit switch. Each R-Pi corresponds a secondary controller of a DG, which runs the proposed algorithm. The server desktop works to monitor and manage the R-Pi cluster and OPAL-RT.

18.5.1 Case 1: step response

In *Case 1*, the performance of the proposed controller is validated with step load changes. In the beginning, DGs 1–4 and Load-1 are connected to the MG. Load-2 is connected into the MG at 20 s, while Load-1 is switched off from the MG at 40 s. The CHIL experimental results of real power output and frequency are shown in

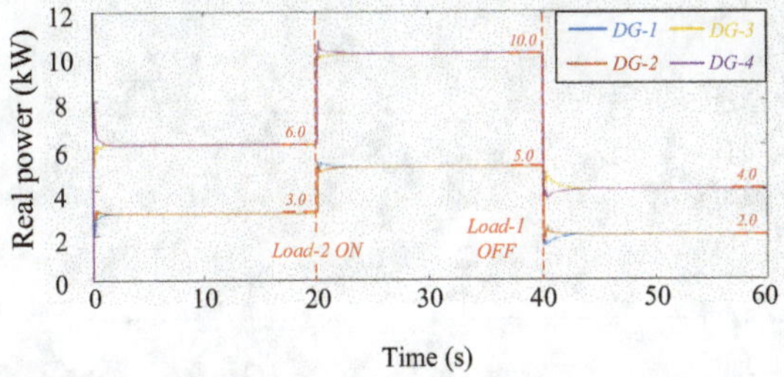

Figure 18.8 The real power output of each DG in Case 1

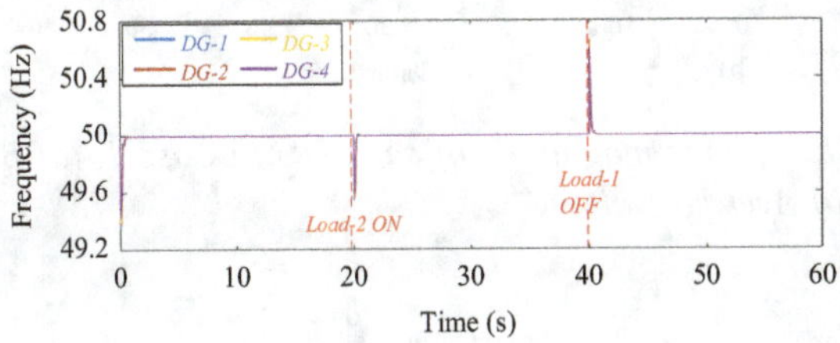

Figure 18.9 The frequency of each DG in Case 1

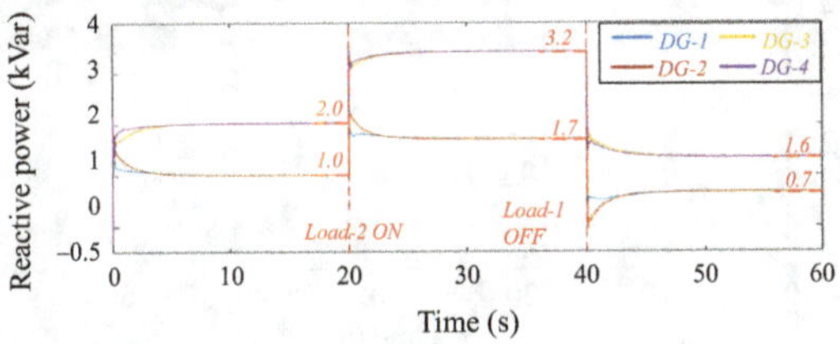

Figure 18.10 The reactive power of each DG in Case 1

Figures 18.8 and 18.9, respectively. It is observed that the real power is shared inversely proportional to the droop coefficients, as (P_1:P_2:P_3:P_4 = 1:1:2:2). The frequency can be restored to reference value (50 Hz) after each load change. The frequency restoration ($t \approx 0.4$ s) responds much faster than the real power sharing ($t \approx 4$ s) process, which verifies the effectiveness of the timescale decoupling. The CHIL experimental results of reactive power output and voltage magnitude are shown in Figures 18.10 and 18.11, respectively. The reactive power is also shared inversely proportional to the droop coefficients as (Q_1:Q_2:Q_3:Q_4 = 1:1:2:2). The estimated average bus voltages can be restored to the reference value ($230\sqrt{2}V$) as shown in Figure 18.11(b), and the actual bus voltages are also restored to some extent in Figure 18.11(a). The voltage restoration ($t \approx 1.4$ s) also operates much faster than reactive power sharing ($t \approx 14$ s).

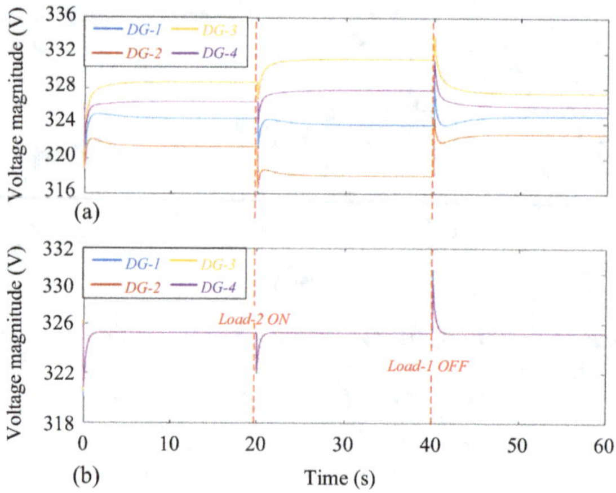

Figure 18.11 The voltage magnitude of each DG in Case 1. (a) Bus voltage. (b) Average voltage

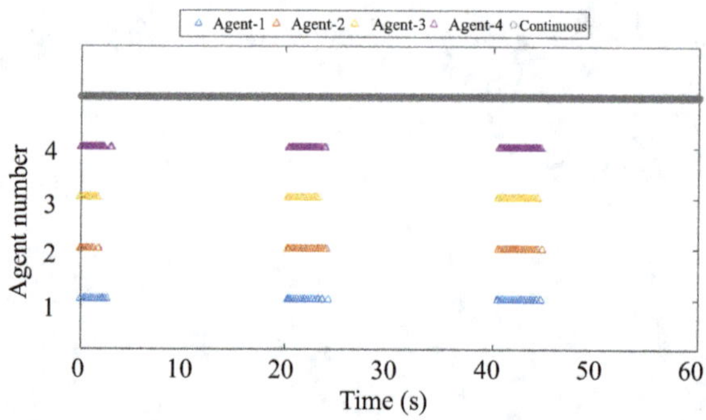

Figure 18.12 The event-triggered time of each agent (DG) under ETC-1 (continuous: conventional method)

18.5.2 Effectiveness of the event-triggered control

The effectiveness of the proposed event-triggered control is discussed in this subsection. The event-triggered time sequence of each agent (DG) in *Case 1* is shown in Figures 18.12 and 18.13. The secondary control on each agent is triggered asynchronously according to its own ETCs. Here, we take Agent-1 as an example. In Figure 18.12, Agent-1 is triggered by ETC#1 last for 4 s. In Figure 18.13, Agent-1 is triggered by ETC#2 last for 14 s. The total communications of Agent-1 are governed by two ETCs together. As compared to conventional control with continuous communications in Figures 18.12 and 18.13, the proposed method can significantly reduce communications, both during the event and after the event. In the meantime, the details of the total event-triggered time for Agent-1 during 20 to

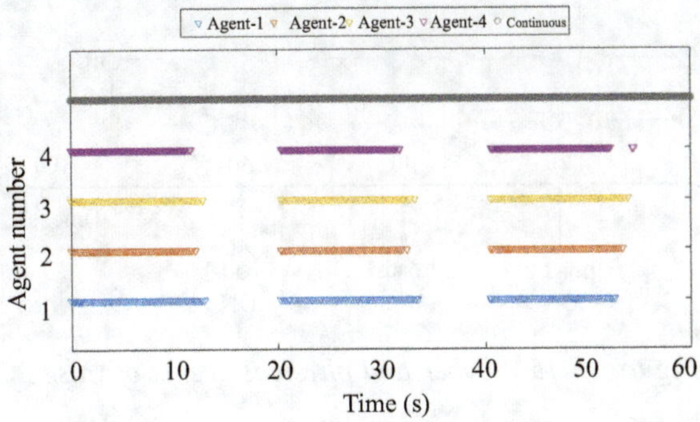

Figure 18.13 The event-triggered time of each agent (DG) under ETC-2 (continuous: conventional method)

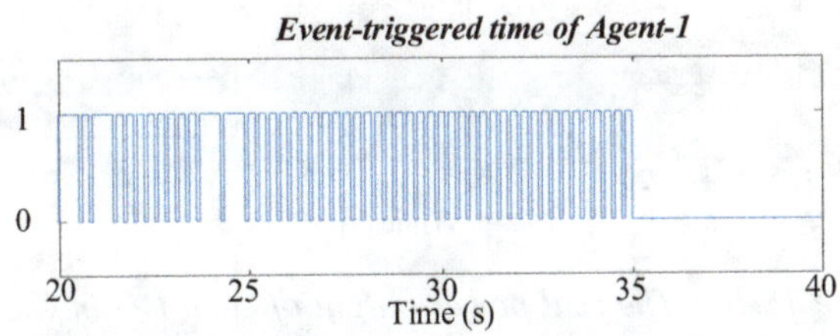

Figure 18.14 The total event-triggered time of Agent-1 (1, triggered; 0, not triggered)

40 s are shown in Figure 18.14. It can be observed that the frequency of communication is significantly reduced. The data flows are turned to idle when the system comes into the steady state.

18.5.3 Case 2: Communication failures and topology change

In *Case 2*, the performance of the proposed method is validated under communication failures and topology change. The physical and cyber events in *Case 2* are shown in Figure 18.15. In the physical system, there are loads switched on and off events. Load-2 is switch into the MG at 20 s. Load-1 is disconnected from MG at 40 s, and then reconnected at 60 s. In the cyber system, the topologies are changed to emulate communication failures. The cyber link 1–2 fails at 15 s, followed by the loss of cyber link 1–4 at 35 s. Afterwards, the cyber link 1–2 is restored at 55 s. The list of neighbours of each agent (DG) is configured initially corresponding to the communication topology in Figure 18.6 (Agent-1: [2,4], Agent-2: [1,3], Agent-3 [2,4], Agent-4: [1,3]). Each agent is able to adapt the alteration of the network topology by online self-adjusting the list of neighbours. For example, the communication of each DG is reconfigured as (Agent-1: [4], Agent-2: [3], Agent-3: [2,4], Agent-4: [1,3]) at 15 s.

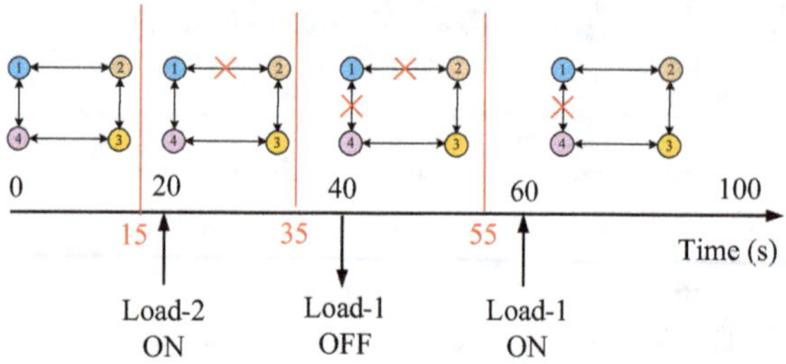

Figure 18.15 Cyber and physical events of Case 2

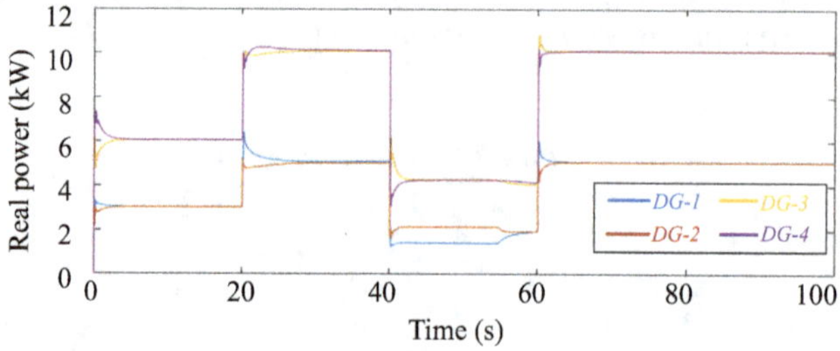

Figure 18.16 The real power output of each DG in Case 2

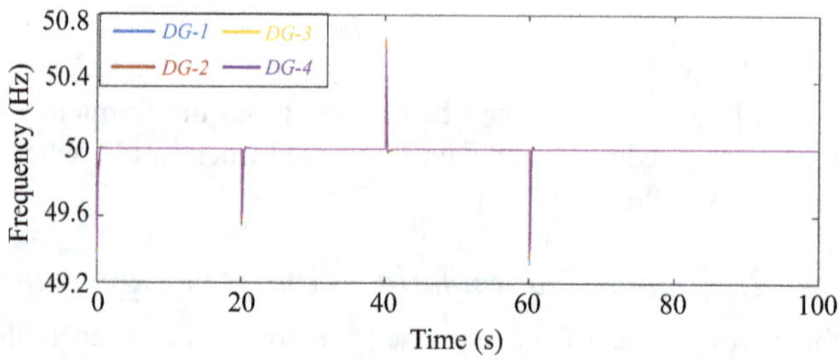

Figure 18.17 The frequency of each DG in Case 2

The CHIL experimental results of real power, frequency, reactive power, and bus voltage of each DG are shown in Figures 18.16–18.19, respectively. It can be found that the system converges slower (15–100 s) compared to the original graph (0–15 s) due to the loss of cyber links. The proportional power sharing is ensured as long as the communication graph is still completed. From 35 to 55 s, the DG-1 is isolated without any cyber links; thus, it is controlled by only droop control. As a result, the proportional real and reactive power sharing cannot be maintained during this period as shown in Figures 18.16 and 18.18.

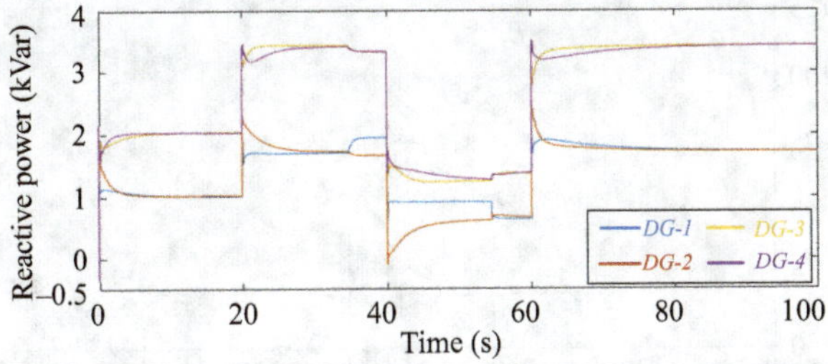

Figure 18.18 The reactive power of each DG in Case 2

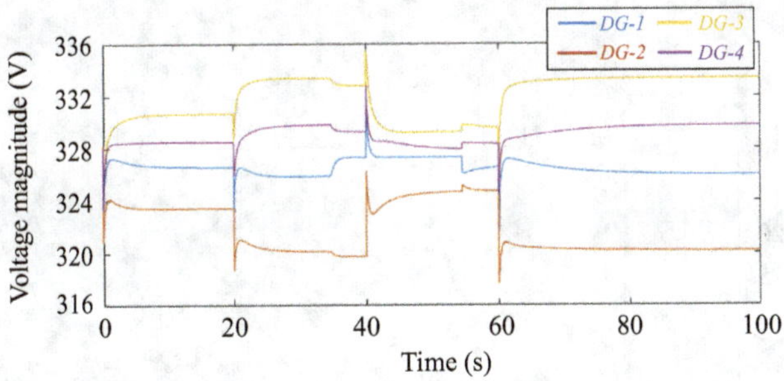

Figure 18.19 The voltage magnitude of each DG in Case 2

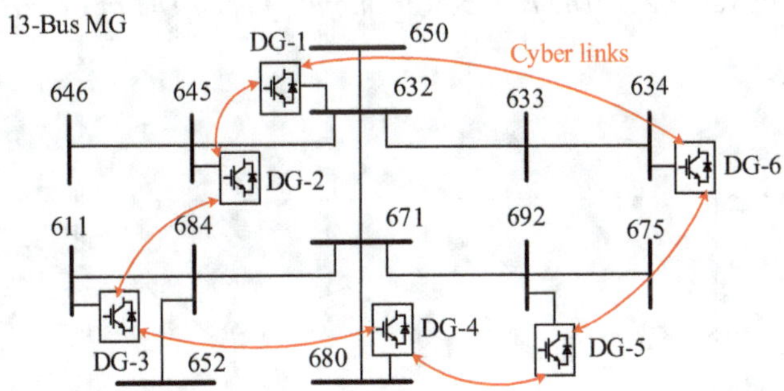

Figure 18.20 Electrical and communication topology of the 13-bus MG in Case 3

18.5.4 Case 3: scalability test

In *Case 3*, the performance of the proposed method is further validated in a larger system with realistic PV and load profiles. The electrical and communication topology of the tested MG is shown in Figure 18.20, which is modified from the IEEE 13-bus system [7]. The droop control coefficients of DGs 1–6 are considered with the ratio of 4:4:2:2:1:1. The PV and load profiles considered in this case are given in Figure 18.21.

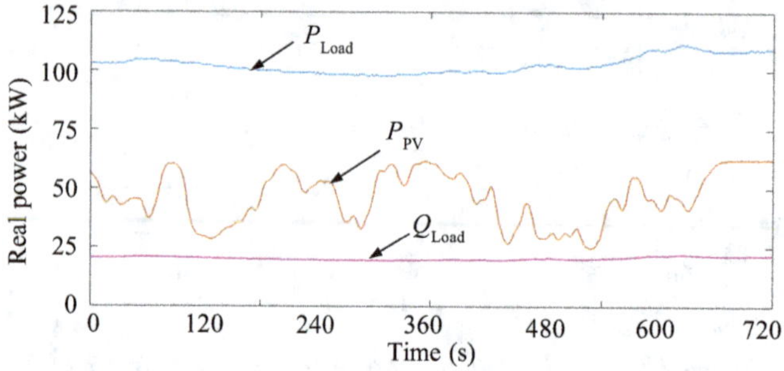

Figure 18.21 The PV and load profiles used in Case 3

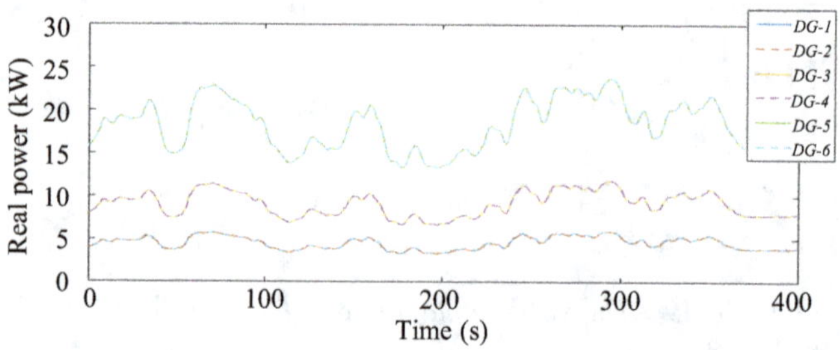

Figure 18.22 The real power output of each DG in Case 3

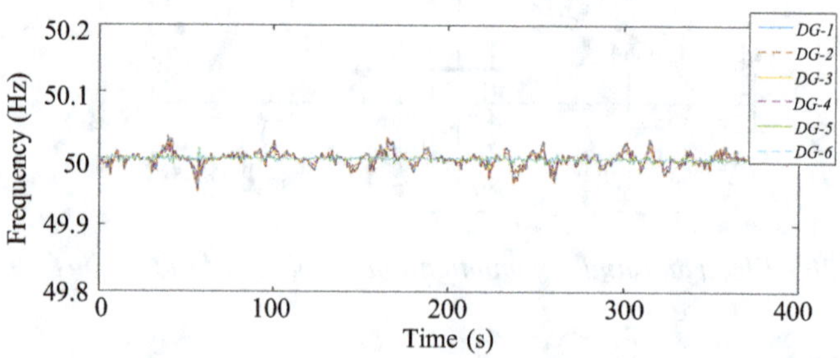

Figure 18.23 The frequency of each DG in Case 3

The OPAL-RT test results of real power, frequency, reactive power, bus voltage, and estimated average voltage for each DG are shown in Figures 18.22–18.26, respectively. It can be found that, regardless of PV and load variations, the real and reactive power can be shared reversely proportional to the droop coefficients as

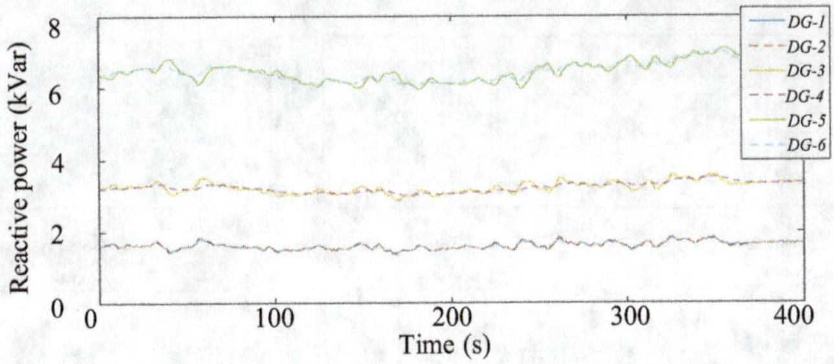

Figure 18.24 The reactive power of each DG in Case 3

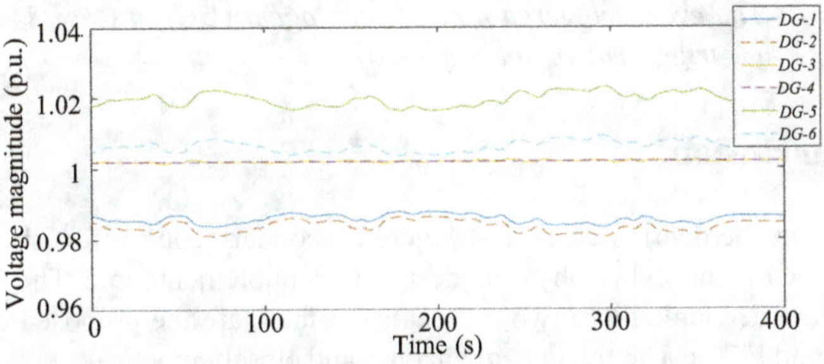

Figure 18.25 The voltage magnitude of each DG in Case 3

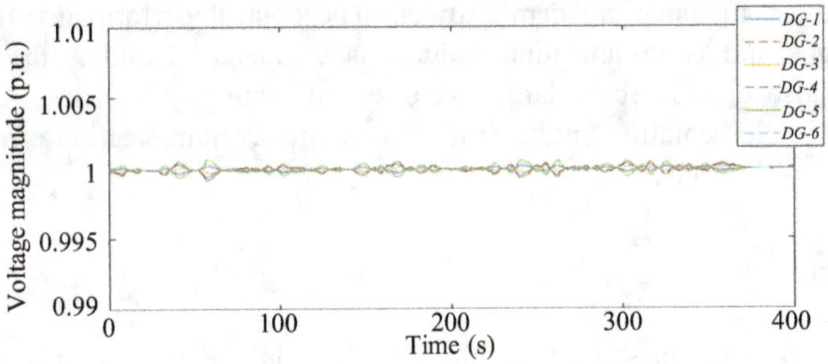

Figure 18.26 The estimated average voltage of each DG in Case 3

1:1:2:2:4:4 for DGs 1–6 in Figures 18.22 and 18.24. The frequency and estimated average voltage can be regulated around the reference values in Figures 18.23 and 18.26. In addition, the event-triggered time of each agent during 100 to 105 s are shown in Figure 18.27. It can be observed that the frequency of communication is reduced by the proposed approach.

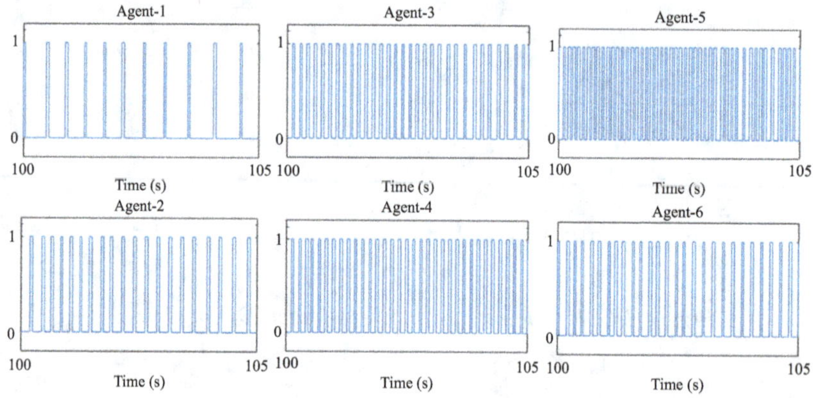

Figure 18.27 The event-triggered time of each agent (DG) in Case 3 (1, triggered; 0, not triggered)

18.6 Conclusion

In this chapter, the distributed event-triggered secondary control for islanded MGs is proposed with the cyber-physical design and implementation. The secondary controllers are decoupled into two timescales to facilitate the proposed distributed controllers and ETCs. The theoretical analysis and closely practical implementation of the proposed method are discussed.

The proposed secondary controllers are implemented and tested on the cyber-physical MG platform. The CHIL experimental results validate the effectiveness of the proposed method. The advantages of the proposed approach over periodical communication methods are demonstrated. The control performance under step load changes and communication failures is evaluated. Besides, the proposed method is also applied for a larger-scale system with real profiles. A practical design and implementation guidance is provided for communication efficient secondary control in MGs.

References

[1] D.V. Dimarogonas, E. Frazzoli, and K.H. Johansson, 'Distributed event-triggered control for multi-agent systems', *IEEE Trans. Automatic Control*, vol. 57, no. 5, pp. 1291–1297, 2012.

[2] M. Zhao, C. Peng, W. He, and Y. Song, 'Event-triggered communication for leader-following consensus of second-order multiagent systems', *IEEE Trans. Cybernetics*, vol. 48, no. 6, pp. 1888–1897, 2018.

[3] S. Weng, D. Yue, S. Member, C. Dou, J. Shi, and C. Huang, 'Distributed event-triggered cooperative control for frequency and voltage stability and power sharing in isolated inverter-based MG', *IEEE Trans. Cybernetics*, 2018.

[4] Y. Fan, G. Hu, and M. Egerstedt, 'Distributed reactive power sharing control for MGs with event-triggered communication', *IEEE Trans. Control Syst. Technol.*, vol. 25, no. 1, pp. 118–128, 2017.
[5] J.W. Simpson-Porco, Q. Shafiee, F. Dörfler, J.C. Vasquez, J.M. Guerrero, and F. Bullo, 'Secondary frequency and voltage control of islanded MGs via distributed averaging', *IEEE Trans. Industrial Electronics*, vol. 62, no. 11, pp. 7025–7038, 2015.
[6] F.L. Lewis, H. Zhang, K. Hengster-Movric, and A. Das, *Cooperative Control of Multi-agent Systems: Optimal and Adaptive Design Approaches*, Springer Science & Business Media, 2013.
[7] T. Morstyn, B. Hredzak, and V.G. Agelidis, 'Distributed cooperative control of microgrid storage', *IEEE Trans. Power Systems*, vol. 30, no. 5, pp. 2780–2789, 2014.

Chapter 19

Three-level coordinated voltage control of DERs and power hardware-in-the-loop validation

The reactive power of the photovoltaic (PV) inverters has great potential for voltage regulation in distribution networks. In this chapter, we propose a new three-level coordinated control method for PV inverters to deal with network voltage fluctuation and violation issues. A ramp-rate control in Level I is developed to smooth the network voltage fluctuations. A droop control in Level II is developed to alleviate the network voltage deviations locally. When local compensation from Levels I and II is not enough to regulate the network voltages within the required limits, a dynamic average consensus control in Level III is developed to fairly share the reactive power requirement among all inverters. The proposed control method functions to smooth the voltage profiles, restrain the voltage rise/drop problem, and coordinate PV inverters in the system when the voltage problem cannot be handled by individual PV inverter. The system model under the proposed three-level coordinated control is derived and the stability analysis is provided. The proposed control method is validated by means of power hardware-in-the-loop (PHIL) experiment.

19.1 Introduction

As introduced in previous chapters, the existing works on voltage control of power distribution networks have the following limitations: (i) The real-time voltage fluctuations and voltage limit violations are not well addressed. (ii) The distributed voltage control scheme relies on a leader (not fully distributed). (iii) The stability problems of multi-level coordinated voltage control are not systematically analysed. (iv) The distributed voltage control algorithms lack experimental validation to prove their applicability in the real world. This chapter addresses the above challenges and introduces a three-level coordinated voltage control approach by utilising reactive power from PV inverters to deal with both voltage fluctuations and limit violations in power distribution networks. The proposed control scheme is implemented and validated on a PHIL experimental platform. As compared to software simulation results, the test platform provides a realistic environment to evaluate the proposed control scheme. The features of this chapter are listed as follows:

- The proposed three-level fully distributed voltage control scheme includes ramp-rate control, droop control, and distributed control. The control autonomously activates based on the network voltage conditions. The guidelines for control gain selection of each control level are presented.

- The system's closed-loop stability under the proposed three-level control method is analysed. The system stable conditions under different voltage ranges are derived.
- The proposed three-level control is validated by means of the PHIL experiment, where a real hardware inverter, individual distributed controller design, and realistic communications are incorporated. The experimental results demonstrate the performance of the proposed control under various operating conditions, which validate the practicability of the proposed method.

19.2 Preliminaries

The preliminaries including the models of distribution networks, PV inverters, and communication networks are introduced in this section.

19.2.1 Power distribution networks

Considering a distribution network with $N+1$ buses denoted by $i = 0, 1, ..., N$, bus $i = 0$ is defined as the slack bus whose voltage can be adjusted by a substation transformer. The remaining N buses are PQ buses connected with PV inverters and loads, injecting or absorbing real and reactive power. The relationship between real power ΔP and reactive power ΔQ injection concerning bus voltage magnitude ΔV and angle $\Delta\theta$ can be obtained from the Jacobian matrix by solving the power flow equations [1,2]:

$$\begin{bmatrix} \Delta P(t) \\ \Delta Q(t) \end{bmatrix} = \begin{bmatrix} \boldsymbol{J}_1 & \boldsymbol{J}_2 \\ \boldsymbol{J}_3 & \boldsymbol{J}_4 \end{bmatrix} \begin{bmatrix} \Delta\theta(t) \\ \Delta V(t) \end{bmatrix} \tag{19.1}$$

where $\Delta P = \mathrm{col}\{\Delta P_1, \Delta P_2, ..., \Delta P_N\}$, $\Delta Q = \mathrm{col}\{\Delta Q_1, \Delta Q_2, ..., \Delta Q_N\}$, $\Delta V = \mathrm{col}\Delta\{V_1, \Delta V_2, ..., \Delta V_N\}$, and $\Delta\theta = \mathrm{col}\{\Delta\theta_1, \Delta\theta_2, ..., \Delta\theta_N\}$. The elements of matrix $\boldsymbol{J}_1$, $\boldsymbol{J}_2$, $\boldsymbol{J}_3$, and $\boldsymbol{J}_4$ are as follows:

$$J_{1,ij} = \left[\frac{\partial P_i}{\partial\theta_j}\right], J_{2,ij} = \left[\frac{\partial P_i}{\partial V_j}\right], J_{3,ij} = \left[\frac{\partial Q_i}{\partial\theta_j}\right], J_{4,ij} = \left[\frac{\partial Q_i}{\partial V_j}\right]$$

Assuming that $\boldsymbol{J}_1$ is invertible, the deviation in bus voltage magnitude can be derived from (19.1) as follows:

$$\Delta V(t) = (\boldsymbol{J}_4\left[\Delta Q(t) - \boldsymbol{J}_3\boldsymbol{J}_1{}^{-1}\Delta P(t)\right] = \boldsymbol{S}_{\mathbf{VQ}}\Delta Q(t) + \boldsymbol{S}_{\mathbf{VP}}\Delta P(t) \tag{19.2}$$

where the voltage sensitivity matrices are derived as $\boldsymbol{S}_{\mathbf{VQ}} = \left(\boldsymbol{J}_4 - \boldsymbol{J}_3\boldsymbol{J}_1{}^{-1}\boldsymbol{J}_2\right)^{-1}$, $\boldsymbol{S}_{\mathbf{VP}} = -\left(\boldsymbol{J}_4 - \boldsymbol{J}_3\boldsymbol{J}_1{}^{-1}\boldsymbol{J}_2\right)^{-1}\boldsymbol{J}_3\boldsymbol{J}_1{}^{-1}$. It is noted that although $\boldsymbol{S}_{\mathbf{VQ}}$ and $\boldsymbol{S}_{\mathbf{VP}}$ are functions of t, the variations of them with respect to t are relatively small for a wide range of operating conditions [2,3]. Equation (19.2) presents a linearised representation of ΔQ and ΔV near the equilibrium point, which is used to perform the system stability analysis. In distribution networks, bus voltages are influenced by both real and reactive power changes due to the high resistance to reactance ratio (R/X). The PV inverters are operated to provide the maximum power output. Therefore, ΔP from PV inverters is not controllable that is just considered as a disturbance. A control approach to effectively adjust ΔQ from PV inverters is investigated to mitigate the impact of ΔP on system voltages. The negative sign to power injection while the positive sign refers to power absorption.

19.2.2 PV inverters

PV inverters are usually operated in current control mode and function as current sources in power distribution networks. It is considered that PV inverters are all in maximum power point tracking mode, while the reactive power output is controlled by the designed control method.

The capacity of PV inverters can be oversized to make PV inverters provide certain reactive power even when rated real power is being produced [4]. The available reactive power capacity of a PV inverter can be represented as

$$\overline{Q}_i(t) = \sqrt{S_i^2 - P_i(t)^2} \tag{19.3}$$

$$S_i = (1 + \alpha_i)\overline{P}_i \tag{19.4}$$

where $\overline{Q}_i(t)$ is the reactive power capacity of ith PV inverter when the real power output is $P_i(t)$. S_i and $\overline{P}_i$ are the total capacity and rated real power of ith PV inverter, respectively. α_i is the percentage by which ith PV inverter is oversized in the deployment stage [5].

19.2.3 Communication network

The communication network of PV inverters in the proposed control scheme can be described by a graph as $\mathscr{G} = (\mathscr{V}, \mathscr{E}, \mathscr{A})$, where $\mathscr{V} = \{v_1, v_2, \ldots, v_N\}$ is a set of N nodes, and $\mathscr{E} = \mathscr{V} \times \mathscr{V}$ is a set of edges. Each node corresponds to a communication agent of a PV inverter while the edges represent the communication links for data exchange. If communication links are bidirectional, $(v_i, v_j) \in \mathscr{E} \Rightarrow (v_j, v_i) \in \mathscr{E}\ \forall i, j$ the graph is undirected, otherwise, it is directed. The adjacency matrix $A = [a_{ij}]$ is associated with the edges; here a_{ij} represents the weight for information exchange between agents i and j, where $a_{ij} = 1$ if agents i and j are connected through an edge $(v_i, v_j) \in \mathscr{E}$, otherwise, $a_{ij} = 0$. The in-degree matrix is defined as $\boldsymbol{D} = \mathrm{diag}(d_i)$, where $d_i = \sum_{i=1}^{N} a_{ij}$. The Laplacian matrix is defined as $\boldsymbol{L} = [l_{ij}]$, where l_{ij} in $\boldsymbol{L}$, and $\boldsymbol{L} = \boldsymbol{D} - \boldsymbol{A}$. In real-world implementation, both cables, optical fiber, and wireless communications like Wi-Fi or Zigbee technology can be used to set up a communication network for control signal exchange.

19.3 Three-level coordinated voltage control

As illustrated in Figure 19.1, the proposed control approach comprises three control hierarchies at different voltage ranges. The Level I ramp-rate control is designed to smooth the voltage fluctuations and operates at all voltage ranges. The Level II droop control is designed to mitigate voltage deviations when the bus voltage is out of the desirable range $[\underset{\sim}{V}, \tilde{V}]$. The Level III distributed control functions will provide corrective voltage control when the bus voltage is beyond the allowable range $[\underline{V}, \overline{V}]$. It is noted that the Levels I and II controls are executed with only local voltage measurement, while Level III control requires the voltage signals from adjacent inverters via a communication network.

The objective of the proposed control approach is to determine an appropriate utilisation ratio for each inverter. The utilisation ratio $u_i(t)$ of ith PV inverter is defined as the reactive power output $\Delta Q_i(t)$ divided by the reactive power capacity $\overline{Q}_i(t)$ i.e. $u_i(t) = \Delta Q_i(t)/\overline{Q}_i(t)$. Figure 19.2 shows the overall architecture of the proposed control method. Each control level will generate one utilisation ratio based on the network voltage conditions, i.e. $u_i^I(t), u_i^{II}(t)$, and $u_i^{III}(t)$. The utilisation ratio $u_i(t)$ of ith PV inverter is the sum of these three levels utilisation ratios. The total reactive power output of ith PV inverter is calculated by

$$\begin{cases} u_i(t) = u_i^I(t) + u_i^{II}(t) + u_i^{III}(t) \\ \Delta Q_i(t) = u_i(t) \cdot \overline{Q}_i(t) \\ u_i(t), u_i^I(t), u_i^{II}(t), u_i^{III}(t) \in [-1, 1] \end{cases} \tag{19.5}$$

The details of each individual control level are explained in the following subsections.

19.3.1 Level I: ramp-rate control

The objective of Level I control is to locally regulate voltage ramp rate within a limit, i.e. $|\frac{V_i(t_1)-V_i(t_2)}{t_1-t_2}| \le VRR_{\text{lim}}$, where VRR_{lim} is the limit of voltage ramp rate.

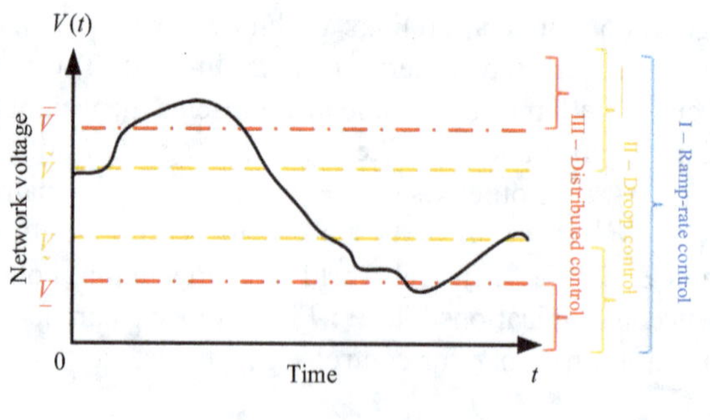

Figure 19.1 The operational ranges of the proposed three-level control

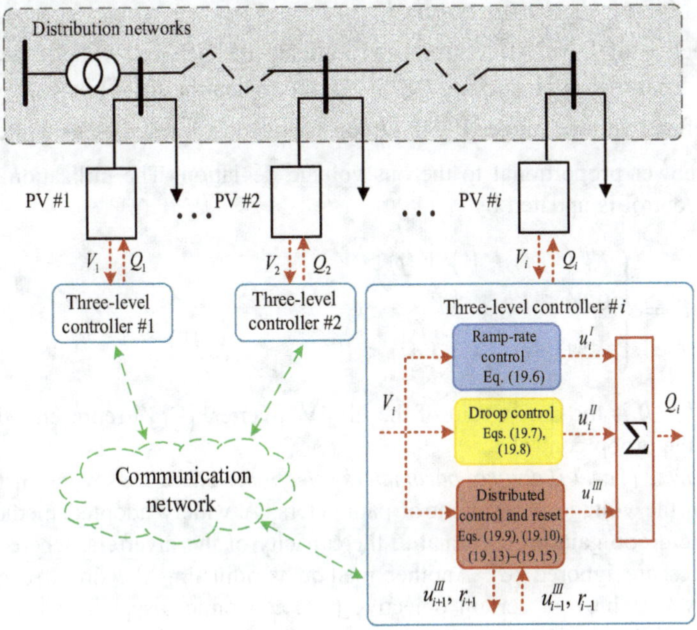

Figure 19.2 The overall architecture of the proposed control scheme

The moving clouds (shading effect) can lead to significant voltage fluctuations, which can be locally smoothed by appropriate reactive power injection [6]. A moving average filter-based ramp-rate controller is designed for each PV inverter. The high-frequency components of each bus voltage are extracted by subtracting the filtered voltage from the measured voltage and compensated with the inverter reactive power output determined by ramp-rate control. The utilisation ratio of Level I control is obtained by

$$u_i^I(t) = K_i^I \left[V_i(t) - \frac{\sum_{j=t-\omega}^{t} V_i(j)}{T(t) - T(t-\omega)} \right] \tag{19.6}$$

where $\omega > 0$ is the length of the moving average window and $K_i^I > 0$ is the ramp-rate control gain.

Remark 1 (Level I control parameters selection): Generally, increasing the moving average filter window length ω will lead to more low-frequency components being compensated. The ramp-rate control gain K_i^I is proportional to the compensation effect of the ramp-rate control. The boundaries of control gain selection can be obtained from the stability analysis in Section 19.4. The criterion for selection of the ramp-rate control parameters is to guarantee that even under a certain level change of load demand, the network voltage fluctuations still satisfy the industrial requirement [7].

19.3.2 Level II: droop control

The objective of Level II control is to mitigate the network voltage deviations based on local measurement, i.e. $V_i'(t) \rightarrow V_i(t)$, when the local bus voltage is out of the defined voltage range $[\underset{\sim}{V}, \tilde{V}]$. Droop control is activated to generate/absorb reactive power proportional to the bus voltage deviation. The utilisation ratio of Level II control is updated by

$$u_i^{II}(t) = \begin{cases} K_i^{II}\left(V_i(t) - \tilde{V}\right), V_i(t) > \tilde{V} \\ 0, \underset{\sim}{V} \leq V_i(t) \leq \tilde{V} \\ K_i^{II}\left(V_i(t) - \underset{\sim}{V}\right), V_i(t) < \underset{\sim}{V} \end{cases} \tag{19.7}$$

where $K_i^{II} > 0$ is the droop gain of the ith PV inverter. $[\underset{\sim}{V}, \tilde{V}]$ represents the dead band for the droop control.

Remark 2 (Level II control parameters selection): There are several methods to determine the voltage droop control parameters. A widely adopted method is to choose the droop gain proportional to the capacity of the inverters, where the line impedances are ignored [8]. Another method is adjusting the droop coefficient adaptively to achieve a certain objective (e.g. economic dispatch and loss minimisation) [9]. In distribution networks, to avoid the instability caused by improper droop settings, it is suggested to choose the droop settings following voltage sensitivity analysis, especially when the distribution network is large [2,10].

In this chapter, the droop control gain of each inverter is selected based on the voltage sensitivity analysis as follows:

$$K_i^{II} = \beta_i \frac{1}{s_{ii}} \tag{19.8}$$

where $s_{ii} = \partial V_i / \partial Q_i \in \boldsymbol{S}_{\mathrm{QV}}$ is obtained from (19.1) and (19.2), and β_i is the scaler of the droop control gain. The boundaries of control gain selection will follow the stability analysis in Section 19.4.

19.3.3 Level III: distributed control

In Level III control, a distributed voltage control with a distributed reset algorithm to terminate the distributed controllers is proposed. The objective of Level III control is to coordinate all available PV inverters when network voltages cannot be regulated within $[V, \overline{V}]$ locally, where the steady-state conditions can be represented as follows:

$$\begin{cases} \lim\limits_{t \to \infty} \|V_i(t) - \overline{V}\| = 0, V_i(t) > \overline{V} \\ \lim\limits_{t \to \infty} \|V_i(t) - V\| = 0, V_i(t) < V \end{cases} \tag{19.9}$$

In such circumstances, all PV inverters with available capacity will increase their reactive power output until all network voltages are back to the allowable range. This control will be deactivated when the network voltages are recovered within a safe range, which is achieved by a distributed reset algorithm presented in this subsection.

The utilisation ratio of Level III control is updated based on the following equations:

$$\dot{u}_i^{III}(t) = G_i^{III}\left[\sum_{i=1}^{N} a_{ij}\left(u_j^{III}(t) - u_i^{III}(t)\right)\right] + e(t) \tag{19.10}$$

$$e(t) = \begin{cases} K_i^{III}\left(V_i(t) - \overline{V}\right), V_i > \overline{V} \\ 0, \underline{V} \leq V_i \leq \overline{V} \\ K_i^{III}\left(V_i(t) - \underline{V}\right), V_i < \underline{V} \end{cases} \tag{19.11}$$

where $K_i^{III} > 0$ and $G_i^{III} > 0$ are the distributed control gains.

The proposed distributed control reaches consensus (i.e. steady state), for any initial state, when the following conditions are met:

$$\lim_{t\to\infty} \|e(t)\| = 0, \lim_{t\to\infty} \|u_i^{III}(t) - u_j^{III}(t)\| = 0 \tag{19.12}$$

Equation (19.12) ensures the control objectives in (19.9) are met. The proposed control can guarantee that the voltage deviations out of $[\underline{V}, \overline{V}]$ are eliminated in steady state with the utilisation ratios of Level III control for each inverter equalised.

Remark 3 (Level III control gain selection): K_i^{III} and G_i^{III} determine the convergence speed of the proposed distributed control. Generally, the larger K_i^{III} and G_i^{III} are chosen, the faster the convergence speed will be. However, the trade-off between stability and convergence speed should also be considered when choosing K_i^{III} and G_i^{III}. The stability criteria for Level III control gain selection will be further discussed in Section 19.4.

Furthermore, integrators in (19.9) are needed to be reset when all network bus voltages are back to the safe operating range. To be consistent with the ideology of distributed control, the integrators also need to be reset in a fully distributed manner. It motivates us to design a distributed reset algorithm for the proposed distributed control. To this end, a flag signal of each inverter is defined as

$$F_i(t) = \begin{cases} 0, \dfrac{\underline{V} + \underset{\sim}{V}}{2} \leq V_i(t) \leq \dfrac{\tilde{V} + \overline{V}}{2} \\ 1, \text{otherwise} \end{cases} \tag{19.13}$$

where $(\tilde{V} + \overline{V})/2$ and $(\underline{V} + \underset{\sim}{V})/2$ are the upper and lower bounds to trigger the flag value, respectively. The flag signal will turn from 0 to 1 when the voltage is out of the range defined in (19.13). The reset signal $R_i(t)$ of each inverter is updated based on the following equations:

$$\dot{r}_i(t) = \chi_i \sum_{i=1}^{N} a_{ij}\big(r_j(t-\tau(t)) - r_i(t)\big) + \gamma_i(F_i(t) - r_i(t)) \tag{19.14}$$

$$R_i(t) = \text{sign}[r_i(t)] \tag{19.15}$$

where $r_i(t)$ and $R_i(t)$ are the transmitted signal and the reset signal of the ith inverter, respectively. $\chi_i > 0$ and $\gamma_i > 0$ are two coefficients that influence the convergence speed of the algorithm. It should be noted that $r_i(t)$ shares the same communication network as of (19.9). The integrators for the distributed control will be reset when $R_i(t)$ falls from 1 to 0. According to the distributed reset algorithm (19.13)–(19.15), $\forall V_i(t) \notin [(\underset{\sim}{V} + V)/2, (\tilde{V} + \overline{V})/2] \Rightarrow \forall F_i(t), R_i(t) \neq 0$, which ensures the integrators are reset in a distributed way when the voltages return to the defined voltage range in (19.13).

19.4 Stability analysis

To analyse the closed-loop system stability under the proposed control, the proposed three-level control (19.5) can be rewritten as

$$\Delta Q_i = u_i \cdot \overline{Q}_i + \overline{Q}_i^d \tag{19.16}$$

It is noted that $\overline{Q}_i^d = 0$ when there is no PV generation. During the daytime, the reactive power from PV inverter contains both controllable term $u_i \cdot \overline{Q}_i$ and a disturbance term $\overline{Q}_i^d$. $\overline{Q}_i$ is a constant indicating the available reactive power capacity of PV inverter for a certain period. In this section, the symbol of time t is omitted for brevity.

In matrix form, (19.16) can be written as

$$\Delta Q = \overline{\boldsymbol{Q}}u + \overline{Q}^d \tag{19.17}$$

where vectors $u = \text{col}\{u_1, u_2, ..., u_N\}$, $\overline{Q}^d = \text{col}\{\overline{Q}_1^d, \overline{Q}_2^d, ..., \overline{Q}_N^d\}$, and matrix $\overline{\boldsymbol{Q}} = \text{diag}\{\overline{Q}_1, \overline{Q}_2, ..., \overline{Q}_N\}$.

Combining (19.2) and (19.17), the system can be rearranged in matrix form as

$$\Delta V = \boldsymbol{S}_{\mathbf{VQ}}\overline{\boldsymbol{Q}}u + \boldsymbol{S}_{\mathbf{VQ}}\overline{Q}^d + \boldsymbol{S}_{\mathbf{VP}}\Delta P = \boldsymbol{S}_{\mathbf{VQ}}\overline{\boldsymbol{Q}}u + \boldsymbol{W} \tag{19.18}$$

where $\boldsymbol{W} = \boldsymbol{S}_{\mathbf{VQ}}\overline{\boldsymbol{Q}}^d + \boldsymbol{S}_{\mathbf{VP}}\boldsymbol{\Delta P}$ is the uncontrollable term, which can be viewed as a disturbance. The system with the proposed control can be derived into three operational ranges, which is illustrated in Figure 19.1.

A. Analysis for $\boldsymbol{V}_i \in [\underset{\sim}{\boldsymbol{V}}, \tilde{\boldsymbol{V}}]$

In this operational range, only the ramp-rate control will take effect. The ramp-rate control functions like derivative control, which generates a control signal u_i^l based on the derivate of voltage ($\mathrm{d}V/\mathrm{d}t$). As the ramp-rate control based on moving average filter is not a linear controller, the approximation dynamics can be obtained from

$$u_i = K_i^I \frac{s}{s+\widehat{\omega}} V_i \tag{19.19}$$

where $\widehat{\omega}_i$ is a parameter corresponding to the window length of the moving average filter.

Substituting (19.19) into (19.18), we can obtain the system dynamics in matrix form as

$$\Delta\dot{V} + \widehat{\boldsymbol{\omega}}\Delta V = \overline{\boldsymbol{Q}}\boldsymbol{S}_{\mathrm{VQ}}\boldsymbol{K}^{\boldsymbol{I}}\Delta\dot{V} + \dot{\boldsymbol{W}} + \widehat{\boldsymbol{\omega}}\boldsymbol{W} \tag{19.20}$$

$$\Delta\dot{V} = -\left(\boldsymbol{I} - \overline{\boldsymbol{Q}}\boldsymbol{S}_{\mathrm{VQ}}\boldsymbol{K}^{\boldsymbol{I}}\right)\widehat{\boldsymbol{\omega}}\Delta V + \boldsymbol{W}^{\boldsymbol{I}} \tag{19.21}$$

where $\widehat{\boldsymbol{\omega}} = \mathrm{diag}\{\widehat{\omega}_1, \widehat{\omega}_2, ..., \widehat{\omega}_N\}$, $\boldsymbol{K}^{\boldsymbol{I}} = \mathrm{diag}\{K_1^I, K_2^I, ..., K_N^I\}$. $\boldsymbol{W}^{\boldsymbol{I}} = \dot{\boldsymbol{W}} + \widehat{\boldsymbol{\omega}}\boldsymbol{W}$, $\boldsymbol{I}$ is an identity matrix of the appropriate dimension.

If the system is stable, the system matrix in (19.21) should be *Hurwitz*, i.e. $-\left(\boldsymbol{I} - \overline{\boldsymbol{Q}}\boldsymbol{S}_{\mathrm{VQ}}\boldsymbol{K}^{\boldsymbol{I}}\right)\widehat{\boldsymbol{\omega}} \prec 0$. As $\widehat{\boldsymbol{\omega}}$ of the is positive definite, thus, it requires $\boldsymbol{I} - \overline{\boldsymbol{Q}}\boldsymbol{S}_{\mathrm{VQ}}\boldsymbol{K}^{\boldsymbol{I}} \succ 0$. Therefore, the chosen value of $\boldsymbol{K}^{\boldsymbol{I}}$ need to satisfy all the eigenvalues $\mathrm{Re}\left[\lambda_i\left(\boldsymbol{I} - \overline{\boldsymbol{Q}}\boldsymbol{S}_{\mathrm{VQ}}\boldsymbol{K}^{\boldsymbol{I}}\right)\right] > 0$.

B. Analysis for $V_i \in [\underline{V}, \underset{\sim}{V}] \cup [\tilde{V}, \overline{V}]$

In this operational range, both ramp-rate control and droop control take effect. The droop control in (19.7) can be rewritten as

$$u_i^{II} = K_i^{II}\left[(V_i - V^*) - \left(V_i^{II} - V^*\right)\right] = K_i^{II}\left(\Delta V_i - \Delta V_i^{II}\right) \tag{19.22}$$

where $V_i^{II} = \begin{cases} \tilde{V}, V_i(t) > \tilde{V} \\ V_i, \underset{\sim}{V} \le V_i(t) \le \tilde{V} \\ \underset{\sim}{V}, V_i(t) < \underset{\sim}{V} \end{cases}$, and V^* is the equilibrium points of ΔV in the system model, which is set as 1 p.u.

Substituting the control law of (19.19) and (19.22) into (19.18) and rearranging in matrix form:

$$\Delta\dot{V} + \widehat{\boldsymbol{\omega}}\Delta V = \overline{\boldsymbol{Q}}\boldsymbol{S}_{\mathrm{VQ}}\left(\boldsymbol{K}^{\boldsymbol{I}}\Delta\dot{V} + \boldsymbol{K}^{\boldsymbol{II}}\Delta\dot{V} + \boldsymbol{K}^{\boldsymbol{II}}\widehat{\boldsymbol{\omega}}\left(\Delta V + \Delta V^{II}\right)\right) + \dot{\boldsymbol{W}} + \widehat{\boldsymbol{\omega}}\boldsymbol{W} \tag{19.23}$$

$$\Delta\dot{V} = \left[\boldsymbol{I} - \overline{\boldsymbol{Q}}\boldsymbol{S}_{\mathrm{VQ}}\left(\boldsymbol{K}^{\boldsymbol{I}} + \boldsymbol{K}^{\boldsymbol{II}}\right)\right]^{-1}\left(\overline{\boldsymbol{Q}}\boldsymbol{S}_{\mathrm{VQ}}\boldsymbol{K}^{\boldsymbol{II}} - \boldsymbol{I}\right)\widehat{\boldsymbol{\omega}}\Delta V + \boldsymbol{W}^{II} \tag{19.24}$$

where $\Delta V^{II} = \mathrm{col}\{\Delta V_1^{II}, \Delta V_2^{II}, ..., \Delta V_N^{II}\}$, $\boldsymbol{K}^{\boldsymbol{II}} = \mathrm{diag}\{K_1^{II}, K_2^{II}, ..., K_N^{II}\}$, and $\boldsymbol{W}^{\boldsymbol{II}}$ is the disturbance term that can be derived from the above equations.

For the dynamic system in (19.24), the system stability requires the following matrix to be *Hurwitz*:

$$\boldsymbol{\Psi} = \left[\boldsymbol{I} - \overline{\boldsymbol{Q}}\boldsymbol{S}_{\mathrm{VQ}}\left(\boldsymbol{K}^{\boldsymbol{I}} + \boldsymbol{K}^{\boldsymbol{II}}\right)\right]^{-1}\left(\overline{\boldsymbol{Q}}\boldsymbol{S}_{\mathrm{VQ}}\boldsymbol{K}^{\boldsymbol{II}} - \boldsymbol{I}\right)\widehat{\boldsymbol{\omega}} \prec 0 \tag{19.25}$$

For a certain control gain $\boldsymbol{K}^I$ in Level I control, the chosen value of K^{II} should satisfy all the eigenvalues $\mathrm{Re}[\lambda_i(\boldsymbol{\Psi})] < 0$.

C. Analysis for $V_i \notin [\underline{V}, \overline{V}]$

In this operational range, all three control levels take effect. In matrix form, the distributed controller in level III is represented as

$$\begin{aligned} \dot{u}^{III} &= -\boldsymbol{G}^{III}\boldsymbol{L}u^{III} + \boldsymbol{e} = -\boldsymbol{G}^{III}\boldsymbol{L}u^{III} + \boldsymbol{K}^{III}(V - V^{III}) \\ &= -\boldsymbol{G}^{III}\boldsymbol{L}u^{III} + \boldsymbol{K}^{III}[(V - V^*) - (V^{III} - V^*)] \\ &= -\boldsymbol{G}^{III}\boldsymbol{L}u^{III} + \boldsymbol{K}^{III}(\Delta V - \Delta V^{III}) \end{aligned} \tag{19.26}$$

where $V_i^{III} = \begin{cases} \overline{V}, V_i(t) > \overline{V} \\ V_i, \underline{V} \le V_i(t) \le \overline{V} \\ \underline{V}, V_i(t) < \underline{V} \end{cases}.$

Substituting all three-level control laws into (19.18), it obtains

$$\begin{aligned} \Delta\dot{V} = {} & \left[(\boldsymbol{I} - \overline{\boldsymbol{Q}}\boldsymbol{S}_{\mathrm{VQ}}(\boldsymbol{K}^I + \boldsymbol{K}^{II})\right]^{-1}[(\overline{\boldsymbol{Q}}\boldsymbol{S}_{\mathrm{VQ}}\boldsymbol{K}^{II} - \boldsymbol{I})\widehat{\omega} + \overline{\boldsymbol{Q}}\boldsymbol{S}_{\mathrm{VQ}}\boldsymbol{K}^{III}]\Delta V \\ & + \left[(\boldsymbol{I} - \overline{\boldsymbol{Q}}\boldsymbol{S}_{\mathrm{VQ}}(\boldsymbol{K}^I + \boldsymbol{K}^{II})\right]^{-1}\left[\overline{\boldsymbol{Q}}\boldsymbol{S}_{\mathrm{VQ}}(-\boldsymbol{G}^{III}\boldsymbol{L} + \widehat{\omega})u^{III}\right] + \boldsymbol{W}^{III} \end{aligned} \tag{19.27}$$

where $\boldsymbol{K}^{III} = \mathrm{diag}\{K_1^{III}, K_2^{III}, \ldots, K_N^{III}\}$, $\boldsymbol{G}^{III} = \mathrm{diag}\{G_1^{III}, G_2^{III}, \ldots, G_N^{III}\}$, and $\boldsymbol{W}^{III}$ is the disturbance term.

Combining (19.26) and (19.27) into the state–space form, it has

$$\begin{bmatrix} \Delta\dot{V} \\ \dot{u}^{III} \end{bmatrix} = \begin{bmatrix} \boldsymbol{\Gamma} & \Xi \\ \boldsymbol{K}^{III} & -\boldsymbol{G}^{III}\boldsymbol{L} \end{bmatrix}\begin{bmatrix} \Delta V \\ u^{III} \end{bmatrix} + \begin{bmatrix} \boldsymbol{W}^{III} \\ -\boldsymbol{K}^{III}(V^{III} - V^*) \end{bmatrix} \tag{19.28}$$

where $\boldsymbol{\Gamma} = [\boldsymbol{I} - \overline{\boldsymbol{Q}}\boldsymbol{S}_{\mathrm{VQ}}(\boldsymbol{K}^I + \boldsymbol{K}^{II})]^{-1}[(\overline{\boldsymbol{Q}}\boldsymbol{S}_{\mathrm{VQ}}\boldsymbol{K}^{II} - \boldsymbol{I})\widehat{\omega} + \overline{\boldsymbol{Q}}\boldsymbol{S}_{\mathrm{VQ}}\boldsymbol{K}^{III}]$ and $\boldsymbol{\Xi} = [\boldsymbol{I} - \overline{\boldsymbol{Q}}\boldsymbol{S}_{\mathrm{VQ}}(\boldsymbol{K}^I + \boldsymbol{K}^{II})]^{-1}[\overline{\boldsymbol{Q}}\boldsymbol{S}_{\mathrm{VQ}}(-\boldsymbol{G}^{III}\boldsymbol{L} + \widehat{\omega})]$.

The system stability requires the following matrix to be *Hurwitz*:

$$\begin{bmatrix} \boldsymbol{\Gamma} & \boldsymbol{\Xi} \\ \boldsymbol{K}^{III} & -\boldsymbol{G}^{III}\boldsymbol{L} \end{bmatrix} \prec 0 \tag{19.29}$$

Therefore, based on the derived closed-loop systems in (19.21), (19.25), and (19.29), the trace of eigenvalues can be plotted to examine the stability boundaries of the studied system.

19.5 Power hardware-in-the-loop experimental tests

19.5.1 Test system

The proposed three-level coordinated voltage control method is validated by a PHIL test in a seven-bus distribution network [2]. The single-line diagram of the test system is shown in Figure 19.3 and the PV capacity and load demand at each

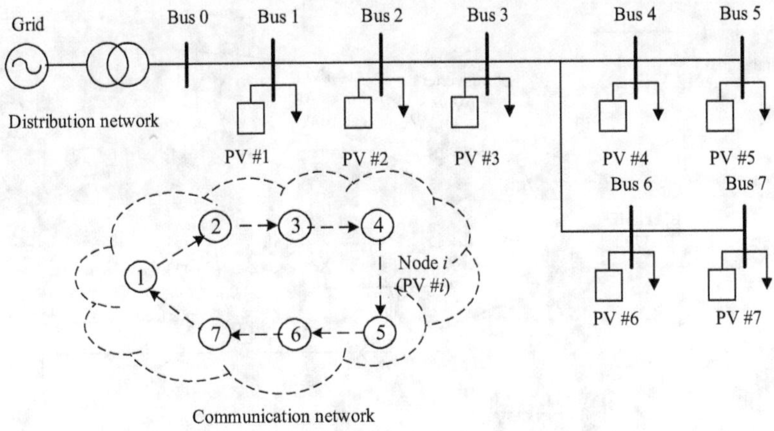

Figure 19.3 The configuration of the seven-bus distribution network and the communication graph of PV inverters

Table 19.1 The PV capacity and load demand in the seven-bus network

Bus ID	**PV capacity (kVA)**	**Load demand (kVA)**
Bus 1	850	700 + 100j
Bus 2	600	850 + 250j
Bus 3	1,200	600 + 150j
Bus 4	900	1,250 + 500j
Bus 5	800	900 + 300j
Bus 6	900	100 + 100j
Bus 7	1,000	1,000 + 350j

bus are modified as in Table 19.1. One-way communication among PV inverters are considered, which forms a directed circle communication graph in Figure 19.3. As all the PV inverters are connected by a directed communication graph, the convergence of the proposed distributed control is guaranteed. The tap change of the on-load tap changer is not considered and the voltage at bus 0 is kept at 1.0 p.u. during the operation. The distribution networks are usually allowed a maximum voltage deviation of ±5%–±10%, considering the national standards of different countries [11]. In this chapter, the voltage ranges to trigger distributed control and droop control are selected as $\left[\underline{V}, \overline{V}\right] = [0.95, 1.05]$ and $\left[\underset{\sim}{V}, \tilde{V}\right] = [0.97, 1.03]$.

The PHIL experiment is performed to test the proposed control method in a more realistic environment [12]. Figure 19.4 shows the real-time PHIL test platform. The PV inverter and load demand at bus 5 are realised with real laboratory equipment. A 15-kVA inverter and a 10-kW resistive load bank with a 90-kVA interfacing converter are used. The 90-kVA interfacing converter reproduces the voltages from the simulation point of common coupling within the laboratory while

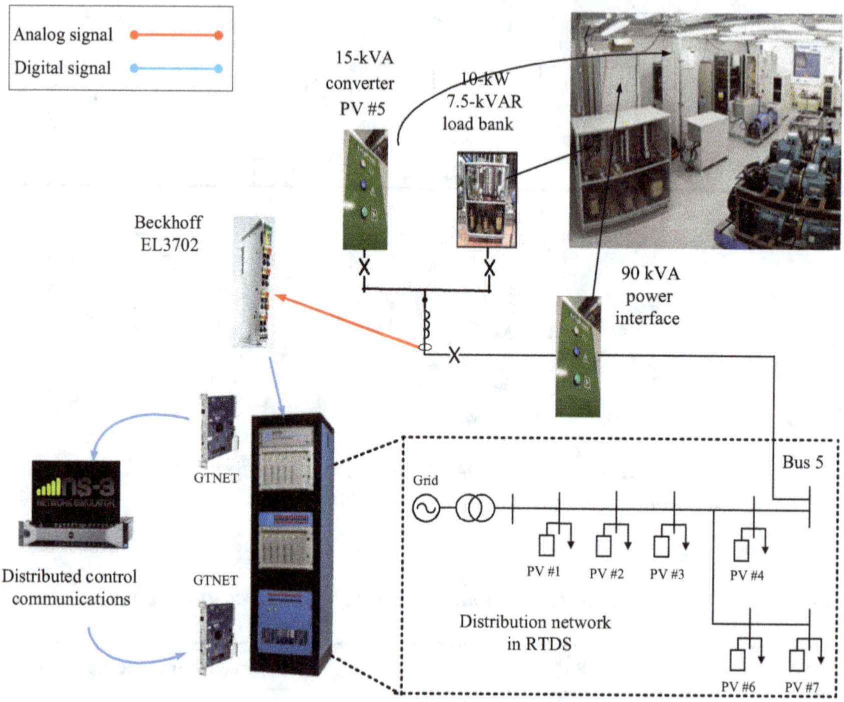

Figure 19.4 PHIL test platform

at the same time measures the hardware response (currents) to be fed in at the simulation point of common coupling. The voltage and current signals are scaled up and down between the real-time simulation and the hardware equipment. To increase the accuracy of the PHIL implementation, time delay compensation technique presented in [13] has been used. The remainder of the seven-bus distribution network is simulated in a real-time digital simulator (RTDS). It should be noted that the inner voltage and current control loop dynamics for the hardware PV inverter are considered as in [14], while the remaining six inverters are simplified into controlled current sources.

The proposed control scheme is embedded individually for each of the PV inverters in the RTDS. The distribution network model in RTDS is simulated at a time step of 50 μs. The network voltages are sampled with a slower rate of 10 ms, which is acquired by the proposed three-level controllers to generate control signals. Furthermore, all the communications required for the distributed control algorithm are implemented through a practical communication system. The control signal from each of the dedicated controller instances is independently routed outside of the RTDS using Giga-Transceiver Network Communication Card (GTNET) cards employing user datagram protocol into a communication emulator by NS-3. It will go back into the neighbouring controller instance in the RTDS. Such an experiment set-up represents close to real-world implementation allowing for the feasibility of the proposed approach to be validated.

Table 19.2 Parameters of the proposed three-level control

Level I	$K_i^I = 15$ and $\omega_i = 10$
Level II	$K_i^{II} = 10$
Level III	$K_i^{III} = 1$, $G_i^{III} = 1$, $\chi_i = 2$, $\gamma_i = 1$

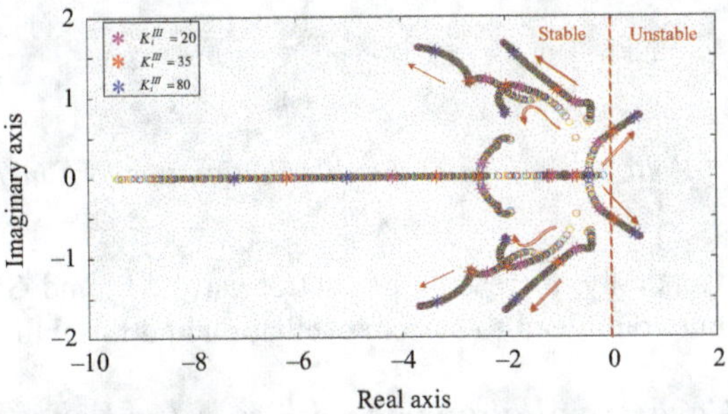

Figure 19.5 The eigenvalue traces of closed-loop system (19.28) as controller gains $G_i^{III} = 1$, K_i^{III} varying from [1, 100]

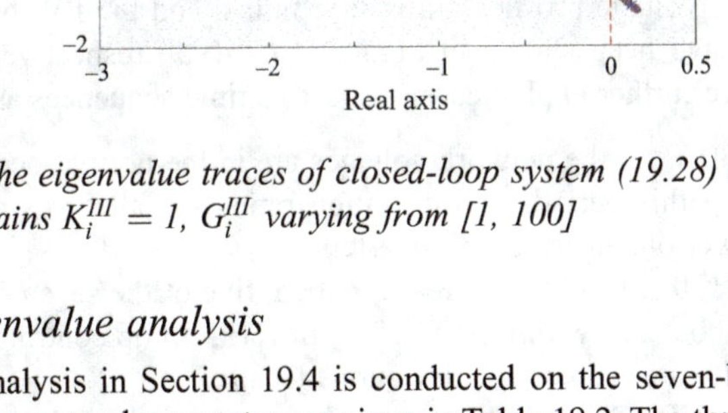

Figure 19.6 The eigenvalue traces of closed-loop system (19.28) as controller gains $K_i^{III} = 1$, G_i^{III} varying from [1, 100]

19.5.2 Eigenvalue analysis

The stability analysis in Section 19.4 is conducted on the seven-bus distribution network, with the control parameters as given in Table 19.2. The three-level control for $V_i \notin [\underline{V}, \overline{V}]$ in Section 19.4.3 is studied as an example. The trace of eigenvalues as control gains K_i^{III} and G_i^{III} varying are shown in Figures 19.5 and 19.6. Arrows indicate the direction of increase in control gains. Significant negative eigenvalues are omitted for clarity. It can be observed that increasing the control gains in Level III leads to system instability. In Figure 19.5, $G_i^{III} = 1$ keeps unchanged and eigenvalue traces cross the negative half-plane of the real axis when $K_i^{III} > 35$, which indicates system is unstable. Similarly, in Figure 19.6, $K_i^{III} = 1$ keeps unchanged and the eigenvalue traces cross the negative half-plane of the real axis

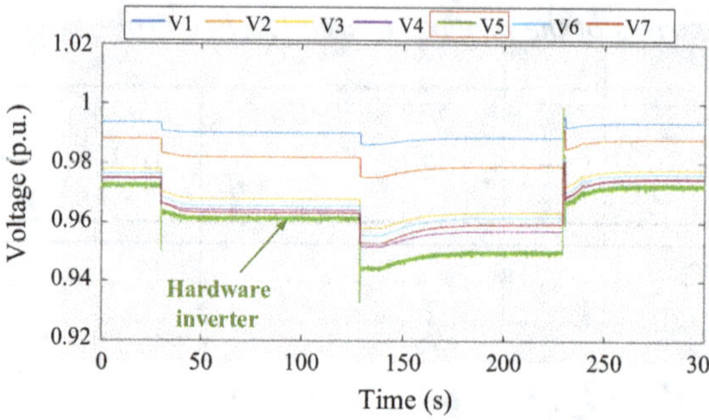

Figure 19.7 Voltage profiles of the distribution network in Test Case 1

when $G_i^{III} > 70$. Generally, increasing of control gains K_i^{III} and G_i^{III} will speed up the convergence but could lead to system oscillation and instability.

19.5.3 Test Case 1: voltage drop with step load change

In Test Case 1, the effectiveness of the proposed control approach to mitigate sudden voltage drops due to step changes in load is tested. Such a situation would normally occur in the evening in distribution grids when load demand is high but there is a lack of PV generation. As there is no PV production, all capacity of PV inverters can be utilised to supply reactive power for network voltage control. The communication delay in the distributed control is not considered in Test Case 1. Three step changes in load are emulated, in terms of percentage representable as follows: (i) an increase from 35% to 65% at $t = 30$ s; (ii) a further increase from 65% to 100% at $t = 130$ s; and (iii) a decrease from 100% to 35% at $t = 230$ s.

The network voltage profiles, utilisation ratios, and the PV inverters reactive power outputs have been shown in Figures 19.7–19.9, respectively. The experimental results are further explained in respective time sequences as follows:

1. 0 s $\leq t <$ 30 s. As the network voltages are in the normal operating range of $[\underset{\sim}{V}, \tilde{V}]$ during this period, the utilisation ratios are all zeros and there is no reactive power output from PV inverters.
2. 30 s $\leq t <$ 130 s. With the increase in load, five of the seven bus voltages are out of $[\underset{\sim}{V}, \tilde{V}]$ but still within $[\underline{V}, \overline{V}]$. The Level II droop control operates in this period to mitigate the voltage deviations locally.
3. 130 s $\leq t <$ 230 s. With the further increase in load, the voltage of bus 5 is out of the range $[\underline{V}, \overline{V}]$, thereby triggering the distributed control. Each inverter generates the same additional utilisation ratio for reactive power injection, as shown in Figure 19.6(c).
4. 230 s $\leq t <$ 300 s. With the decrease in load, the network voltages are back to the voltage range of [0.97, 1.03]. The effect of ramp-rate control mitigates the rate of voltage changes. The integrators of distributed control are reset according to the distributed reset algorithm presented in (19.13)–(19.15). Then all the utilisation ratios and reactive power outputs of PV inverters will be back to zero.

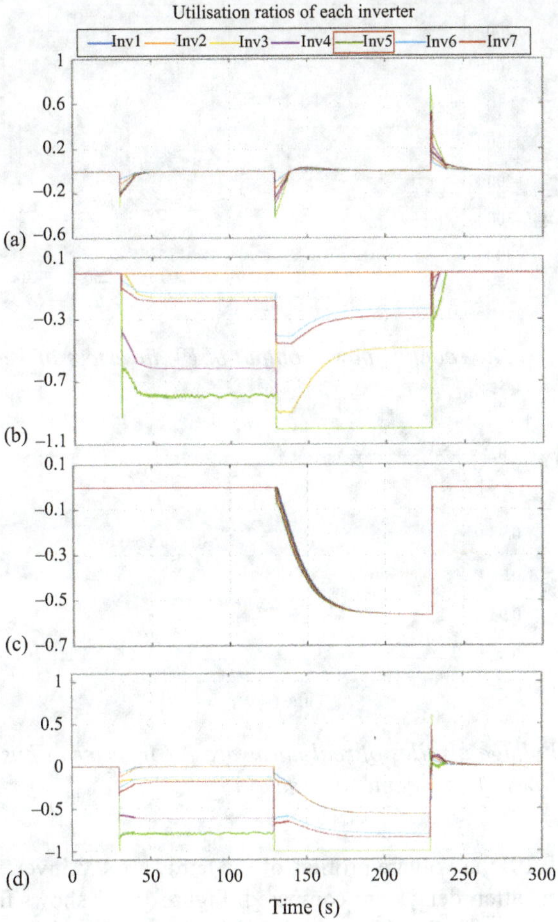

Figure 19.8 The utilisation ratios of each inverter in Test Case1. (a) u_i^I. (b) u_i^{II}. (c) u_i^{III}. (d) u_i

19.5.4 Test Case 2: communication delays

In Test Case 2, the performance of the proposed control method with communication delays is investigated. The test system conditions remain the same as in Test Case 1. Two types of communication delays are considered: (i) a constant time delay $\tau_1(t) = 0.5$ s, and (ii) a variable time delay representable by Gaussian distribution as

$$\tau_2(t) = \frac{1}{\sigma\sqrt{2\pi}} e^{-\frac{1}{2}\left(\frac{x-\mu}{\sigma}\right)^2} s$$

where mean value $\mu = 0.15$ and variance $\sigma^2 = 0.05$. These two time delays are bounded and satisfy $\tau(t) \leq \overline{\tau}$.

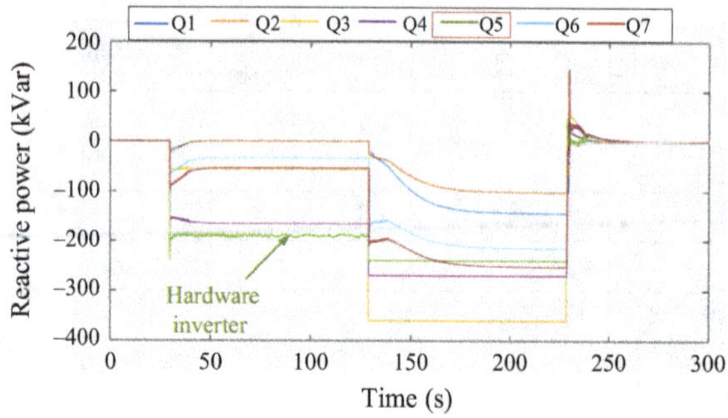

Figure 19.9 The reactive power output of PV inverters in Test Case 1

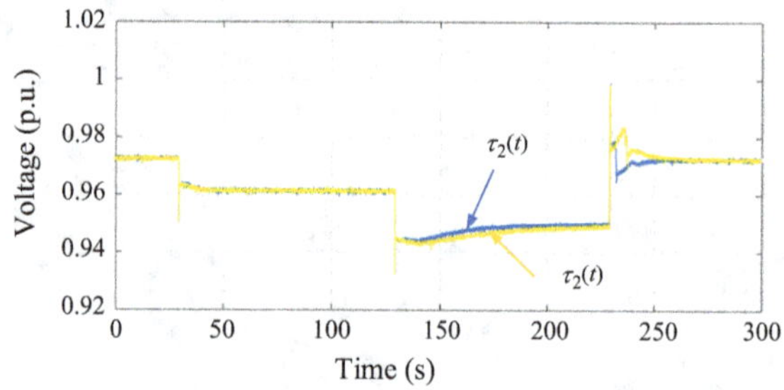

Figure 19.10 Voltage profiles of real hardware PV inverter at bus 5 with two types of communication delays

In Figure 19.10, the voltage profiles of the hardware PV inverter (bus 5) with the two communication delays are compared. Figure 19.11 shows the comparison of utilisation ratios $u_i^{III}(t)$of each inverter with these two communication delays. As shown in Figures 19.10 and 19.11, the proposed control approach works normally under both constant delay $\tau_1(t)$ and variable delay $\tau_2(t)$. This validates the effectiveness of the proposed approach under different communication delay conditions. It can also be noticed that the increase in communication delay impacts the convergence speed of the distributed control. Therefore, it takes a longer time to control the bus voltage back to 0.95 p.u., as shown in Figure 19.10. Furthermore, with the increase in communication delay, the reset time of the integrators also increases, as shown in Figure 19.11(a) and (b).

19.5.5 Test Case 3: voltage rise with real PV data

In Test Case 3, the ability of the proposed control approach to mitigate voltage rise conditions arising from high PV penetration is tested. Such a situation usually occurs during the daytime, when the residential load demand is low while PV generation is at its peak. Ten-minute PV data with 1-s resolution as shown in Figure 19.12, measured by EPRI in June 2012, is used here. As per the profile

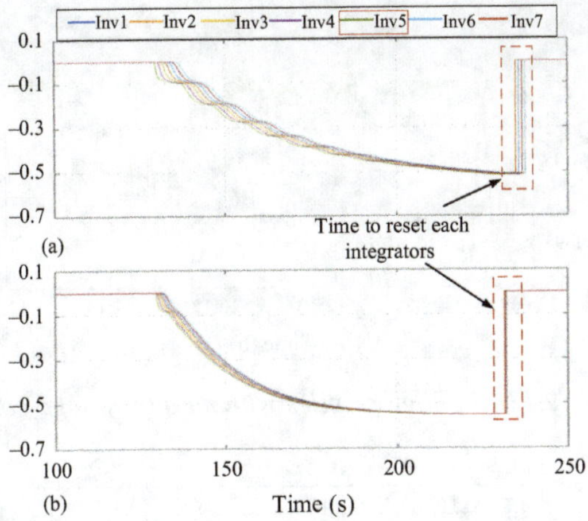

Figure 19.11 The comparison of utilisation ratios $u_i^{III}(t)$ with (a) τ_1(t) and (b) τ_2(t)

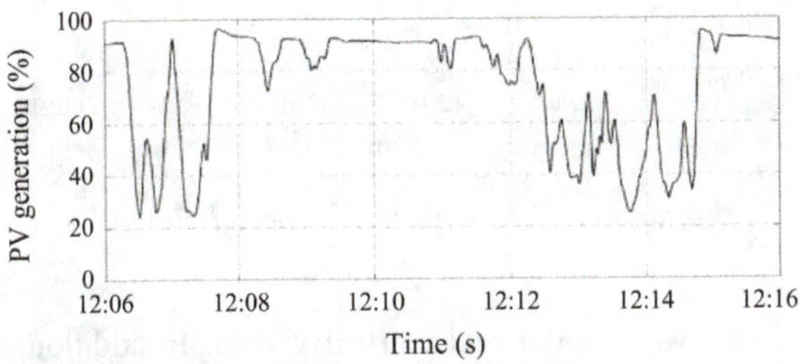

Figure 19.12 PV profiles in Test Case 3

shown in Figure 19.12, the PV real power output ramps up and down from 20% to 95% of its capacity. Therefore, the PV capacity is slightly oversized ($\alpha_i = 0.1$) to enable sufficient reactive power output support even under peak PV generation. The line resistance is increased by ten times (compared to the previous studies, Test Cases 1 and 2) to enable emulation of voltage rise issues experienced in a weaker, more resistive distribution grid. The communication delay is considered as $\tau(t) =$ 100 ms. The load percentage is set to 40% while variations in load are ignored. The voltage profiles without any voltage control are shown in Figure 19.13 as a reference case for comparison.

The network voltage profiles, utilisation ratios, and reactive power outputs are shown in Figures 19.14–19.16, respectively. As can be observed from Figure 19.14, the network voltages can be mostly regulated within [0.95, 1.05], except for very short transient periods. The PHIL results reinforce the verification of the proposed control approach to (i) mitigate voltage rise issues, (ii) mitigate the voltage

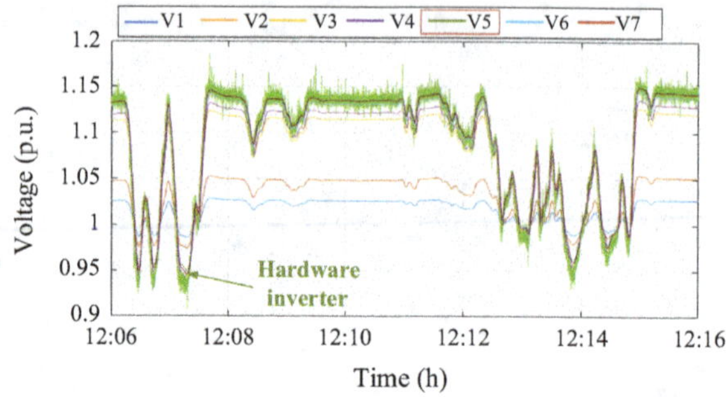

Figure 19.13 Voltage profiles without any voltage control

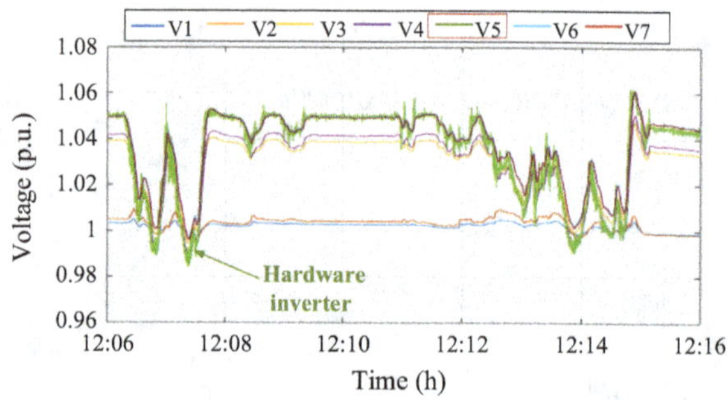

Figure 19.14 Voltage profiles with the proposed method in Test Case 3

fluctuations, and (iii) work under real-world PV data. In addition, the integrator reset control stops the Level III control when the network voltages are back within the safe operational range [0.96, 1.04], as shown in Figures 19.14 and 19.15(c).

19.5.6 Test Case 4: comparison study

In Test Case 4, the proposed method is compared with the method in [15] to further demonstrate its advantages. In [15], a two-level voltage var control scheme incorporating a ramp-rate control and distributed leader-follower control is proposed. The distinctive differences with the proposed control, i.e. the droop control and integrator reset, are not considered. The performance of the two methods is compared concerning the control objective of regulating the voltage within [0.95, 1.05]. The network voltage profiles, utilisation ratios, and reactive power outputs with the method in [15] are shown in Figures 19.17–19.19, respectively.

It can be seen that the proposed method offers the following two advantages:

1. The proposed three-level voltage control is fully distributed, while the method in [15] relies on the measurement of the voltage at critical buses. Thus, the proposed control structure is more resilient than the method in [15], as the critical bus measurement and leader control are not required.

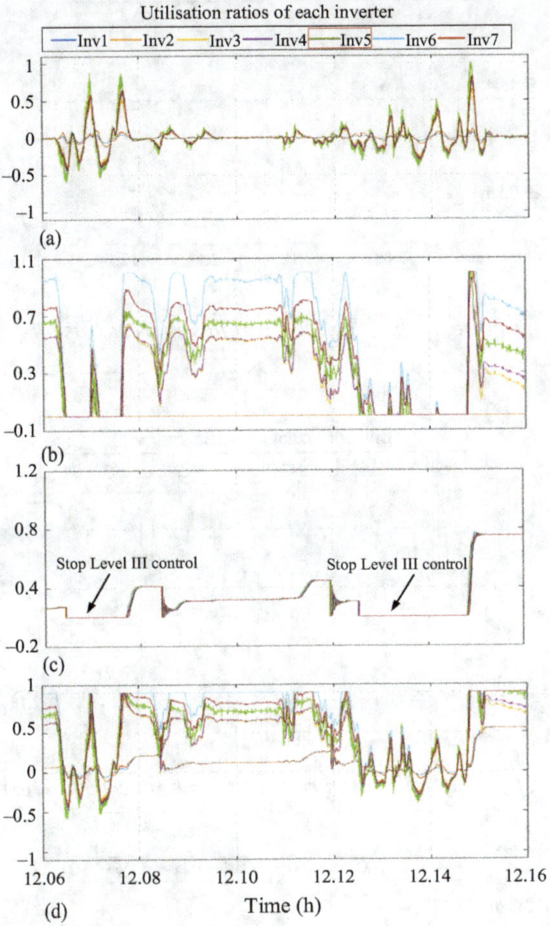

Figure 19.15 The utilisation ratios of each inverter in Test Case 3. (a) u_i^I. (b) u_i^{II}. (c) u_i^{III}. (d) u_i

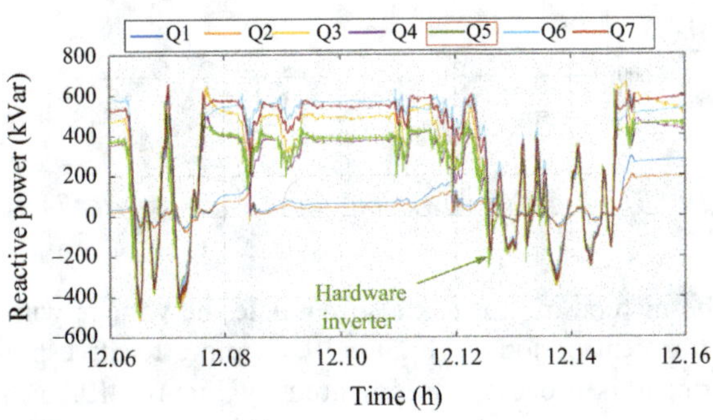

Figure 19.16 The reactive power output of each inverter in Test Case 3

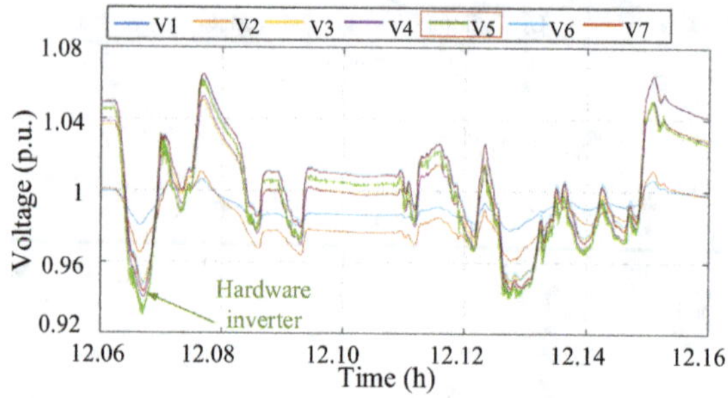

Figure 19.17 Voltage profiles in Test Case 4

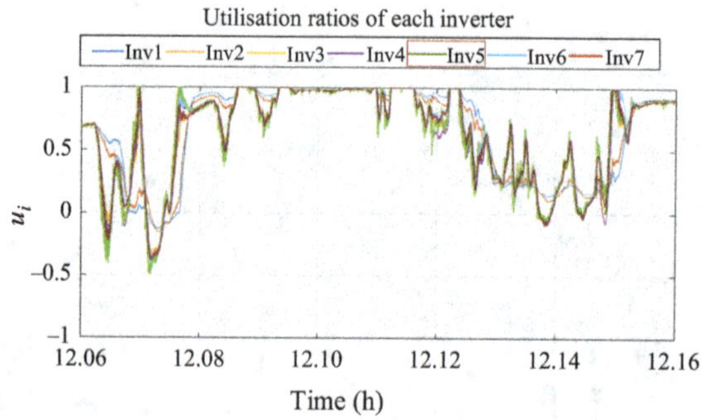

Figure 19.18 The utilisation ratios of each inverter in Test Case 4

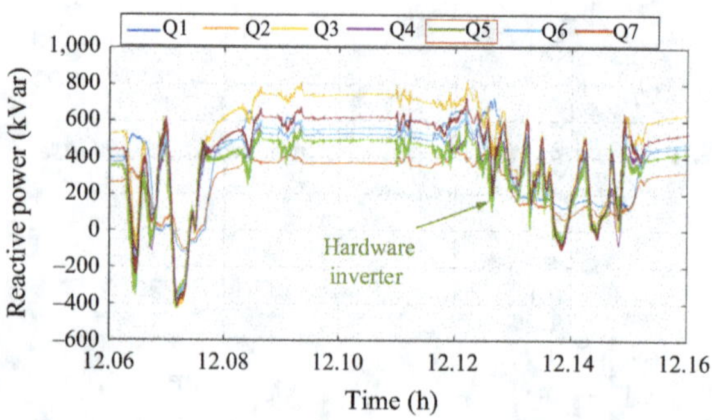

Figure 19.19 The reactive power output of each inverter in Test Case 4

2. Although the method in [15] can also regulate the voltage within [0.95, 1.05] except for transient periods in Figure 19.17, more reactive power is actually utilised during this process, as compared by Figures 19.16 and 19.19. The usage of reactive power by the method in [15] is 35% higher than the proposed method. The utilisation ratio of each inverter also indicates more use of

reactive power, as compared by Figures 19.15(d) and 19.18. The main reason is that the droop control and integrator reset are not considered in [15].

19.6 Conclusion

In this chapter, a three-level coordinated control method for PV inverters has been proposed for voltage control in distribution networks. The proposed three-level control can achieve both local voltage control and distributed voltage control by using the reactive power from PV inverters. The stability of the proposed three-level coordinated voltage control is analysed for different network voltage ranges.

The effectiveness of the proposed control approach has been validated by means of PHIL experiment. The proposed controllers are separately designed and communication links among them are realistically implemented. The results have demonstrated that the three-level control can determine the utilisation ratio based on the network voltage conditions. The corresponding reactive power can be provided by each PV inverter to deal with both the voltage fluctuations and voltage limits violation issues. Besides, it has been validated that the proposed control method performs effectively under bounded communication delays. The comparison study demonstrates the advantages of the proposed three-level voltage control method.

References

[1] M.J.E. Alam, K.M. Muttaqi, and D. Sutanto, 'A multi-mode control strategy for var support by solar PV inverters in distribution networks', *IEEE Trans. Power Systems*, vol. 30, no. 3, pp. 1316–1326, 2015.

[2] B.A. Robbins, C.N. Hadjicostis, and A. Domínguez-Garcíz, 'A two-stage distributed architecture for voltage control in power distribution systems', *IEEE Trans. Power Systems*, vol. 28, no. 2, pp. 1470–1482, 2013.

[3] H. Zhu, and H.J. Liu. 'Fast local voltage control under limited reactive power: optimality and stability analysis', *IEEE Trans. Power Systems*, vol. 31, no. 5, pp. 3794–3803, 2016.

[4] S. Alyami, Y. Wang, C. Wang, J. Zhao and B. Zhao, 'Adaptive real power capping method for fair overvoltage regulation of distribution networks with high penetration of PV systems', *IEEE Trans. Smart Grid*, vol. 5, no. 6, pp. 2729–2738, 2014.

[5] Y. Zhang, Y. Xu, H.M. Yang, and Z.Y. Dong, 'Voltage regulation-oriented co-planning of distributed generators and battery storage in distribution networks', *Int. J. Electrical Power and Energy Systems*, vol. 105, pp. 79–88, 2019.

[6] M.J.E. Alam, K.M. Muttaqi, and D. Sutanto, 'A novel approach for ramp-rate control of solar PV using energy storage to mitigate output fluctuations caused by cloud passing', *IEEE Trans. Energy Conversion*, vol. 29, no. 2, pp. 507–518, 2014.

[7] *Pacific Power Engineering Handbook*, Available: www.pacificpower.net/content/dam/pacific_power/doc/Contractors_Suppliers/Power_Quality_Standards/1C_5_1.pdf.

[8] Y. Li and C.N. Kao, 'An accurate power control strategy for power-electronics-interfaced distributed generation units operating in a low-voltage multibus microgrid', *IEEE Trans. Power Electronics*, vol. 24, no. 12, pp. 2977–2988, 2009.

[9] K. Baker, A. Bernstein, E. Dall'Anese, and C. Zhao, 'Network-cognizant voltage droop control for distribution grids', *IEEE Trans. Power Systems*, vol. 33, no. 2, pp. 2098–2108, 2018.

[10] C. Zhang, Y. Xu, Z.Y. Dong, and J. Ravishankar 'Three-stage robust inverter-based voltage/var control for distribution networks with high PV', *IEEE Trans. Smart Grid*, vol. 10, no. 1, pp. 782–793, 2019.

[11] R. Tonkoski, L.A.C. Lopes, and T.H.M. El-Fouly, 'Coordinated active power curtailment of grid connected PV inverters for overvoltage prevention', *IEEE Trans. Sustainable Energy*, vol. 2, no. 2, pp. 139–147, 2011.

[12] P.C. Kotsampopoulos, F. Lehfuss, G.F. Lauss, B. Bletterie, and N.D. Hatziargyriou, 'The limitations of digital simulation and the advantages of PHIL Testing in studying distributed generation provision of ancillary services', *IEEE Trans. Industrial Electronics*, vol. 62, no. 9, pp. 5502–5515, 2015.

[13] E. Guillo-Sansano, A.J. Roscoe and G.M. Burt, 'Harmonic-by-harmonic time delay compensation method for PHIL simulation of low impedance power systems', *2015 International Symposium on Smart Electric Distribution Systems and Technologies (EDST)*, Vienna, pp. 560–565, 2015.

[14] J. Rocabert, A. Luna, F. Blaabjerg, and P. Rodríguez 'Control of power converters in AC microgrids', *IEEE Trans. Power Electronics*, vol. 27, no. 11, pp. 4734–4749, 2012.

[15] Y. Wang, Y. Xu, Y. Tang, M.H. Syed, E. Guillo-Sansano, and G.M. Burt, 'Decentralised-distributed hybrid voltage regulation of power distribution networks based on power inverters', *IET Gen., Trans. & Dist.*, vol. 13, no. 3, pp. 444–451, 2019.

Index

www.ingramcontent.com/pod-product-compliance
Lightning Source LLC
LaVergne TN
LVHW011553200926
848484LV00001B/311
* 9 7 8 1 8 3 9 5 3 2 6 8 9 *